THE PHYSICS OF
IONIZED GASES

To learn more about the AIP Conference Proceedings, including the Conference Proceedings Series, please visit the webpage **http://proceedings.aip.org/proceedings**

THE PHYSICS OF IONIZED GASES

23rd Summer School and International Symposium on the Physics of Ionized Gases

Invited Lectures, Topical Invited Lectures and Progress Reports

National Park Kopaonik, Serbia
28 August – 1 September 2006

EDITORS
Ljupčo Hadžievski
Vinča Institute of Nuclear Sciences, Belgrade

Bratislav P. Marinković
Institute of Physics, Belgrade

Nenad S. Simonović
Institute of Physics, Belgrade

SPONSORING ORGANIZATIONS
Ministry of Science and Environmental Protection, Republic of Serbia
European Physical Society (EPS)

Melville, New York, 2006
AIP CONFERENCE PROCEEDINGS ■ VOLUME 876

Editors:

Ljupčo Hadžievski
Vinča Institute of Nuclear Sciences
P. O. Box 522
1001 Belgrade
Serbia

E-mail: ljupcoh@vin.bg.ac.yu

Bratislav P. Marinković
Nenad S. Simonović

Institute of Physics, Belgrade
P. O. Box 68
11080 Belgrade
Serbia

E-mail: bratislav.marinkovic@phy.bg.ac.yu
simonovic@phy.bg.ac.yu

L.C. Catalog Card No. 2006937292
ISBN 978-0-7354-0377-2
ISSN 0094-243X
Printed in the United States of America

CONTENTS

SECTION 1
ATOMIC COLLISION PROCESSES

Invited Lectures

Topical Invited Lectures

Progress Reports

v

SECTION 2
PARTICLE AND LASER BEAM INTERACTION WITH SOLIDS

Invited Lectures

Topical Invited Lectures

Progress Reports

SECTION 3
LOW TEMPERATURE PLASMAS

Invited Lectures

Topical Invited Lectures

Progress Reports

SECTION 4
GENERAL PLASMAS

Invited Lectures

Topical Invited Lectures

Progress Reports

PREFACE

This volume contains the Invited lectures, Topical invited lectures and Progress reports presented at the 23rd Summer School and International Symposium on the Physics of Ionized Gases – SPIG2006. The conference was held in National park Kopaonik, Serbia, from August 28 to September 1, 2006.

This conference, as the most recent in the series of biannual meetings which started more than forty years ago, provided a forum for 180 active researchers from 27 countries to discuss current advances in the physics of ionized gases and related fields. The program of the invited talks covered all traditional scientific fields of the conference that are grouped in the present book in the following sections: Atomic Collision Processes, Particle and Laser Beam Interactions with Solids, Low Temperature Plasmas and General Plasmas. We hope that this book will be a valuable source of information to graduate students as well as to active scientist in the field.

We are indebted to lecturers for presenting their contributions at the conference and for preparing the manuscripts for publication. We also express our gratitude to the members of the Scientific and Organizing committees for their efforts in organizing the meeting. Special gratitude is due to the Ministry of Science and Environmental Protection of the Republic of Serbia for providing the financial support to the conference. The conference is sponsored by the European Physical Society (EPS) and we are grateful for grants that had been provided trough East West Task Force scheme.

September 2006   *The Editors*

ACKNOWLEDGMENTS

23ʳᵈ SUMMER SCHOOL AND INTERNATIONAL SYMPOSIUM ON THE
PHYSICS OF IONIZED GASES

is organized by the

**Institute of physics
Belgrade, Serbia**

under the auspices and with support of the

Ministry of Science and Environmental Protection, Republic of Serbia

and also sponsored by:

- **European Physical Society**
- **Kryooprema d.o.o. Pančevo**
- **EUnet d.o.o. Belgrade**
- **"Knjaz Miloš" a.d. Arandjelovac**

SPIG 2006

SCIENTIFIC PROGRAM

Section 1.   ATOMIC COLLISION PROCESSES
  1.1.   Electron and Photon Interactions with Atomic Particles
  1.2.   Heavy Particle Collisions
  1.3.   Swarms and Transport Phenomena

Section 2.   PARTICLE AND LASER BEAM INTERACTION WITH SOLIDS
  2.1.   Atomic Collisions in Solids
  2.2.   Sputtering and Deposition
  2.3.   Laser and Plasma Interaction with Surfaces

Section 3.   LOW TEMPERATURE PLASMAS
  3.1.   Plasma Spectroscopy and Other Diagnostics Methods
  3.2.   Gas Discharges
  3.3.   Plasma Applications and Devices

Section 4.   GENERAL PLASMAS
  4.1.   Fusion Plasmas
  4.2.   Astrophysical Plasmas
  4.3.   Collective Phenomena

SPIG 2006

INTERNATIONAL SCIENTIFIC COMMITTEE

Lj. Hadžievski (Chairman) Serbia

N. Bibić, Serbia
S. Buckman, Australia
Z. Donko, Hungary
V. Guerra, Portugal
M. Kuraica, Serbia
D. Jovanović, Serbia
J.J. Jureta, Serbia
K. Lieb, Germany
T. Makabe, Japan
G. Malović, Serbia

A. Maluckov, Serbia
S.T. Manson, USA
N.J. Mason, UK
E. Mediavilla, Spain
Z. Mijatović, Serbia
K. Mima, Japan
N. Nedeljković, Serbia
L. Popović, Serbia
Y. Serruys, France
N. S. Simonović, Serbia

ADVISORY COMMITTEE

D. S. Belić
N. Konjević
J. Labat
B. P. Marinković
B. Milić
M. Milosavljević

B. Perović
Z. Lj. Petrović
M. M. Popović
J. Purić
B. Stanić
M. Škorić

ORGANIZING COMMITTEE

N. S. Simonović (Co-chairman)
B. P. Marinković (Co-chairman)
D. Šević (Secretary)
A. R. Milosavljević (Secretary)

S. Jovićević
S. Ćirković
P. Kolarž
S. Milisavljević

M. Parđovska
D. Pavlović
D. Radosavljević
M. Maksimović

SECTION 1

ATOMIC COLLISION PROCESSES

Invited Lectures
Topical Invited Lectures
Progress Reports

Probing Radiation Damage at the Molecular Level

N J Mason [1], M A Smialek [1], S A Moore [1], M Folkard [2]
and S V Hoffmann [3]

[1] *Centre of Atomic and Molecular Engineering. The Open University, Walton Hall, Milton Keynes, MK7 6AA, United Kingdom.*
[2] *Gray Cancer Institute, Mount Vernon Hospital, Northwood, Middlesex HA6 2JR, United Kingdom.*
[3] *Institute of Storage Ring Facilities, University of Aarhus, Ny Munkegade, DK 8000 Aarhus, Denmark*

Abstract. Radiation damage of DNA and other cellular components has traditionally been attributed to ionisation via direct impact of high-energy quanta or by complex radical chemistry. However recent research has shown that strand breaks in DNA may be initiated by secondary electrons and is strongly dependent upon the target DNA base identity. Such research provides the fascinating perspective that it is possible that radiation damage may be described and understood at an individual molecular level introducing new possibilites for therapy and perhaps providing an insight into the origins of life.

Keywords: Molecular physics, Radiation chemistry; DNA damage; Electron interactions.
PACS: 34.50, 34.80, 82.50, 87.14, 87.15, 87.50.

INTRODUCTION

Advances in cancer radiation therapy are most likely to arise through a deeper understanding of the mechanisms by which radiation interacts with cells, and the means by which such mechanisms can be manipulated. Furthermore, although it is well known that ionizing radiation is also a carcinogen, the risks associated with typical occupational and environmental exposures are still not well understood. As measurements of biological effects are difficult at low doses, much reliance is placed on the construction of 'damage' models to predict such effects, models that ultimately must allow legislation to be developed that will mediate any perceived risks. In order to develop models of the effects of ionizing radiation a mechanistic model of radiation action is required that is based on sound experimental data. Since damage to DNA within the cell is the major cause of cell death and mutation we particularly need to investigate the effect of ionizing radiations on DNA.

Ionizing radiations can produce a range of structural and chemical modifications of the DNA helix. Of these, double-strand breaks (dsb), where both strands of the helix are broken within a few base pairs, can lead to lasting damage via the production of chromosome aberrations, mutations and ultimately cell killing. It is now known that the effectiveness of different ionizing radiations is critically dependent on the patterns of ionizations they produce on a nanometre scale, comparable with the diameter of the DNA helix. Theoretical track structure modelling is therefore being used with

CP876, *The Physics of Ionized Gases: 23rd Summer School and International Symposium*,
edited by L. Hadžievski, B. P. Marinković, and N. S. Simonović
© 2006 American Institute of Physics 978-0-7354-0377-2/06/$23.00

increasing sophistication to simulate the distinctive patterns of ionizations produced by a wide range of ionizing radiations.

Such models show us that penetrating primary radiation (i.e. energetic photons, electrons or ions) produce a significant number of nanometer sized clusters of ionization at the low energy track ends liberating a large number of secondary electrons and ions. Although the effects of high energy photons and electrons on DNA have been studied for many years, much less work has been published on the effects of such low energy electron and ions. Low energy atomic and molecular ions can be produced either directly by the ionizing radiation from the DNA or from the surrounding environment along the ionization track. Such ions undergo electron capture collisions and can also displace atoms from surrounding molecules resulting in bond-breaking and the production of molecular cations in excited states which dissociate into reactive fragments. Recent observations[1] of damage to DNA components by $0.25 - 1.75$ eV Da^{-1} Ar^{+} ions have shown that such low energy ions can induce structurally complex strand-breaks in DNA, which are less easily repaired than the predominantly 'clean' breaks produced directly by energetic radiation. Similarly recent pioneering research of Sanche and co-workers[2,3] have shown that strand breaks in DNA may be initiated by secondary electrons at sub-ionization energies and are dependent upon the target DNA base identity, DNA sequence, and incident electron energy[4]. They also showed that the probability of inducing strand breaks is one to two orders of magnitude larger for electrons than for photons of the same energy.

It is therefore important to fully characterize the mechanisms by which different types of radiation damage DNA. This will allow models to be formulated that can predict not just the patterns of ionizations, but also the spectra of damage complexity that different types of radiation can induce in DNA. Hence we require knowledge of the relationship between the amounts of energy deposited within a given region of the DNA helix and the type and severity of damage that is produced.

In this brief overview we will explore how recent research in the study of electron interactions with the biomolecules and DNA itself has developed to provide new insights into the possible mechanisms of DNA damage within the cell. We will also discuss some of the rapidly developing areas of research which seek to explore the energetics of biomolecular excitation and fragmentation that have consequences for our understanding of how the 'molecules of life' may have been assembled as the origins of life itself.

Such a brief review can only provide a guide to a research field that has grown rapidly in the last five years and for further details and a deeper discussion readers are invited to consult several of the recent reviews referenced in this text. European research in this area has been led by several major collaborative networks; The Framework VI Research Training Network Electron and Positron Induced Chemistry (EPIC); The COST Action on Radiation Damage (RADAM) and the European Science Foundation Programme 'Electron Induced Processing At the Molecular level (EIPAM) details of which and further references can be found at www.isa.au.dk/epic; www.isa.au.dk/cost; www.isa.au.dk/eipam respectively.

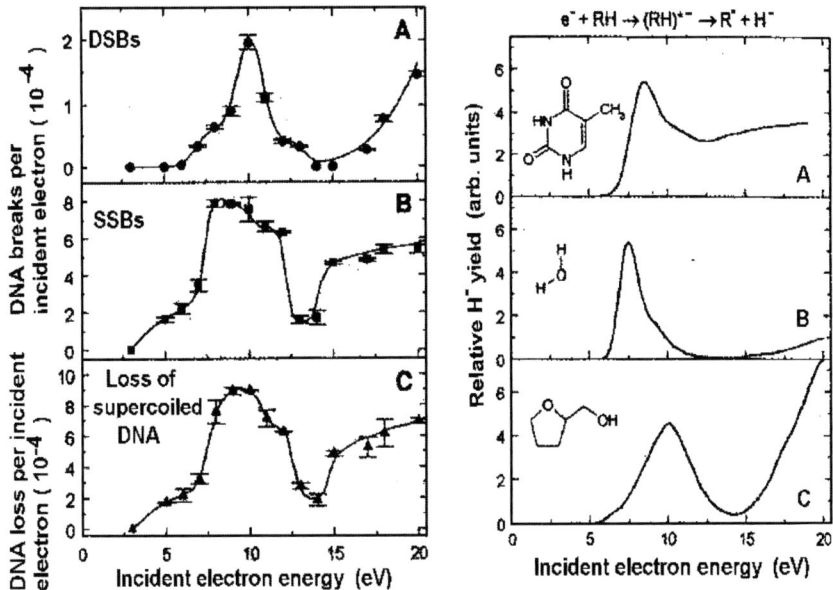

FIGURE 1. Double and single strand breaks observed in electron irradiation of DNA compared with formation of H⁻ ions formed by dissociative electron attachment of some typical biomolecules; thymine, water and tetrahydrofurfuryl alcohol (From Boudiaffa et al[5])

Electron Induced DNA Damage

The recent explosion in the study of electron interactions with biomolecules was largely pioneered by the publication of a pioneering paper by Sanche and co-workers at the University of Sherbrooke in Science in 2000[5]. Samples of plasmid DNA with 3199 base pairs, extracted from *Escherichia coli,* were irradiated with low energy electrons and electrogelphoresis used to analyse the irradiated samples for damage i.e. loss of supercoiled DNA and creation of both single and double strand breaks. In all cases they observed that low energy electrons were able to induce DNA damage at energies far below that those commonly believed to be necessary to rupture the DNA helix through ionization of its constituent species (Figure 1). Indeed further work has shown that single strand beaks may be induced with electrons with energies of less than 1 eV- far below those energies needed to rupture any of the chemical bonds of the constituent molecules[6]. Such results were in contradiction with prediction (and many radiation damage models) and could not be explained by any direct fragmentation process. However in the same experiment Sanche and co-workers provided an explanation observing that the damage yields seemed to be enhanced over some energy regions. Such enhancements, often entitled resonances, are common in electron induced dissociation of molecules through a process known as 'dissociative electron attachment'.

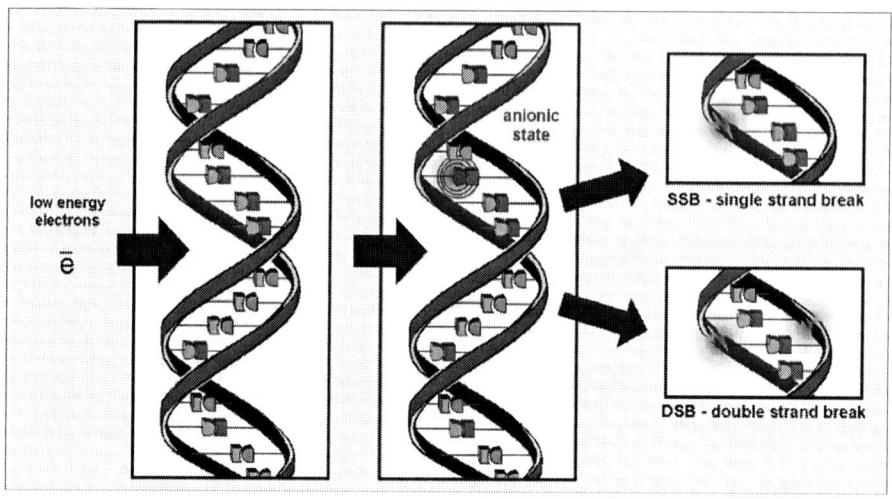

FIGURE 2. Schematic summary of strand breaks induced in DNA by low energy electrons.

Dissociative Electron Attachment; A New Route Of Molecular Dissociation and DNA Damage

In the process of Dissociative Electron Attachment (DEA) the incident electron is captured by the target molecule (ABC) to form a temporary negative ion (anion) ABC^- which subsequently decays to produce a negative fragment accompanied by one or more associated neutral counterpart(s), the process being summarised as

$$e^- + ABC \rightarrow ABC^{\#--} \rightarrow B^- + AC \text{ or } B^- + A + C$$

The product ions and neutral (often radical) products may subsequently initiate new new chemistry, a process that is now being exploited to effect chemical transformations of thin films[7] and is used in the plasma etching industry[8]. Sanche and co-workers compared the patterns of DNA damage with the formation of anions in simple biomolecules (e.g. water, thymine and tetrahydrofurfuryl alcohol) by DEA. Resonances in the DEA cross sections of these biomolecules were found to coincide with those energy regions in which enhanced DNA damage was observed using low energy electrons (Figure 1). This observation led Sanche et al to suggest that *"from a radiobiological perspective...the abundant low-energy electrons and possibly their ionic and radical reaction products play a crucial role in the nascent stages of cellular DNA radiolysis"*(Figure 2).

Subsequently there has been an extremely active research programme to investigate DEA of those molecular systems (M) that comprise DNA, e.g. the nucleotide bases (thymine, cytosine, uracil) and sugar (deoxyribose) as well as so-called DNA analogues such as tetrahydrofuran (THF)and tetrahydrofurfuryl alcohol (THFA). In the case of nucleobases, studies have also been performed using isotopically labelled molecules where hydrogen atoms are replaced by deuterium atoms at some specific sites of the target molecule. In all these targets the most common DEA process at the lowest energies (<5eV) is that of hydrogen atom abstraction (Figure 3) with the simultaneous formation of an anion (M-H)⁻. Furthermore detailed experimental studies have revealed that such DEA is both bond (C-H versus C-N) and site selective (N1-H versus N3-H).

These results provide an fascinating insight into the possible mechanism of DNA damage within the cell in which low energy secondary electrons form temporary anions at specific sites within the DNA helix the decay of which subsequently leads to strand breaks for example through the reactive H atom. Thus, in contrast to most modern radiation models, DNA damage may be described at an individual molecular level ! For example such a molecular picture may be used to explain the well known correlation between electron attachment rates of many molecules and their carcinogenicity and may be used to suggest new compounds to be adopted in radiation therapy as treatment enhancing sensitizers. The high DEA attachment rates common in halogenated species therefore makes them a good candidate for radiosensitizers, Recent experiments have shown that bromouracil (a ready replacement for thymine and commonly used in therapy) has a DEA cross section over two orders of magnitude higher than that of thymine[10].

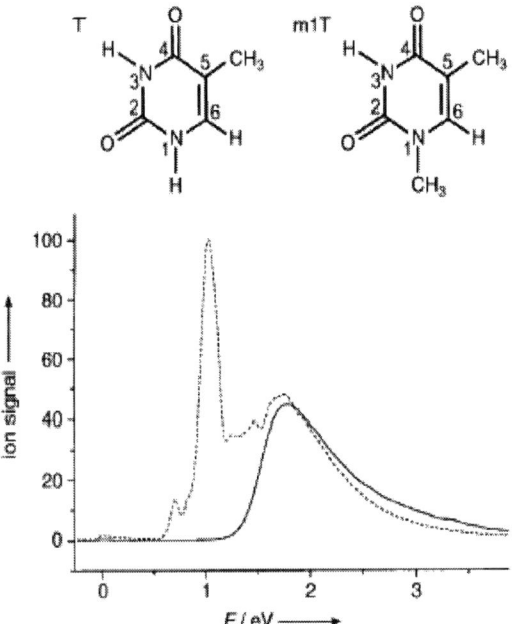

FIGURE 3. Plot of ion yield of the (M-H)⁻ anions (indicating loss of a hydrogen atom) from methylated thymine (m1T) at the N1 position (----) with that of thymine T (————) (From Ptasinska et al[9]) indicating that DEA is bond selective.

Electron Impact Studies of Biomolecular Spectra and Cross Sections

Developing a model of DNA damage at the molecular level requires detailed knowledge of the spectroscopy and dissociation dynamics of many biomolecular systems. Many of the higher lying anion resonances (>5eV) are formed with neutral electronic states of the target molecule as 'parents' and may decay leaving the molecule excited in electronic and ro-vibrational states. Therefore it is vital to understand the electronic sate spectroscopy of such molecular systems as the DNA bases, simple sugars etc. Unfortunately due to the difficulty in preparing such molecules in the gaseous phase there is currently little or no information on the spectroscopy of such systems. Only in the last two years have 'traditional' methods such as Electron Energy Loss Spectroscopy (EELS) been adopted to study such targets. EELS has an advantage over optical spectroscopy in that by varying the incident electron energy and the scattering angle at which spectra are measured both allowed and forbidden transitions may be observed. Figure 4 shows an EELS in 5-bromouracil[11] with low lying forbidden transitions apparent below 4 eV excited using low energy electrons while allowed transitions are more apparent at high energies. However such EELS have insufficient resolution to reveal any ro-vibrational structure in these excited states and hence can not reveal if such states are bound (capable of supporting vibronic excitation) or unbound. Hence in the near future it is hoped that optical techniques (using synchrotron radiation) will be extended to such targets[12].

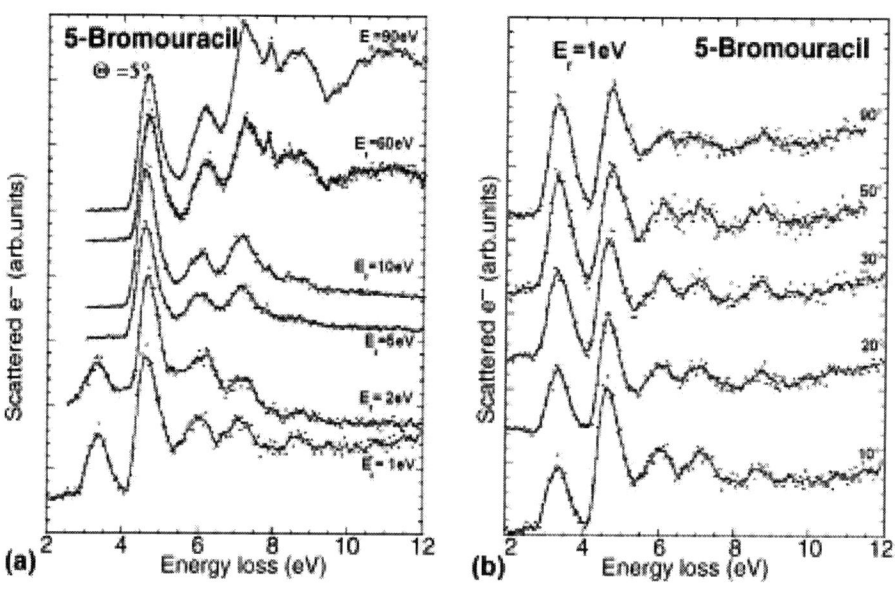

FIGURE 4. Electron Energy loss spectrum of bromouracil revealing low lying triplet states (< 4eV).

These first EELS spectroscopy experiments have recently been followed by the first experiments measuring elastic and inelastic scattering cross sections. These measurements may in turn be compared with derived theoretical cross sections with such theory in turn providing an insight into the dynamics of electron interactions with biomolecules. Inelastic and angular cross sections are necessary if both the energy deposition and spatial profiles along radiation tracks are to be modelled accurately. Figure 5 shows a recently measured differential cross section in THFA, a molecule that has often be used as a mimic of DNA, these experimental results are compared with a simple independent atom model[13]. This data is now providing data for input into a new generation of refined ionization track codes such as the GEANT 4 code now being adopted to study to nanoscale dosimetry[14]. Such codes may ultimately provide a full molecular representation of DNA damage and may be used to predict radiation doses used in therapy.

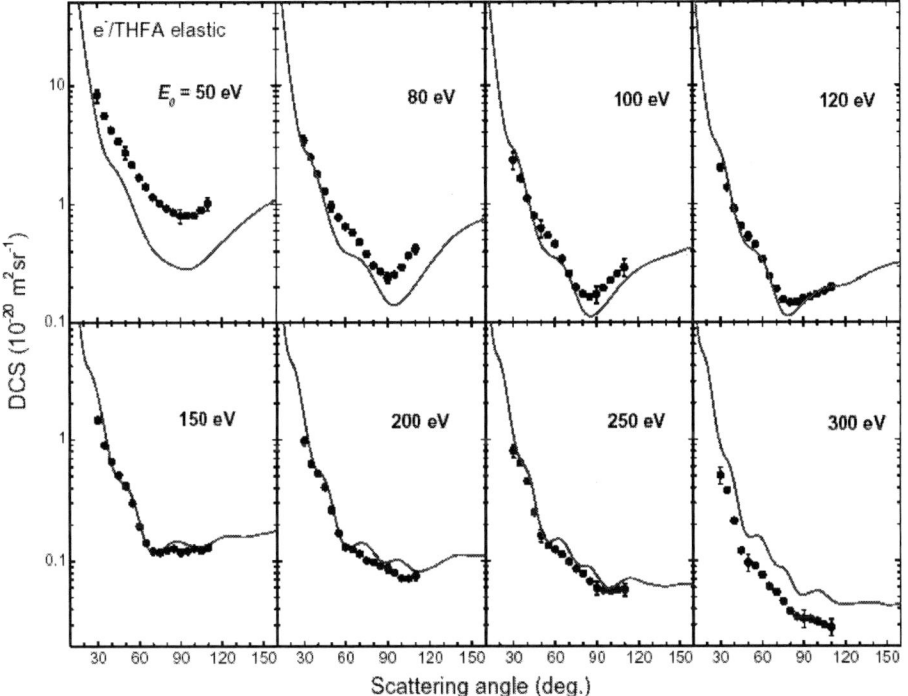

FIGURE 5. Differential cross section for elastic scattering from THFA compared with calculations using independent atom approximation[13].

However whilst a great deal of data has been assembled in recent years on electron interactions with biomolecules and the induced dissociation and energy deposition processes there remain many unanswered questions and the relevance of electron induced processes in cellular radiation damage is still somewhat controversial.

First it must be stressed that most of the experiments have investigated gaseous phase targets with only a few comparable DNA experiments being performed in the condensed phase[15,16] which while showing similar features to those observed in the gas phase also show notable differences. For example DEA resonances are shifted in energy (usually lower) and hence some fragmentation pathways are no longer accessible. Some electronic states are no longer capable of being excited in the condensed phase (e.g. Rydberg states) and the excitation energies of others are shifted sometimes by 100's of meV. Consider water, photon absorption experiments have revealed a large blue shift in the condensed phase (Figure 6) this in turn suggests that the threshold for dissociation of water within a cell should not (as in common practice) be modelled using gas phase data.

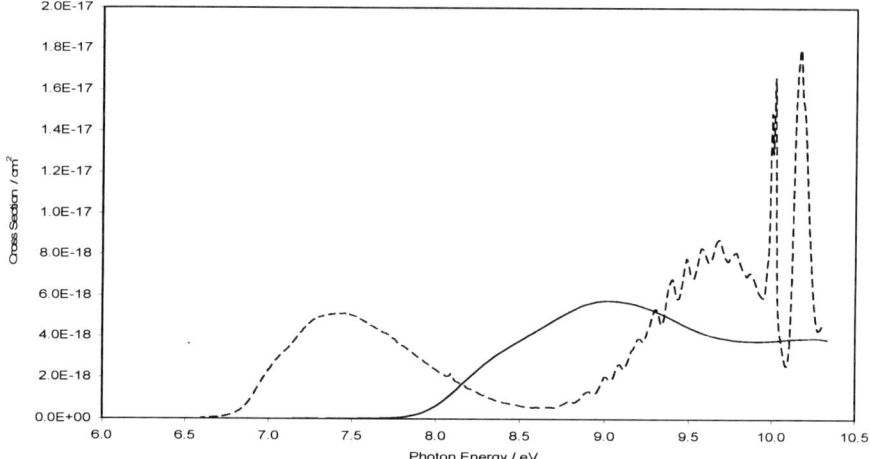

FIGURE 6. Photon absorption spectra recorded in gas phase (----) and in water ice (——). A strong blue shift is observed in the solid phase and quenching of the excitation of Rydberg states[17].

The role of water is of course central to any true study of radiation damage in cellular systems since much of the cell is filled with water and all DNA contains entrapped structural water. During irradiation water is fragmented releasing OH free radicals which are commonly believed to be largely responsible for inducing strand breaks. Such OH induced damage is often entitled 'indirect damage' in contrast to direct damage induced by DNA impact from ions and electrons. Whether it is better to model the water embedded within DNA helix and the cell as gaseous or condensed phase is still unclear with a recent experiment[17] seeking to study water in cellular membrane mimics surprisingly suggesting that a gaseous model is more accurate in determining the electronic state spectroscopy of water under cellular conditions.

Indeed our knowledge of electron interactions with water incredibly remains remarkably poor[18], in part due to a reluctance be experimental groups to put water into their apparatus and the difficulties that theoreticians have in modelling the strong dipole moment of water. Water molecules also like to form dimers and larger complexes such that under biological conditions much of the water may be found in larger clusters. Only recently have theoreticians started to model electron interactions with water dimers and to date there are no experiments on water dimers. However recently DEA in water in both the gaseous and condensed phase has finally been explored experimentally and previous discrepancies resolved[19,20].

Future Studies In Radiation Damage And Its Consequences For Exploring the Origins of Life

The study of radiation induced damage remains an extremely active topic of modern research requiring transdisciplinary collaborations. Whilst considerable progress has been made since 2000 in the study of electron interactions with DNA and several of its component biomolecules much remains to be done. An immediate priority is repeat and extend the study of DNA strand breaks undertaken by Sanche and his co-workers in Sherbrooke to other plasmids and under a wider variety of experimental conditions (e.g. different degrees of hydration to explore the role of OH damage). Experiments on smaller oglionucleotides will allow site specific reaction dynamics to be explored whilst allowing more reproducible samples to be prepared.

The standard method of monitoring DNA damage is gel electrophoresis. This is in reality a somewhat indirect way of monitoring strand breaks and once a sample has been analysed it can not be reused. Using modern Atomic Force Microscopes (AFM) it is possible to image DNA strands (Figure 7) and directly view damage induced after irradiation with the advantage that the same sample may then be irradiated once again and the effects of successive irradiations explored.

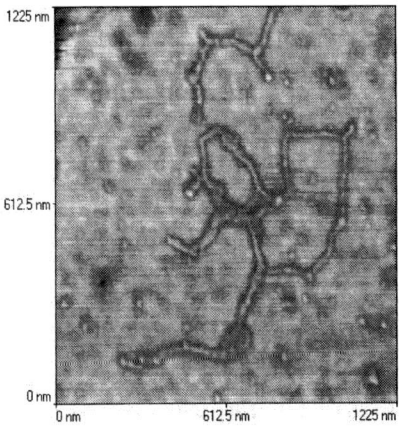

FIGURE 6. Atomic Force Microscope image of strands on DNA deposited on a mica sheet

In studying DNA damage alone we do not gain a complete picture of the effects of radiation within live tissue. Tissues are composed of intercommunicating cells of one or more types, hence damage in one irradiated cell in a tissue culture may be transmitted to a neighbouring unirradiated cell in the same or another tissue culture dish. This intriguing "bystander effect"[21-22] is reminiscent of the well described "abscopal effect" in radiation therapy and to date the physical/chemical processes of both such effects are largely unknown. Thus in obtaining a fuller understanding of radiation damage it is necessary to study a broader range of structures in the cell than just those linked to DNA.

All cells whether bacterial, plant or animal are enclosed by membranes, indeed membranes make up about 80% of the total dry-matter content of animal cells. How cells generate and maintain their internal structures and integrity depends upon how the cell membrane controls the transport of material in and out of the cell. The basic structural components of any cell membrane are lipid bylayers. Lipid bylayers are semi-permeable such that small uncharged molecules can pass more or less freely across the membrane but for charged species and macromolecules such as proteins and DNA the membrane is a major obstacle to diffusion. Thus the membrane essentially keeps 'the inside in and the outside out' and ultimately defines what constitutes a cell and what constitutes the 'rest of the world' in a biological sense (often called intracellular and extracellular space). However cells do need to transport proteins, DNA and ions into and out of the cell, across the lipid bilayer. The methods of transfer across membranes are therefore one of the defining mechanisms of life and our understanding of these processes is fundamental to our understanding of all other aspects of cellular physiology. To date there have only been a few studies probing the effect of radiation on the cell membrane[23]. A body of evidence has begun to emerge to suggest that ionizing radiation may have both direct (i.e. direct energy deposition within the target tissue) and indirect (through the production of chemically active secondary species such as OH radicals) effects on lipid membranes[24]. Radiation induced changes within the lipid bilayers of the membrane may alter ionic pumps leading to either increased or decreased permeability. This may be due to changes in the viscosity of intracellular fluids associated with disruptions in the ratio of bound to unbound water[25]. Such changes would result in an impairment of the ability of the cell to maintain metabolic equilibrium and can be very damaging even if the shift in equilibrium is quite small. The most extreme effects may lead to rupture of the cell membrane and thence cell death.

Experiments are now being prepared to study the effects of radiation on cellular membranes by preparing membrane mimics in the laboratory using a layer-by-layer films technique. In this technique alternating cationic and anionic layers are deposited onto a solid substrate, according to the following procedure: i) the solid substrate is immersed in a polycationic (or polyanionic) solution for a set period of time, after which the substrate and adsorbed film is washed in pure water or in an aqueous solution with the same pH of the polyelectrolyte solution. ii) the substrate is then immersed in the polyanionic (or polycationic) solution to complete a bilayer, with the substrate and bilayer being again washed to remove weakly adsorbed material. Repetition of the steps above may be performed to deposit the desired number of bilayers (Figure 7). These bilayers are a good mimic of the structure of a cellular membrane but one that is controllable within the laboratory. Water and proteins may be placed within the bilayers as can a radiation sensitive chemical sensor such that any induced effects may be probed as a function of exposure dose,

A study of radiation damage reveals the fragility of life and the narrow 'envelope' within which it has evolved. For example DNA does not absorb light in the visible part of the spectrum but strongly absorbs at wavelengths less than 350nm, wavelengths which are filtered from our atmosphere by the ozone layer. Thus without an ozone layer any DNA based life would struggle to survive since the absorption of UV light would lead to fragmentation of the component biomolecules.

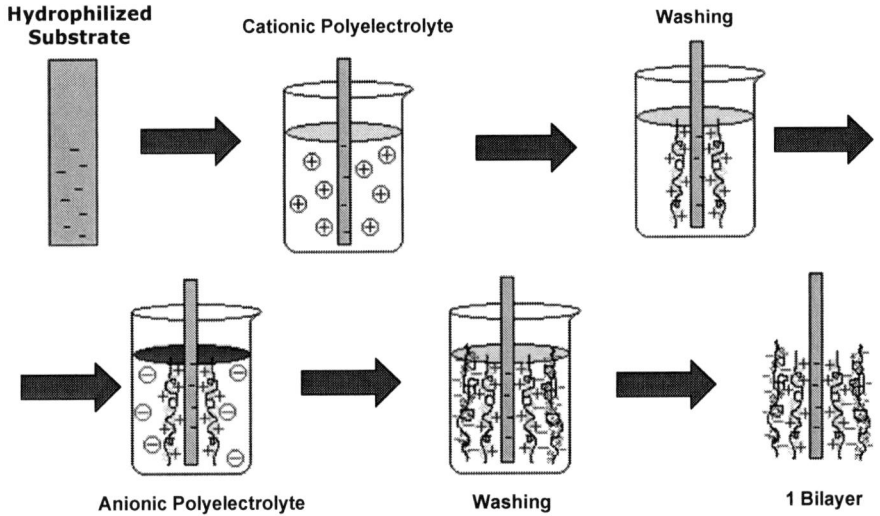

Figure 7; Schematic of the layer-by-layer(LBL) technique used for synthetic membrane production.

Recent research has shown that many of the molecular building blocks of life (amino acids, bio-sugars, simple lipids) may be readily formed as a consequence of organic chemistry both on the early Earth and as a byproduct of the process of stellar and planetary formation, indeed many of these compounds are found in meteorites, themselves residues of such processes[26]. But how are such compounds assembled to form cellular structures and self replicating molecules such as DNA ? To date we do not have an answer to this most fundamental scientific question, however for the first time, laboratory studies are being conducted to explore such phenomena. Indeed a new scientific discipline known as 'Astrobiology' has emerged in order to provide an interdisciplinary forum for both the origins of life on Earth and to evaluate the probability of life having evolved on some of the exoplanets we are now able to detect in other solar systems[27].

Hence our recent recognition that radiation damage may ultimately be described at a molecular level may in turn be a consequence of the way in which biomolecular systems were first assembled from 'prebiotic building blocks. Thus in understanding how radiation damages our own DNA and cellular structures we may unravel the mechanisms that led to origins of life.

ACKNOWLEDGMENTS

We are pleased to recognize support of EU EPIC network HPRN-C-2002-00179, the EU/ESF COST Action P9 on Radiation Damage and the European Science Foundation EIPAM Programme and the support of the EU I3 project IA-SFS, RII3-CT-2004-506008 for the use of the ISA facilities at University of Aarhus, Denmark. M A Smialek thanks the Open University for the support of a postgraduate studentship.

REFERENCES

1. Z Deng, I Bald, E Illenberger and M A Huels Phys Rev Letters **95** (2005) 153201
2. L Sanche, Mass Spec. Rev. **21** (2002) 349
3. M A Huels B Boudaiffa, P Cloutier, D hunting and L Sanche., J. Am. Chem. Soc. **125** (2003) 4467
4. H Abdoul-Carime and L Sanche, Int. J. Radiat. Biol. **78** (2002) 89
5. B Boudaiffa, P Cloutier , D Hunting, M A Huels and L Sanche, Science **287** (2000) 1658
6. R. Panajotovic, F. Martin, P. Cloutier, D. Hunting and L. Sanche, Rad. Res, **165** (2006),452
7. A.Lafosse, M.Bertin, D.Caceres, C.Jäggle, P.Swiderek, D.Pliszka, and R.Azria, Eur.Phys.J. D **35**, (2005) 363
8. S Samukawa, K Sakamoto and K Ichiki, J Vacuum Science and Technology A **20** (2002) 1566
9. S. Ptasińska, S. Denifl, V. Grill, T. D. Märk, E. Illenberger and P. Scheier Phys. Rev. Lett. **95**, (2005) 093201
10. H Abdoul-Carime, P Limao-Vieira, I Petrushko, N J Mason, S Gohlke & E Illenberger Chem Phys Lett.**393** (2002) 442
11. R Abouaf , J Pommier and H Dunet Chem Phys Lett **381** (2003) 486
12. A. Giuliani , I. C. Walker ,J. Delwiche, S. V. Hoffmann, P. Limão-Vieira,N. J. Mason, B. Heyne, M. Hoebeke and M. -J. Hubin-Franskin J. Chem. Phys. **119** (2003) 7282.
13. A.Milosavljević, F.Blanco, D.Sevic, G.Garcia and B.Marinkovic Eur J Phys D **40** (2006) 107
14. A J Wroe, R Schulte, V Bashkirov, A B Rosenfeld, B Keeney, P Spradlin, H F W Sadrozinski and B Grosswendt IEEE Transactions on Nuclear Science **53** (2006) 532
15. P L Levesque, M Michaud and L Sanche J Chem Phys **122** (2005) 094701
16. P L Levesque, M Michaud, W Cho and L Sanche J Chem Phys **122** (2005) 224704
17. R Mota, R Parifita, A Guiliani, MJ Hubin-Franskin, J MC Lourenco, S V Hoffmann , N J Mason, PA Ribeiro, M Raposo and P Limao-Vieira Chem Phys Lett **416** (2005) 152
18. Y Itikawa and N J Mason J. Phys. Chem. Ref. Data **34** (2005) 1
19. B C Garrett et al Chemical Reviews **105** (2005) 355
20. J. Fedor, P. Cicman, B. Coupier, S. Feil, M. Winkler, K.G luch, J. Husarik, D. Jaksch, B. Farizon, N.J. Mason,P.Scheier,T.D. Mark J. Phys B (2006) in press
21. O Belyakov, A Malcolmson, M Folkard, K M Prise and B D Michael Br. J. Cancer **84** (2001) 674.
22 G Schettino, M Folkard, B D Michael and K M Prise Radiation Research **163**: (2005) 332
23. M B Yatvin, J J Gipp and E A Werts E A 'Membrane Aspects of Radiation Biology' Berlin-New-York. W. de Crugter Co. (1984)
24. K P Mishra J Environ Pathol Toxicol Oncol. **23** (2004) 61
25. M Weik, U Lehnert and G Zaccai, Biophys. J. **89** (2005) 3639
26. P Ehrenfreund and H J Fraser, in Solid State Astrochem. NATO ASI series, Eds Pirronello & Krelowski (2002)
27. The Astrophysical Context of Life Committee on the Origins and Evolution of Life, National Research Council USA ISBN: 0309096278 (2005)

Jumps And Bi-stability Effects In Low Temperature Plasmas

N. A. Dyatko[1], Y. Z. Ionikh[2], A.V. Meshchanov[2], A. P. Napartovich[1]

[1] *State Research Center Troitsk Institute for Innovation and Fusion Research, Troitsk, Moscow region, 142190, Russia*
[2] *Fock Research Institute of Physics, St. Petersburg State University, Ul'yanovskaya ul. 1, Petrodvorets, St. Petersburg, 198504, Russia*

Abstract. In present paper the two kinds of bi-stability in low temperature plasmas are considered. The first kind is the bi-stability of the electron energy distribution function (EEDF) and related plasma parameters. Qualitative explanation of the possibility of the effect is given using Maxwellian EEDF approach. The review of physical conditions is done under which EEDF bi-stability was predicted using Boltzmann equation analysis and known experimental data interpreted in terms of bi-stability are described. The possible experiments are also discussed. The second kind is the bi-stability of the form of the positive column of glow discharge. The possibility of the existence of diffuse and constricted forms of positive column under the same discharge current is considered. It is found that under certain experimental conditions constricted and diffuse modes can simultaneously exist in the same discharge tube, i.e. steady state partially-constricted discharge can be realized. The problems of modeling of such kind of discharge are finally discussed.

INTRODUCTION

Plasma in is a classical sample of a non-linear system. One of the most pronounced manifestations of this non-linearity is the bi-stability effect, i.e. the appearance of two different stable states. In present paper two different kinds of bi-stability are considered: bi-stability of the electron energy distribution function (EEDF) and bi-stability of the glow discharge properties.

The first kind is that in homogeneous low temperature plasma under the same conditions (reduced electric field value, excitation degree, ionization degree, external ionization source intensity and so on) two stable states are possible with different EEDFs and, respectively, with different plasma parameters such as mean electron energy, electron drift velocity and others. At present EEDF bi-stability effect was studied mainly theoretically using Boltzmann equation analysis and only implicit experimental evidences of the possibility of this effect were obtained. Below there is a review of physical conditions under which EEDF bi-stability effect was studied. Possible experiments for the observation of effects caused by the transition between two states are also discussed. The second kind is that the positive column of glow discharge can exist in two forms (diffuse and constricted) under the same discharge current. It is well known that the increase of the discharge current leads to the constriction of the diffuse glow discharge. In most cases it looks like step-wise

CP876, *The Physics of Ionized Gases: 23rd Summer School and International Symposium*, edited by L. Hadžievski, B. P. Marinković, and N. S. Simonović
© 2006 American Institute of Physics 978-0-7354-0377-2/06/$23.00

transition after the discharge current exceeds some critical value. It is also known that in some cases hysteresis effect takes place, i.e. transition from constricted to diffuse form occurs at lower current value than transition from diffuse to constricted form. Respectively, there is the range of current values where glow discharge can exist in diffuse form as well as in constricted form. Most of all surprisingly that under certain experimental conditions constricted and diffuse modes can simultaneously exist in the same discharge tube, i.e. steady state partially-constricted glow discharge can be realized. The results of the experimental studies of such kind of discharge are presented in the paper. The problems of modeling this phenomenon are also discussed.

EEDF BI-STABILITY IN LOW TEMPERATURE PLASMAS

General Remarks

The solution of Boltzmann equation (BE) for the electrons in homogeneous plasma under a steady electric field action is usually based on the expansion of the electron velocity distribution function $f(\vec{v})$ in Legendre polynomials $P_n(\cos\Theta)$:

$$f(\vec{v}) = \sum_{0}^{\infty} f_n(v)P_n(\cos\theta),\tag{1}$$

where $\vec{v}$ is the electron velocity, θ is the angle between electron velocity and electric field and functions f_n depend only on the absolute value of electron velocity. If the electric field applied to the plasma is not too high, then the distribution function is almost isotropic and the expansion (1) can be restricted by two terms [1-2]:

$$f(\vec{v}) = f_0(v) + \cos\theta f_1(v).\tag{2}$$

In this case, the steady state Boltzmann equation for the symmetrical part, $f_0(u)$, of the distribution function can be expressed in the following form (further on we use an energy $u = 0.5mv^2$ instead of v):

$$0 = \frac{1}{3}\left(\frac{eE}{N}\right)^2 \frac{d}{du}\left(\frac{u}{Q_m(u)}\frac{df(u)}{du}\right) + I_{e-M} + I_{in} + I_{sup} + I_{e-e},\tag{3}$$

where $f(u) \equiv f_0(u)$, e and m are the charge and mass of electron, E is the electric field strength, $Q_m(u)$ is the momentum transfer cross section and N is the number of atoms and molecules per unit volume. The first term on the right-hand side of equation (3) describes the heating of electrons by the electric field and I_{e-M}, I_{in}, I_{su} and I_{e-e} are the collision integrals for the elastic and inelastic collisions, superelastic collisions, and electron-electron (e-e) collisions, respectively. A detailed description of all terms in equation (3) is given in [3-4]. The normalization condition for the function $f(u)$ is

$$\int_0^\infty \sqrt{u} f(u)du = 1.\tag{4}$$

The equation (3) is usually solved numerically and it is assumed that for a given set of parameters (gas mixture, gas pressure, gas temperature, electric field strength, excitation degree, ionization degree and so on) there is a single solution to this

16

equation. Really, in most cases this is true. But there are conditions under which BE has two essentially different stable solutions.

Qualitative Explanation Of The Possibility Of The Bi-stability Effect

To understand the origin of the bi-stability effect let us consider a simplified situation when the EEDF is Maxwellian: $f(u) \sim exp(-u/T_e)$, where T_e is the electron temperature. Then, the time-dependent equation for the electron temperature can be presented in the following form:

$$\frac{d(T_e)}{dt} = \Phi(T_e) = H(T_e) - L(T_e),$$ (5)

where $H(T_e)$ represents heating of electrons by electric field and $L(T_e)$ describes the losses of electrons energy due to elastic and inelastic collisions. The function $\Phi(T_e)$ can be easily calculated by multiplying terms on the right-hand side of equation (3) by u and integrating over energy.

As it is well known, the number of steady state stable solutions to equation (5) depends on the shape of the $\Phi(T_e)$ function. In particular, if this function has the shape shown in Fig. 1, then there are three steady state solutions to equation (5) ($T_e = T_1$, $T_e = T_1$ and $T_e = T_1$), two of which ($T_e = T_1$, $T_e = T_3$) are stable.

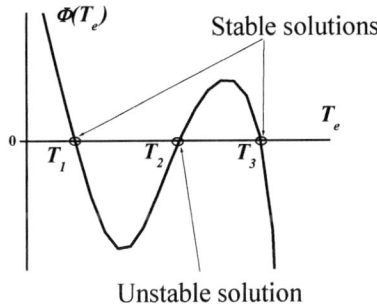

FIGURE 1. The shape of $\Phi(T_e)$ function at which there are two stable solutions to equation (5).

The shape of $\Phi(T_e)$ function depends on the shape of the cross sections for the electron scattering from atoms and molecules and on the value of the reduced electric field, E/N. It is rather obvious that in order to obtain $\Phi(T_e)$ function similar to that shown in Fig. 1 there should be some peculiarities in the shape of cross sections involved in calculations.

EEDF Bi-stability In Plasma Of Heavy Rare Gases

Maxwellian EEDF Approach

Using the Maxwellian EEDF approach, the bi-stability in plasma of heavy rare gases with the applied electric field was predicted in ref. [5]. It was shown that, in a certain range of E/N values, right-hand part of (5) has two stable roots. The result is attributed to the specific shape of the momentum transfer cross sections [6] (see Fig. 2). As an example, in Fig. 3 $H(T_e)$, $L(T_e)$ and $\Phi(T_e)$ functions calculated for pure Xe at $E/N = 0.03$ Td and gas temperature $T_g = 300$ K are presented. It follows from Fig. 3 that the shape of $\Phi(T_e)$ function is similar to that shown in Fig. 1. The detailed examination of $H(T_e)$ and $L(T_e)$ terms has shown that two low-side solutions ($T_e = T_1$ and $T_e = T_2$) are coupled with a rapid decrease in the momentum transfer cross-section $Q_m(u)$ in the energy interval 0–0.6 eV (see Fig. 2). The rate of energy losses of electrons in elastic collisions is proportional to $Q_m(u)$ while the rate of heating is inversely proportional to $Q_m(u)$ (see equation (3)). The exact interplay between these two terms provides the bi-stability. It should be noted that bi-stability effect takes place at low electric fields.

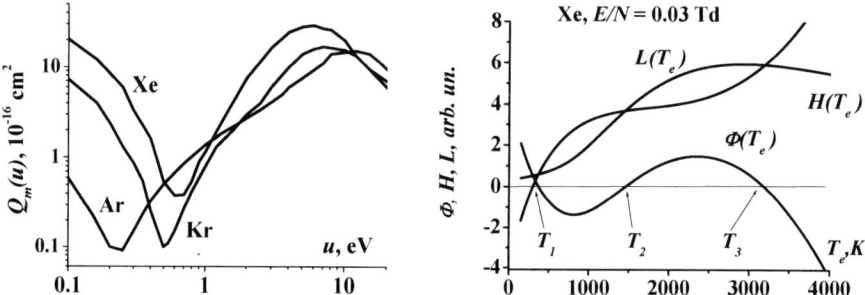

FIGURE 2. Momentum transfer cross sections for the electron scattering from Ar, Kr and Xe atoms.
FIGURE 3. $H(T_e)$, $L(T_e)$ and $\Phi(T_e)$ functions calculated for pure Xe at $E/N = 0.03$ Td and $T_g = 300$ K.

In addition, using Maxwellian distribution function approach the bi-stability effect was studied in [7] for positrons swarms in He.

Boltzmann Equation Analysis (E/N And n_e/N Parametric Studies)

In reality, the EEDF is not Maxwellian in the whole energy range, therefore, the results described in the previous section may serve as an indication the possibility of two stable solutions to the Boltzmann equation in the presence of e-e collisions, since exactly e-e collisions lead to the Maxwellization of the distribution function. Besides, when e-e collisions are taken into account, equation (3) is nonlinear and, thus, can have two or more solutions.

The possible way to find numerically different stable solutions to BE is to solve time-dependent BE (relaxation method) with different initial $f(u)$. If the iteration method is used to solve steady state BE then different initial $f(u)$ should also be exploited. For example, the Maxwellian distribution function with a given electron temperature T_{e0} can be used in calculations as an initial $f(u)$, in this case electron temperature T_{e0} should be varied. The bi-stability effect in plasma of heavy rare gases was studied in [8] by solving the simplified BE taking into account e-e collisions. Recently this effect was revisited by numerically solving the BE [9] with the use of modern data on momentum transfer cross sections. It follows from [8-9] that, in a certain range of parameters (namely, the reduced electric field, E/N, and the degree of ionization, n_e/N, where n_e is the electron concentration), the BE has two stable solutions. It means that in such plasma two different steady states with different plasma parameters can be realized.

The EEDFs in Xe plasma calculated at $E/N = 0.035$ Td and different n_e/N values are shown in Fig. 4. According to calculations, for $n_e/N > 10^{-10}$ there are two stable solutions to the BE. Let us note that for the ionization degree $n_e/N = 10^{-9}$ two calculated $f(u)$ functions (curves 3 and 3' in Fig. 4) differ markedly from Maxwellian function, while for $n_e/N = 10^{-7}$ EEDFs are almost Maxwellian.

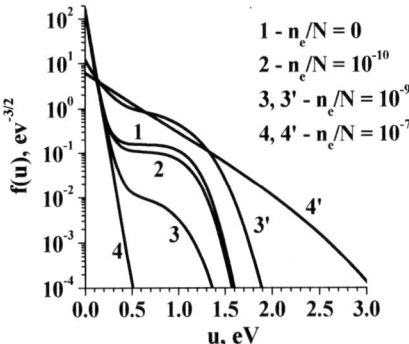

FIGURE 4. EEDFs in Xe plasma calculated at E/N = 0.035 Td and different ionization degrees.

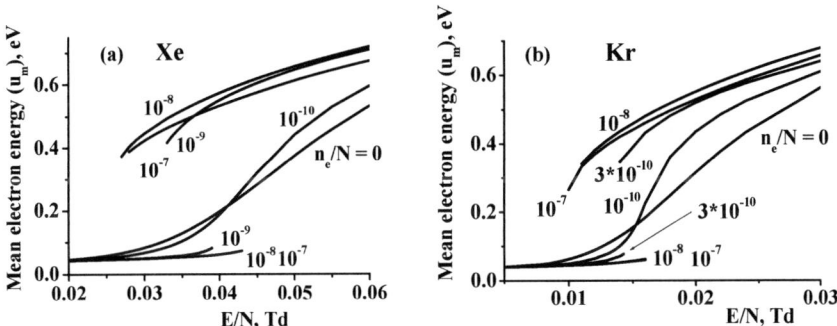

FIGURE 5. Mean electron energy in Xe (a) and Kr (b) plasma as functions of E/N calculated at different ionization degrees.

Mean electron energies, u_m, in pure Xe and Kr calculated at low electric fields are shown in Figs 5a and 5b. As it follows from these Figs, bi-stability effect takes place for $n_e/N > 10^{-10}$ in the E/N intervals $E/N \approx 0.027 \div 0.043$ Td (Xe) and $E/N \approx 0.01 \div 0.015$ Td (Kr). Let us also note that, in the case of bi-stability, two values of u_m are essentially different.

Boltzmann Equation Analysis (Self-Consistent Studies)

In studies described above the electron number density was considered as an independent parameter. Since the range of the electric field strengths where the bi-stability effect has been found lies at rather low electric fields (i.e. electron impact ionization is negligible) the situation considered corresponds to the conditions of the decaying plasma. Besides, the conditions required for the observation of the bi-stability can be realized in non-self-sustained discharge. This opportunity was examined in our papers [10-11], in which the EEDF bi-stability in plasma of heavy rare gases under steady state conditions was studied. The conditions corresponding to a discharge sustained by a beam of fast electrons were considered.

The process of ionization of gas by a steady state fast electron beam (including cascade processes) is characterized by the production rate, q, of the secondary electrons, i.e. the electrons with energies lower than the ionization potential of atoms or molecules. For a given gas mixture, the value of q depends on the e-beam current density and energy of fast electrons. In [10-11] q was taken as an independently varied parameter. The secondary electrons lose their energy in elastic and inelastic collisions and disappear when recombining with positive ions. Under constant e-beam current conditions, a steady concentration of electrons with an appropriate energy spectrum is established in the plasma.

The relevant Boltzmann equation for the EEDF can be written as follows:
$$I_E(n_e f) + qS(u) + St(n_e f) = 0, \qquad (6)$$
where the term $I_E(n_e f)$ describes heating of plasma electrons by electric field, $qS(u)$ is the source of secondary electrons with a spectrum $S(u)$, $St(n_e f)$ is the collision integral (inclusive, in particular, e-e collisions and electron-ion recombination processes). An equation for the electron concentration can be derived by integrating equation (6) over energy:
$$q - n_e^2 \alpha_r = 0, \qquad (7)$$
where α_r is recombination rate coefficient calculated by solving (6). In [10-11] it was assumed that under considered conditions ($P = 760$ Torr) molecular ions dominate in quasineutral plasma.

It should be noted that physical situation under consideration significantly differs from that discussed in previous section. In e-beam sustained plasma the electron number density is found self-consistently. More important is the fact that electrons gain energy not only from the electric field but also from the e-beam.

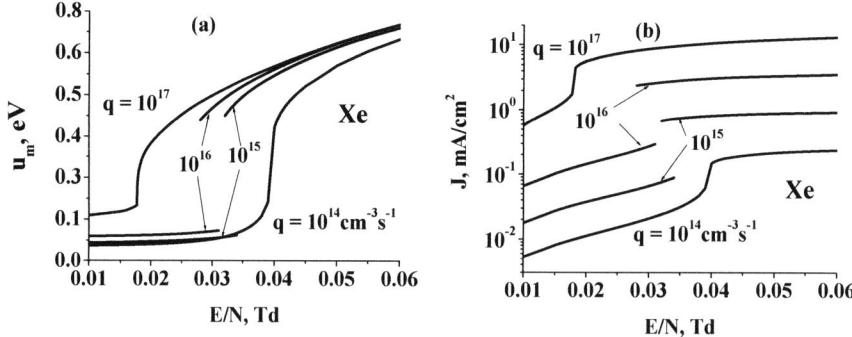

FIGURE 6. Plasma parameters in e-beam sustained discharge in Xe ($P = 760$ Torr, $T = 300$ K) [10]. (a) – mean electron energy, (b) – electron current density.

Equations (6) and (7) were solved in parallel, i.e. $f(u)$ and n_e were calculated in a self-consistent manner. Respectively, the electron drift velocity and plasma electron current density are also determined from this calculation. It was found for Xe [10] and Kr [11] that in a certain range of q and E/N values two different stable plasma states exist, the effect is more pronounced in the case of Xe. For an Ar plasma [11], the BE has a unique solution over the entire parameter range over study. In Fig. 6 the calculated mean electron energy and plasma electron current density in Xe are demonstrated [10]. As can be seen from Fig. 6, bi-stability effect in non-self-sustained discharge in Xe takes place in the interval of q values 10^{14} cm^{-3}s$^{-1} < q < 10^{17}$ cm^{-3}s^{-1}. The difference in discharge currents in bi-stability range is almost ten-fold, and can be easily measured by common diagnostics. As it was estimated in [10], production rate $q = 10^{16}$ cm^{-3}s^{-1} is provided by the e-beam current density $j_b \approx 2$ μA/cm^2 with the energy of fast electrons $\sim$ 300 keV, the e-beam with such characteristics can be easily realized.

Possible Experimental Studies

Theoretical results presented in [10] (see previous sub-section) indicate clearly conditions for the bi-stability effect existence: e-beam sustained discharge, for example, in xenon ($P = 760$ Torr, $T = 300$ K, $j_b \approx 2$ μA/cm^2, $E/N \approx 0.03$ Td). Measuring current-voltage discharge characteristic $I(V)$, could give an evidence of the bi-stability effect. The magnitudes of the voltage applied across the discharge gap should be chosen in such a way to provide the variations of the electric field strength in the range $0 \div 0.06$ Td. In the case of bi-stability, the hysteresis effect should be observed and $I(V)$ function will exhibit a stepwise changes at some V values.

Another possible approach is to investigate the effects associated with transition between two states of the system. For example, if the alternating electric field with a certain amplitude and frequency is applied across discharge gap, then the current will behave differently in response to an increase and decrease in the applied voltage [9].

EEDF Bi-stability In Ar : N$_2$ Afterglow Plasmas

EEDF in nitrogen (and nitrogen containing gas mixtures) afterglow has been a subject of many experimental and theoretical studies [12-18]. A specific feature of EEDF formation in such plasma is that heating of electrons is due to the superelastic collisions of electrons with vibrationally excited molecules and, as a consequence, there is the strong coupling between the vibrational excitation of the nitrogen molecules and the EEDF. It was also shown [16] that e-e collisions strongly influence EEDF in nitrogen afterglow. The degree of the vibrational excitation is characterized by the vibrational temperature T_v, assuming that the population of the lower vibrational levels is close to the Boltzmann distribution. One of the EEDF characteristics is an effective electron temperature $T_e = 2/3 \times u_m$. Theoretical studies of the EEDF in Ar:N$_2$ afterglow plasma [17] were initiated to understand the peculiarities in time variation of the electron temperature observed in experiments [17].

21

Ar : N₂ Afterglow Experiments

Measurements were made in Ar:N₂ = 100:1 mixture at a gas pressure P of 0.5 and 1 Torr. The discharge ran in the pulse repetition mode with pulse current I = 200 – 700 mA. Pulse duration was 40 μs and the pulse repetition frequency was 1 kHz. In experiments the EEDF, as well as the vibrational temperature of N₂ molecules were measured at different delay times after the discharge pulse. The measured EEDF was characterized by electron temperature. Besides, the time dependence of the electron concentration was also estimated. Results of measurements are shown in Fig. 7.

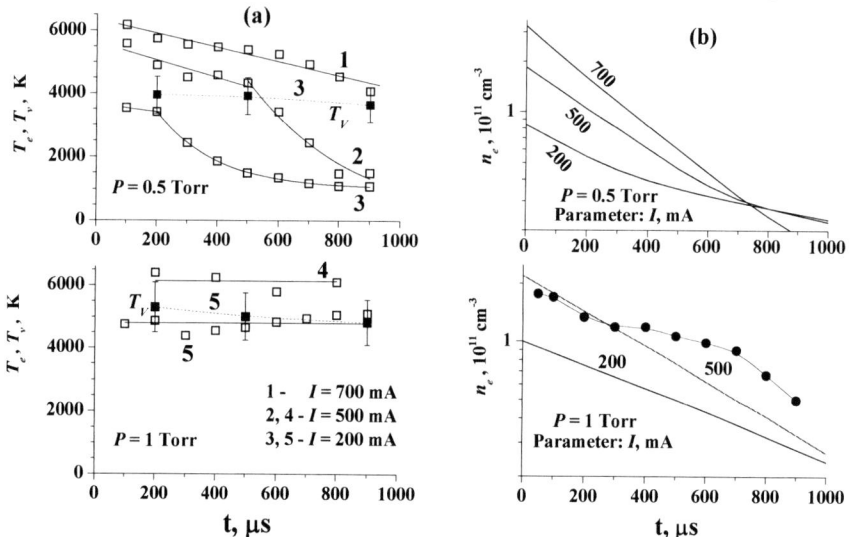

FIGURE 7. (a) Experiment [17]. Electron (open squares) and vibrational (solid squares) temperatures in the afterglow. (b) Experimental (points) and calculated (curves) time dependence of electron density.

It was found that in all cases no noticeable decay of T_v occurs during the post discharge period. On the other hand, rather sharp knee appears in the $T_e(t)$ dependence in the post-discharge at lower pressure and the pulse current. The conditions under study are evidently quasistationary for the formation of the EEDF. Therefore the origin of $T_e(t)$ changing is the varying of plasma parameters influencing it. Of these parameters only electron density shows noticeable time dependence in the afterglow.

Ar : N₂ Afterglow Calculations

Calculations were made by numerical solving the steady state BE for the EEDF:

$$0 = I_{e-M} + I_{in} + I_{sup} + I_{e-e}, \tag{8}$$

Let us note that the steady state BE is appropriate for EEDF calculations in Ar : N₂ afterglow since the electron thermalization time is considerably shorter than the time variations of plasma parameters such as populations of metastable and vibrational levels and electron concentration. In calculations the vibrational distribution has been supposed to be a Boltzmann distribution with a given temperature T_v. The electron concentration was also considered as an independent parameter. The population of all electronically excited states except $A^3\Sigma_u^+$ was neglected. The concentration of $A^3\Sigma_u^+$ state (n_A,) was taken as 10^{10} cm^{-3} and 10^{11} cm^{-3}. Results of calculations [17] for Ar : N$_2$ = 100 : 1 gas mixture are shown in Fig. 8.

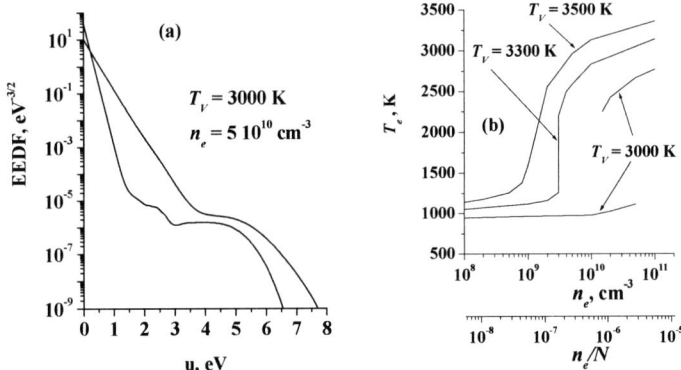

FIGURE 8. Theory [17]. (a) - Two different EEDFs in a post discharge, $P = 0.5$ Torr. (b)-Electron temperature in a post discharge as a function of n_e, $P = 0.5$ Torr, $n_A = 10^{10}$ cm^{-3}.

It appeared that for a given T_v the calculated T_e depends on the electron concentration. In the case of high electron concentrations T_e is close to T_v and slightly decreases as n_e subsides. Then the sharp transition from high to low T_e is observed at some critical n_e value (n_e^*). The lower is the vibrational temperature the higher is the n_e^* value and more sharp is the transition. Moreover, for lower T_v (see Fig. 8) values the range of n_e exists, in which there are two different stable solutions to BE. The bi-stability is provided by the resonance structure of the cross sections for the excitation (and, de-excitation) of nitrogen vibrational levels by electron impact.

Electron density decays in the afterglow. Then the jump-like $T_e(t)$ variation observed in experiments can be explained as a transition from high to low T_e values after the electron density will subside down to some critical value. Described experimental results can be considered as an implicit evidence of the possibility of the bi-stability effect in Ar:N$_2$ afterglow plasma.

The bi-stability effect was also theoretically studied in pure nitrogen afterglow plasma [18]. Besides, it is noteworthy a theoretical work of Hurle [19] where the similar effect was mentioned.

BI-STABILITY OF THE GLOW DISCHARGE PROPERTIES

It is well known that the increase of the discharge current leads to the constriction of the diffuse glow discharge and formation of one or several filaments, transverse sizes of which are small with respect to the discharge chamber diameter (see, for example, [20]). The transition scenario from diffuse to constricted mode depends on the conditions, in particular, on the gas mixture and pressure. In some cases constriction of plasma occurs gradually, i.e. the visible diameter of plasma in the whole of positive column diminishes gradually with current increase. But in most cases it looks like step-wise transition after the discharge current I exceeds some critical value I^*. In this case the hysteresis effect can be observed, i.e. the reverse transition from constricted to diffuse form may occur at slightly lower current value $I^{**} < I^*$ [20]. Therefore in the range of current values $I^{**} < I < I^*$ glow discharge can exist in diffuse form as well as in constricted form, i.e. the bi-stability effect takes place. Most of all surprisingly that under certain experimental conditions constricted and diffuse modes can simultaneously exist in the same discharge tube.

In the case of step-wise transition the time evolution of constriction process can be different. For example, in rare gases rapid change of discharge current from $I_1 < I^*$ to $I_2 \gg I^*$ leads to practically uniform constriction of the positive column along the current. If the discharge current only slightly exceeds critical value, $I_2 \geq I^*$, then the constriction of the positive column arises near the one of electrodes and the boundary between constricted and diffuse parts moves to the other electrode. Respectively, during some time interval constricted and diffuse modes exist in the same discharge volume. In paper [21] discharge current in neon glow discharge was modulated at about I^* value with an appropriate frequency, so that the boundary between constricted and diffuse mode cannot reach the electrode during half a cycle. Visually glow discharge in this case looked like steady state partially constricted discharge. In 1966 steady-state partially constricted glow discharge without discharge current modulation was observed in argon [22], but no special studies of this effect were made. About the observation of such kind of discharge in argon-nitrogen mixture it was briefly reported in [23].

Experimental Observations Of the Steady State Partially Constricted Glow Discharge

Recently we have started systematic studies of the steady state partially constricted glow discharge [24-25]. Preliminary experiments were made using U-shaped discharge tube, the length of the horizontal part of which was about 12 cm and with 2.7 cm i.d. [24].

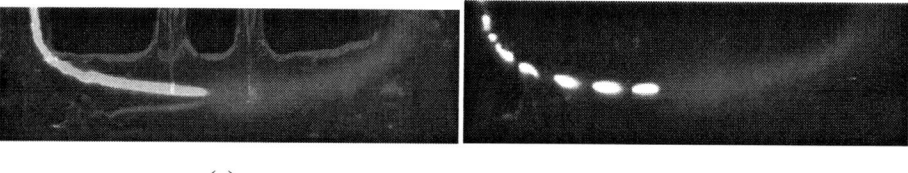

(a)   (b)

FIGURE 9. (a) - An image (obtained with long-time exposition) of the partially constricted steady-state glow discharge in Ar:N₂ mixture. P~30 Torr, I ~100 mA [24]. The cathode is on the left. (b) -Time-resolved (with 150 μs time resolution) image of the discharge under the same conditions.

Fig. 9(a) shows the image recorded at long-time exposition of the discharge under conditions when constricted and diffuse forms are observed simultaneously. The constricted part of the discharge is on the left side of the tube. It is connected to the cathode and looks like narrow bright cord. Such a discharge turned out to be very stable and could exist more than ~ 10 min with no adjusting of parameters. A very small variation of discharge current (in the limit of ~ 1 mA) forced the boundary between the constricted and diffuse parts to shift at some distance along the tube. Switching polarity of the electrodes resulted in changing locations of the two parts. This fact, in particular, means that effect of partial constriction is not governed by the gas flow.

Fig. 9(b) presents the time-resolved image of the discharge under the same conditions with 150 µs exposure. Bright white sports in constricted part of the discharge are the striations (moving striations), this is in agreement with the well known fact that the constricted medium-pressure glow discharge is usually stratified, too. It is noteworthy that no striations are observed in diffuse part of discharge.

It became clear that effect will be more pronounced it the case of the long discharge tube. For this reason the experimental setup was redesigned to do experiments with the discharge tube of 110 cm long (and 2.7 cm i.d.). The first results of studies using new discharge tube are presented in [25].

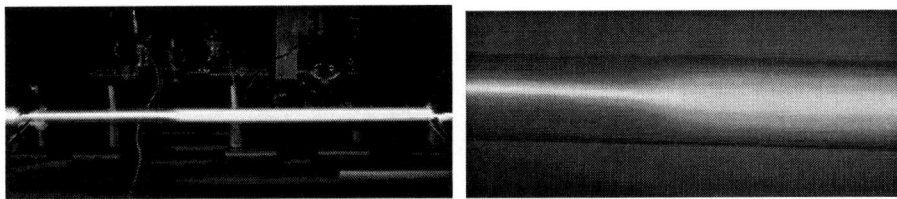

FIGURE 10. L_{tube} = 110 cm. Ar:N$_2$ mixture with 0.07% of N$_2$, P = 39 Torr, $I \approx$ 90 mA. An image of the partially constricted steady-state glow discharge (on the left) [25]. The enlarged view of the discharge region with transition from diffuse to constricted form (on the right).

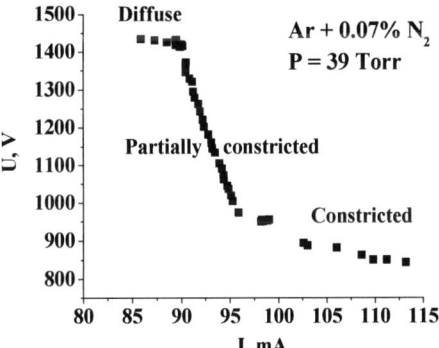

FIGURE 11. L_{tube} = 110 cm. Volt-Ampere characteristic of the glow discharge in Ar:N$_2$ mixture with 0.07% of N$_2$. P = 39 Torr [25]. Measurements were made by variation of discharge current from high to low values.

Fig. 10 shows the image recorded at long-time exposure of the discharge under the conditions when constricted and diffuse forms are observed simultaneously. Fig. 11 presents the measured volt-ampere characteristic of the glow discharge in Ar:N$_2$ mixture. It follows from our measurements that for a given gas mixture steady-state partially constricted glow discharge exists in the narrow interval of discharge current values: 88 mA ÷ 94 mA.

Problems Of Modeling

The very interesting problem is to model partially constricted discharge. In fact, this problem is extremely complicated and requires 2D modeling. As a first step, the possibility of the existence of diffuse and constricted forms of glow discharge under the same discharge current could be analyzed. In this case 1D axial-symmetric model can be used.

A few physical effects are known [20], which provide non-linear dependence of the ionisation rate on the electron concentration and provoke the constriction of diffuse glow discharge: EEDF Maxwellization due to electron-electron collisions, step-wise ionisation and gas heating. Besides, it is known that in pure rare gases radiation transfer processes can noticeably influence on the characteristics of discharge plasma [26].

At present we are developing 1D axial-symmetric discharge model for pure argon in which all mentioned effects are taken into account. This model includes a system of kinetic equations for populations of electronic states, balance equations for the charged particles, the equation for the gas temperature and the equation for the electric circuit. Radiation transfer effects are taken into account in terms of Holstein-Biberman equation. Rate coefficients for electron-induced processes are calculated from solution to the electron Boltzmann equation (with taking into account electron-electron collisions) solved in parallel with the system of kinetic equations.

ACKNOWLEDGMENTS

This work was partially supported by the Russian Foundation for Basic Research project # 06-02-17272.

REFERENCES

1. Shkarofsky, I. P., Johnston, T. W., and Bachinskii, M. P., *The Particle Kinetics of Plasmas*, Reading: Addison-Wesley, 1966.
2. Golant, V. E., Zhilinskii, A. P., and Sakharov, S. A., *Principles of Plasma Physics*, Moskow: Atomizdat, 1977 (in Russian).
3. Huxley, L. G., and Crompton, R. W., *The Diffusion and Drift of Electrons in Gases*, New York: Wiley, 1974.
4. Aleksandrov, N. L. and Son, E. E., "Electron energy distribution and electron kinetic coefficients in gases in the electric field" in *Plasma Chemistry*, edited by B. M Smirnov, Moscow: Energoatomizdat, 1980,Vol. 7, pp 35-75 (in Russian).
5. Gerasimov, G. N., Maleshin, M. N., and Petrov, S. Ya., *Opt. Specrosc.*(USSR) **59**, 562 (1985).
6. Pack, J. L., Voshall, R. E., Phelps, A. V., and Kline, L. E., *J. Appl. Phys.*, **71**, 5363-5371 (1992).
7. Robson, R. E., *J. Chem. Phys.* **85**, 4486 –4501 (1986).
8. Ivanov, V. A., and Prikhod'ko, A. S., *Sov. Phys.-Tech. Phys.* **31**, 1202 (1986).
9. Dyatko, N. A., and Napartovich, A. P., *Proceedings of 15-th ESCAMPIG* (Grenoble, France, 2002), 1, 215-216 (2002).

10. Dyatko, N. A., and Napartovich, A. P., *J. Phys. D: Appl.Phys.* **36**, 2096-2101 (2003).
11. Dyatko, N. A., and Napartovich, A. P., *Plasma Physics Reports* **30**, 953-961 (2004).
12. Capitelli, M., Gorse, C., and Ricard, A., in *Nonequilibrium Vibrational Kinetics,* edited by M Capitelli, Springer-Verlag Berlin, 1986.
13. Gorbunov, N., A., Kolokolov, N. B. and Kudryavtsev, A. A. *Sov.Phys.Tech.Phys.* **36**, 616 (1991).
14. Kudryavtsev, A. A. and Ledyankin, A. I., *Physica Scripta* **53**, 597-602 (1996).
15. Dilecce, G., and Benedictis, S., *Plasma Sources Sci.Technol.* **2**, 119-122 (1993).
16. Dyatko, N. A., Kochetov, I. V., and Napartovich, A. P., *J. Phys. D: Appl.Phys.* **26**, 418-423 (1993).
17. Dyatko, N. A., Ionikh, Yu. Z., Kolokolov, N. B., Meschanov, A. V., and Napartovich, A. P., *J. Phys. D: Appl.Phys.* **33**, 2010-2018 (2000).
18. Dyatko, N. A., Kochetov, I. V., and Napartovich, A. P., *Plasma Physics Reports* **30**, 953-961 (2004).
19. Hurle, I. R., *J.Chem.Phys.* **41**, 3592-3603 (1964).
20. Raizer, Yu. P., *Gas Discharge Physics,* Moscow: Nauka, 1987 (Berlin: Springer-Verlag, 1991).
21. Golubovskii, Yu. B., Kudryavtsev, A. A., Nekuchaev, V. O., Porohova, I. A., Tsendin, L. D., *Electron kinetics in non-equilibrium gas discharge plasmas,* St. Petersburg: St. Petersburg State University, 2004.
22. Garscadden, A. and Lee, D. A., *Int. J. Electronics* **20**, 567 (1966).
23. Ionikh, Y. Z., Chernysheva, N. V., Macheret, S. O., Miles, R. B., *Bull. Am. Phys. Soc.* **44**, 77 (1999).
24. Dyatko, N. A., Ionikh, Y. Z., Meshchanov, A. V., Napartovich, A. P., Yuretskiy, A. V. *Abstracts of Invited Lectures and Contributed Papers* (XVIII-th ESCAMPIG, Lecce, Italy, 2006), pp. 209-210.
25. Dyatko, N. A., Ionikh, Y. Z., Meshchanov, A. V., Napartovich, A. P., Petrov, F. B. *Contributed papers of 23-th SPIG* (Kopaonik, Serbia, 2006), (accepted).
26. Starostin, S. A., Peters, P. J. M., Van der Poel, G., Udalov, Yu. B., Witteman, W. J., Kochetov, I. V., Napartovich, A. P., *Plasma Physics Reports* **27**, 432 (2001).

Coherent Atom Optics with fast metastable rare gas atoms

J. Grucker, J. Baudon [c], J.-C. Karam, F. Perales, V. Bocvarski*, G. Vassilev, and M. Ducloy

Laboratoire de Physique des Lasers, Université Paris 13
Avenue J.B. Clement, 93430-Villetaneuse, France
(*) Institute of Physics, Pregrevica 118, 11080 – Belgrade-Zemun, Serbia
(c) Corresponding author : baudon@galilee.univ-paris13.fr

Abstract. Coherent atom optics experiments making use of an ultra-narrow beam of fast metastable atoms generated by metastability exchange are reported. The transverse coherence of the beam (coherence radius of 1.7 μm for He*, 1.2 μm for Ne*, 0.87 μm for Ar*) is demonstrated *via* the atomic diffraction by a non-magnetic 2μm-period reflection grating. The combination of the non-scalar van der Waals (vdW) interaction with the Zeeman interaction generated by a static magnetic field gives rise to "vdW-Zeeman" transitions among Zeeman sub-levels. Exo-energetic transitions of this type are observed with Ne*($^3P_2$) atoms traversing a copper micro-slit grating. They can be used as a tunable beam splitter in an inelastic Fresnel bi-prism atom interferometer.

Keywords: Matter waves, Metastable Atoms, Scattering from surfaces, Coherent Atom Optics

PACS: 39.10.+j Atomic and molecular beam sources and techniques, 03.75.-b Matter waves, 68.49.Bc Atom scattering from surfaces

I – INTRODUCTION

Manipulation of matter waves at a mesoscopic or a macroscopic scale is the domain of coherent atom optics (CAO). Collisions are generally excluded from this domain, in spite of the fact that they exhibit a number of coherent matter wave phenomena such as interference, because of an incomplete control of the target (ion, atom, molecule). CAO experiments, i.e. essentially diffraction and interference of atoms hitting material obstacles of a micrometric or sub-micrometric size, such as slits [1], double slits [2], gratings [3], started in the late eighties with thermal velocity (a few 100m/s) species (ground-state alkali, alkaline earth, metastable rare gas atoms). Immaterial structures like resonant light standing waves, light pulses, inhomogeneous magnetic field [4] have been used a short time later in place of material ones, delicate to make and expensive at sub-micrometric sizes. Nevertheless the major change in the domain has been the introduction of cold atoms. These are released from traps with velocities ranging from 10^{-2} m/s up to a few m/s [5]. More recently, other powerful tools have been developed, as Bose-Einstein condensates [6] and, probably in a near future, "atom lasers" [7].

CP876, *The Physics of Ionized Gases: 23rd Summer School and International Symposium,*
edited by L. Hadžievski, B. P. Marinković, and N. S. Simonović
© 2006 American Institute of Physics 978-0-7354-0377-2/06/$23.00

It is obviously a great advantage in CAO experiments to deal with de Broglie wavelengths much larger than those existing at thermal velocities (up to a few μm instead of a few tens of pm) and large coherence lengths (several μm up to a few mm). However these are essentially practical points, that do not really affect the general principles. For instance, the fact that matter wavelengths reach today the visible-light range is not a fundamental advantage, in so far as it makes the CAO instrument less accurate than its counterpart at wavelengths of a few tens of pm. It is worthwhile noticing however that the extremely short wavelengths associated with fast atoms represent in fact an ultimate limit, which is, up to now, far from being accessible. This should appear as a strong encouragement to slow down the atoms. Nevertheless there is a wide gap between thermal velocities (e.g. 580 m/s for Ar*) and cold atom velocities (~ 1 m/s or less). Hence the intermediate velocity range (~ 10 m/s to 100 m/s) represents a compromise worth to be investigated whilst it has not been much explored, except recently, e.g. for He* metastable atoms [8] and K atoms [9]. Actually to take benefit of high and medium velocity range has a cost, which is the great difficulty to fulfil with a fast atom beam the criteria imposed by CAO. These criteria deal with velocity and angular resolutions and coherence properties. As it will be seen, a new type of fast rare gas metastable atom beam will allow us to satisfactorily reach at the same time these three criteria, and perform elastic and inelastic diffraction experiments using reflection and transmission gratings.

The paper is organised as follows. In part II the production of an ultra-narrow and coherent metastable atom beam is described. Part III is devoted to the elastic diffraction of He* and Ne* atoms by a 1D micrometric reflection grating, which gives a direct evidence for the transverse coherence of the beam. Part IV reports a special inelastic-scattering experiment realised with a nano-slit transmission grating. It is known that the vdW interaction alters significantly the diffraction patterns [10]. Here we combine the *non-scalar* part of this interaction, with the Zeeman interaction within a static magnetic field, to induce vdW-Zeeman transitions among the atomic magnetic sub-levels [11]. Such transitions can be used as tunable beam splitters. A new scheme of atom interferometer, based upon an inelastic Fresnel bi-prism configuration, is described [12].

II- METASTABILITY EXCHANGE, PRINCIPLE AND BENEFITS

II-1. Principle

Metastability exchange is a resonant process occurring in a collision between two atoms A_1, A_2 having the same atomic number, but not necessarily the same mass number. It can be schematised as follows :

$$A_1^* + A_2 \to A_1 + A_2^* \tag{1}$$

where (*) stands for the metastable state. The process is due to the invariance of the electronic Hamiltonian of the (A_1A_2) molecule as regards the permutation $A_1 \leftrightarrow A_2$. In the Born-Oppenheimer approximation (largely valid in our case), electronic states are either symmetric, and called g as *gerade*, or anti-symmetric and called u as *ungerade*, under this permutation. For sake of clarity, let us consider the simple case of two atoms in states S, e.g. $He^*(2^3S_1) + He(1^1S_0)$, where only two electronic states, $|g\rangle = |^3\Sigma_g\rangle, |u\rangle = |^3\Sigma_u\rangle$, are involved. Obviously neither the incoming state $A_1^* + A_2$ nor the outgoing one, $A_1 + A_2^*$, has a g or u character. The former state is in fact a linear combination :

$$2^{-1/2} (\,|g\rangle + |u\rangle) \qquad (2)$$

It is readily verified that the latter one is :

$$2^{-1/2} (\,|g\rangle - |u\rangle) \qquad (3)$$

For given collision parameters (incident energy, impact parameter or scattering angle), two different potentials $V_g(R)$, $V_u(R)$, which simply are the electronic energies of states $|g\rangle$ and $|u\rangle$, act on the nuclear motion. Each potential induces a specific phase shift [13] (resp. α_g, α_u), which transforms (2) into :

$$|4\rangle = \exp(i\alpha)\,[\,\cos(\delta\alpha)\,|2\rangle + i\sin(\delta\alpha)\,|3\rangle\,] \qquad (4)$$

where $\alpha = (\alpha_g + \alpha_u)/2$ and $\delta\alpha = (\alpha_g - \alpha_u)/2$. $|2\rangle$ being the incoming state, the first term in (4) corresponds to the elastic, so-called *direct*, collision, whereas the second one containing $|3\rangle$ corresponds to the *exchange* process. Notice that this last term is present provided that V_g is not identical to V_u. When P states are involved, such as in the collision $Ar^*(^3P_{0,2}) + Ar(^1S_0)$, more molecular states ($\Sigma_{g,u}$, $\Pi_{g,u}$) are involved, but the description of the exchange process remains basically the same.In   atom-atom collisions at thermal and sub-thermal energy, the direct scattering amplitude is generally peaked at angle $\theta_C = 0$ in the centre-of-mass (CM) frame, whereas, just because of the exchange, the exchange amplitude is peaked around $\theta_C = \pi$ [14]. This suggests that, in a first approximation, the two partners simply exchange their momentum.

II-2. Production of a metastable atom nozzle beam

If one excepts very few cases in which metastable atoms are produced by two-photon transitions, these species are generated in most cases by a DC discharge or by electron bombardment. In the former case, the intensity is high but both velocity spread ($\delta v/v \sim 30\%$) and angular aperture ($\delta\theta \sim 10\text{-}20°$) are too wide to be readily used in a CAO experiment. Then velocity selection and drastic angular collimation (both signal-consuming) are needed to fulfil the CAO criteria [10a]. The bombardment of a Campargue-type nozzle beam of ground-state atoms by electrons the energy of which is far above the excitation threshold (for efficiency reasons) strongly damage the natural properties of such a beam, because of the electron-atom momentum transfer. It makes the velocity spread $\delta v/v$ passing from 1% to 7% or more, and multiplies the original angular aperture ($\delta\theta \approx 0.5$ mrad) by about the same factor. In the present experiment, a nozzle beam of ground-state rare gas atoms (A = He, Ne, Ar) is bombarded by 150 eV contra-propagating electrons, within an axial magnetic field of 150 G, making the electrons spiralling around the field lines and converging towards the skimmer. In spite of that, the relative efficiency of the metastable atom A* production remains small (a few 10^{-4}) but sufficient owing to the initial high flux of ground-state atoms (10^{14} at/s/strd) and to the high detection efficiency ($\approx$ 30%). Nevertheless, as it is, this "primary" A* beam is not suitable for CAO experiments.

Precisely because most of the atoms are not excited, A* atoms propagate within an almost intact A- nozzle beam, at an average velocity slightly lower than that of the ground-state atoms (momentum electron transfer). Then low energy (1-2 meV) collisions occur between these species, giving rise to the metastability exchange process. In fig.1a the Newton (velocity) diagram of the collision is shown, in the lab. frame (initial velocities v_1, v_2 for A and A* atoms respectively, final velocities v'_1, v'_2) and in the CM frame (velocities φ_1, φ_2 and φ'_1, φ'_2). In modulus, v_2 is slightly lower than v_1 and the lab. angle $\gamma = (v_1, v_2)$ is relatively small (< 0.05 rad). Because of the kinetic-energy conservation, all CM velocities have the same modulus and, because of the momentum conservation rule, $\varphi_2 = - \varphi_1$ and $\varphi'_2 = - \varphi'_1$. Let us define a scattering direction in the CM frame by standard spherical co-ordinates : θ_C = angle (φ_1, φ'_1) , ϕ_C azimuthal angle around φ_1. The final velocity distribution of metastable atoms in the lab. frame is derived by standard kinematics relationships from the CM elastic differential cross section (DCS), $d\sigma_C/d\omega_C$, where $d\omega_C$ is the CM elementary solid angle. It includes direct and exchange processes. CM elastic differential cross sections have been measured and calculated in the past for all systems considered here [15, 16], at energies above 10 meV. The general behaviour as a function of θ_C at 1-2 meV is expected to be similar to that of DCS-s at higher energy, with a maximum at $\theta_C = 0$ (direct process), another one at $\theta_C = \pi$ (exchange process) and oscillations (u-g interference) in between. Actually, as we shall see, the details of this behaviour are not of a great importance for our purpose. Indeed we are interested in the metastable atom distribution in the *lab. frame*, which is governed by the lab. DCS : $d\sigma_L/d\omega_L$, $d\omega_L$ being the elementary lab. solid angle. As the elementary atom flux is the same one in both frames, one has : $d\sigma_L = d\sigma_C$, then $d\sigma_L/d\omega_L = d\sigma_C/d\omega_C * R_\omega$, where $R_\omega = (d\omega_C/ d\omega_L)$. This solid angle ratio has been calculated in the case where $\gamma = 0.1$ rad, $v_2 = 0.9 \, v_1$. Fig.1b shows $R_\omega(\theta_L)$ in the vicinity of $\theta_L = 0.1$ rad, i.e. in the region where the exchange is expected. It is seen that the exchange process will be tremendously enhanced and narrowed at $\theta_L = 0.099505$ rad, within a width of the order of 5 µrad. Therefore it can be said that, to a high accuracy, exchanged ("secondary") metastable atoms simply take the velocity of ground-state atoms. In other words the secondary A* atom distribution is almost exactly that of the nozzle beam. One may notice that this quite favourable issue is due to the fact that the two initial velocities (v_1, v_2) are close in modulus and make with each other a rather small angle γ, which makes the direction of φ_1 almost perpendicular to that of v_1.

These theoretical predictions are confirmed by experiment [10b]. A circular diaphragm of diameter 80 µm, is placed on the nozzle beam axis, just beyond the electron gun, at 116 mm from the nozzle and 1280 mm from a position-sensitive detector (a multi-channel plate followed by a phosphor screen and a CCD camera). As primary metastable atoms are emitted from a somewhat extended source (the electron-bombardment zone), the diaphragm gives on the detector an inverted image, of the type *camera obscura*. On another hand straight line atom trajectories within the nozzle beam, and hence within the secondary-metastable atom beam, are emitted from the small sized nozzle (effective diameter, estimated from *umbra-penumbra* considerations, of 15 μm), with a small angular aperture ($\approx$ 0.5 mrad). The corresponding narrow profile is superimposed on the detector to the wide primary metastable profile. Finally, by adding another diaphragm (100 μm in diameter) at 395 mm from the nozzle, one can isolate the secondary metastable nozzle beam (plus a contribution of about 10% of the primaries). Fig.1c shows an example of the final profile for He* atoms. In fact the technique works as well for Ne* and Ar*, with similar properties of the beam, namely (i) a narrow angular aperture (directly measured on the detector) of 0.35 mrad FWHM; (ii) a relative velocity spread $\delta v/v$ (measured by time-of-flight) of about 2% FWHM. In the special case of Ar* it has been possible to measure the ratio of exchanged $^3P_0$ and $^3P_2$ populations, by using a laser beam (closed transition $^3P_2$-$^3D_3$, λ = 811.5 nm) pushing $^3P_2$ atoms out of the initial profile. It turns out that this ratio is *not* the statistical-weight ratio [1:5] but rather [3.38 : 6.62], because the exchange probability is stronger (by a factor of about 2.55) for state $^3P_0$ than for state $^3P_2$. A similar behaviour is expected for Ne*.

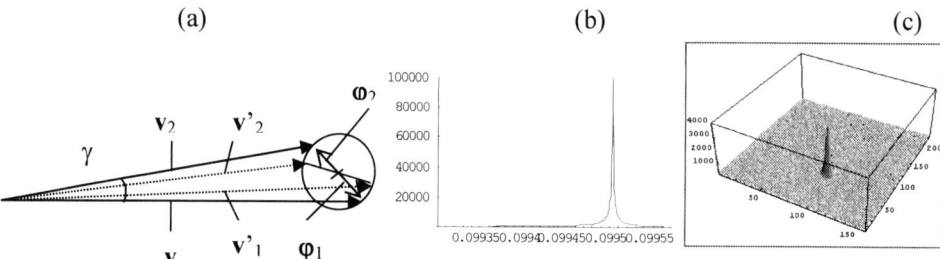

(a)   (b)   (c)

FIGURE 1: (a) Newton diagram: v_1, v_2 (v'_1, v'_2) are the initial (final) velocities in the lab. frame; φ_i, φ_2 (φ'_1, φ'_2) are initial (final) velocities in the CM frame. (b) Ratio of elementary solid angles R_ω = $d\omega_C/d\omega_L$ as a function of θ_L. (c) Experimental 2D profile of the exchanged beam (co-ordinates in pixels)

The longitudinal coherence length $\ell_{//}$ is readily derived from the velocity spread: $\ell_{//}$ = $(v/\delta v)\,\lambda_{//}$, where $\lambda_{//}$ is the longitudinal de Broglie wavelength (0.057 nm, 0.025 nm, 0.018 nm for He*, Ne* and Ar* respectively). For a 2% velocity spread, one gets : $\ell_{//}$ = 2.85 nm for He*, 1.25 nm for Ne* and 0.9 nm for Ar*. The transverse coherence radius $\ell_\perp$ (the beam profile is circular), at a distance d from the nozzle, is given by the van Cittert-Zernike theorem: $\ell_\perp = \dfrac{x_m}{2\pi}\,\lambda_{//}\,\dfrac{d}{a}$, where a is the effective nozzle radius (a = 7.5 μm) and x_m a dimensionless parameter fixing, somewhat arbitrarily, the fraction of the maximum beyond which one considers as negligible the intensity diffracted by the circular aperture. Taking d = 597 mm (cf. §

III) and the limiting value 0.2, one gets: $0.2 = 4 (J_1(x_m)/x_m)^2$, where J_1 is the 1st order Bessel function, then $x_m = 2.36$. One finally obtains the following transverse coherence radii: $\ell_\perp(He^*) = 1660$ nm, $\ell_\perp(Ne^*) = 1236$ nm, $\ell_\perp(Ar^*) = 873$ nm.

All characteristics given above make the secondary metastable atom beam directly usable in CAO experiments, namely elastic and inelastic diffraction by reflection and transmission gratings, without any need of additional angular collimation nor velocity selection.

This conversion from ground-state to metastable atoms may find applications also in *ground-state atom optics*. Indeed it is well known that the detection of such atoms is difficult (ionisation by photons or electrons, followed by a mass analysis) and poorly efficient. The conversion to metastables at the output of an optical device would provide good detection efficiency and the possibility of imaging the beam profile.

III- ATOM DIFFRACTION BY A 1D - REFLECTION GRATING

III-1. Reflection gratings : a problem with fast metastable atoms

An important and general problem in the use of reflection gratings with fast metastable rare gas atoms is the quenching process (*via* Auger effect in most cases). This effect is experienced by atoms approaching a solid at distances short enough to make overlap the atomic outer electronic orbital with the solid electronic cloud, namely at distances shorter than 0.5 nm. This quenching is very efficient at thermal energies since the reflection factor is only of a few 10^{-5}. This is obviously not dramatic for transmission gratings. On the other hand, reflection gratings need a reflection of atoms. This means (for non-magnetic solids) they approach the surface close enough to be repelled by the repulsive part of the interaction (which is otherwise of the van der Waals type), the range of which is precisely that of the quenching. A noticeable exception exists at extremely low normal velocities, obtained by use of falling cold atoms and grazing incidence (quantum reflection) [17]. Grazing incidence angles (e.g. $\theta_i = 7$ mrad) are also interesting in our case, in so far as they reduce the normal velocity (e.g. 12.2 m/s for He* atoms, instead of a longitudinal velocity of 1750 m/s), which results into a reflection factor of about 10^{-4}. It makes sense to choose such small incidence angles since the angular aperture of the beam is 0.35 mrad. It is also worth to notice that, because of the small size of the atom source (effective nozzle radius a = 7.5 μm), an almost perfectly defined velocity vector is present at each point in the beam, i.e., in particular, at each point of the grating lighted by this beam.

III-2. Diffraction by a 1D-non magnetic reflection grating

In the present experiment, carried out with $He^*(2^1S_0, 2^3S_1)$, $Ne^*(3s, ^3P_{0,2})$ and $Ar^*(4s, ^3P_{0,2})$ atoms, the reflection grating is made of parallel aluminium thin stripes (thickness 30 nm) deposited on a silicon substrate through a resist mask realised by electron-beam lithography in a scanning electron microscope (lift-off technique). These stripes are 500 nm wide. They are separated by 1500 nm, leading to a grating pitch $\Lambda = 2$ μm. The grating is placed in the horizontal plane at 597 mm from the nozzle and then rotated by the incident angle θ_i (a few mrad), the stripes being perpendicular to the incident beam axis. Provided that a single incident wave-packet reaches several stripes, an atomic diffraction is expected in the vertical plane. As the transverse coherence radius is $\ell_\perp (He^*) = 1660$ nm, the number of stripes coherently lighted is $\ell_\perp / (\Lambda \, \theta_i) = 118$, for $\theta_i = 7$ mrad. The observed direction, referred to the incident beam axis (angle β), makes an angle $\theta = \beta - \theta_i$ with the grating plane. The condition on θ under which all stripes diffract in-phase amplitudes is:

$$k_{//} \, \Lambda \, (\cos \theta - \cos \theta_i) = - 2\pi \, N \qquad (5)$$

where $k_{//} = 2\pi/\lambda_{//}$ is the longitudinal wave number, and N an integer, the diffraction order (the minus sign is introduce to take in account the fact that here $\theta \geq \theta_i$). As all angles are small (θ itself is limited to 15 mrad by the detector size), Eq.(5) can be approximated by:

$$\theta^2 = \theta_i^2 + A \, N \qquad (6)$$

where $A = 2 \, (\lambda_{//} / \Lambda)$. Numerically $A(He^*) = 57$ mrad^2, $A(Ne^*) = 25$ mrad^2, $A(Ar^*) = 18$ mrad^2.

   The grating being removed, one readily gets the incident axis direction, as being the centre of the beam profile on the position-sensitive detector. This provides us with a direct determination of β angles. On another hand, the inclination θ_i of the grating with respect to the atom beam axis (not necessarily horizontal) cannot be directly measured with a good accuracy. A satisfactory method to obtain θ_i is to compare the diffraction patterns $\beta(N)$ obtained with He*, Ne* and Ar* atoms. Indeed only zero-order peaks (specular reflection) are allowed to coincide. It is true that, strictly speaking, N is not exactly known a priori (uncertainty ±1). In practice however, the $N = 0$ peaks can be unambiguously located, at an angle $\beta(0) = 2 \, \theta_i = 14$ ±1 mrad, then $\theta_i = 7 \pm 0.5$ mrad. Fig.2a shows the diffraction pattern observed with He* atoms. In fig.2b the squared scattering angle, θ^2, is plotted as a function of the diffraction order N, for He*, Ne* and Ar* atoms. As expected from eq.(6), straight lines are obtained. As mentioned before they approximately cross at point $N = 0$, $\theta^2 = \theta_i^2 \approx 49$ mrad^2. Their slopes are about 56, 27 and 23 mrad^2, i.e. rather close to theoretical values (57, 25, 18 mrad^2) if one takes in account the uncertainty on the angle values ($\delta\theta = \pm 0.5$ mrad), which leads to a large uncertainty on θ^2 : $\delta(\theta^2) = \pm 18$ mrad^2 at $\theta^2 = 324$ mrad^2.

   As it was mentioned at the end of § II, the longitudinal coherence lengths $\ell_{//}$ may appear as short at the grating scale, in spite of the fact they cover 50 longitudinal wavelengths or more. Strictly speaking, the in-phase condition (eqs. 5, 6) is valid for an incident plane wave. One may wonder about its validity for an incident wave packet of a finite longitudinal extension $\ell_{//}$. To answer this question, one has to consider the overlap of wave packets scattered by 2 subsequent stripes. Ignoring the natural wave-packet spreading (negligible in the present case), the time delay between the two wave-packet centres is given by : $\Delta t = \Lambda (\cos \theta - \cos \theta_i) / v_g$, where $v_g = \hbar k_{//} / m$ is the group velocity, which corresponds to the spatial shift $\Delta \ell = \Lambda (\cos \theta - \cos \theta_i)$.Then an overlap exists provided that $\Delta \ell \le \ell_{//}$. This leads to the condition : $k_{//} \Lambda (\cos \theta - \cos \theta_i) \le 2\pi \, v/\delta v$, i.e. $N \le v/\delta v$. This simply means that the maximum observable diffraction order is $v/\delta v \approx 50$, a value far above our experimental range since $\theta_{50} \approx 54$ mrad for He*, 36 mrad for Ne* and 31 mrad for Ar*. Exactly the same result would be obtained by considering the incident (longitudinal) wave-packet as a superposition of plane waves, over the spectral range $k_{//} [1-\delta v/(2v)]$, $k_{//} [1+\delta v/(2v)]$.

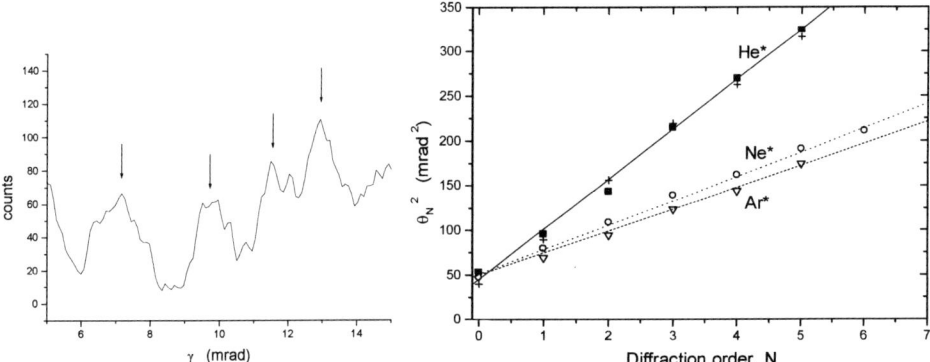

FIGURE 2 : Atom diffraction by a 1D non magnetic reflection grating (period $\Lambda = 2$ μm). (a) profile of the diffracted He* beam (vertical direction, γ is the observation angle, see text). (b) Plot of the squared scattering angle, θ^2 (mrad^2), as a function of the diffraction order N: full line, solid squares and crosses, 2 experiments with He* ; dotted line, open circles, Ne* ; dashed line, open triangles, Ar* (from Ref. [20]).

III-3. Magnitude of diffraction peaks

The angular location of diffraction peaks only involves the grating period Λ. On another hand, the peak magnitudes are not so easy to measure (because of the residual background signal) nor to predict. *Nevertheless it is worthwhile to examine this point since it represents the link between diffraction and interaction.* Exactly as in the case of a transmission grating, the diffraction pattern is given by :

$$I(\theta) = i(\theta) \left(\sin \left[(p+1) \varphi / 2 \right] / \sin (\varphi / 2) \right)^2 \qquad (7)$$

where p is the number of active stripes and $\varphi = k_{//} \Lambda (\cos \theta - \cos \theta_i)$. In this formula, $i(\theta)$ is the diffraction pattern produced by a single stripe. In the case of a transmission grating, the single-slit diffraction pattern is rather easily calculated, even when the phase-shift effect due to the vdW interaction with the slit walls, is taken into account [10a]. The reason for this relative simplicity is that straight-line trajectories can be considered as a starting "zero order approximation". No such a reference exists in the case of reflection, even in the semi-classical approximation. An exception is the situation in which the distance of closest approach is zero, i.e. when the potential $V(z)$ is modelled by an infinitely high wall at $z = 0$, preceded ($z > 0$) by some attractive potential. The effect of the wall is simply to transform any trajectory into its symmetric counterpart with respect to the plane $z = 0$. Consequently the calculation of phase-shifts can be made in a configuration where the wall is absent, $V(z)$ being replaced by $W(z)$, an even function of z. One has then to consider an incident trajectory making the angle θ_i with the grating plane, and the emerging trajectory issue from an element $[(y, y+dy), (Z, Z+dZ)]$ and inclined by some prescribed angle θ with respect to the grating plane. For sake of simplicity, let us consider the case where $W(z)$ is a square-shaped well of depth ($- E_m$) and extension 2ε. The phase-shift between amplitudes scattered at $z = Z$ and $z = 0$ is :

$$\Phi(Z) = k_{//}\, Z\, (v - 1)\, \left((\sin\beta)^{-1} - (\sin\beta_i)^{-1}\right) \tag{8}$$

where v is the index given in the semi-classical approximation by :

$$v = (1 + E_m / E_0)^{1/2} \approx 1 + E_m / (2E_0) \tag{9}$$

provided that $E_m \ll E_0$ (the incident kinetic energy). Angles β and β_i are given by the Descartes-Snell law : $\cos\theta_{(i)} = v \cos\beta_{(i)}$. Setting $\xi = E_m / (2E_0)$ and assuming that θ and θ_i are small angles, one gets from (8) :

$$\Phi(Z) = k_{//}\, Z\, (v - 1)\, (\xi/2)^{1/2}\, [(1 - \tfrac{1}{2}\theta^2/\xi)^{-1/2} - (1 - \tfrac{1}{2}\theta_i^2/\xi)^{-1/2}] \tag{10}$$

The total scattered amplitude for a given incident ray is then :

$$u = \int_{-\varepsilon}^{+\varepsilon} dZ \exp[i\,\Phi(Z)] = 2\varepsilon\, \sin[\Phi(\varepsilon)] / \Phi(\varepsilon) \tag{11a}$$

The overall scattered amplitude is simply the product of u by the width b of a stripe. If the potential well is sufficiently deep so that $\xi \gg \theta^2$ and θ_i^2, then :

$$u \approx (\varepsilon/8)\, (k_{//}^2/k_m)\, (\theta^2 - \theta_i^2) \tag{11b}$$

where $k_m = \hbar^{-1} \sqrt{2m\,E_m}$. Under such a condition the argument in $i(\theta) = b^2\, |u|^2$ is $\theta^2 - \theta_i^2$, i.e. that present in the grating factor. Choosing $k_m = k_{//} / 100$, i.e. $E_m \approx 0.64$ μeV (a value compatible with the condition $\xi \gg \theta^2$ and θ_i^2), the first zero of i is obtained for $\Phi(\varepsilon) = \pi$, i.e. $(\theta^2 - \theta_i^2)_0 \approx 0.04\, \lambda_{//} / \varepsilon$. As an example, for He*, with $\lambda_{//} = 0.057$ nm and $\varepsilon = 4$ nm, one gets $(\theta^2 - \theta_i^2)_0 \approx 570$ mrad^2, i.e. ten times the constant $A(He^*)$ (eq. 6) governing the location of the grating diffraction peaks. This is an indicative but experimentally compatible result, at least if one assumes that the effect of quenching is just to reduce the intensity by a constant factor.

IV- VAN DER WAALS-ZEEMAN TRANSITIONS: ATOMIC BEAM SPLITTERS AND INTERFEROMETER

IV-1. Origin of van der Waals-Zeeman (vdW-Z) transitions

It is well known that the van der Waals (vdW) interaction between a metastable atom and a planar metallic surface, the conductivity of which is assumed infinite, is given by $V_v(d) = -(4\pi\varepsilon_0)^{-1}(16d^3)^{-1}(D^2 + D_z^2)$ where d is the atom-surface distance and $\mathbf{D}$ the atomic dipole operator. The latter factor can be expanded as $(4/3)$ $\mathbf{D}^2 + (D_z^2 - \mathbf{D}^2/3)$, i.e. as a sum of two irreducible tensors $T_0^{(0)}$ (scalar) and $T_0^{(2)}$ (quadrupolar). This interaction can be expressed as well with the atomic angular momentum $\mathbf{J}$ as : $V_v = -C_3/d^3 - \eta[(\mathbf{J}.\hat{n})^2 - \mathbf{J}^2/3]/(16d^3)$. Here $\hat{n}$ is the normal to the surface, C_3 and η are constants that can be evaluated from spectroscopic data.

Let us now consider the collision of a thermal energy metastable Ne* atom (velocity v_0 = 780 m/s) with a copper grating immersed in a static magnetic field $\mathbf{B}$. The parallel bars of the grating are modelled by cylindrical rods of radius a = 1 μm. The atomic trajectory, assumed to be a straight line (x-axis) perpendicular to the rod axis (z-axis), passes at a distance ρ (the impact parameter) from the surface (see fig.3a). Compared to the vdW potential range (a few nm up to a few 10 nm) the rod radius is large enough to make locally valid the above expression of V_v. The magnetic field (a few hundred G) lies in the x,y plane and makes an angle β with the y-axis. As ρ << a, the factor d^{-3} where $d = [(a+\rho)^2 + x^2]^{1/2} - a$, is a rapidly varying function of x, sharply peaked at x = 0 (FWHM $\Delta x \approx 1.02 (a\rho)^{1/2}$ << a). Therefore (i) the normal to the surface $\hat{n}$ rotates by a small angle $\Delta x/a$ along the interaction zone Δx, (ii) the interaction time $\Delta t = \Delta x/v_0$ is much smaller than the Larmor period T_L in $\mathbf{B}$, e.g. $\Delta t \approx$ 0.1 T_L for B = 100G. Under such conditions a *sudden approximation* can be used to treat the effect of the surface on the spin evolution and, during this interaction, $\hat{n}$ remains practically aligned with the y-axis. This leads to an analytic expression of the transition probability from a Zeeman sub-level, m, to another one, m' [18] :

$$P_{m,m'}(v_0, \rho) = \left| \langle m | \text{Exp}[-i\,\alpha(v_0,\rho)J_y^2] | m' \rangle \right|^2 \qquad (12)$$

where $\alpha(v_0, \rho) = 3\pi\eta(2a)^{1/2}(128\,\hbar v\,\rho^{5/2})^{-1}$. It may be noticed that only the non-scalar part of the vdW interaction (coefficient η) participates into these transitions. For a non polarised beam, one gets the averaged probability :

$$\overline{P}_{\Delta m}(v_0, \rho) = (2J+1)^{-1} \sum_{m'=m+\Delta m} P_{m,m'}(v_0, \rho) \qquad (13)$$

A numerical calculation carried out with Ne* atoms (v_0 = 780 m/s) for an angle β = 12° (the experimental value) shows that $\overline{P}_{-1}(v_0,\rho)$ exhibits, as a function of ρ, an upper threshold at $\rho_{th} \approx$ 3 nm and oscillates for $\rho < \rho_{th}$, between 0 and several %.

IV-2. Experiment

Owing to this relatively large transition probability, we have undertaken an experiment to evidence exo-energetic vdW-Z transitions ($\Delta m = -1, -2, -3, -4$) with Ne* ($^3P_2$) metastable atoms traversing an inclined copper grating, in the presence of a static magnetic field ranging from 0 up to 600G. This experiment has been described in details elsewhere [11] and we only give here a summary of the results. It is readily verified that, because of energy and momentum conservation rules, exo-energetic transitions ($\Delta m < 0$) induce deflections of the atomic trajectory by angles

$$\gamma = (\Delta E/E_0)^{1/2} = 0.360 \ 10^{-3} \ B^{1/2}(G^{1/2}) \ |\Delta m|^{1/2} \tag{14}$$

In our case, these angles are a few mrad. They can be resolved owing to both the narrow angular spread of the incident beam and the good angular resolution of the detector. A clear signature of vdW-Z transitions is the dependence of γ on B : in a diagram [γ, $B^{1/2}$], inelastic peak locations are indeed aligned along straight lines the slope of which are, according to (14), $0.360 \ 10^{-3} \ |\Delta m|^{1/2}$ (fig.3b). As the transverse coherence radius ($\ell_\perp = 1236$ nm for Ne*) is largely above the range of the vdW interaction, coherent atomic wave packets are scattered at various γ angles, for a given magnitude of B. Hence the surface behaves as a *multiple beam splitter*, the separation angles of which are tunable, simply by changing B. This inelastic effect is readily applicable to atom interferometry. Nevertheless we shall examine now a similar effect occurring at the two edges of a slit, which results, as it will be seen, into a new and very simple scheme of atom interferometer.

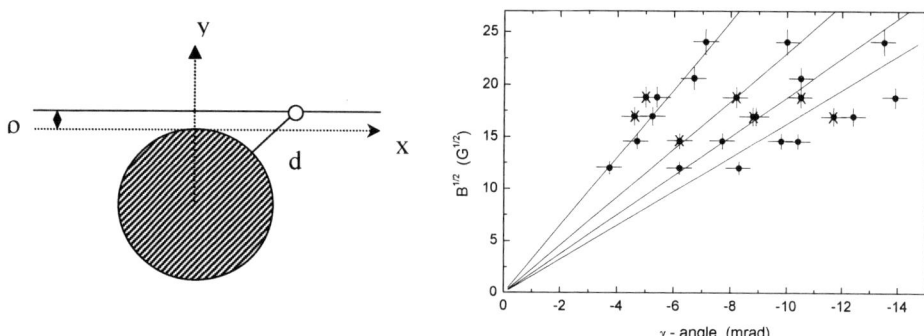

FIGURE 3 : Left side : Scheme of atom-solid collision. The solid is modelled by a cylindrical rod, the impact parameter is ρ, the distance atom-surface is d. Right side : diagram (γ (mrad), $B^{1/2}$ ($G^{1/2}$)) where γ is the location of inelastic peaks; full lines: theoretical prediction for $\Delta m = -1, -2, -3, -4$.

IV-3. Inelastic Fresnel bi-prism

Let us consider a single slit of width w = 5μm, the two edges of which are modelled, as before, by two parallel cylindrical metallic rods of radius a = 1μm. With a slowed down spin-polarised metastable atom beam, e.g. He* (2^3S_1, m = +1) at a velocity $v_0 = 17.5$ m/s ($E_0 \approx 6.5$ μeV), the transverse coherence radius ($\ell_\perp \approx 6$μm) can

38

easily be made larger than w. In such a case the two slit borders are coherently shined. Then they produce two inelastically scattered wave packets, e.g. with $\Delta m = -1$ (final state : m = 0), accompanied by two opposite repulsive deflections ($+ \gamma$ and $- \gamma$). The elastically scattered atoms remain in state m = +1. They are, in principle, easily pushed out, using a resonant laser beam.

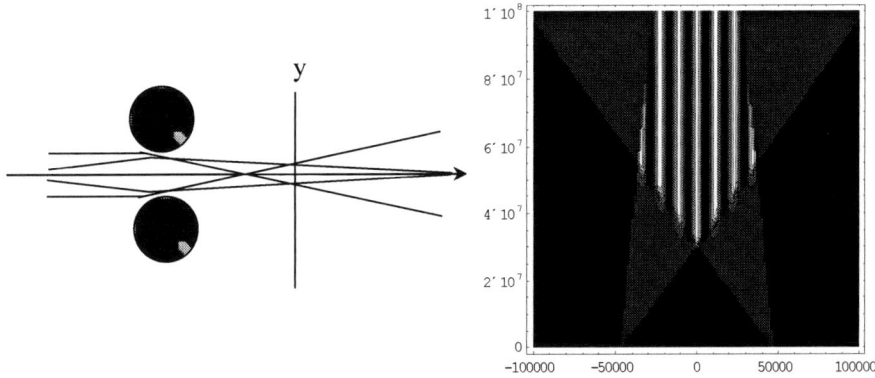

FIGURE 4. Bi-prism interferometer. Left side : 2 slit borders are modelled by 2 cylindrical rods. Trajectories (rays) are shown for a single incident wave packet experiencing a vdW-Zeeman transition ($\Delta m = -1$) on both sides (angles have been amplified). The vertical line indicates where a detector must be placed to observe an interference pattern along the vertical axis (see text). Right side: interference pattern in a diagram y (horizontal pattern axis), x (vertical pattern axis). Distances are in atomic units.

Using an appropriate value of the magnetic field B, the deflection angle γ can be chosen in such a way that the two wave packets make an overlap at a rather large distance x_{ov} from the slit (see fig 4a). For example $x_{ov} = 5$ mm corresponds to $\gamma = 0.5$ mrad. This overlap gives rise to a so-called non-localised interference. The initial width of the wave packets in the direction $\hat{y}$ perpendicular to the incident axis $\hat{x}$ is relatively small because of the short range of the transition probability $P(\rho)$: under the present conditions it can be estimated to be about 10 nm. However, because of the angular aperture of the incident (coherent) beam (≈ 0.35 mrad), this width increases as the wave packets propagate along $\hat{x}$, becoming as large as the slit itself. It is worth noticing that all trajectories involved in the process experience the same deviation, independently of the incidence angle on the surface. This makes each border work as a stigmatic device, exactly as a small-angle prism should do. The whole system actually works as a Fresnel bi-prism. In this geometry, under the above conditions, an approximate straightforward calculation of the interference pattern along $\hat{y}$ gives a spacing between 2 fringes of 0.57µm, the envelope of the interference pattern covering a range of several mm around x_{ov} (fig.4b). Standard position-sensitive detectors do not reach such a spatial resolution, but this difficulty can be overcome in the following way : a secondary-emission plate is placed at x_{ov}, slightly inclined with respect to x-axis. For an inclination of 1°, the fringe spacing on the plate becomes 32.6 µm. When the metastable-atom impacts on this plate are imaged by an electron microscope of magnification 50-100 [19] the spacing becomes 1.63-3.26 mm. A series of slits (grating) would give repeated interference patterns. Such a device could be readily exploited as a very compact two-arm atom interferometer. In the present example the area in between the 2 arms is 0.025 mm^2, which leads to a reasonable sensitivity to inertial effects as gravity or rotation.

We use here a continuous beam and the signal is time-independent. Nevertheless, as already discussed in § III.2, some attention must be paid to the length of the wave packets along $\hat{x}$, namely $\ell_{//} \approx 50 \lambda_{//} = 285$ nm. Indeed as soon as the difference of paths exceeds this length the two wave packets no longer overlap, i.e. no longer interfere. Obviously this makes somewhat critical the positioning of the two slit edges in front of each other. Nevertheless this positioning should be successfully achieved at an accuracy of 0.1 μm, by a small rotation around the 3rd axis ($\hat{z}$), since 0.1 μm corresponds to a rotation of 20 mrad.

REFERENCES

1. O. Nairz, M. Arndt, A. Zeilinger, Phys.Rev. A **65** (2002) 32109
2. O. Carnal and J. Mlynek, Phys. Rev. Lett. **66**, 2689 (1991)
3. D W. Keith, C. R. Ekstrom, Q. A. Turchette, and D. E. Pritchard, Phys.Rev.Lett. **66**, 2693 (1991)
4. S. Nic Chormaic et al., Phys.Rev.Lett., **72**, 1 (1994) ; M. Boustimi et al., Phys.Rev.A, **61**, 033602 (2000)
5. D.S. Weis, B.C. Young and S. Chu, Phys.Rev.Lett., **70**, 2706 ; F. Shimizu, K. Shimizu and T. Takuma, Phys.Rev.A, **46**, R17 (1992)
6. T. Schumm, S. Hofferberth, L. M. Andersson, S. Wildermuth, S. Groth, I. Bar-Joseph, J. Schmiedmayer, P. Kruger, Nature Physics **1**, 572005 (2005)
7. T. Lahaye and D. Guéry-Odelin, Phys. Rev. A **73**, 063622 (2006)
8. A.E.A. Koolen et al., Phys.Rev.A, **65**, 041601 (R) (2002)
9 J. Catani et al., Phys.Rev.A, 73, 033415 (2006)
10. (a) R.Brühl et al., Europhys. Lett., **59**, 357 (2002); (b) J.-C. Karam et al. J.Phys.B, At. Mol. Opt.Phys.,**38**, 2691 (2005)
11. J.-C. Karam, J. Grucker, M. Boustimi, F. Perales, V. Bocvarski, J. Baudon, G. Vassilev, J. Robert and M. Ducloy,
    Europhys.Lett., **74**, 36 (2006)
12. J. Grucker et al. 38th EGAS Conference, Ischia, Italy (2006), book of abstracts, p.155
13. In this simplified description, only electronic states are considered. Actually the relative motion of the 2 atoms, involving impact parameter, scattering angle, etc., makes the situation more complex. Nevertheless the essential features of the exchange mechanism are given by indiscernibility this description.
14. For identical isotopes, because of atom y, an additional interference appears, combining scattering amplitudes at angles θ_C and $\pi - \theta_C$. This effect is especially important around $\theta_C = \pi/2$, an angular domain not considered here.
15. F.T. Smith, D.C. Lorents, W. Aberth, R.P. Marchi, Phys.Rev.Lett., **15**, 742 (1965)
16. B. Brutschy, H. Haberland, J.Phys.E **10**, S90 (1979) ; I. Colomb de Daunant, G. Vassilev, J. Baudon, B. Stern, J.Physique, **43**, S591 (1982)
17. F. Shimizu, J. Fujita, Phys.Rev.Lett., **88**, 123201 (2002)
18. J.-C. Karam, Thesis, University Paris 13 (2005)
19. A.R. Milosavljevic., V Bocvarski, J. Jureta., B. Marinkovic. J.-C.Karam, J. Gruker, F. Perales., Vassilev, J.
    Reinhardt, J. Robert, J. Baudon, Meas. Sci. Technol., **16**, 1997-2004, (2005).
20. J. Grucker et al., submitted to publication, Europhys. Lett. (2006)

Role of Excited Nitrogen In The Ionosphere

L. Campbell[1], M. J. Brunger[1], D. C. Cartwright[2] and M. A. Bolorizadeh[3]

[1]ARC Centre for Antimatter-Matter Studies, SoCPES, Flinders University,
GPO Box 2100, Adelaide 5001 Australia
[2]formerly of Theoretical Divisions, B285, LANL, Los Alamos, NM 87545, USA
[3]Physics Department, Shahid Bahonar University of Kerman, Kerman, Iran

Abstract. Sunlight photoionises atoms and molecules in the Earth's upper atmosphere, producing ions and photoelectrons. The photoelectrons then produce further ionisation by electron impact. These processes produce the ionosphere, which contains various positive ions, such as NO^+, N^+, and O^+, and an equal density of free electrons. $O^+(^4S)$ ions are long-lived and so the electron density is determined mainly by the density of $O^+(^4S)$. This density is dependent on ambipolar diffusion and on loss processes, which are principally reactions with O_2 and N_2. The reaction with N_2 is known to be strongly dependent on the vibrational state of N_2 but the rate constants are not well determined for the ionosphere. Vibrational excitation of N_2 is produced by direct excitation by thermal electrons and photoelectrons and by cascade from the excited states of N_2 that are produced by photoelectron impact. It can also be produced by a chemical reaction and by vibrational-translational transitions. The vibrational excitation is lost by deexcitation by electron impact, by step-wise quenching in collisions with O atoms, and in the reaction with $O^+(^4S)$. The distribution of vibrational levels is rearranged by vibrational-vibrational transitions, and by molecular diffusion vertically in the atmosphere. A computational model that includes these processes and predicts the electron density as a function of height in the ionosphere is described. This model is a combination of a "statistical equilibrium" calculation, which is used to predict the populations of the excited states of N_2, and a time-step calculation of the atmospheric reactions and processes. The latter includes a calculation of photoionisation down through the atmosphere as a function of time of day and solar activity, and calculations at 0.1 s intervals of the changing densities of positive ions, electrons and N_2 in the different vibrational levels. The validity of the model is tested by comparison of the predicted electron densities with the International Reference Ionosphere (IRI) of electron density measurements. The contribution of various input parameters can be investigated by their effect on the accuracy of the calculated electron densities. Here the effects of two different sets of rate constants for the reaction of vibrationally excited N_2 with $O^+(^4S)$ are investigated. For reference, predictions using the different sets are compared with laboratory measurements. Then the effect of using the different sets in the computational model of the ionosphere is investigated. It is shown that one set gives predictions of electron densities that are in reasonable agreement with the IRI, while the other set does not. Both sets result in underestimation of the electron density at the height of the peak electron density in the atmosphere, suggesting that either the amount of vibrational excitation or the rate constants may be overestimated. Our comparison is made for two cases with different conditions, to give an indication of the limitations of the atmospheric modeling and also insight into ways in which the sets of rate constants may be deficient.

Keywords: ionosphere, electron impact, vibrationally excited nitrogen
PACS: 82.33.Tb; 34.80.Gs

CP876, *The Physics of Ionized Gases: 23rd Summer School and International Symposium*,
edited by L. Hadžievski, B. P. Marinković, and N. S. Simonović
© 2006 American Institute of Physics 978-0-7354-0377-2/06/$23.00

INTRODUCTION

The electron density is one of the most important parameters of the Earth's ionosphere. It is determined mainly by the density of $O^+(^4S)$ ions, which are relatively long-lived. They are removed in reactions with O_2 and N_2, and redistributed with height by ambipolar diffusion. The reaction rate with N_2 is strongly dependent on the vibrational level of N_2. We have developed an atmospheric model which includes detailed calculations of electron impact excitation of the vibrational levels of N_2 and the subsequent decay processes of these levels. Applying this model for different sets of reaction rates for the $O^+(^4S) + N_2$ reaction is therefore a useful test of both the model and the reaction rates. We do this for two cases with different latitude, longitude and solar activity, to given an indication of the uncertainty in the atmospheric modeling process and some insight into discrepancies between predictions and measurements.

COMPUTATIONAL MODEL

We build on two empirical models [1] which are fits to a wide range of measurements of atmospheric parameters. The MSIS-E-90 model gives the densities of neutral components, while the IRI-2001 (International Reference Ionosphere) gives the neutral and electron temperatures and the electron density, the latter being the "measured" value to which our predictions will be compared.

Photoionisation rates are calculated down through the atmosphere, giving production rates of positive ions and photoelectrons. This calculation is based on a standard model of sunlight intensities for 37 EUV (5–105 nm) lines or wavelength ranges, appropriately scaled for solar activity, and corresponding photoionisation cross sections [2]. Intensites for wavelengths less than 20 nm are increased by factors suggested in more recent work [3]. The ionisation of O atoms and ionisation and dissociation of N_2 and O_2 are calculated in each 1-km layer of the atmosphere and the sunlight intensity exiting the layer is correspondingly reduced. The solar zenith angle (as a function of time of day) and curvature of the Earth are taken into account. Rates are calculated for production of $O^+(^4S)$, $O^+(^2D)$, $O^+(^2P)$, O_2^+, N_2^+, NO^+, N^+, $N(^4S)$ and $N(^2D)$. It is assumed that the photoelectrons lose energy in a series of subsequent collisions, ending up as thermal electrons with a density equal to that of all the positive ions.

Further ionisation and dissociation by the photoelectrons is calculated by multiplying a photoelectron spectrum [4] by the electron impact cross sections for ionisation of O [5], O_2 [6] and N_2 [6,7] and dissociation of N_2 [8]. The photoelectron spectrum is adjusted for solar activity and zenith angle by multiplying by the ratio of the calculated photoionisation rate for the particular zenith angle and solar activity to the photoionisation rate calculated for the conditions of the given photoelectron spectrum.

The main reactions and decay processes [9–12] of the ions and N atoms produced by photoionisation and electron impact are then accounted for in a time-step procedure [13], where the density of the reactants and products is calculated at 0.1 s intervals. For each time step the densities of all species are calculated at each 10-km mark in the altitude range 130–600 km. Ambipolar diffusion of $O^+(^4S)$ [14] and corresponding electrons is applied between adjacent heights at each time step.

The excitation rates of N_2 excited to various vibrational levels are calculated by considering both direct electron impact (by thermal plus photoelectrons, producing both excitation and deexcitation) [15] and radiative decay from the excited states of N_2, the latter being produced by electron impact on N_2 [16] by photoelectrons. For thermal electrons, it is assumed that they have a Maxwell-Boltzmann distribution appropriate to the electron temperature from the IRI. This distribution is converted to an isotropic flux, for compatibility with the photoelectron spectrum. The total flux at each energy is then multiplied by the electron-impact excitation cross section at that energy, with the product summed over all energies to give the excitation rate.

Due to the large number of different transitions between the excited states of N_2 and the wide range of transition probabilities, the densities of the excited states are determined in a "statistical equilibrium" calculation [17]. In this the gain due to electron impact and radiative transitions into each level is equated with the loss due to radiative transitions out of the level and quenching in collisions. This gives an estimate of the population density of the N_2 molecules in each state and level. The calculation for all states and levels is repeated until the calculated densities reach equilibrium. These equilibrium densities are then used in the time-step procedure to determine their contribution to the ground electronic state vibrational levels.

Vibrationally excited N_2 is also produced by the reaction:

$$N(^4S) + NO \rightarrow N_2(v'' = 4) + O$$

where v'' is the vibrational level [18]. The distribution among vibrational levels is reconfigured by VV (vibrational-vibrational) and VT (vibrational-translational) interactions [19]. Excited N_2 can drop one vibrational level at a time in collisions with O atoms [20]. It disappears in the reaction with $O^+(^4S)$ [12], and also moves downwards in the atmosphere by molecular diffusion [21,18]. The population densities of N_2 in different vibrational levels are calculated for all of these processes at each 0.1-s time step.

VIBRATIONALLY EXCITED NITROGEN

Two sets of rate constants for the reaction:

$$O^+(^4S) + N_2(v'') \rightarrow NO^+ + N(^4S) \tag{R1}$$

are investigated. One is the set advanced by Van Zandt and O'Malley [22], and the other is a combination of the rate for $N_2(v'' = 0)$, determined by St. Maurice and Torr [23] for ionospheric conditions, with the values for $v'' > 0$ given by multiplying by rate coefficients determined by Schmeltekopf et al. [24]. Reaction rates as a function of v'' are plotted for 5 temperatures in Figure 1 for each set.

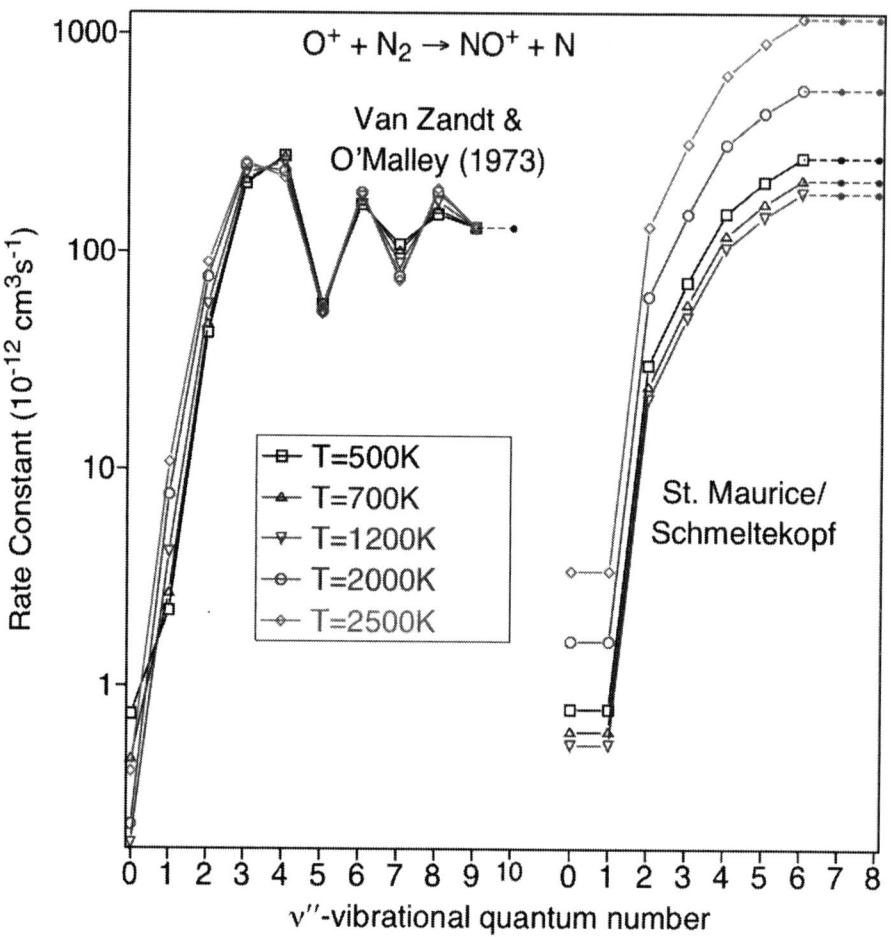

FIGURE 1. Rates for the reaction $O^+(^4S) + N_2 (v'') \rightarrow NO^+ + N(^4S)$ as a function of v'' as predicted using the theory of Van Zandt and O'Malley (left) and the theory of St. Maurice and Torr with the rate coefficients of Schemltekopf *et al.* (right). The rates are given for temperatures of 500, 700, 1200, 2000 and 2500 K. Extrapolation ($\bullet--\bullet--\bullet--\bullet$) is used above the highest level defined.

It can be seen in Figure 1 that there is a large enhancement of the rate constant for vibrationally excited N_2. The most significant difference between the two sets is the ratio of the rate for $v'' = 1$ to that for $v'' = 0$, which is a greater than 10 at higher temperatures for the rates of Van Zandt and O'Malley, but unity using the enhancement coefficients of Schemltekopf *et al.*

The actual reaction rate in a gas is calculated by weighting the rates in Figure 1 by the fraction of N_2 molecules in each vibrational level and summing over all levels. For a Boltzmann distribution of the N_2 levels this gives the reaction rates plotted in Figure 2, where they are compared with laboratory measurements by Hierl *et al.* [25].

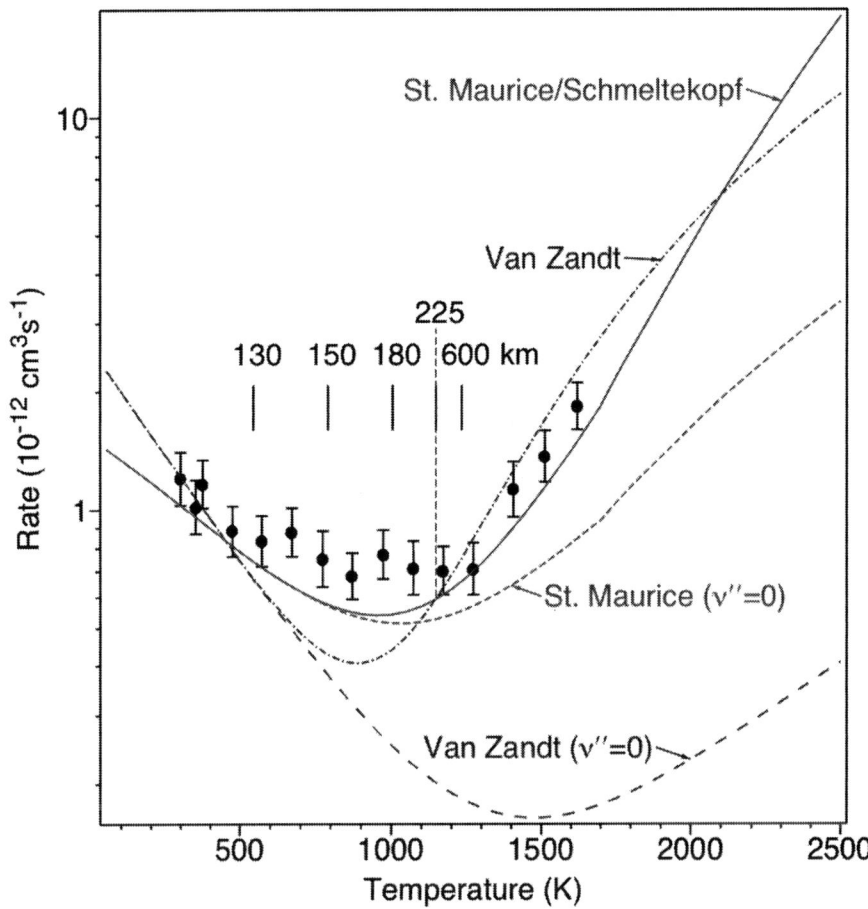

FIGURE 2. A comparison of predicted and measured rates for $O^+(^4S) + N_2 (v'') \rightarrow NO^+ + N(^4S)$ as a function of temperature. The measurements of Hierl *et al.* (●) are shown with error bars. Predictions according to Van Zandt and O'Malley are shown for $N_2(v'' = 0)$ (– – – –) and for a Boltzmann distribution of the vibrational levels of N_2 (– - – - – -). The predictions of St. Maurice and Torr are shown for $v'' = 0$ (-----) and for a Boltzmann distribution with the enhancement coefficients of Schmeltekopf *et al.* (———) applied. The temperature range relevant to the atmosphere is indicated at neutral temperatures for the heights 130, 150, 180, 225 and 600 km (|).

ATMOSPHERIC CALCULATIONS

Altitude markers are placed in figure 2 to indicate the temperature range corresponding to the temperature of neutral components in the atmosphere. It can be seen that in the altitude range 135–225 km both sets underestimate the measured reaction rates, while for 225–600 km both sets give a rate within the error range of the measurements. These rates for the two sets are similar (*e.g.* at about 1150 K), despite the large differences in the enhancement seen in Figure 1, because the higher $v'' = 0$ contribution of St. Maurice and Torr is cancelled by the lower enhancement with vibrational quantum number (particularly for $v'' = 1$) of Schmeltekopf *et al.* The similarity of the rates for the altitude range 180–600 km suggests that atmospheric calculations of the electron density will not be significantly different for the two different sets. This is confirmed in Figure 3, where the calculated electron density is compared with the IRI for the two sets of rate constants applied to a Boltzmann distribution of N_2 vibrational levels.

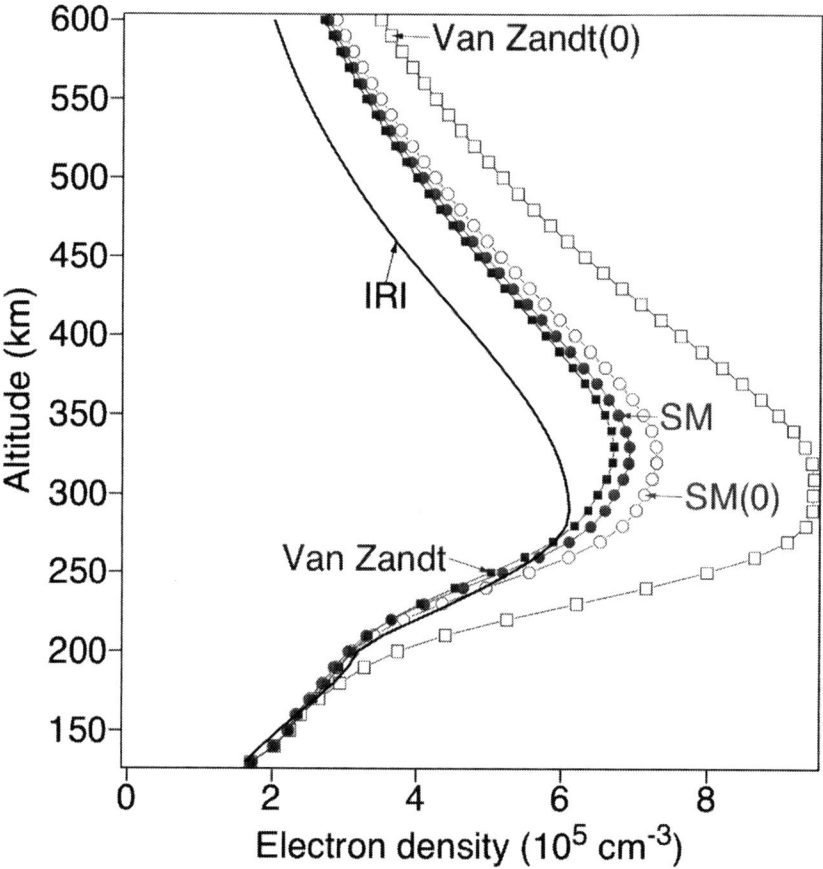

FIGURE 3. Calculated electron densities, as a function of altitude (on the vertical axis), using the rate constants of Van Zandt and O'Malley, for no vibrational excitation (□—□—□—□) and for a Boltzmann distribution of N_2 vibrational levels (■—■—■—■). The same calculation is applied using the rate constants of St. Maurice and Torr for no vibrational excitation (○—○—○—○), and using these with the enhancement coefficients of Smeltekopf *et al.* for a Boltzmann distribution (●—●—●—●). "Measured" electron densities are indicated by the predictions of the IRI empirical model (——).

Figure 3 shows calculated values of the electron density at noon, produced by running the time-step calculation at 0.1 s intervals from 4am until noon, for solar maximum conditions on 21st June 2001 at 0°E, 60°N. In Figure 3 it can be seen that using the $N_2(v'' = 0)$ rate of St. Maurice and Torr gives good agreement with the IRI up to 260 km. There is only a slight change when vibrationally excited N_2 is taken into account, using the rate enhancement factors of Schemltekopf *et al.* Using the $v'' = 0$ rate of Van Zandt and O'Malley leads to much higher values of the electron density but this is mitigated (as explained above for Figure 2) by their higher rates for vibrationally excited N_2, so that the results are very similar to those using the St. Maurice and Torr plus Schmeltekopf *et al.* values.

Above 250 km the $v'' = 0$ rates of both sets yield electron densities above the IRI value. In the range 250–270 km inclusion of the enhanced rates for vibrationally excited N_2 leads to agreement with the IRI, but above this both sets of rate constants give an electron density in excess of the IRI. This disagreement could be because the population of the vibrational levels of N_2 is not a Boltzmann distribution in the ionosphere. This is investigated below by applying the two sets of rate constants to the distributions of vibrational levels predicted by the detailed model described above *i.e.* as produced by the combined effects of electron impact excitation and deexcitation, chemical excitation, VV and VT transitions, chemical reactions and molecular diffusion.

This detailed model is run with and without vibrational enhancement of reaction (R1) for each set of reaction rates, for the situation above and also for 49°N, 238°E on June 28th 1980. In Figure 4 the electron density profiles at noon, calculated with the rates of Van Zandt and O'Malley, with and without applying the enhanced rates for $v'' > 0$, are plotted for comparison with the IRI values. The two different atmospheric situations are labeled with the latitude. The "no enhancement" electron densities at 60°N are the same as in Figure 3, because the same reaction rate for all vibrational levels is used and so the overall rate is not affected by a different distribution of levels. In the 49°N case the densities calculated without the enhanced rates are higher than in the 60°N case. This can be explained if the $v'' = 0$ reaction rate is too low. As the neutral temperatures are higher in the 49°N case (due mainly to higher solar activity), any underestimation of the $v'' = 0$ reaction rate has a greater effect, in that a lower rate produces a lesser reduction in the calculated electron density.

With vibrational enhancement there is good agreement in Figure 4 between the calculated and IRI electron densities only for 130–200 km and near 600 km in the 60°N case and for 130–150 km in the 49°N case. The lack of agreement in the lower altitude range at 49°N can be explained by the higher neutral temperatures coupled with an underestimated $v'' = 0$ rate. As there is no difference between the electron densities predicted with and without enhancement at these altitudes, it is evident that reaction (R1) here is dominated by the $v'' = 0$ component. As the $v'' = 0$ rate decreases with increasing temperature, the effect of an underestimated rate becomes apparent at a lower altitude in the 49°N than in the 60°N case, as the neutral temperatures in the 49°N case are higher.

Above 200 km the vibrational excitation of N_2 has a significant effect on the rate of reaction (R1) and hence on the calculated density. When the enhancement is excluded, the electron density is far higher than the IRI at all altitudes for both atmospheric situations. When it is included, the calculated densities are too low, suggesting that the $v'' > 0$ rate constants for (R1) are too high. An exception is the altitude range 200–250 km at 49°N, which can be explained as being due to the reaction rate for $v'' = 0$ being too low at higher temperatures, with the compensation for this being by the vibrationally excited levels not becoming sufficiently large until 250 km.

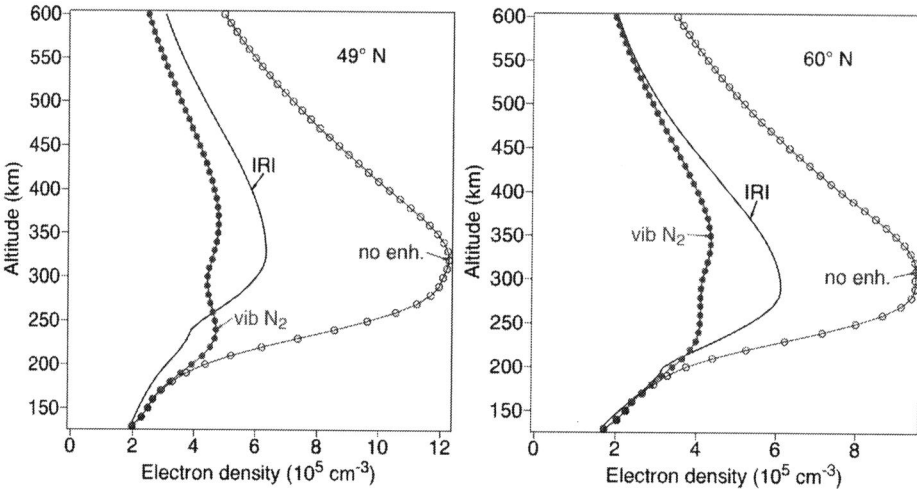

FIGURE 4. Calculated electron densities (as a function of altitude vertically) with (●—●—●—●) and without (O—O—O—O) the enhancement for excited levels of N_2, plus the densities from the IRI empirical model (————). The rate constants of Van Zandt and O'Malley are used.

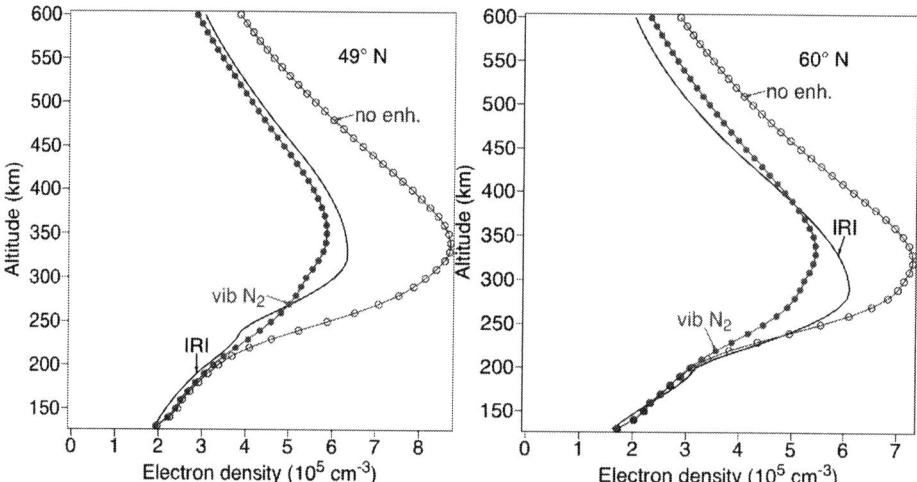

FIGURE 5. As in Figure 4, except that the $v'' = 0$ rate of St. Maurice and Torr and the rate enhancement coefficients of Schmeltekopf *et al.* are used.

The results for repeating the same calculations using the $v'' = 0$ rate of St. Maurice and Torr with the enhancement coefficients of Schmeltekopf *et al.* are shown in Figure 5. Here the agreement with the IRI is good below 220 km for both atmospheric situations, suggesting that the $v'' = 0$ rate of St. Maurice and Torr is appropriate at

48

lower altitudes. Above this the electron density is overestimated when the enhancement due to vibrationally excited nitrogen is not included. When it is included, the agreement with the IRI densities is much better than in Figure 4, suggesting that this second set of rate constants is more accurate. In both atmospheric cases the electron density is somewhat underestimated in the region of the peak electron density, suggesting that enhancement coefficients are still too high, or that the $v'' = 0$ rate of St. Maurice and Torr is too high at higher altitudes. The deviations could also be due to the cumulative effect of inaccuracies in electron impact rates and quenching rates for vibrationally excited N_2, and in the IRI and MSIS models. The deviations from the IRI densities above the peak may not be significant as they are related to the density at the peak by ambipolar diffusion.

In comparing Figure 5 with Figure 3, it can be seen that a Boltzmann distribution of the N_2 vibrational levels gives a good prediction of the electron density up to 260 km using either set of rate constants, while using the calculated distribution leads to an underestimation of the density in the altitude range 210–340 km. A number of factors could account for this, including overestimation of electron impact, underestimation of VV rates, or overestimation of the enhancement coefficients. Also, it is possible that the agreement for the Boltzmann distribution is fortuitous, as there are uncertainties in many other reactions involved in the calculation. For example, the recombination rates of NO^+, O_2^+ and N_2^+ with electrons depend on the vibrational excitation level of the ions, but the distribution of these vibrationally excited levels is not known [26].

Above 260 km, the use of the Boltzmann distribution gives predicted electron densities that are too high, while the detailed calculation gives values that are a little low. Thus it is shown that it is necessary to include the non-Boltzmann distribution of N_2 vibrational levels in calculations of the electron density, but the effect appears to be overestimated in the current calculation.

CONCLUSIONS

A computational model of the ionosphere is used to investigate the effect of vibrationally excited N_2. The reaction of N_2 with $O^+(^4S)$ reduces the electron density, and the reaction rate and density reduction are enhanced if N_2 is vibrationally excited. This is investigated by running the calculation with and without the enhancement for two sets of rate constants for the reaction. If a Boltzmann distribution for the vibrational levels of N_2 is assumed, both sets of rate constants lead to predictions of electron densities which are in agreement with measurements in the lower ionosphere, but are too high at greater altitudes. The analysis is repeated for a non-Boltzmann distribution of the N_2 levels, which is predicted in a detailed calculation that includes electron impact and other excitation, VV and VT transitions, chemical reactions and molecular diffusion. It is found for both sets of reaction rates that the predicted electron density is always too high if the enhancement is not included. When it is included, the predicted density is generally too low if the rate constants of Van Zandt and O'Malley are used. Much better agreement with the IRI empirical model is obtained using the rate constant of St. Maurice and Torr with the enhancement coefficients of Schmeltekopf *et al.* This second set is also better for predictions of the electron density in the lower ionosphere in a second atmospheric situation, where the temperatures are higher. However, the predicted densities are still a little lower than measurements in the regions of the peak electron density, suggesting that this second set of rate constants are also somewhat on the high side. Other possible sources of the discrepancies are noted.

ACKNOWLEDGMENTS

This work was supported by an Australian Research Council Grant.

REFERENCES

1. http://modelweb.gsfc.nasa.gov/
2. P. G. Richards, J. A. Fennelly and D. G. Torr, *J. Geophys. Res.* **99**, 8981-8992 (1994).
3. S. C. Solomon, S. M. Bailey and T. N. Woods, *Geophys. Res. Lett.* **28**, 2149-2152 (2001).
4. D. J. Strickland (private communication).
5. Y. Itikawa and A. Ichimura, *J. Phys. Chem. Ref. Data* **19**, 637-651 (1990).
6. B. G. Lindsay and M. A. Mangan, "Cross sections for ion production by electron collisions with molecules" in *Landolt–Börnstein, 17/C,* edited by Y. Itikawa, Berlin: Springer, 2003, pp. 5-1–114.
7. D. C. Cartwright (private communication).
8. P. C. Cosby, *J. Chem. Phys.* **98**, 9544-9553 (1993).
9. C. A. Barth, *Planet. Space Sci.* **40**, 315-336 (1992).
10. D. J. Strickland, J. Bishop, J. S. Evans, T. Majeed, P. M. Shen, R. J. Cox, R. Link and R. E. Huffman, *J. Quant. Spectrosc. Radiat. Transfer,* **62** 689-742 (1999).
11. D. G. Torr, K. Donahue, D. W. Rusch, M. R. Torr, A. O. Nier, D. Kayser, W. B. Hanson and J. H. Hoffman, *J. Geophys. Res.,* **84** 387-392, (1979).
12. A. V. Pavlov, *Ann. Geophysicae,* **16** 589-601 (1998).
13. L. Campbell, D. C. Cartwright, M. J. Brunger and P. J. O. Teubner, *J. Geophys. Res.* (in press).
14. M. J. Buonsanto, D. P. Sipler, G. B. Davenport and J. M. Holt, *J. Geophys. Res.* **102**, 17267-17274 (1997).
15. L. Campbell, M. J. Brunger, D. C. Cartwright and P. J. O. Teubner, *Planet. Space Sci.* **52**, 815-822 (2004).
16. L. Campbell, M. J. Brunger, A. M. Nolan, L. J. Kelly, A. B. Wedding, J. Harrison, P. J. O. Teubner, D. C. Cartwright and B. McLaughlin, *J. Phys. B: At. Mol. Opt. Phys.* **34**, 1185-1199 (2001).
17. D. C. Cartwright, *J. Geophys. Res.* **83**, 517-531 (1978).
18. B. Jenkins, G. J. Bailey, A. E. Ennis and R. J. Moffett, *Ann. Geophysicae* **15**, 1422-1428 (1997).
19. A. S. Kirrilov, *Ann. Geophysicae* **16**, 838-846 (1998).
20. A. V. Pavlov and A. A. Namgaladze, *Geomag. Aeron* **28**, 607-620 (1988).
21. A. G. Kolesnik, *Geomag. Aeron.,* **22**, 601-607 (1982).
22. T. E. Van Zandt and T. F. O'Malley, *J. Geophys. Res,* **78**, 6818-6820 (1973).
23. J. P. St.-Maurice and D. G. Torr, *J. Geophys. Res.* **83**, 969-977 (1978).
24. A. L. Schmeltekopf, E. E. Ferguson and F. C. Fehsenfeld, *J. Chem. Phys.* **48**, 2966-2973 (1968).
25. P. M. Hierl, I. Dotan, J. V. Seeley, J. M. Van Doren, R. A. Morris and A. A. Viggiano, *J. Chem. Phys.* **106**, 3540-3544 (1997).
26. C. H. Sheehan and J.-P. St.-Maurice, *J.Geophys Res.* **109**, A03302, doi:10.1029/2003JA010132 (2004).

Non-equilibrium electron transport in gases: Influence of magnetic fields on temporal and spatial relaxation

R. D. White[*], B. Li[†], S. Dujko[*], K. F. Ness[*] and R. E. Robson[**,*]

[*]*School of Mathematics, Physics and IT, James Cook University, Townsville, QLD Australia*
[†]*School of Physics, University of Sydney, NSW 2006, Australia*
[**]*Research School of Physical Sciences, Australian National University, ACT 2600, Australia*

Abstract. The ability to control the temporal and spatial relaxation of electron swarms in gases through application of an orthogonal magnetic field is examined via solutions of Boltzmann's equation. Multi-term solutions of Boltzmann's equation are presented for two specific applications: temporal relaxation in the time-dependent hydrodynamic regime, and spatial relaxation in the steady state non-hydrodynamic regime. We highlight the commonality of methods and techniques for handling the velocity dependence of the phase-space distribution function as well as their point of departure for treating the spatial dependence. We present results for model and real gases highlighting the explicit influence of the magnetic field on spatial and temporal relaxation characteristics, including the existence of transiently negative diffusion coefficients.

Keywords: electron swarms, Boltzmann equation, transient phenomena, spatial relaxation
PACS: 51.10.+y, 52.25.Dg, 52.25.Fi, 52.25.Xz, 52.20.Fs

INTRODUCTION

Non-equilibrium, low-temperature plasma discharges sustained and controlled by electric and magnetic fields are widely used in a variety of scientific and industrial fields [1]. Within these discharges the electric and magnetic fields can vary in space, time and orientation depending on the type of discharge. Although there has been a tremendous amount of research into temporal and spatial non-locality of electron transport for electric fields only (see e.g. the review [2]), the study of such effects where magnetic fields are included explicitly has not developed to such a level (see e.g. the reviews [3, 4]). This may be due in part to the unavoidable additional complexity associated with introducing the magnetic field into the theories and simulations. When a magnetic field is present, in general transport can not be accurately predicted from d.c. *E* only data. There is no easy way out when high accuracy is required. The effects of the magnetic field must be included explicitly in the kinetic and/or simulation equations and approximations such as equivalent/effective fields avoided. The present paper will emphasize this and focus on two main areas:

- *Temporal non-locality*: In recent times, the transient kinetic behaviour of electron swarm transport coefficients under the combined action of electric and magnetic fields has been systematically addressed by the Petrović group (see e.g. the review [4]) and the JCU group (see e.g. the review [3]). Also of note is the work by the group at Griefswald (see e.g. [5]). These studies have unearthed a variety

CP876, *The Physics of Ionized Gases: 23ʳᵈ Summer School and International Symposium*,
edited by L. Hadžievski, B. P. Marinković, and N. S. Simonović
© 2006 American Institute of Physics 978-0-7354-0377-2/06/$23.00

of new and important phenomena - the existence of transient negative diagonal diffusion tensor elements in rf E and B fields being particularly noteworthy [6]. This temporal non-locality means that many of these new kinetic phenomena are not predictable from d.c. electric-field (or combined d.c. electric and d.c. magnetic field) swarm data and they highlight important kinetic information that may need to be included or accounted for in plasma models. In this paper we focus on the extension of the two-term work in [5] to include spatially inhomogeneous transport coefficients (diffusion tensor elements) in a multi-term framework. We demonstrate that transiently negative diagonal diffusion elements can also be achieved through switching on a d.c. magnetic field of sufficient strength.

- *Spatial relaxation*: The explicit influence of a magnetic field under non-hydrodynamic conditions has been addressed in [7, 8, 9]. In this paper, we study an idealised steady-state Townsend experiment including a superimposed orthogonal magnetic field and demonstrate its ability to control the spatial re-laxation characteristics of the electron swarm. Through tuning of the orthogonal magnetic field strength, we demonstrate that we can suppress or induce oscillatory relaxation behaviour (Franck-Hertz oscillations) as well as enhance or retard spatial relaxation lengths.

In Theory section we outline the general multi-term solution of Boltzmann's equation highlighting the requirements for treating the space dependence of the phase-space distribution function under hydrodynamic and non-hydrodynamic conditions. In Results section we present results highlighting the explicit influence of an orthogonal magnetic field on the temporal relaxation under hydrodynamic conditions and the spatial relax-ation under non-hydrodynamic conditions.

THEORY

The behaviour of electrons in gases under the influence of electric and magnetic fields is described by the phase-space distribution function $f(r, c, t)$ representing the solution of the Boltzmann equation

$$\frac{\partial f}{\partial t} + c \cdot \frac{\partial f}{\partial r} + \frac{q}{m} [E + c \times B] \cdot \frac{\partial f}{\partial c} = -J(f, f_0), \qquad (1)$$

where r and c denote the position and velocity co-ordinates, q and m are the charge and mass of the swarm particle and t is time. The electric and magnetic fields are assumed spatially homogeneous and orthogonal with magnitudes E and B respectively. Swarm conditions are assumed to apply and $J(f, f_0)$ denotes the rate of change of f due to binary particle-conserving collisions with the neutral molecules only. The original Boltzmann collision operator [10] and its semiclassical generalization [11] are used for elastic and inelastic processes respectively. In what follows, we employ a co-ordinate system in which qE is in the z-direction, while qB is in the y-direction.

Representation of the velocity dependence

The velocity dependence of f is represented in terms of a combined spherical harmonic and Sonine polynomial expansion:

$$f(r,c,t) = w(\alpha,c) \sum_{v=0}^{\infty} \sum_{l=0}^{\infty} \sum_{m=-l}^{l} F(vlm|r,\alpha,t) R_{vl}(\alpha c) Y_m^{[l]}(\hat{c}), \tag{2}$$

where

$$w(\alpha,c) = \left(\frac{\alpha^2}{2\pi}\right)^{3/2} \exp\left\{\frac{-\alpha^2 c^2}{2}\right\} \tag{3}$$

$$R_{vl}(\alpha c) = N_{vl} \left(\frac{\alpha c}{\sqrt{2}}\right)^l S_{l+1/2}^{(v)}\left(\frac{\alpha^2 c^2}{2}\right) \tag{4}$$

$$N_{vl}^2 = \frac{2\pi^{3/2} v!}{\Gamma(v+l+3/2)}, \tag{5}$$

$Y_m^{[l]}(\hat{c})$ are spherical harmonics, $\hat{c}$ denotes the angles of c, $S_{l+1/2}^{(v)}\left(\frac{\alpha^2 c^2}{2}\right)$ are Sonine polynomials and $\alpha^2 = \frac{m}{kT_b}$. The modified Sonine polynomials satisfy the orthonormality relation

$$\int_0^{\infty} w(\alpha,c) R_{v'l'}(\alpha c) R_{vl}(\alpha c) c^2 dc = \delta_{v'v}\delta_{l'l}. \tag{6}$$

The various properties of the moments due to symmetry and reality considerations carry over from the steady state theory and are described in [12].

Using the appropriate orthogonality relations the following system of coupled differential equations for the moments $F(vlm;r,t,\alpha)$ is generated:

$$\sum_{v'=0}^{\infty} \sum_{l'=0}^{\infty} \sum_{m'=-l'}^{l'} \left[\left(\frac{\partial}{\partial t}\delta_{vv'} + NJ_{vv'}^l(\alpha)\right) \delta_{l'l}\delta_{m'm} + i\frac{qE}{m}\alpha(l'm10|lm) < vl||K^{[1]}||v'l' > \delta_{m'm} \right.$$

$$+ \frac{1}{2}\frac{qB}{m}\left\{ \sqrt{(l-m)(l+m+1)}\delta_{m'm+1} - \sqrt{(l+m)(l-m+1)}\delta_{m'm-1} \right\}\delta_{l'l}\delta_{v'v}$$

$$\left. - i\frac{1}{\alpha}(l'm10|lm) < vl||\alpha c^{[1]}||v'l' > \delta_{m'm}\nabla \right] F(v'l'm';r,t,\alpha) = 0, \tag{7}$$

$$(v,l) = 0,1,2,\ldots,\infty, \quad m = -l,-l+1,\ldots,l-1,l,$$

where N is the neutral gas number density. The reduced matrix elements $J_{vv'}^l(\alpha)$, $< vl||\alpha c^{[1]}||v'l' >$ and $< vl||K^{[1]}||v'l' >$ of the collision operator, velocity and velocity derivative are given by (11), (12a) and (12b) of Ref. [13], respectively. For further details the reader is referred to [3].

Representation of the spatial dependence

The treatment of the spatial dependence of the phase-space distribution function is dependent on the conditions under which the experiment is performed.

Hydrodynamic regime: In carefully controlled swarm experiments, spatial gradients are designed to be small so that the *hydrodynamic regime* in general prevails, and the space-time dependence can be projected onto the number density [14]. Transport coefficients unfolded from swarm experiments in this manner are essentially independent of the geometry of the experiment, since all spatial dependence is accounted for by (functionals) of $n(r,t)$ which is in turn found through the solution of the diffusion equation. For studies of transport in the hydrodynamic regime the spatial dependence is projected onto the number density through a (time-dependent) density gradient expansion:

$$F(vlm|r,t,\alpha) = \sum_{s=0}^{\infty}\sum_{\lambda=0}^{\infty}\sum_{\mu=-\lambda}^{\lambda} F(vlm|s\lambda\mu;t,\alpha)G_{\mu}^{(s\lambda)}n(r,t), \tag{8}$$

where $G_{\mu}^{(s\lambda)}n(r,t)$ is the irreducible gradient tensor operator. Substituting into (7) and equating coefficients of $G_{\mu}^{(s\lambda)}n(r,t)$ yields the following hierarchy of equations for the calculation of time-dependent transport coefficients:

$$\sum_{v'=0}^{\infty}\sum_{l'=0}^{\infty}\sum_{m'=-l'}^{l'}\left[\left(N\frac{d}{dt}\delta_{vv'}+NJ_{vv'}^{l}(\alpha)\right)\delta_{l'l}\delta_{m'm}\right.$$
$$+i\frac{qE}{m}\alpha(l'm10|lm)\langle vl||K^{[1]}(\alpha)||v'l'\rangle\delta_{m'm}+\frac{1}{2}\frac{qB}{m}\left\{\sqrt{(l-m)(l+m+1)}\delta_{m'm+1}\right.$$
$$\left.-\sqrt{(l+m)(l-m+1)}\delta_{m'm-1}\right\}\delta_{l'l}\delta_{v'v}\Bigg]F(v'l'm'|s\lambda\mu;t,\alpha) = X(vlm|s\lambda\mu), \tag{9}$$

where

$$X(vlm|000) = 0 \tag{10}$$

$$X(vlm|11\mu) = \sum_{v'=0}^{\infty}\sum_{l'=0}^{\infty}\left[\left(-\frac{1}{\alpha}\right)(l'm-\mu1\mu|lm)\langle vl||\alpha c^{[1]}||v'l'\rangle F(v'l'm-\mu|000)\right]$$
$$-\frac{(-1)^{\mu}}{\alpha}F(01-\mu|000)F(vlm|000). \tag{11}$$

Explicit expressions for the reduced matrix elements are given by [13]. Discretising in time using an implicit finite difference scheme converts the hierarchy of systems of coupled differential equations into a hierarchy of coupled matrix equations. To establish the transport coefficient of interest we are required to solve the following members of the hierarchy $(s,\lambda,\mu) = (0,0,0),(1,1,0),(1,1,1)$. The transport properties can be expressed in terms of the calculated moment via:

$$\varepsilon = \frac{3}{2}kT_b\left[1 - \sqrt{\frac{2}{3}}\text{Re}\{F(100|000)\}\right] ,$$

$$W_x = \frac{1}{\alpha}\sqrt{2}\text{Im}\{F(011|000)\} , \qquad W_z = -\frac{1}{\alpha}\text{Im}\{F(010|000)\} ,$$

$$D_E = -\frac{1}{\alpha}F(010|110) , \qquad D_{E\times B} = -\frac{1}{\alpha}\left[\text{Re}\{F(011|111)\} - \text{Re}\{F(01-1|111)\}\right] ,$$

$$D_B = -\frac{1}{\alpha}\left[\text{Re}\{F(011|111)\} + \text{Re}\{F(01-1|111)\}\right] .$$

We note Re{} and Im{} respectively represent the real and imaginary parts of the moments.

Non-hydrodynamic regime: In most discharges, large spatial gradients are formed in response to sources, boundaries or spatially varying fields and so-called *non-hydrodynamic* conditions are assumed to prevail. In this case, the spatial dependence must be treated explicitly. Here we use a second-order finite differencing scheme with appropriate modifications at the boundaries. The quantities of interest in terms of the calculated moments are:

$$v_{E\times B}(z) = v_x(z) = \frac{\sqrt{2}}{\alpha}\frac{\text{Im}[F(011;\alpha,z)]}{F(000;\alpha;z)} , \qquad v_E(z) = v_z(z) = -\frac{1}{\alpha}\frac{\text{Im}[F(010;\alpha,z)]}{F(000;\alpha;z)} ,$$

$$\varepsilon(z) = \frac{3}{2}kT_b\left[1 - \sqrt{\frac{2}{3}\frac{F(100;\alpha,z)}{F(000;\alpha,z)}}\right] , \qquad n(z) = F(000;\alpha,z).$$

For further details the reader is referred to [9].

RESULTS

Temporal relaxation

In this section we consider the response of macroscopic transport properties to the application of a magnetic field under hydrodynamic conditions. This work is an extension of the work in [5] to consider spatially inhomogeneous transport coefficients. The initial conditions represent the steady state magnetic field free case where the electron swarm is acted on solely by a d.c. electric field ($E/N=12\text{Td}$, $B/N=0\text{Hx}$; $1\text{Td} = 10^{-21}$ V m^2, $1\text{Hx} = 10^{-27}$ T m^3). At time $t=0$, a crossed magnetic field is switched on (electric field is unaltered) and the relaxation properties of the swarm are monitored as a function of time (normalised time Nt). The results are displayed in Figure 1 for electrons in CO_2 at 293K. The steady state results are in general well known and we refer the reader to [15, 16, 17, 4] for such discussions. In this article we will focus entirely on the transient behaviour. The theory and associated code have been benchmarked against an independent time-resolved Monte-Carlo simulation at the University of Belgrade. In

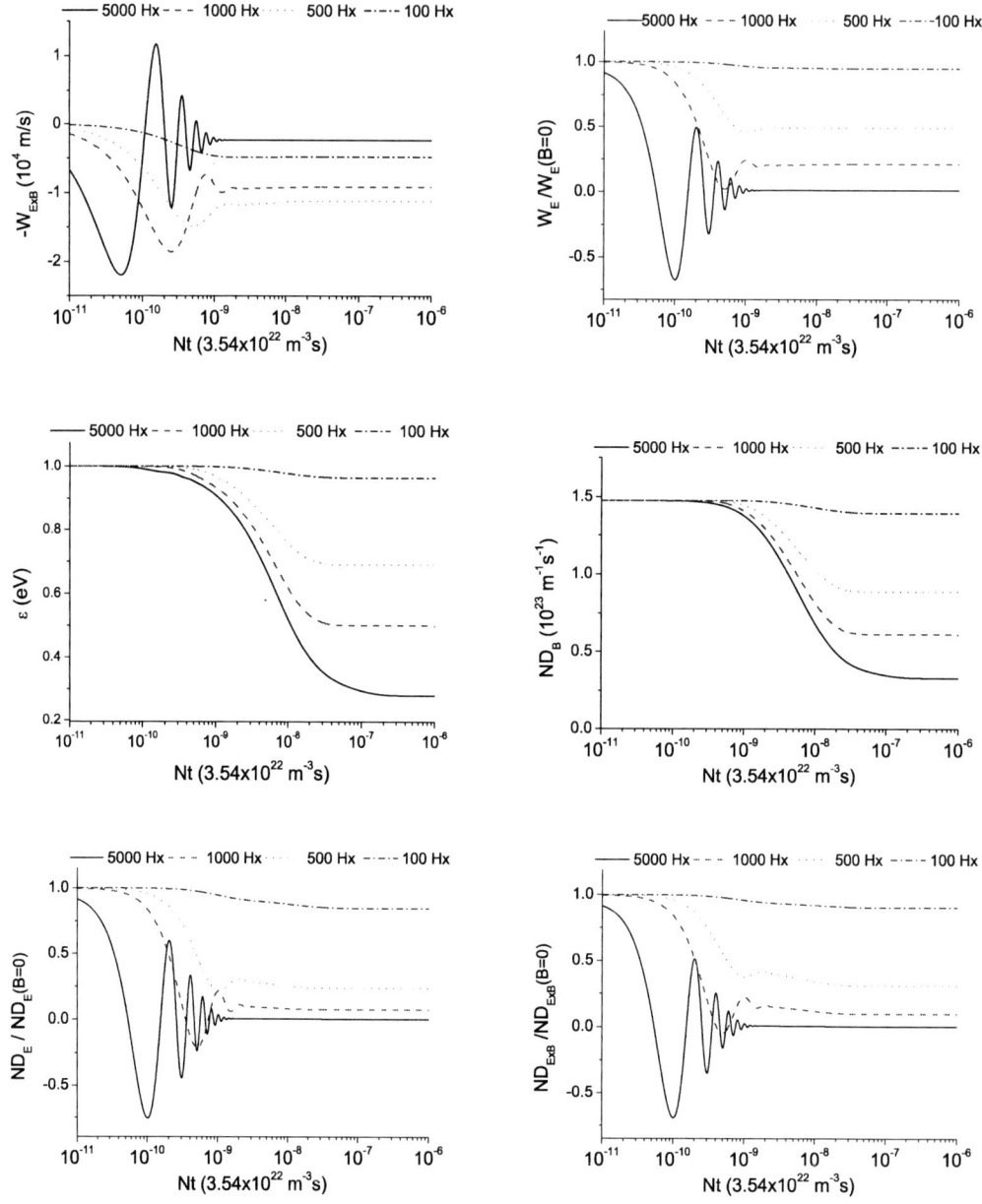

FIGURE 1. Temporal relaxation of the mean energy, drift velocity and diagonal elements of the diffusion tensor for various applied magnetic fields for electrons in CO_2 (E/N=5Td).

the relaxation profiles we observe the existence of three distinct timescales: (i) the gyro-period of the electrons τ, (ii) the momentum relaxation time τ_m, and (iii) the energy relaxation time τ_e. The latter two timescales are functions of energy. The various transport properties display profiles that are either monotonic relaxation or damped period relaxation. For quantities like the ε and ND_B, relaxation is in general always monotonic and occurs on the timescale governed by τ_e. In contrast, the relaxation profiles of the drift and diffusion in the E and $E \times B$ directions exhibit a transition from monotonic decay to damped periodic decay as the magnetic field strength is increased to values where $\tau \leq \tau_m$. For the damped periodic profiles, the oscillations are on the timescale of the gyro-orbits τ and the envelope decays on a timescale of τ_m together with a further relaxation on the timescale of τ_e. The existence of the additional oscillatory behaviour in the relaxation profiles is an imprint of the collective gyrations of the electrons damped by collisions that exchange momentum and energy. Perhaps the most striking phenomena is the existence of transiently negative excursions of the diffusion tensor elements in both the E and $E \times B$ directions. The existence of transiently negative diagonal diffusion elements in swarms was initially observed by the Petrović group [6] for $ND_{E \times B}$ when considering radiofrequency electric and magnetic fields. This was independently verified by the JCU group [3] and shown to exist in ND_E as well. The results in Figure 1 indicate that transiently negative diffusion in both the E and $E \times B$ directions can be achieved through the abrupt application of a d.c. magnetic field of sufficient strength.

The behaviour demonstrated for the drift and diffusion in the E and $E \times B$ directions is in general not predictable from steady state d.c. results. The manifestation of these complex relaxation profiles when considering time-dependent fields (e.g. rf and/or pulsed rf) is behaviour which is distinctly non-local in time. Contemporary understanding of field frequency effects (viz. reduction in amplitude and increase in phase-lag with respect to the field) fail or have a limited range of validity when the relaxation is not monotonic. Understanding such effects requires recourse to a systematic investigation of relaxation profiles as presented here [18].

Spatial relaxation

In this section we extend previous work on the idealised steady-state Townsend experiment [19, 20, 21] to include the explicit influence of an orthogonal magnetic field. The system is schematically represented in Figure 2 where charged particles are emitted at a constant rate from an infinite plane source at $z = z_0$ and interact with the neutral gas under the influence of static uniform electric and magnetic fields. The boundary conditions on the distribution function are detailed in [19, 9]. The theory and associated code have been benchmarked against an (i) analytic model and (ii) an independent eigenfunction expansion technique [9]. In this section we present results for a model gas and CH_4. In the absence of magnetic fields, the spatial relaxation characteristics are controlled by an interplay between the collisional energy loss mechanisms. Elastic collisional processes are essentially continuous energy loss processes while inelastic collisional processes are discrete energy loss processes. In the absence of elastic collisions, electrons would experience energy accumulation from the field followed by a discrete energy loss due to an

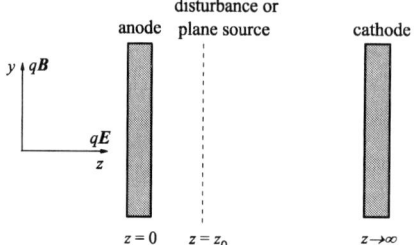

FIGURE 2: Schematic representation of an idealised steady-state Townsend experiment. We scale the length in terms of a representative mean free path $\lambda = 1/\sqrt{2}N\sigma_0$ where $\sigma_0 = 10^{-20}\text{m}^2$.

inelastic collision. This would result in a periodic spatial profile with a wavelength (period) inversely proportional to the field strength and proportional to the energy threshold for the process. The presence of elastic collisions tends to damp this oscillatory behaviour and broaden the peaks. In summary and broadly speaking (i) if the collisional energy loss is governed essentially by 'continuous' energy loss processes then we have monotonic decay (e.g. when the mean swarm energy is much less than the lowest energy threshold and elastic collisional processes are dominant, or when mean swarm energies are much greater than the lowest energy threshold), (ii) if the collisional energy loss is dominated by 'discrete' energy loss processes then we have damped periodic decay (e.g when the mean swarm energy is less than lowest energy threshold). Spatial relaxation can thus be characterised by a spatial relaxation length and a relaxation period if oscillatory behaviour exists. Under the influence of an electric field only, certain gases only exhibit oscillatory spatial relaxation over a limited range of E/N (window) [19].

We now highlight in Figures 3-4 some important implications on the spatial relaxation of macroscopic transport properties associated with the application of an orthogonal magnetic field. In Figure 4, we have chosen the electric field such that the mean energy lies in the vicinity of the Ramsauer minimum in the CH_4 elastic cross-section. The relaxation length is consequently very long. The steady state behaviour is well known and the reader is referred to [15] for details.

Ability to control oscillatory relaxation: The well known cooling action of an orthogonal magnetic field allows control of the distribution function and hence control of the energy loss processes. In Figure 3, for the conditions specified, we observe in the absence of a magnetic field the relaxation profile is monotonic. Application of an orthogonal field cools the swarm, enhancing the discrete nature of energy loss processes and generating oscillatory relaxation. Further enhancement of the magnetic field cools the swarm to a state such that continuous energy loss elastic collisional processes are dominant and relaxation is again monotonic. The corollary to this is the application of a magnetic field will in general increase the window of electric field strengths for oscillatory behaviour.

Modification of relaxation properties: If the relaxation type (monotonic/oscillatory) remains unchanged on application of a magnetic field, then relaxation is faster, and the period (if oscillatory) is longer (see Figure 4). If the relaxation type changes however, the relaxation occurs slower/faster if the original magnetic field free relaxation is monotonic/periodic. For a sufficiently high magnetic field, relaxation becomes

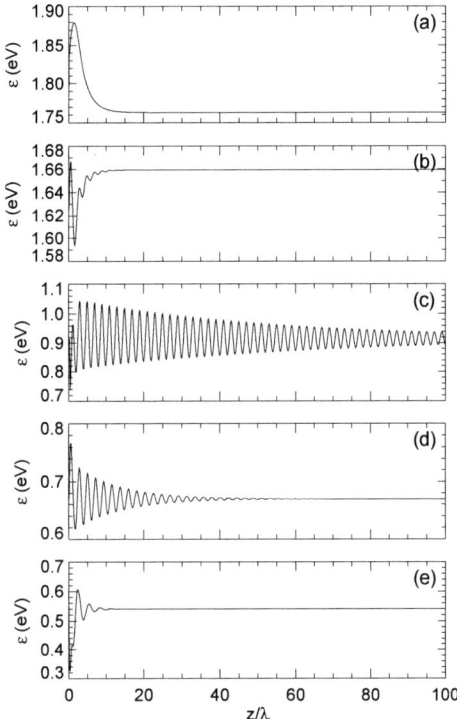

FIGURE 3. Spatial relaxation of ε for various applied magnetic fields for electrons in the step function collision model [19]. The disturbing source properties: $T=8000K$ and $v=2000$ ms^{-1}. ($E/N=15$Td; (a) $B/N=0$Hx; (b) $B/N=100$Hx; (c) $B/N=500$Hx; (d) $B/N=1000$Hx; (e) $B/N=2000$Hx).

monotonic and the relaxation distance is reduced over the magnetic field free distance.

CONCLUDING REMARKS

The explicit influence of a magnetic field on (i) hydrodynamic temporal relaxation, and (ii) non-hydrodynamic steady-state spatial relaxation of electron swarms was considered using time- and space-dependent multi-term solutions of Boltzmann's equation respectively. The magnetic field was found to induce damped oscillatory temporal relaxation profiles for the drift and diffusion elements in the E and $E \times B$ directions. Most strikingly, the existence of transiently negative diffusion tensor elements in these directions was observed for sufficiently high magnetic fields. Likewise, the application of a magnetic field to the idealised steady-state Townsend experiment was found to significantly modify the spatial relaxation profiles. The magnetic field was shown to be able to both suppress and generate Frank-Hertz type oscillations as well as modify the spatial relaxation distance. Importantly, it was demonstrated that the application of a magnetic

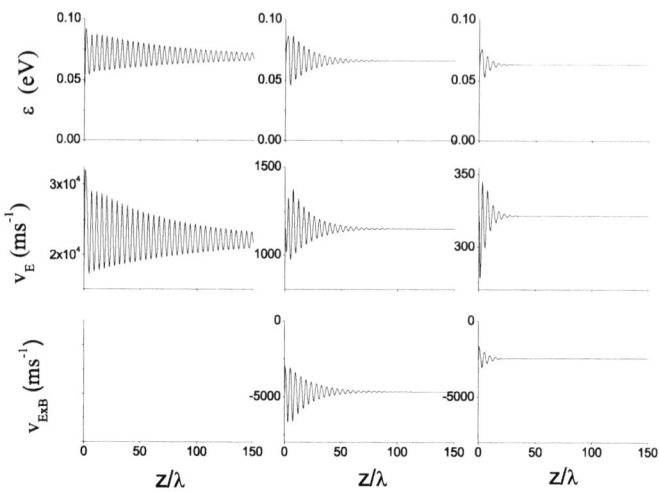

FIGURE 4. Spatial relaxation of ε, $v_x = v_{E \times B}$ and $v_z = v_E$ for various applied magnetic fields for electrons in CH_4. The disturbing source properties: $T=200K$ and $v=2000$ ms^{-1}. ($E/N=0.5$Td; Column 1: $B/N=0$Hx; Column 2: $B/N=100$Hx; Column 3: $B/N=200$Hx).

field orthogonal to the electric field significantly modified temporal- and spatial- non-locality of the system, and this has important ramifications in the study of time- and space-dependent magnetic fields.

REFERENCES

1. M. A. Lieberman, and A. J. Lichtenberg, *Principles of Plasma Discharges and Materials Processing*, Wiley, New York, 1994.
2. R. Winkler, D. Loffhagen, and F. Sigeneger, *Applied Suface Science* **192**, 50–71 (2002).
3. R. D. White, K. F. Ness, and R. E. Robson, *Appl. Surf. Sci.* **192**, 26–49 (2002).
4. Z. L. Petrović, Z. M. Raspopović, S. Dujko, and T. Makabe, *Appl. Surf. Sci.* **192**, 1–21 (2002).
5. D. Loffhagen, and R. Winkler, *IEEE Plasma Sci. Technol.* **27**, 1262 (1999).
6. Z. Raspopović, S. Sakadžić, Z. L. Petrović, and T. Makabe, *J. Phys. D: Appl. Phys.* **33**, 1298 (2000).
7. R. Winkler, . A. Mairorov, and F. Sigeneger, *J. Appl. Phys.* **87**, 2708 (2000).
8. I. A. Porokhova, B. Yu, B. Golubovskii, J. Bretagne, M. Tichy, and J. F. Behnke, *Phys. Rev. E* **71**, 066407 (2005).
9. B. Li, R. E. Robson, and R. D. White, *Phys. Rev. E* **74**, (accepted) (2006).
10. L. Boltzmann, *Wein. Ber.* **66**, 275 (1872).
11. C. S. Wang-Chang, G. E. Uhlenbeck, and J. De Boer, , in *Studies in Statistical Mechanics*, edited by J. D. Boer, and G. E. Uhlenbeck, Wiley, New York, 1964, vol. II, p. 241.
12. R. D. White, K. F. Ness, R. E. Robson, and B. Li, *Phys. Rev. E* **20**, 2231 (1999).
13. K. F. Ness, and R. E. Robson, *Phys. Rev. A* **34**, 2185 (1986).
14. K. Kumar, H. R. Skullerud, and R. E. Robson, *Aust. J. Phys.* **33**, 343 (1980).
15. K. F. Ness, *J. Phys. D: Appl. Phys.* **27**, 1848 (1994).
16. R. D. White, M. J. Brennan, and K. F. Ness, *J. Phys. D: Appl. Phys.* **30**, 810–816 (1997).

17. S. Bzenić, Z. Raspopović, S. Sakadžić, and Z. L. Petrović, *IEEE. Trans. Plasma Sci.* **27**, 78 (1999).
18. R. D. White, *Phys. Rev. E* **64**, 056409 (2001).
19. R. E. Robson, B. Li, and R. D. White, *J. Phys. B: At. Mol. Opt. Phys.* **33**, 507–520 (2000).
20. B. Li, R. D. White, and R. Robson, *J. Phys. D: Appl. Phys.* **35**, 2914 (2002).
21. R. Winkler, S. Arndt, D. Loffhagen, F. Sigeneger, and D. Uhrlandt, *Contrib. Plasma Phys.* **44**, 437 (2004).

Coherent Spectroscopy of Sodium and Potassium Vapour

G. Alzetta[1], S. Cartaleva[2], S. Gozzini[1], T. Karaulanov[2], A. Lucchesini[1], C. Marinelli[3]

[1]IPCF-CNR, Area della Ricerca, via Moruzzi 1, 56124 Pisa, Italy
[2]Institute of Electronics, BAS, boul. Tzarigradsko shosse 72, 1784 Sofia, Bulgaria
[3]CNISM Department of Physics, University of Siena, Via Roma 56, 53100 Siena, Italy

Abstract. A brief introduction is presented related to three different approaches for Coherent Population Trapping (CPT) observation, namely CPT resonance prepared: (i) in degenerate two-level system, (ii) by coupling pairs of non-degenerate Zeeman sublevels of two ground levels and (iii) utilizing Zeeman sublevels within a single hyperfine transition. As an illustration of the first two approaches recent results concerning the CPT preparation on the first resonance line in Na and K are described. The first, to our best knowledge, CPT resonance observation on the second resonance line of K (violet, 404.4 nm) is reported. A transfer of the CPT resonance due to cascade transitions, without compromising its parameters, is evidenced. The presented results are promising for investigation of processes in atomic vapour and gas discharge.

Keywords: Coherent population trapping, Optical pumping, Atomic collisions
PACS: 32.80.Bx; 32.80.Qk; 42.50.Gy

I. INTRODUCTION

Coherent excitation of atoms at long-living states results in interesting effects with significant importance for fundamental physics and applications. Among others, Coherent Population Trapping (CPT) [1] and the related effect of Electromagnetically Induced Transparency (EIT) are nowadays extensively studied. The EIT effect manifests itself as a dip in the fluorescence or peak in the transmitted through the atomic vapour laser light, which is several orders of magnitude narrower than the natural width of the corresponding optical transition.

In this presentation, a brief introduction will be done related to different approaches for the CPT observation, namely CPT resonance prepared: (i) in degenerate two-level system (so called Hanle configuration), (ii) by coupling pairs of non-degenerate Zeeman sublevels of two ground-state hyperfine (hf) components and (iii) utilizing Zeeman sublevels within a single hf transition. Then in the following Sections, a more detailed description of the recent results obtained in Na and K, based on the first two excitation schemes, will be presented.

The CPT resonances prepared in Hanle configuration have been widely investigated for Cs and Rb atoms [2]. Alkali atoms are irradiated by monochromatic laser field in such way that different polarization components couple the Zeeman sublevels of one

CP876, *The Physics of Ionized Gases: 23rd Summer School and International Symposium*,
edited by L. Hadžievski, B. P. Marinković, and N. S. Simonović

hf ground level to a common excited state and introduce coherence between ground magnetic sublevels at magnetic field B = 0. As has been shown in [3], in absence of depolarizing collisions of the excited state, and depending on the ratio of the degeneracy of the two states involved in the optical transition, sub-natural width resonances of both EIT, dark resonance and Electromagnetically-Induced Absorption (EIA), bright resonance can be observed. EIT is realized when the condition $F_g \rightarrow F_e = F_g - 1, F_g$ is met, while EIA is observed for $F_g \rightarrow F_e = F_g + 1$ kind of transitions. Here, F_g and F_e are the hf quantum numbers of the ground- and excited-state hf levels, respectively. It has been shown [3] that the bright resonance is very sensitive to the cell buffering. It has been found [3] and very recently confirmed [4] that if the cell containing alkali atoms is buffered by some noble gas, the CPT resonance at the $F_g = 4 \rightarrow F_e = 5$ transition on the D_2 line of Cs transforms from bright to the dark one. The theoretical modeling has shown that this reversal of the resonance sign can be attributed to depolarization of the F_e level by collisions of alkali atoms with the buffer gas atoms. Note that the cell buffering does not reverse the sign of the dark resonances observed on the $F_g \rightarrow F_e = F_g - 1, F_g$ type of transitions.

The second approach for the CPT preparation is based on the coupling of pairs of non-degenerate Zeeman sublevels belonging to two ground levels to a common excited-state level by means of bi-chromatic coherent laser fields. The two frequencies required for CPT preparation are most frequently obtained through modulation of the diode laser injection current at frequency in the GHz region (for Rb and Cs).

The third method [5] can be considered as a combination of the first two in that it uses polychromatic laser field coupling non-degenerate Zeeman sublevels of a single ground-state hf level to Zeeman sublevels of the excited state. The needed mutli-frequency coherent fields are again obtained by modulation of the diode laser injection current, but here the modulation frequency is in the kHz range because the ground-state Zeeman sublevels are split only by the applied magnetic field (MF). With this approach the complication of the laser frequency modulation is significantly relaxed but the resonance parameters can be compromised when using buffered or coated cell due to the hf optical pumping to the hf level non-interacting with the laser field.

II. COMPLETE EIT IN SODIUM

In this Section we will briefly concern the realization of the first approach on the D_1 line of Na, in a significantly improved version. As it was mentioned above, the method is based on the alkali atom excitation by monochromatic light field of different polarizations. As only one of the two ground-state levels is excited by the light and involved in the formation of the CPT resonance, a significant portion of the atoms populates the other ground-state hf level. Due to this population loss, the contrast of the resonance reaches a maximum value less than 50%. To overcome this problem we proposed [6] a simple method to make negligible the population loss by involving all atoms of the Na ground state in the formation of the narrow coherent resonances. The basic idea is to excite the D_1 line of Na by a multi-mode dye laser light, whose spectral bandwidth is larger than the D_1 absorption line-width and whose longitudinal mode separation is comparable with the homogeneous line-width. Indeed, when we excite only one of the hf sublevels, the other one, due to hf optical pumping, becomes a population sink and hence, a source of loss. In our case instead both ground-state hf levels are simultaneously excited and the hf optical pumping is quenched. Moreover, when a large number of laser modes is used, all velocity classes of the Na atoms are resonant with the radiation and contribute to the resonance preparation. As a result, dark resonances with a contrast up to 100% have been observed. They are detected looking at the atomic fluorescence when the applied MF is scanned around B=0. The main experimental result is illustrated in Fig.1a. It can be seen that utilizing Na cell with spin anti-relaxation coating or Na cell buffered by noble gas, complete transparency of Na vapour is evidenced in a narrow interval of the MF scan.

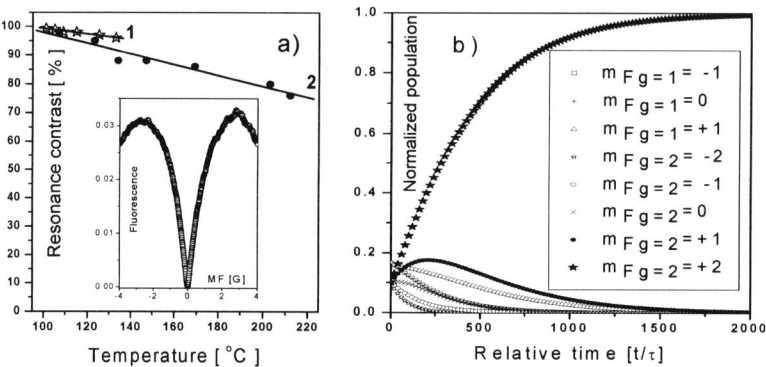

FIGURE 1. (a) Illustration of the complete transparency observation in Na vapour: (1) confined in anti-relaxation coated cell; (2) confined in buffered (by 6 Torr of Ar)cell. In the insert: Na fluorescence vs magnetic field scanned around B = 0. (b) Dynamics of the population of the ground-state Zeman sublevels of the hf levels $F_g = 1$ and $F_g = 2$ (D_1 line of Na) under broadband σ+ excitation. τ is the optical transition lifetime. Light power density is 400 mW/cm^2.

To explain the experimental observation a simple rate equation model is used [6], which describes the evolution of the population of all levels involved in the interaction scheme. Let us consider atom excitation on the D_1 line by circularly polarized light (σ+). It is not difficult to predict that if the hf transitions starting from the $F_g = 1$ or $F_g = 2$ levels are excited, the atoms will be accumulated on the $F_g = 1$, $m_{Fg} = +1$ and $F_g = 2$, $m_{Fg} = +2$ magnetic sublevels. However, our theoretical consideration has shown an interesting result: if both ground-state levels are excited simultaneously, at steady state all atoms will be accumulated only on the ground-state sublevel $F_g = 2$, $m_{Fg} = +2$ (Fig.1b). Consequently, exciting both hf transitions, complete orientation of atoms along a single direction can be achieved. Due to the complete accumulation of atoms on a Zeeman sublevel not interacting with the laser field at B = 0, the application of magnetic field orthogonal to the atomic orientation will lead to a 100% contrast dark resonance. This is the case for coated and buffer gas cells, where the interaction time between atoms and light is large enough for reaching steady state. Hence the theoretical consideration is in good agreement with the experimental results.

The obtained result is of basic importance, showing that it is possible to overcome the population loss due to the hf optical pumping. However, it should be noted that the bulky dye laser is not convenient for broad practical applications. In opposite, a number of practical systems are based on diode lasers which are suitable for excitation of alkali atoms. The alkali atoms spectral properties analysis has shown that the K atom can be considered as a promising candidate for the reduction of atomic population loss while still utilizing the convenient diode lasers. In the following Sections our very recent results on the CPT preparation using K vapour and different configurations for its excitation will be discussed.

III. COHERENT SPECTROSCOPY ON THE D1 LINE OF POTASSIUM

In this section we present the main results related to the CPT preparation on the D_1 line of K. The relevant energy level diagram and hf transitions are presented in Fig.2. Opposite to the case of Cs and Rb atoms that are mostly applied in the EIT experiments due to the widely available commercial narrow-band diode lasers matching their resonance transitions, in K the splitting of the ground-state hf levels is less than the Doppler width (FWHM = 765 MHz at room temperature) of the hf transitions.

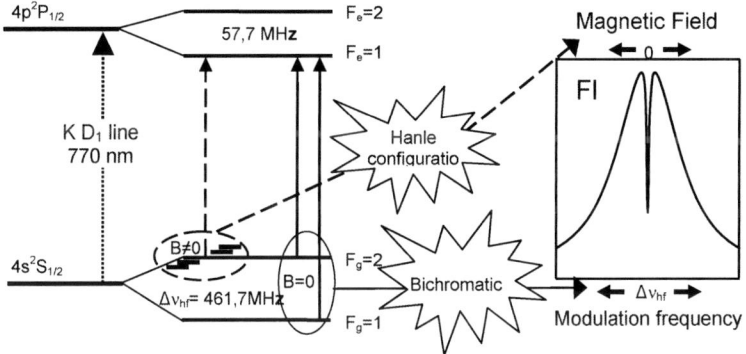

FIGURE 2. Energy levels related to the D_1 line of K. Preparation of the coherent resonance in Hanle configuration (upper branch) and by bi-chromatic excitation (lower branch) is illustrated.

Concerning first the CPT preparation in Hanle configuration, we will limit our discussion to the $F_g = 2 \rightarrow F_e = 1$ transition denoted by dashed line in Fig.2. Suppose that K atoms are excited by circularly polarized mono-mode laser field (at B = 0). Then, due to the laser excitation and the following spontaneous decay, part of the atoms will be accumulated to the $F_g = 2$, $m_{Fg} = 1,2$ Zeeman sublevels, non-interacting with the laser field. As a result of this process the absorption of atoms will be decreased and the angular momentum (and spin) of K atoms will be oriented. If MF is applied in a direction orthogonal to the atomic orientation caused by the light, the atomic population will evolve among the Zeeman sublevels of $F_g = 2$ level, resulting in enhancement of the light absorption (fluorescence from $F_e = 1$ level). In this way, a narrow dark resonance is observed in the fluorescence dependence on the scanned around B = 0 MF (see the upper branch, noted as Hanle configuration, in Fig.2). However, it should be stressed that part of the atoms will decay to the $F_g = 1$ level and will not interact any more with the light. The last atoms contribute to the population loss from the $F_g = 2$ level and are the reason for resonance contrast strong reduction in Rb and Cs. In case of K, however we expect that due to the significant overlapping of the hf transitions starting from $F_g = 1$ and $F_g = 2$ levels, the population loss will be reduced to a great extent.

In order to estimate the contribution of hf optical pumping, K atoms are excited by single-frequency laser light and resonances are registered monitoring the absorption dependence on MF scanned around B = 0. The used experimental set up is shown in Fig.3. Extended Cavity Diode Laser (ECDL, Toptica DL100) is utilized operating in single mode. The laser FWHM is of the order of 1MHz. Different K glass cells are used: evacuated and buffered by 6 Torr of Ar. Both types of cells are built in two options: uncoated and coated by poly-dimethyl-siloxane (PDMS) anti-relaxation coating. K atoms are irradiated by circularly-polarized single-frequency light tuned to the maximum fluorescence of D_1 line. The laser frequency tuning is monitored by means of an auxiliary K reference cell. The main K cell, where the coherence resonance is prepared, is shielded against stray MFs by μ – metal shielding. A pair of Helmholtz coils is situated inside the shielding, producing a MF orthogonal to the laser beam. The cell heating is accomplished by an oven situated outside the shield.

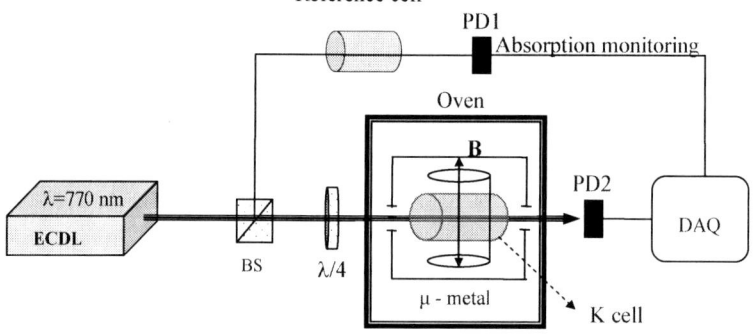

FIGURE 3. Experimental set up for EIT resonance observation on the D_1 line of K, in Hanle configuration.

Our experimental study has shown that in pure dilute K, only EIT resonances are observed. Hence, no contribution of the $F_g = 1 \rightarrow F_e = 2$ transition has been registered as an EIA resonance. Probably, due to the significant overlapping of the four hf transitions, the effect of the three $F_g \rightarrow F_e = F_g - 1, F_g$ transitions predominates.

As an example, the EIT resonance observed in the buffered coated cell (at laser power of 30 mW/cm^2) is shown in Fig.4a. In agreement with the work [7] the resonance profile is of Lorentzian shape. Processing the experimentally observed EIT resonance absorption profiles, their contrast is determined as the ratio of the resonance amplitude (A_0 - A) to the absorption outside the resonance A_0. The resonance contrast dependences on the cell temperature are shown in Fig.4b, for three different cells. It can be seen that in the vacuum coated cell the resonance contrast is significantly smaller than that in the buffered cells. In the vacuum non-coated cell, the resonance contrast is similar to that observed in the vacuum coated cell. In order to explain this result, it should be pointed out that the natural width of the hf transitions (6 MHz) is less than the hf splitting of the ground and the excited states. Hence, in the evacuated cells, the four hf transitions excited by the mono-mode laser light can be considered as independent because different velocity-class atoms are excited at different hf transitions. Due to this hf transition selective excitation, optical pumping occurs to the ground state hf level non-interacting with the laser field.

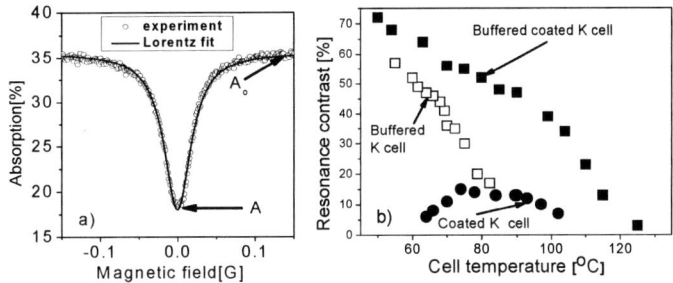

FIGURE 4. EIT resonance in absorption (a) and its contrast dependence on temperature (b), for three different types of cells.

On the contrary, in the case of buffered cells, due to velocity changing collisions of alkali atoms with Ar atoms, almost all atoms interact with the light, thus reducing the hf optical pumping and increasing the resonance contrast. As shown in the work [3], the situation is different in the case of Cs. There, the EIT resonance contrast in the evacuated cell is larger than that in the buffered one. This result also follows from the hf optical pumping. In Cs, where the ground-state hf splitting is 9,2 GHz (much larger than the Doppler broadening, 300 MHz), the hf optical pumping is higher for the buffered cell than for the evacuated one. The buffering prevents Cs atoms from spin changing collisions with the cell walls resulting in hf optical pumping enhancement and EIT resonance contrast reduction. Comparing the results shown in Fig.4b with those of Ref. [3], it can be concluded that the EIT resonance contrast obtained in buffered K cells is significantly larger than that in buffered Cs cells. This result shows the advantages of K medium, since namely buffered cells are largely used because there the resonance width is orders of magnitude lower than that observed in pure K.

Let us go back to Fig.2, where the second approach for CPT preparation is also illustrated (see the lower branch noted as bi-chromatic excitation). To decrease the laser modulation frequency while still involving the two ground-state hf levels in the resonance preparation, it is promising to use K atoms. The aim is to observe a coherent superposition of the two ground-state hf levels of K, resulting in EIT resonance. For the experiment, it was not possible to use the ECDL system applied successfully in the previous one because of the needed laser frequency modulation in the MHz region. Unfortunately, the available diode lasers operating at wavelength matching the D_1 line of K provide output consisting of a great number of modes in a large spectral interval.

Due to these reasons, a solitary multi-mode diode laser (Sacher Lasertechnik Mod. FP-0770-100) is used and one of its modes is tuned in resonance with the D_1 line of K. The frequency modulation of the laser modes is performed by modulation of the laser injection current (see Fig. 5a). The modulation frequency f_m is swept around the ground-state hf level frequency difference $\Delta v_{hf} = 461,7$ MHz. In this way, the two coherent laser fields with frequency difference Δv_{hf}, necessary for the EIT resonance preparation, are provided. The disadvantage of utilizing a multi-mode laser is that the intensity of the used mode is very low and it can not be measured experimentally.

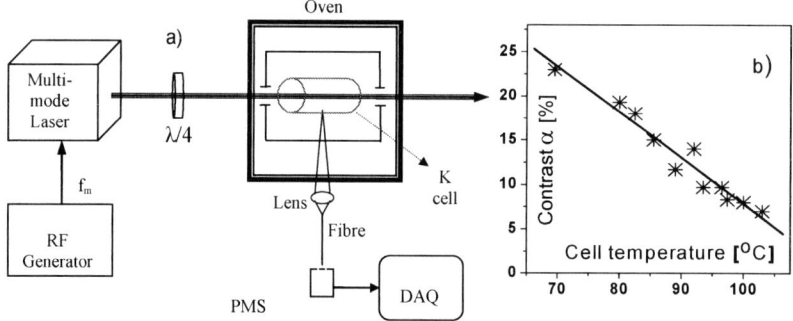

FIGURE 5. Experimental set up for CPT observation based on coherent superposition of $F_g = 1$ and $F_g = 2$ level of K (a), and the observed EIT resonance contrast dependence on cell temperature (b).

In this experiment, the fluorescence of K atoms is measured. The EIT signal is evidenced by monitoring the fluorescence in dependence on the laser modulation frequency. At this stage of our work, we are not able to perform a complete study of the EIT resonance in dependence on the light and cell parameters. Instead we limited ourselves in providing an experimental evidence of the EIT resonances in K based on coherent superposition of the two hf levels of the ground state and estimating the efficiency of the EIT preparation. The resonance contrast, determined as the ratio of the resonance amplitude to the maximum fluorescence outside the resonance, is measured in dependence on the cell temperature. For the cell with pure K, the result is shown in Fig. 5b. Taking into consideration the assumption of low power of the mode used for the EIT resonance preparation, it can be seen that the resonance contrast is quite good, above 20%, for cell temperature around 70°C.

IV. COHERENT RESONANCES OBSERVED ON THE SECOND RESONANCE LINE OF POTASSIUM

The EIT phenomenon has mainly been studied on the D_1 and D_2 lines of alkalis due to the widely available single-mode near-infrared laser diodes. The promising to many applications EIT spectroscopy can now be extended to the violet/blue spectrum due to the progress in the development of diode lasers emitting at these spectral regions.

To our best knowledge, here we present the results of the first observation of the EIT resonance on the second resonance line of Potassium $(4s^2S_{1/2} \rightarrow 5p^2P_{3/2})$ with wavelength of 404.4 nm. The energy levels and optical transitions, related to both resonance lines in K, are shown in Fig.6a. In the experiment presented in this Section, the laser excitation is performed only at the $4s^2S_{1/2} \rightarrow 5p^2P_{3/2}$ transition. Narrow EIT resonance is observed in the 404.4 nm fluorescence dependence on the scanned around $B = 0$ MF (see Fig.6b). As the 404.4 nm transition probability is more than an order of magnitude less than that of the D_1 hf transitions, the observation of coherent resonance at 404.4 nm is of significant importance.

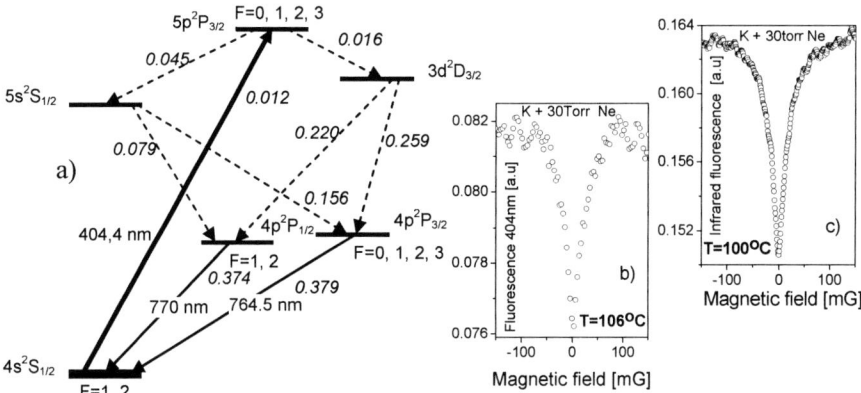

FIGURE 6. (a) Scheme of transitions relevant to the 404.4 nm coherent resonance formation and its transfer by cascade transitions. (b) EIT resonance in the 404.4 nm fluorescence. (c) EIT resonance in the infrared fluorescence.

Moreover, our experimental study has shown that a called by us for simplicity "transfer of the EIT resonance" occurs to the $4p^2P_{1/2}$ and $4p^2P_{3/2}$ excited states due to the spontaneous cascade transitions noted by dashed line in Fig.6a. This transfer is proved by the experimental observation of narrow EIT resonance in the infrared ($4s^2S_{1/2} \rightarrow 4p^2P_{1/2}$ and $4s^2S_{1/2} \rightarrow 4p^2P_{3/2}$ transitions) fluorescence dependence on MF (Fig.6c) when alkali excitation is performed only by the 404.4 nm laser radiation. Note that under similar conditions, the resonance in the infrared fluorescence is not broader than the resonance in the 404.4 nm fluorescence. It should be particularly stressed that the resonance observed in the infrared fluorescence is with superior signal/noise ratio compared to that in the 404.4 nm fluorescence. The superiority is provided by the fact that when observing the EIT resonance in the infrared fluorescence, the exciting laser light at 404.4 nm can be completely filtered, thus avoiding the laser noise. It is well known that the laser light noise is one of the main drawbacks limiting the applications of coherent resonances for precise measurements where the EIT feature is usually with less amplitude than the scattered light.

The experimental setup used for observation of both resonances is presented in Fig.7. Potassium atoms are excited by single-frequency circularly-polarized light at 404.4 nm and resonances are observed in the fluorescence dependence on an orthogonal to the laser beam MF scanned around $B = 0$, in the violet and infrared fluorescence separately. Two types K cells are used: containing pure K and K buffered by 30 Torr of Ne. Appropriate filters are used to perform separate measurements in the violet or infrared fluorescence. For the buffered cell, in the 404.4 nm fluorescence dependence on MF EIT resonance is observed centered at B = 0. The same sign resonance is observed in the infrared fluorescence when K atoms are irradiated by the light at 404.5 nm. Strong narrowing of the 404.5 nm EIT resonance in the buffered cell is registered, which is also transferred through the cascade transitions.

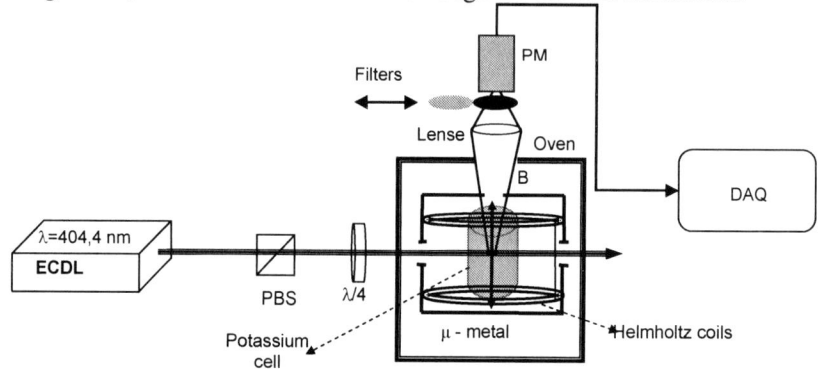

FIGURE 7. Experimental setup for the coherence transfer observation.

In Fig.8, the EIT resonance contrast vs cell temperature is shown for buffered and pure alkali cells.

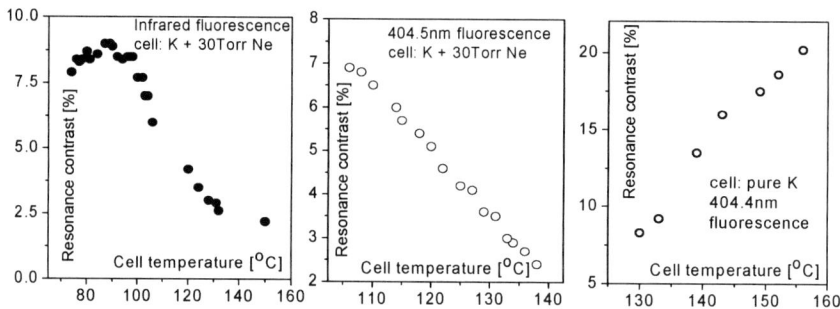

FIGURE 8. EIT resonance contrast dependence on temperature: (a) buffered cell, resonance in the infrared fluorescence; (b) buffered cell, resonance in the 404.4 nm fluorescence, (c) cell containing pure K, resonance in the 404.4 nm fluorescence.

It can be seen that in case of buffered cell, the contrast of both types of resonances [in the infrared (Fig.8a) and 404.4 nm (Fig.8b) fluorescence] reduces with cell temperature. Opposite, the resonance in the 404.4 nm fluorescence observed in the cell containing pure K significantly increases its contrast with temperature. At the same time, starting from cell temperature of 130°C, the EIT resonance in the infrared fluorescence starts transformation to EIA resonance (Fig.9). Below 130°C only EIT resonance is observed. Then a formation of the EIA resonance starts on the bottom of the EIT resonance. The EIA resonance enhances its amplitude with cell temperature.

The richness of physical processes behind the experimental observations presented in this section is evident. Most of the observed dependences are not explained yet.

Theoretical and further experimental investigations are in progress, in order to perform a profound analysis of the coherent resonance formation, as well as its transfer and modification by the processes in K vapour.

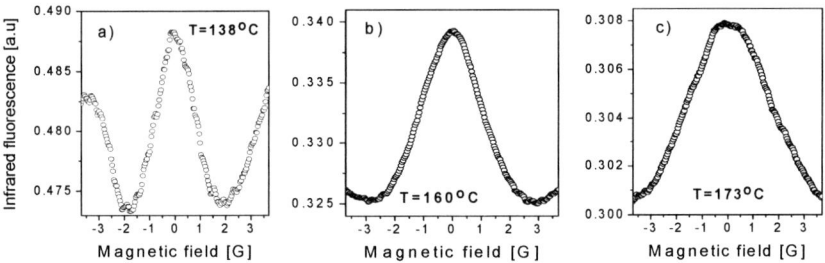

FIGURE 9. (a) Appearance of EIA resonance superimposed on the EIT one; (b) EIA resonance at cell temperature of 160°C; (c) Significantly broader resonance at cell temperature of 173°C.

IV. CONCLUSIONS

The observed complete transparency in Na has shown that the optical pumping to the ground-state hf level non-interacting with the laser field can be suppressed and a perfect orientation of atoms can be realized. This result is applicable for investigation of spin-changing collisions. The results related to the EIT resonance transfer by cascade transitions have potential for study of processes in vapour and gas discharges. The developed methods for CPT-based measurements of MF are with potential for direct application to plasma study. A method for MF diagnostic of plasma, based on the CPT in gas discharge [8] is proposed, which is promising for future expanding of the coherent spectroscopy application for plasma study.

REFERENCES

1. G. Alzetta, A. Gozzini, L. Moi, G. Orriols, *Nuovo Cimento* **36**, 5 (1976); E. Arimondo, *Prog. Opt.* **35** 257 (1996).
2. F. Renzoni, W. Maichen, L. Windholz, E. Arimondo, *Phys. Rev.* **A55**, 3710-3718 (1997); Y. Dancheva, G. Alzetta, S. Cartaleva, M. Taslakov, C. Andreeva, *Opt. Comm.* **178**, 103-110 (2000); G. Alzetta, S. Cartaleva, Y. Dancheva, C. Andreeva, S. Gozzini, L. Botti, A. Rossi, *J. Opt. B: Quantum Semiclass. Opt.* **3**, 181-188 (2001); F. Renzoni, C. Zimmermann, P. Verkerk, E. Arimondo, *J. Opt. B: Quantum Semiclass. Opt.* **3**, S7 (2001); F. Renzoni, S. Cartaleva, G. Alzetta, and E. Arimondo, *Phys. Rev.* A **63**, 065401 (2001); A. Papoyan, A. Auzinsh, K. Bergmann, *Eur. Phys. J.* **D21**, 63-71 (2002).
3. C.Andreeva, S. Cartaleva, Y. Dancheva, V. Biancalana, A. Burchianti, C. Marinelli, E. Mariotti, L. Moi, K. Nasyrov, *Phys. Rev.* **A66**, 012502 (2002).
4. D.V. Brazhnikov, A.M. Tumaikin, V.I. Yudin, A.V.Taichenachev, *J. Opt. Soc. Am.* **B22**, 57-64 (2005).
5. C. Andreeva, G. Bevilacqua, V. Biancalana, S. Cartaleva, Y. Dancheva, T. Karaulanov, C. Marinelli, E. Mariotti, L. Moi, *Appl. Phys.* **B76**, 667-675, (2003).
6. G.Alzetta, S.Gozzini, A.Lucchesini, S.Cartaleva, T. Karaulanov, C. Marinelli, L. Moi, *Phys. Rev.* **A69**, 063815 (2004).
7. F. Levi, A. Godone, J. Vanier, S. Micalizio, and G. Modugno, *Eur. Phys. J.* **D12**, 53-59 (2000).
8. R. Akhmedzhanov, I. Zelenscy, R. Kolesov, E. Kuznetsova, *Phys. Rev.* **A69**, 036409 (2004).

Spin and Electron Correlation Effects in Excitation of $3d$ Metal Atoms

L Pravica, D Cvejanović, S Napier and J F Williams

Centre for Atomic Molecular and Surface Physics
ARC Centre for Antimatter-Matter Studies
Physics Department, University of Western Australia, Perth 6009, Australia

Abstract. Excitation of zinc atoms by electron impact reveals the existence of negative ion states in the autoionizing region of the spectrum where both the configuration interaction and core shielding effects are important. Two resonances are also observed in the energy dependence of the three polarization Stokes parameters when spin polarized electrons are used and the 636.2 nm photons from the $4\,^1D_2$ observed. These observations point to the role of exchange and spin orbit interaction in the respective negative ion states. The statistical accuracy of the polarization measurements permits the precise determination of the energies and widths of the two negative ion states with better sensitivity than offered by integral cross section measurements.

Keywords: Zinc excitation, Polarized electrons, Negative ion resonances, Polarization of fluorescence
PACS: 34.80.Dp, 34.80.Nz

INTRODUCTION

Electron impact experiments have played a crucial role in the study of atomic and molecular structure and in providing the cross sections needed for modelling of different types of plasmas, the developments of lasers and in particular discharges, important examples including fluorescence lamps using Hg and recently Zn. Equally important is the contribution which the field has made toward the understanding of the fundamentals of atomic structure and dynamics in forming a basis for the development of a new generation of theories and their extension to new systems. Theoretical models, such as the Convergent Close Coupling (CCC) [1], R-matrix Close coupling [2], and B-spline method [3] are being now applied to the complex metal atoms and experiments are essential to stimulate their further developments and also to test their predictive accuracy. Here we report on an experimental investigation of electron impact excitation of a $3d$ metal atom, zinc, an important case where electron correlations, configuration interactions and relativistic effects are known to be an important ingredient for accurate modelling.

With a $3d^{10}4s^2$ ground state electron configuration and its Z=30 position in the periodic table the spin orbit interaction in zinc atom is already strong enough to cause perturbations and configuration mixing as observed in autoionization by both photon [4] and electron impact [5] and successfully modeled by Mansfield [6]. Photoionization in zinc and other group II_B atoms, Cd and Hg, is dominated by the single electron inner shell excitation of the form $(n-1)d^{10}ns^2 \rightarrow (n-1)d^9ns^2mp\,^{1,3}P_1\,^3D_1$. A similar situation is observed in electron impact studies on zinc [5] but in addition a number of

CP876, The Physics of Ionized Gases: 23rd Summer School and International Symposium,
edited by L. Hadžievski, B. P. Marinković, and N. S. Simonović

new optically forbidden transitions have been observed at low electron impact energies. The proximity of the $3d$ and $4s$ shells means that the core shielding and electron correlations must be treated in electron scattering calculations beyond the frozen core when the collision process involves one of the atomic $3d$ electrons. So far only one calculation [7] has been published with modelling of the negative ion resonances in the integral cross sections of states below the first ionisation threshold.

Experiments with spin polarized electrons [8, 9, 10] offer an additional strength to experimental studies with the possibility to explore the influence of the spin orbit and exchange interaction as well as momentum couplings. Our investigations in Perth show that temporary negative ions involving a single electron excitation from the $3d$ orbital in zinc are strongly influencing electron scattering. These experiments also present the most stringent tests for theoretical models as the relativistic effects have to be included. Although technologically demanding, experiments with spin polarized electrons allow for a deeper insight into the scattering dynamics using a simpler and more efficient approach as indicated below.

EXPERIMENTAL METHOD - INTEGRATED STOKES PARAMETERS

Work in progress in the laboratory in Perth at UWA encompasses a range of studies including polarization measurements from electron impact excitation and electron impact spectroscopy of zinc atom. Both types of measurements are based on crossed electron and atomic beams and the use of spin polarized electrons.

The measurement of integrated Stokes parameters is an established method [11] based on observation of the polarization characteristics of photons emitted after the excitation of atoms by spin polarized electrons. Here integrated denotes integration over all scattering angles of undetected scattered electrons. The method has been applied to two atoms from the group II_B elements, mercury [12, 13] and zinc whose studies include excitation of the 5^3S_1 state [14] and excitation with ionization of the $3d^94s^2\,^2D_{3/2}$ state [10] and the ongoing experiments in our laboratory. Built upon the fundamental principles of symmetry and conservation of angular momentum, these experiments offer details of the alignment and orientation of the atomic charge cloud to a level comparable with coincidence experiments without spin polarization.

The geometry of the experiment is depicted in figure 1. The electron spin vector $\vec{P}$ oriented along the Y-axis defines a sense of rotation in the X-Z plane and the experiment then has a reflection symmetry in this plane. In contrast, without spin polarization, the experiment has cylindrical symmetry along the axis defined by the linear momentum of the incident electron, $\vec{k}_0$, the Z-axis in figure 1. The polarization of photons detected along the direction $\vec{P}$, when spin polarized electrons are used, is more detailed than observed in the case of cylindrical symmetry of experiments without spin polarization. Generally three Stokes parameters are needed to describe the polarization. The direction of the incident beam, vector $\vec{k}_0$, and the spin vector $\vec{P}$ define the reference axes for polarization measurements. The three Stokes parameters, P_1, P_2 and P_3 are then defined

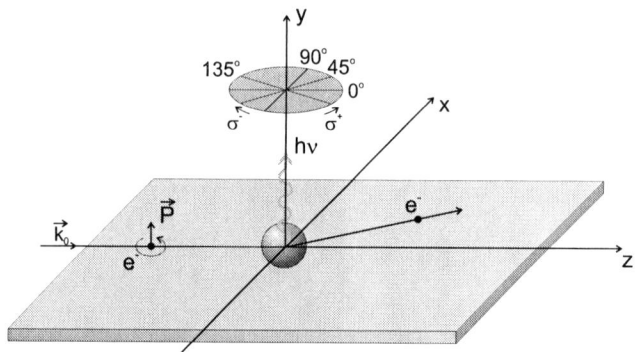

FIGURE 1. Schematic diagram showing experimental geometry relevant for definition of measured Stokes parameters.

as

$$P_1 = \frac{I(0°) - I(90°)}{I(0°) + I(90°)}, \quad P_2 = \frac{I(45°) - I(135°)}{I(45°) + I(135°)}, \text{ and } P_3 = \frac{I(\sigma^-) - I(\sigma^+)}{I(\sigma^-) + I(\sigma^+)}, \quad (1)$$

where $I(\alpha)$ corresponds to the observed photon intensities measured for the linear polarizations P_1 and P_2 with the polarizer transmission axis oriented at an angle α with respect to the incident electron beam direction and for circular polarization P_3 with the positive (σ^+) or negative (σ^-) helicity.

Bartschat and Blum [15] established the theoretical basis relating the integrated Stokes parameters to electron exchange and the spin-orbit interaction. The linear polarization P_1 is not affected by electron spin. Measurement of P_1 can generally give a non-zero value as it probes predominantly electrostatic interactions. A non-zero value of the Stokes parameter P_2 is an indication of spin orbit interaction. In the case of atoms with small and medium values of Z, and this is the case for the zinc atom, only the spin orbit interaction of the atomic electrons (responsible for the fine structure), can be observed. The circular polarization i.e. Stokes parameter P_3, is a direct measure of the momentum transfer and is affected by both the exchange and spin-orbit interaction. If no spin orbit interaction is observed, then circular polarization results from only electron exchange. In the experiments with unpolarized electrons, the measurement of P_2 and P_3 always results in zero values.

THE APPARATUS

The apparatus used for measurement of integrated Stokes parameters is shown in figure 2. Polarized electrons are obtained from photo emission from a GaAs crystal using $\lambda=830$ nm circularly polarized laser light [17]. Ordinary and strained GaAs crystals produce emitted electron polarizations of typically 30% and 66%, respectively. After passing the electron beam through a 90° hemispherical electrostatic deflector, initial longitudinal spin polarization is transformed into transverse. Spin orientation is changed

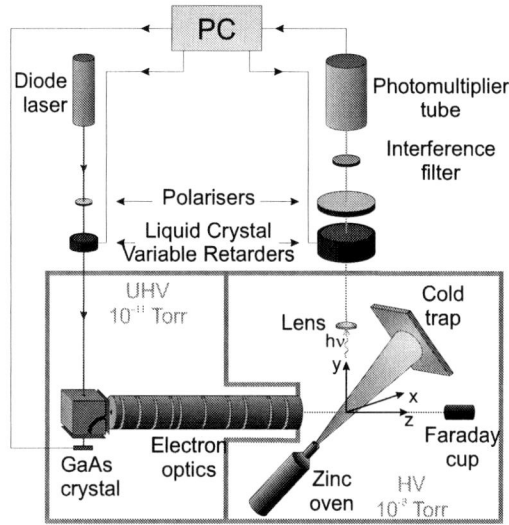

FIGURE 2. Schematic diagram of the apparatus.

by changing the helicity of circularly polarized laser light from right to left using a liquid crystal variable retarder (LCVR).

Zinc atoms are produced by evaporation in a resistively heated oven providing satisfactory intensities for observation of a range of transitions, yet creating negligible magnetic field and thermal background photons. The atomic beam is crossed at right angles with a beam of polarized electrons. Photons emitted from the interaction region are wavelength-selected using an interference filter while the polarization is measured using a combination of a liquid crystal variable retarder and polarizer [18] with easy control from a PC based data acquisition system.

The energy scale was calibrated by measuring the photon excitation function for the $4s4d\,^1D_2 \rightarrow 4s4p\,^1P_1$ decay photons in the threshold region where it exhibits a steep rise. This excitation function was then compared to the theoretical shape of the integral cross section [7] convoluted with a Gaussian apparatus function of appropriate width, as shown in figure 4. The estimated energy calibration error is 50 meV.

RESULTS AND DISCUSSION

The two different sets of experimental results are presented here and will be discussed in more detail at the conference. They illustrate some subtle points of experiments with spin polarized electrons and specific ways how electron correlations and momentum couplings can be explored when high precision polarization measurements are made. Both experiments indicate how polarization can be influenced by negative ion resonances, and how they can be observed with spin-polarised electron experiments.

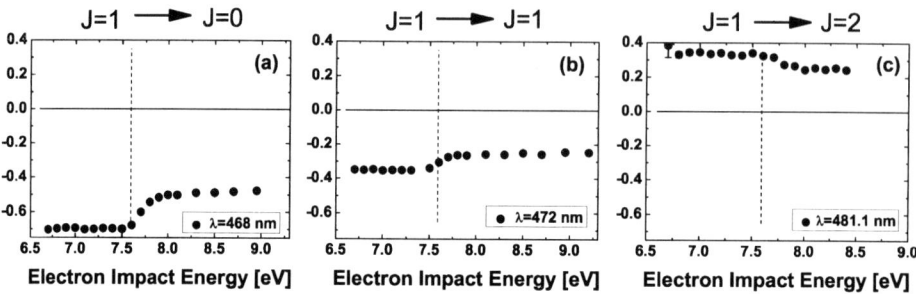

FIGURE 3. Stokes parameter P_3 measured for the three lines emitted in de-excitation process $5^3S\rightarrow$ $5^3P_{J=0,1,2}$ near threshold. Vertical line indicates the threshold for cascades above which the depolarization is observed.

While spin-resolved experiments test the existence, competition and interplay of exchange and spin orbit interaction in the excitation process, the correlation effects are enhanced by excitation via formation of intermediate short-lived negative ion resonances. Combined with the different ways the exchange and spin orbit interaction are manifested via resonance, studies with spin polarized electrons offer an increased sensitivity and observation of specific ways of their formation and decay. Such negative ion states have been observed in the energy region between 10 and 12 eV for all three Stokes parameters for the $3d^{10}4s4d\,^1D_2 \rightarrow 3d^{10}4s4p\,^1P_1$ decay photons. Negative ion resonances can most probably explain non-zero near threshold P_2 values measured for the $4s5s\,^3S_1$ state which is predicted to be well LS coupled.

The near threshold excitation of the 5^3S state - Exchange scattering

Excitation of the 5^3S_1 state in zinc by spin polarized electrons has been proposed earlier [14] and used in the present work to determine the spin polarization of the incident electrons by an optical method. For zinc atoms with a ground state ns^2 configuration, the transition $3d^{10}4s^2\,^1S_0 \rightarrow 3d^{10}4s5s\,^3S_1$ in which no net orbital angular momentum is transfered and excitation proceeds by electron exchange, the polarization of the emitted light is related simply to the photon polarization. Spin transfer in the collision polarizes the atom and if detected in the direction of spin transfer, the Y-axis in figure 1, the subsequently emitted light is circularly polarized. The measure of circular polarization, the Stokes parameter P_3, is related simply to the electron spin polarization P_e [14]. Expressions for each of the fine structure components in the decay $5s\,^3S_1 \rightarrow 4p\,^3P_{0,1,2}$ are

$$P_3 = +P_e\ (\lambda = 468\text{ nm}); \ P_3 = +P_e/2\ (\lambda = 472\text{ nm}); \ P_3 = -P_e/2\ (\lambda = 481\text{ nm}) \quad (2)$$

The measured values of P_3 for each of the fine structure components permit independent determination of the electron polarization as shown in figure 3. The measured value of the electron spin polarization was $P_e=0.66$ with ratios of P_3 for the three different transitions in agreement with equation (2), as can be seen in figure 3.

Several conditions have to be met for this method to be applicable as an optical electron spin-polarimeter [14]. The spin orbit interaction has to be strong enough for the fine structure splitting to be adequate for wavelength separation but not too strong to distort LS coupling. This condition rules out the spin-orbit interaction of the continuum electron for zinc. It is also assumed that the excited state is populated only by direct electron impact and not cascades. Above the threshold energy for excitation of the $4s5p\,^3P$ state cascading can cause significant depolarization as seen in all three measurements in figure 3. This requirement limits the incident electron energy range to a near threshold region indicated in the figure by a vertical line. Also the possible influence of negative ion resonances on polarization has to be checked carefully.

Excitation of the 4^1D_2 state - Characteristics of the 3d orbital

Excitation of zinc atoms by spin-polarised electrons has revealed negative ion resonances overlapping the ionization continuum and in the region of the $3d^94s^24p$ states. These resonances decay into several excited states with $3d^{10}4snl$ electron configuration and their study can probe properties of the atom with a vacancy in the $3d$ orbital. High precision measurements of the integrated Stokes parameters, P_1, P_2 and P_3, of the decay photons, $\lambda = 636.2$ nm, from the decay of the $4s4d\,^1D_2$ state, normalized to electron spin polarization, are shown in figure 4. An ordinary GaAs crystal was used in the P_1 measurement ($P_e=0.30$), and a strained GaAs crystal for P_2 and P_3 measurements ($P_e=0.66$). The energy scale was calibrated by recording the threshold for each measurement and comparing the intensity curve with the theoretical BSRM integral cross sections convoluted with the appropriate Gaussian instrumental function, shown in figure 4. In the narrow energy region close to threshold there are no cascade contributions, the transmission of the electron gun does not change significantly and the photon excitation function is proportional to the integral cross section. The energy resolution of the incident beam, full width at half height, was 180 meV for the ordinary GaAs crystal used in the P_1 measurements, and 250 meV for the strained crystal used in the P_2 and P_3 measurements.

A fit to the Fano profile using each of the Stokes parameter curves in figure 4 enabled precise determination of the energies, (10.98 ± 0.02) eV and (11.33 ± 0.02) eV, and the widths, $\Gamma = (0.25 \pm 0.03)$ eV and $\Gamma = (0.33 \pm 0.05)$ eV, of the two structures respectively. The uncertainties associated with each value originate from the fitting of the resonances. Several different considerations indicate that the observed resonances are associated with the $3d^94s^24p^2$ electron configuration where a pair of electrons in the $4p$ orbital is bound to the $^2D_{5/2}$ (lower energy resonance) and $^2D_{3/2}$ (higher energy resonance) ion core. Consideration of the centre of gravity of the states arising from different configurations indicates the most probable ion core state. Also the energy separation of the two structures is nearly equal to the separation of the those ion states.

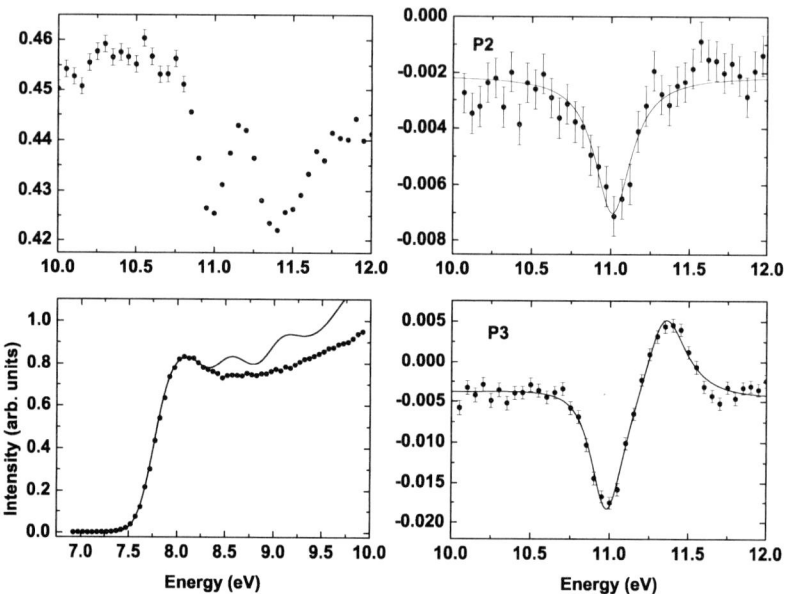

FIGURE 4. Stokes parameters P_1, P_2, P_3 and threshold intensity behavior of the $\lambda=636.2$ nm photons used in the energy calibration. ●, Experimental values; Full line, Fit to Fano profile for P_1, P_2, P_3 and to BMSR calculations [7] convoluted with the Gaussian apparatus function appropriate to each particular measurement.

In addition, both structures are, within experimental uncertainties, coincident with the two optically forbidden autoionizing states with the same configuration, $3d^94s^24p\,^3P$, J=2,0 as observed in the ejected electron spectra [5] and with the calculated assignments [6].

A similar situation is observed in the excitation of the neutral states of Cd and Hg [19], where negative ion resonances, associated with autoionizing states involving excitation of one of the least bound d-electrons, significantly modify the cross section for production of metastable atoms. The only previous photon excitation functions indicated the existence of the resonances [20, 21] but did not have the sensitivity for precise characterisation in terms of the energies and widths. As indicated by our experiments, photon excitation functions are not a very sensitive observation channel in this particular case.

The behavior of the spin-dependent Stokes parameters P_2 and P_3 in figure 4 permits some interesting conclusions about the role of exchange and spin-orbit interaction in the two negative ion states. As P_2 is a direct indication of spin-orbit interaction, there is a measurable but small spin-orbit interaction in the $^2D_{5/2}4s^24p^2$ negative ion state but none in the $^2D_{3/2}\,4s^2 4p^2$ state. In contrast P_3 shows the effects of both resonances indicating that exchange is effective at least for the latter state. As P_3 is directly proportional to angular momentum transfer, in this case transfer of spin, P_3 should have the same sign as the spin polarization, here positive, if there is no spin orbit interaction. On the other hand, a negative sign of P_3 is an indication of separate and competing roles of both the spin orbit interaction and exchange while the observation of the lower energy resonance in P_2 we already know that spin orbit interaction is observable in this resonance state.

ACKNOWLEDGMENTS

This work was supported by the Australian Research Council and the University of Western Australia. LP and SN are supported by Postgraduate Research Scholarships.

REFERENCES

1. D. V. Fursa, and I. Bray, *J. Phys. B: At. Mol. Opt. Phys.* **30**, 5895–5913 (1997).
2. K. A. Berrington, W. B. Eissner, and P. H. Norrington, *Comput. Phys. Commun.* **92**, 290 (1995).
3. O. Zatsarinny, and C. F. Fischer, *J. Phys. B: At. Mol. Opt. Phys.* **33**, 313 (2000).
4. D. W. O. Heddle, R. G. W. Keesing, and J. M. Kurepa, *Proc. Roy. Soc., Ser. A* **359**, 389–410 (1978).
5. C. G. Back, M. D. White, V. Pejčev, and K. J. Ross, *J. Phys. B: At. Mol. Phys.* **14**, 1497–1507 (1981).
6. M. W. D. Mansfield, *J. Phys. B: At. Mol. Phys.* **14**, 2781–2792 (1981).
7. O. Zatsarini, and K. Bartschat, *Phys. Rev. A* **71**, 022716 (2005).
8. G. F. Hanne, *Phys. Rep.* **95**, 95–165 (1983).
9. N. Andersen, K. Bartschat, J. T. Broad, and I. V. Hertel, *Phys. Rep.* **279**, 252–396 (1997).
10. D. H. Yu, L. Pravica, J. F. Williams, N. Warrington, and P. Hayes, *J. Phys. B: At. Mol. Opt. Phys.* **34**, 3899–3908 (2001).
11. P. A. Hayes, D. H. Yu, J. Furst, M. Donath, and J. F. Williams, *J. Phys. B: At. Mol. Opt. Phys.* **29**, 3989–4000 (1996).
12. K. Bartschat, G. F. Hanne, and A. Wolcke, *Z. Phys. A* **304**, 89–94 (1982).
13. J. Goeke, G. F. Hanne, J. Kessler, and A. Wolcke, *Phys. Rev. Lett.* **51**, 2273–2275 (1983).
14. M. Eminyan, and G. Lampel, *Phys. Rev. Lett.* **45**, 1171–1174 (1980).
15. K. Bartschat, and K. Blum, *Z. Phys. A* **304**, 85–88 (1982).
16. S. Napier, D. Cvejanovic, J. F. Williams, and L. Pravica, *J. Phys. B: At. Mol. Opt. Phys.* (2006).
17. P. A. Hayes, D. H. Yu, and J. F. Williams, *Rev. Sci. Instrum.* **68**, 1708–1713 (1997).
18. J. E. Furst, D. H. Yu, P. A. Hayes, C. M. D'Souza, and J. F. Williams, *Rev. Sci. Instrum.* **67**, 3813–3817 (1996).
19. S. J. Buckman, and C. W. Clark, *Rev. Mod. Phys.* **66**, 539–655 (1994).
20. O. B. Spenik, I. P. Zapesochnyi, V. V. Sovter, E. E. Kontrosh, and A. N. Zavilopulo, *Sov. Phys. JETP* **38**, 898–902 (1974).
21. E. E. Kontrosh, I. V. Chernishova, L. Sovter, and O. B. Spenik, *Optics and Spectroscopy* **90**, 339–343 (2001).

Electron Interaction with DNA Deoxyribose Analogue Molecules

A. R. Milosavljević

Institute of Physics, Pregrevica 118, 11080 Belgrade, Serbia

Abstract. The experimental results on elastic and inelastic low(medium)-electron interaction with tetrahydrofuran (THF: C_4H_8O) and tetrahydrofurfuryl alcohol (THFA: $C_5H_{10}O_2$) are presented. The differential cross sections (DCSs) were measured both as a function of scattering angle and incident electron energy. The experimental DCSs for elastic electron scattering by THF and THFA at the energies of 50, 100 and 250 eV are compared with the available recent theoretical data. The applicability of different theoretical approaches to describe elastic electron scattering from THF and THFA is discussed. Furthermore, electron energy loss spectra for THF and THFA are presented at the incident electron energy of 100 eV and the scattering angle of $10°$. Differences in both elastic and inelastic scattering processes upon substitution of one H atom (THF) by CH_2OH group (THFA) are discussed.

Keywords: tetrahydrofuran, THF, tetrahydrofurfuryl alcohol, THFA, elastic electron scattering, electron energy loss.
PACS: 34.80.Bm, 34.80.Gs

INTRODUCTION

The low- and medium-energy electron interaction with tetrahydrofuran (THF: C_4H_8O) and tetrahydrofurfuryl alcohol (THFA: $C_5H_{10}O_2$) has attracted considerable attention in recent years. Majority of the recent studies were prompted by investigations of radiation damage to biological systems and a need to understand elementary electron driven processes in living cell, which are believed to be to a great extent responsible for most genotoxic effects. THF and THFA were considered to be simple prototypes for investigations of electron interactions with deoxyribose molecules and further with the DNA sugar backbone. The existing results cover: VUV absorption and electron energy loss spectra of THF [1,2] (and references therein), resonant enhanced vibrational excitation of THF [3], electron-stimulated desorption yields of H⁻ from thin films [4,5], electron-induced damage of solid THF films [6], total cross sections for electron(positron)-THF scattering [7,8], absolute DCSs for elastic electron-THF(THFA) scattering [9,10], calculations of integral and differential elastic and ionization cross sections using the independent atom method (IAM) [10,11], calculations of low-energy electron interaction with THF [12,13] and dissociative electron attachment (DEA) to gas-phase THF [14]. Much more results have been obtained for the simplest THF molecule, while only few have been reported for THFA and none of them for electron energy loss spectra.

CP876, *The Physics of Ionized Gases: 23ʳᵈ Summer School and International Symposium*,
edited by L. Hadžievski, B. P. Marinković, and N. S. Simonović

The most recent studies [14] suggest that, actually, THF represent a poor surrogate for sugar molecules, considering DEA processes at low incident electron energies (below 5 eV). However, the present work deals with elastic DCSs and electron energy loss spectra at higher incident electron energies. These results are of interest in energy deposition modelling that is based on Monte Carlo simulation of a single scattering process. Experimental and theoretical absolute DCSs for elastic electron scattering from gaseous THF and THFA have only recently been published. In the present paper, we compare the experimental elastic DCSs with the most recent calculations and discuss the applicability of different theoretical approaches to describe elastic electron scattering from THF and THFA. Also, we investigate the change of elastic DCSs upon substitution of one α-H atom in THF by the CH_2OH group in THFA, in order to learn the applicability of present elastic DCSs for the more complex deoxyribose molecule (the comparison of structural formulae for THF, THFA and deoxyribose can be find in reference [10]). Finally, we present new electron energy loss spectra for THF and THFA at relatively high incident electron energy of 100 eV and a small scattering angle of 10^0, the conditions that are more favorable to electric dipole interaction, so the results can be compared with photoabsorption spectra. The results for THF are compared with available previous VUV and electron energy loss data, while similarities in electron energy loss spectra of THF and THFA (for which no results have been reported, according to our knowledge) are discussed.

EXPERIMENTAL

A more detailed description of the experimental set-up and the measurement procedure has been given recently [15]. Briefly, an electron gun produces a nonmonochromated, well collimated incident electron beam, which is crossed perpendicularly by a molecular beam produced by a stainless still needle (see Figure 1). The gun can be rotated around the needle in the angular range from about -40^0 to 130^0. The scattered electrons are retarded and focused by a four-element cylindrical electrostatic lens into a double cylindrical mirror analyzer, followed by three-element focusing lens and a single channel electron multiplier. The base pressure was about 4×10^{-7} mbar (turbo-molecular pump). The working pressure was usually about 5×10^{-6} mbar. The uncertainty of the incident energy scale was determined by observing a threshold for He^+ ions yield to be less than ± 0.4 eV. The best energy resolution was about 0.5 eV (limited by thermal spread of primary electrons). The resolution was lowered (1-1.5 eV) for measurements of DCSs as a function of incident electron energy in order to reduce energy dependence of the transmission function [15]. It should be noted that even with the best applied resolution, the present elastic DCSs inevitably include rotational and vibrational excitations. The latter, however, should not distort significantly elastic DCSs at the presented incident energies. The angular resolution was determined to be better than $\pm 2^0$. The experimental procedure was checked according to benchmark DCSs for elastic electron scattering by Kr, as a function of both scattering angle and incident electron energy, which were measured directly before and after electron-THFA measurements. The anhydrous both THF and THFA were purchased from Merck KGaA with declared purities of >99.9% and >98%, respectively, and were used after several cycles of freeze-thaw under vacuum.

Because of a rather low vapor pressure of THFA (more than two orders of magnitude lower at 20 °C than that of THF), the sample container was heated during a measurement at the temperature of about 70 °C. The calibration of the relative elastic DCSs to the absolute scale has been obtained according to relative flow measurements for THF (see [9]) or according to calculated cross sections for THFA (see [10]).

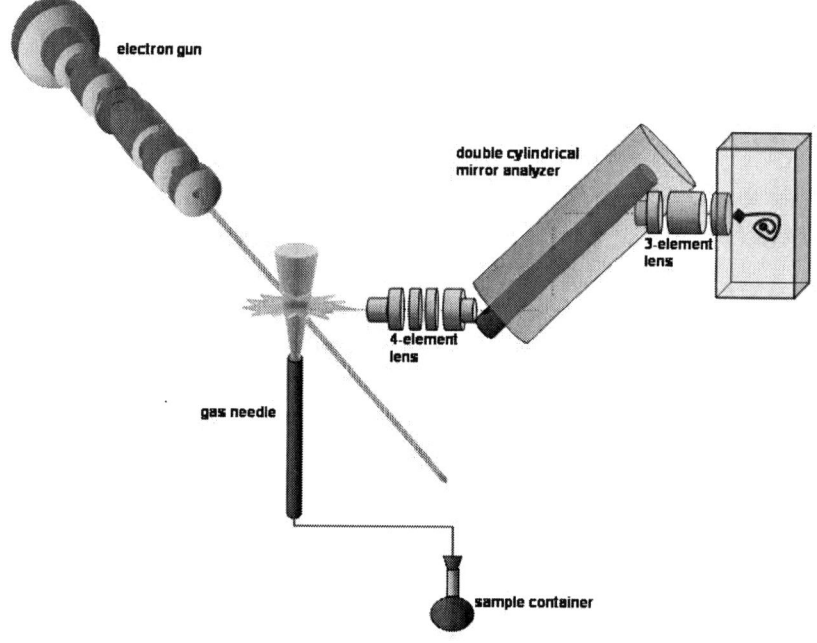

Figure 1. Schematic view of the experimental set-up.

RESULTS AND DISCUSSION

Elastic scattering

DCSs for elastic electron scattering from THF and THFA, at representative incident energies of 50, 100 and 200 eV, are shown in Figures 2a and 2b, respectively. The most recent calculated DCSs [10,11] are presented, as well. For better comparison of the shape of DCSs, the theoretical curves are normalized to the experimental results at the scattering angle of 50°. Absolute DCSs are compared for incident electron energy of 100 eV in Figure 3. The theoretical results of Možejko and Sanche [11] (dashed curves) have been obtained using independent atom method (IAM), which is based on the following assumptions: electrons scatter independently from each atom of the molecule, redistribution of atomic electrons due to molecular binding is unimportant and multiple scattering within the molecule is negligible (see [11] and references therein). The relative behavior of calculated cross sections is generally in good agreement with the experiment, except in the region around DCS minima where

82

theoretical results [11] show deeper minima. This deviation is even more pronounced at higher incident electron energies where IAM should be more accurate.

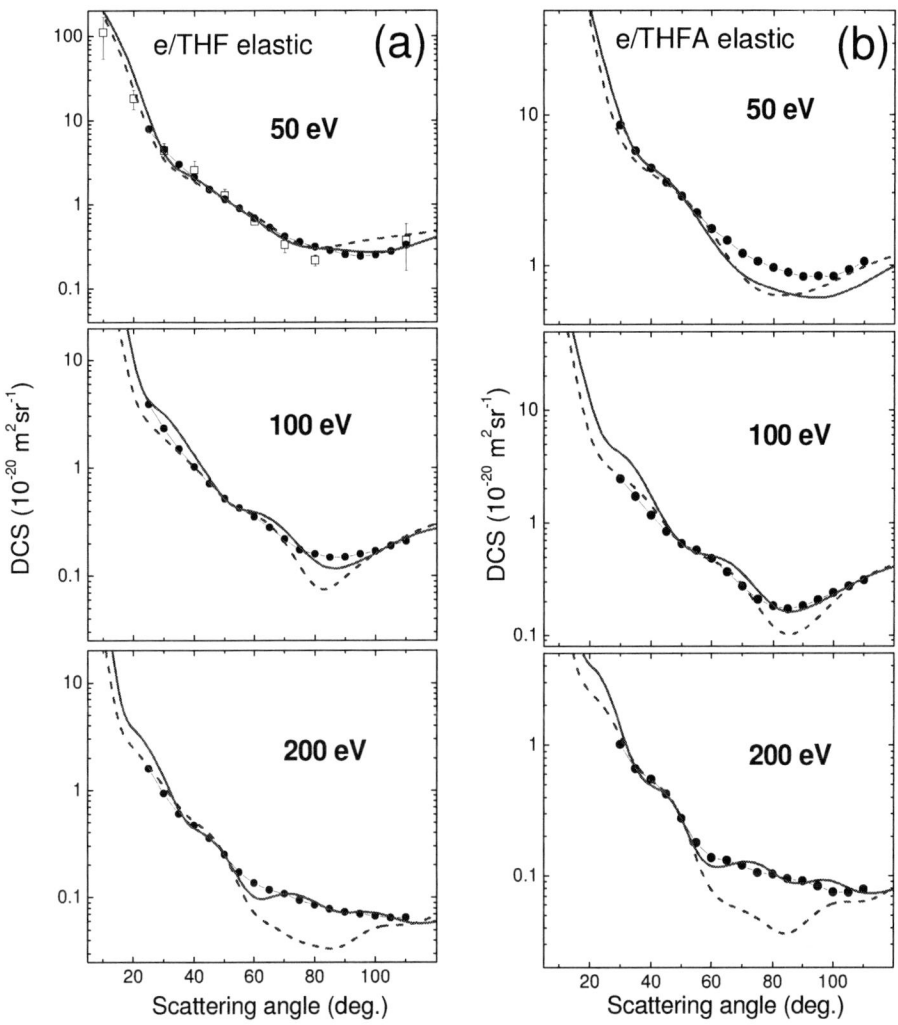

FIGURE 2. Absolute differential cross sections for elastic electron scattering from THF (a) and THFA (b) at the incident electron energies of 50, 100 and 200 eV: •, present experiment [9,10]; □, high-energy-resolution experiment [9]; —, theory (SCAR) [10]; --, theory (IAM) [11]. For better comparison of DCS shapes, the theoretical DCSs are normalized to the experimental results at the scattering angle of 50°. The experimental points are connected by strait lines.

The calculations of Blanco and Garcia [10] (full green line) are based on a corrected form of the IAM, known as the SCAR (Screen Corrected Additivity Rule) procedure (see [10] and references therein). In this procedure, screening corrections are introduced to account for partial overlapping of geometrical atomic cross sections as seen by the incident electrons. The calculations of atomic cross sections use an improved quasifree absorption model potential to account for inelastic processes (see [10]). The relative behavior of the calculated DCSs is in a very good agreement with the experiment, even in the region around minima (see Figure 2 – full line).

Absolute experimental DCSs for THF have been obtained according to relative flow measurements (see [9] for details). Since there are no absolute measurements for THFA, these DCSs were normalized to the absolute scale according to SCAR calculations (see [10] for details). In Figure 3, experimental and theoretical absolute DCSs for both THF and THFA are compared at the incident electron energy of 100 eV. DCSs for THF and THFA have similar absolute magnitudes according to both experimental and theoretical results. The DCSs for THF are somewhat lower in comparison to DCSs for THFA, which is to be expected considering sizes of the molecules. The calculated DCSs generally agree well with the experiment, although the DCSs of Možejko and Sanche [11] somewhat overestimate, while DCSs calculated by Blanco and Garcia [10] somewhat underestimate experimental DCSs. Still, later calculations [10] are closer to experimental results.

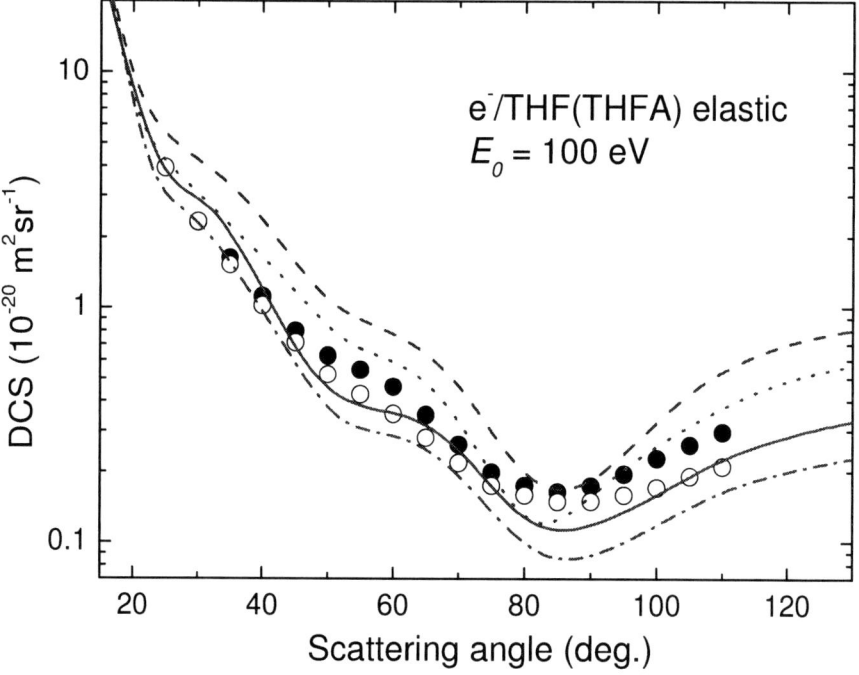

FIGURE 3. Absolute differential cross sections for elastic electron scattering from THF and THFA at the incident electron energy of 100 eV: $\bigcirc$, experiment THF [9]; $\bullet$, experiment THFA [10]; --, theory (IAM) THF [11]; ....., theory (IAM) THFA [11]; ---, theory (SCAR) THF [10]; —, theory (SCAR) THFA [10].

Electronic excitations

Electron energy loss spectra for THF and THFA molecules, obtained at the incident electron energy of 100 eV and the scattering angle of $10°$, are shown in Figures 4a and 4b, respectively. The marked electronic levels for THF (bars – above the spectrum), which are described as excitation to Rydberg levels, are taken from Bremner et al. [1] and are obtained according to VUV measurements. The levels obtained with high resolution electron energy loss measurements by Tam and Brion [2] are shown, as well (bars – below the spectrum).

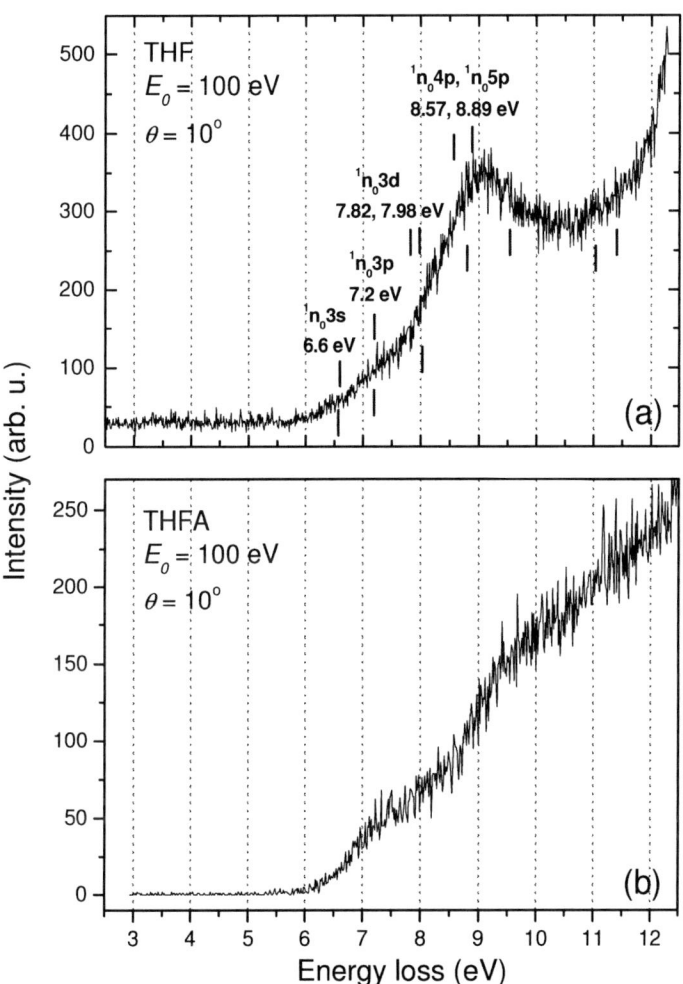

FIGURE 4. Electron energy loss spectra of THF (a) and THFA (b) recorded at the incident electron energy of 100 eV and the scattering angle of $10°$. The energy resolution was about 0.5 eV (FWHM). The marked electronic states for THF are taken from VUV measurements [1] (above the spectrum) and high resolution electron energy loss measurements [2] (below the spectrum).

The present spectrum for THF (Figure 4a) is generally in good agreement with previous results. According to our knowledge, no results for electronic excitation of THFA have been published yet. The present spectrum for THFA (Figure 4b) shows similar excitation bands as the spectrum for THF, in the covered energy loss region.

CONCLUSION

The elastic and inelastic electron collisions with tetrahydrofuran (THF) and tetrahydrofurfuryl alcohol (THFA) are investigated using crossed-beam experimental set-up and recent published results. For elastic scattering, the most recent experimental [9,10] and calculated [10,11] DCSs for THF and THFA are compared. It is shown that SCAR procedure improves the agreement of the shape of calculated DCSs with the experimental results upon IAM, mostly in the regions around DCS minima. Furthermore, the DCSs for THF and THFA appear to be very similar both in shape and on the absolute scale. Therefore, the substitution of one α-H atom in THF by the CH_2OH group in THFA does not influence significantly the elastic scattering process in the incident energy region between 50 and 300 eV.

The low resolution electron energy loss spectra for THF and THFA, taken at the incident electron energy of 100 eV and the scattering angle of 10° are shown, as well. The spectrum for THF is in a good agreement with previous published results. According to our knowledge, the first electron energy loss spectrum for THFA is presented in this paper. The first several bands in electronic excitation of THFA (in the energy loss region up to about 10 eV) seem to have similar position as for THF and can be also described as Rydberg excitations.

ACKNOWLEDGMENTS

I am very grateful to Dr B. P. Marinković from Belgrade for critical reading and suggestions. Also, I am very grateful to Dr P. Možejko from Gdansk and to Dr F. Blanco and Prof. G. García from Madrid for sending the calculated cross sections in numerical form, as well as for useful discussion. I am also grateful to Prof. M.-J. Hubin-Franskin from Liége for useful discussion on electronic excitation of THF and THFA. This work has been supported by Ministry of Science and Environmental Protection of Republic of Serbia under project 141011 and motivated by research within COST Action P9 "Radiation Damage in Biomolecular Systems".

REFERENCES

1. L. J. Bremner, M. G. Curtis, I. C. Walker, *J. Chem. Soc. Faraday Trans.* **87**, 1049 (1991)
2. W.-C. Tam and C. E. Brion, *J. Electron Spectrosc. Relat. Phenom.* **3**, 263 (1974)
3. M. Lepage, S. Letarte, M. Michaud, F. Motte-Tollet, M.-J. Hubin-Franskin, D. Roy, L. Sanche, *J. Chem. Phys.* **109**, 5980 (1998)
4. D. Antic, L. Parenteau, M. Lepage, L. Sanche, *J. Phys. Chem. B* **103**, 6611 (1999)
5. D. Antic, L. Parenteau, L. Sanche, *J. Phys. Chem. B* **104**, 4711 (2000)
6. S.-P. Breton, M. Michaud, C. Jäggle, P. Swiderek, L. Sanche, *J. Chem. Phys.* **121**, 11240 (2004)
7. A. Zecca, C. Perazzoli, M. J. Brunger, *J. Phys. B: At. Mol. Opt. Phys.* **38**, 2079 (2005)
8. P. Možejko, E. Ptasińska-Denga, A. Domaracka, C. Szmytkowski, *Phys. Rev. A* **74**, 012708 (2006)

9. A. R. Milosavljević, A. Giuliani, D. Šević, M.-J. Hubin-Franskin, B. P. Marinković, *Eur. Phys. J. D* **35**, 411 (2005)

10. A. R. Milosavljević, F. Blanco, D. Šević, G. García and B.P. Marinković, *Eur. Phys. J. D* **40**, 107 (2006)

11. P. Možejko, L. Sanche, *Radiat. Phys. Chem.* **73**, 77 (2005)

12. D. Bouchiha, J. D. Gorfinkiel, L. G. Caron, L. Sanche, *J. Phys. B: At. Mol. Opt. Phys.* **39**, 975 (2006)

13. C. S. Trevisan, A. E. Orel, T. N. Rescigno, *J. Phys. B: At. Mol. Opt. Phys.* **39**, L255 (2006)

14. P. Sulzer, S. Ptasińska, F. Zappa, B. Mielewska, A. R. Milosavljević, P. Scheier, T. D. Maerk, I. Bald, S. Gohlke, M. A. Huels, and E. Illenberger, *J. Chem. Phys.* **125**, 044304 (2006), also selected to *Virtual Journal of Biological Physics Research* **12(3)** (2006)

15. A. R. Milosavljević, S. Madžunkov, D. Šević, I. Čadež, and B. P. Marinković, *J. Phys. B: At. Mol. Opt. Phys.* **39**, 609 (2006)

Electron impact excitation of the 3s3p $^1P_1$ state in magnesium

Branko Predojević

Faculty of Natural Sciences, University of Banja Luka, Republic of Srpska, Bosnia and Herzegovina

Abstract. Differential cross sections (DCSs) for electron-impact excitation of the 3s3p $^1P_1$ resonance state of magnesium have been measured at 10, 15, 20, 40, 60, 80 and 100 eV incident electron energies (E_o). Scattered-electron intensities were measured over wide range of scattering angles from 2° to 150°. The absolute DCS scale for the $^1P_1$ state was determined through normalizations of its relative DCSs to optical oscillator strength using forward scattering function method, except at $E_o \leq 15$ eV where the excitation function of the 3s3p $^1P_1$ state experimentally obtained by Leep and Gallagher (*1976 Phys. Rev. A* **13** 148) was utilized for normalization. These absolute DCSs were extrapolated to 0° and 180° and numerically integrated to yield integral, momentum transfer and viscosity cross sections. Our results are compared with available experimental and theoretical data.

Keywords: differential cross section, magnesium, scattering.
PACS: number 34.80.Dp

INTRODUCTION

Following our investigation of electron scattering by Yb, Hg, Cd, Zn and Ca [1-6], we have extended our interest of two valence electron atoms to Mg. As Mg is a relatively light atom (Z=12), the ground state and excited states can be well described within the LS-coupling formalism. The large optical oscillator strength and small difference between ground $3s^2$ $^1S_0$ state and the 3s3p $^1P_1$ resonance state suggest a strong coupling between these two states. For these reasons, Mg as target in electron-atom collision processes shows different behavior compared with He, a much studied target in experimental and theoretical works concerning the electron-atom collisions processes.

Magnesium has been the subject of a number of experimental electron atom scattering studies, but only limited information is available on the 3s3p $^1P_1$ excitation cross sections. Williams and Trajmar [7] measured DCSs for excitation of the 3s3p $^1P_1$ state at impact energies (E_o) of 10, 20 and 40 eV and scattering angles (θ) from 10° to 130°. Their normalized DCSs were extrapolated to 0° and 180° and then the integrated; integral (Q_I) and momentum transfer cross sections (Q_M) were calculated. For the normalization of relative DCSs they chose the optical excitation function of Leep and Gallagher [8]. The same normalization procedure has been applied by Brunger *et al* [9] to measured DCSs for the 3s3p $^1P_1$ state at energies of 10 eV and 20 eV (from 5° to 130°) and at 40 eV (from 3° to 130°). Recently, Brown *et al* [10] reported relative

CP876, *The Physics of Ionized Gases: 23rd Summer School and International Symposium*,
edited by L. Hadžievski, B. P. Marinković, and N. S. Simonović
© 2006 American Institute of Physics 978-0-7354-0377-2/06/$23.00

DCSs for excitation of the resonance state in magnesium by 20 eV incident electrons, adding these results to the previously reported DCSs at 40 eV [11], both measurements were performed in angular range from $10°$ to $140°$.

Calculations of the DCS and Q_I cross section for magnesium have been conducted by Fabrikant [12] for the $3s^2$ $^1S_0 \rightarrow 3s3p$ $^1P_1$ transition at $E_o = 10$ and 20 eV, using the two-state close-coupling (CC2) approximation. Mitroy and McCarthy [13] computed DCSs for the excitation of the four lowest singlet states ($3s3p$ $^1P_1$, $3s4s$ $^1S_0$, $3s4d$ $^1D_2$ and $3s4p$ $^1P_1$) and the elastic scattering at $E_o = 10$, 20, 40 and 100 eV, using the five-state close-coupling (CC5) approximation. They also calculated the integral cross sections for excitation of the resonance transition including cascade contributions at all energies studied. McCarthy et al [14] calculated DCSs and Q_I cross sections for the inelastic (excitation of the $3s3p$ $^1P_1$ and $^3P_1$ states) and elastic scattering at $E_o = 10$, 20 and 40 eV using the six-state close-coupling (CC6 and optical CCO6) methods. Meneses et al [15] used the first-order many-body theory (FOMBT) to calculate DCSs for excitation of the $3s3p$ ($^1P_1$ and $^3P_1$) states at $E_o = 20$, 30, 40, 50 and 100 eV. Clark et al [16] used both FOMBT and distorted-wave (DW) approximation to study the influence of target-state wave functions on DCS (at 10 and 40 eV) and Q_I (from 10 to 100 eV) results for the $3s3p$ $^1P_1$ state. Kaur et al [17] applied the relativistic distorted-wave (RDW) approximation for calculation of DCSs for the $3s3p$ ($^1P_1$ and $^3P_{0,1,2}$), $3s3d$ ($^1D_2$ and $^3D_{1,2,3}$) and $3s4p$ ($^1P_1$ and $^3P_{0,1,2}$) states in magnesium at $E_o = 10$, 20 and 40 eV. Fursa and Bray [18] calculated DCSs for electron-impact excitation of the $3s3p$ $^1P_1$ state at $E_0 = 10$, 20 and 40 eV using the 27-state (CC27) and CCC approach.

EXPERIMENT AND PROCEDURE

The apparatus used in the measurements is a conventional cross-beam electron spectrometer. Hemispherical energy selectors are used both in monochromator and analyzer. The spectrometer can be operated in three different modes: to record electron energy-loss spectra, to scan the incident energy and to measure the angular distribution of inelastically (or elastically) scattered electrons. Doppler broadening of lines in energy-loss spectra is avoided because the electron beam is perpendicular to the atom beam and because of the narrow electron beam geometry. All cylindrical electrostatic lenses are made of OFHC copper, which is gold plated, while hemispherical energy selectors and diaphragms are made of molybdenum. The analyzer can be positioned from $-30°$ to $150°$. A channel electron multiplier is used for single electron counting.

The metal-vapour beam source consists of a tubular stainless-steel crucible. The crucible is placed in a stainless-steel cylinder co-axially wrapped with two resistive bifilar heaters (top and bottom), enabling the top of the source to be at approximately 100 K hotter then the bottom. This prevents clogging and minimizes dimmer production. The temperature is monitored by two thermocouples, on the top and bottom of the crucible. To attained higher temperature and more stable operating conditions, titanium foils around the cylinder and on its top end are mounted to reflect thermal radiation back to the crucible. An additional outer copper cylinder served as a holder for the helical tube of the water-cooler.

The angular resolution of the spectrometer is estimated to be $1.5°$. The real zero scattering angle was determined on the basis of the symmetry of the scattered electron intensity with respect to the mechanical zero, within $0.2°$ uncertainty. Overall energy resolution (as FWHM) of about 120 meV was estimated in these measurements. The energy scale was calibrated against the 3s3p $^1P_1$ excitation threshold of Mg at 4.346 eV. Our measurements were performed at temperature of 780 K for magnesium of 99.9% purity. The working temperature mentioned above corresponds to the metal-vapour pressure approximately of 9.5 Pa. The beam of magnesium effuses through a cylindrical channel (aspect ratio 0.075) in the cap of the crucible.

The experiment has been conducted in two steps. In the first series of measurements, we have determined the relative DCSs at small scattering angles (up to $14°$). Briefly, for a given E_o the position of analyzer was changed in steps of $2°$ and the angular distribution of scattered electrons was measured. A correction of the scattering intensity was made due to the angular dependence of the effective interaction volume [19]. In order to put results for $E_o \geq 20$ eV on an absolute scale, relative DCSs were normalized using known values of the optical oscillator strength (OOS) [20] by utilizing the forward scattering function (FSF) method [21]. The FSF method may not be used for normalization of the relative DCS at $E_o = 10$ eV because the necessary condition $E_o \geq 2.5\omega$ (where $\omega = 4.346$ eV is the excitation energy) is not satisfied. At energies close to this limit ($E_o \leq 15$ eV) we normalized our experimental DCSs to optical excitation function measured by Leep and Gallagher [8].

In the second series of measurements, we have determined the relative DCSs at scattering angles from $10°$ to $150°$ (in steps of $10°$) in the same way as in previous case. These relative DCSs were normalized using the absolute values of the DCSs obtained in first step of measurements at scattering angle of $10°$. Absolute DCSs were extrapolated to $0°$ using polynomial functions and to $180°$ using the shape of the DCSs calculated by Srivastava and Stauffer [22] (MCGS RDW multi-configuration ground-state relativistic distorted-wave) and than integrated to obtain integral, momentum transfer and viscosity cross sections.

Contributions to the total error of the absolute DCSs come from: a) uncertainties in our experimental values, and b) uncertainties in normalization procedure. Uncertainties in our experimental values reflect the uncertainty of relative DCS. These arise from statistical errors, uncertainty of the effective path-length correction factor, and uncertainties of energy and angle. Uncertainty in the normalization procedure arises from uncertainty of the OOS and fitting of relative generalized oscillator strengths (GOS). Total errors of absolute DCSs are obtained as the square root of the sum of squared particular errors. The total errors of the integrated cross sections arise from the DCS errors mentioned above and errors of the extrapolation of DCS to $0°$ and to $180°$ and numerical integration.

RESULTS

Energy-loss spectrum of magnesium at 40 eV incident electron energy and scattering angle of $4°$ is shown in figure 1, together with the spectrum at 60 eV and $10°$ in the inset. The spectra contain well-resolved features that correspond to the elastic scattering and the 3s3p $^1P_1$ (resonance) state. The inset in the same figure shows details of the spectrum at $E_o = 60$ eV and $\theta = 10°$ in region above of the resonance state.

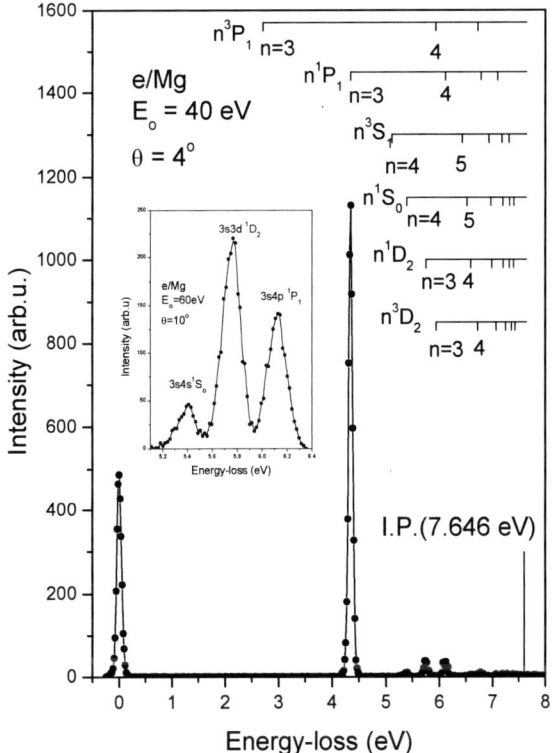

FIGURE 1. Electron energy-loss spectra of magnesium at 40 eV and 4° and 60 eV and 10° (in inset).

We have measured differential cross sections for electron impact excitation of the 3s3p $^1P_1$ state in magnesium at incident impact energies of 10, 15, 20, 40, 60, 80 and 100 eV and scattering angles from 2° to 180°. As mentioned above, at $E_o \geq 20$ eV the relative DCSs were normalized utilizing the forward scattering function method (FSF). The normalized generalized oscillator strength for excitation of the 3s3p $^1P_1$ versus the squared momentum transfer and their linear fits at all energies studied are plotted in figure 2. The absolute DCSs values (total error bars are indicated) at $E_o = 20$, 40, 80 and 100 eV together with several experimental and theoretical results are presented in figures 3 to 5. Calculated 3s3p $^1P_1$ integral cross sections are shown in figure 6, and compared with other available experimental data and theoretical calculations.

DISCUSSION AND CONCLUSION

The normalized generalized oscillator strengths for the excitation of the 3s3p $^1P_1$ states versus squared momentum transfer are presented in fig.2 (a) and (b). As expected [23] both slope and linear fit decreases and region of linearity of the GOS shrinks near the K^2_{min} as E_o increases.

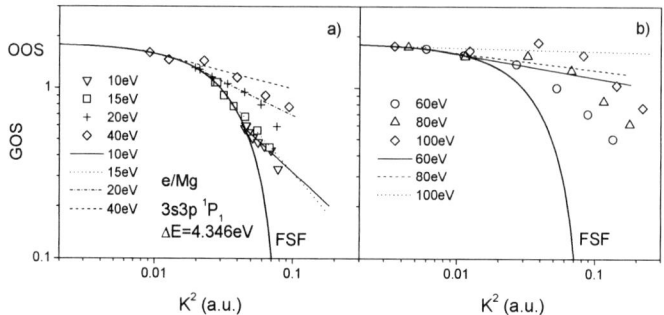

FIGURE 2. Generalized oscillator strength for electron excitation of the 3s3p $^1P_1$ state of magnesium at a) 10, 15, 20 and 40 eV; b) 60, 80, 100 eV incident electron energies.

At 20 eV (fig. 3a), there is disagreement among the theories as to whether there are one or two minima. Our measurements support the existence of two minima around 80° and 130°. The present experimental data exhibit the same general shape as the previous data [7, 9]. The agreement with [9] is better especially at small scattering angles (left inset in figure 3). The CCC [18] calculations yield substantially smaller cross sections at higher angles. At scattering angles $\theta \leq 70^\circ$ good agreement between present and CC5 [13] has been obtained.

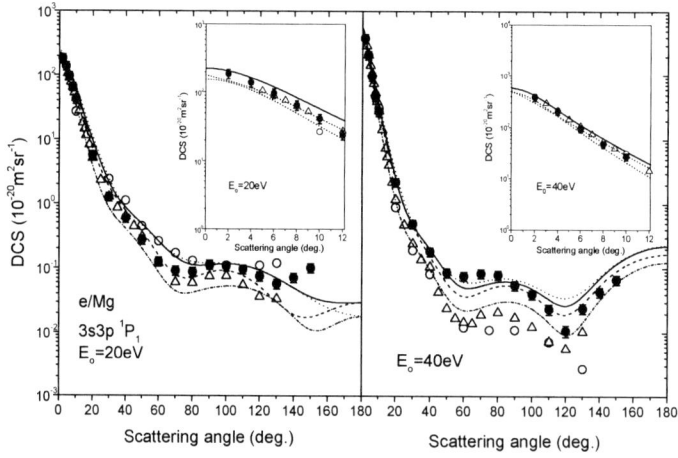

FIGURE 3. Differential cross sections for electron impact excitation of the 3s3p $^1P_1$ state of Mg at 20 and 40 eV. ●, present (total error bars are indicated); ○, Williams and Trajmar [7]; △, Brunger *et al* [9]; —, Srivastava and Stauffer [22]; – – –, Mitroy and McCarthy [13]; ···, Meneses *et al* [17]; –··–, Fursa and Bray [18].

92

At 40 eV, there is good agreement in shape and magnitude among the present data and CC5 [13], MCGS RDW [22] and FOMBT [17] calculations, except in depths of the both minima. Present and Brunger *et al* [9] experimental data and theoretical calculation [13, 18, 22] are in very good agreement at small scattering angles ($\theta \leq 10°$), right inset in figure 3. As one can see in figure 4, our experiment is in good agreement in shape with the recent experimental DCS for 20 and 40 eV [11, 10], normalized to ours at $100°$.

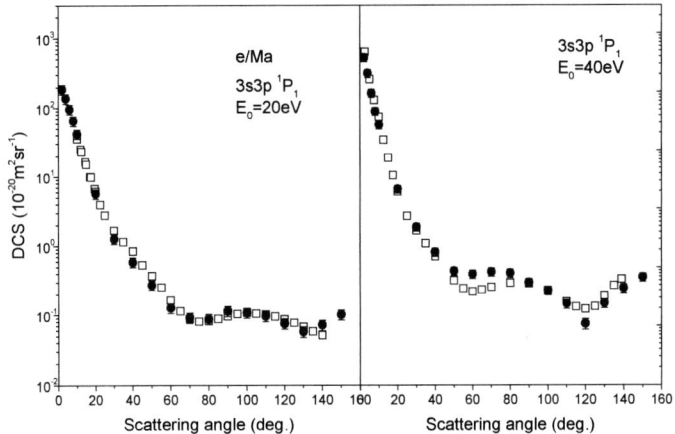

FIGURE 4. Comparison between present ($\bullet$), and Brown *et al* [10, 11] normalized to the present experimental DCS at $100°$ ($\square$).

As can be seen from figure 5, at 80 and 100 eV electron impact energy perfect agreements in shape between the present data measured for scattering angles up to $150°$ and MCGS RDW [22] calculation have been obtained. The excellent agreement, within experimental errors, between present and MCGS RDW [22] calculations at small scattering angles is clearly seen in insets in figure 4.

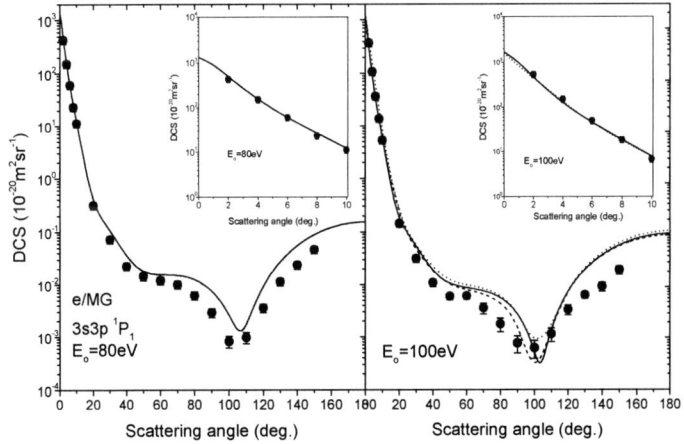

FIGURE 5. Sam as figure 3, but at 80 and 100 eV incident electron energies.

93

The integral cross section for excitation of the resonance state of magnesium together with several different calculations and optical excitation function by Leep and Gallagher [8] is shown in figure 6. The present Q_I are in good agreement in shape and magnitude with [8] at all energies measured. Also, good agreement, within experimental error bars, with CC5 [13] and UDW [16] was obtained. Going toward lower impact energies, the MCGS RDW [22] calculation overestimates the present experiment in contrast to the CCC [18] results that underestimates our measurement by a similarly amount.

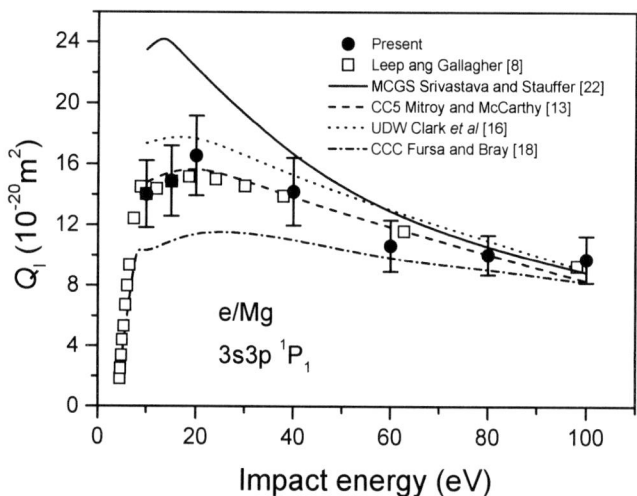

FIGURE 6. Integral cross section for excitation 3s3p $^1P_1$ state in magnesium atom.

We have performed differential cross section measurement for the excitation 3s3p $^1P_1$ at 10, 15, 20, 40, 60, 80 and 100 eV and in a broad range of scattering angles (up to 150°). The DCSs at 15, 60, 80 and 100 eV are determined for the first time. Comparing to the previous measurements, agreement among present DCSs and recent calculations was improved. In order to compare Mg with other atoms studied in our Laboratory, that also have two valence electrons, we considered a general rule that effective cross section for electron/atom impact is lower if the ionization potential is higher as it is noted in paper dealing with Zn and Cd [23]. We found that this rule is also valid in the case of electron-impact excitation of the resonant n ^1P states of Mg and Ca [6].

ACKNOWLEDGMENTS

I thank Dr R Srivastava and Dr A Stauffer for sending their calculated e/Mg DCS data in numerical form. I am very grateful to Dr. D. F. Filipović and other colleagues from The Laboratory of Atomic Collision Processes. This experimental work has been carried out in The Institute of Physics, Belgrade within MNZZS project No.141011 of Republic of Serbia.

REFERENCES

1. Predojević B, Šević D, Pejčev V, Marinković B P and Filipović D M, *J. Phys .B: At. Mol. Opt. Phys.* **38**, 1329 (2005)
2. Predojević B, Šević D, Pejčev V, Marinković B P and Filipović D M, *J. Phys .B: At. Mol. Opt. Phys.* **38**, 3489 (2005)
3. Panajotović R, Pejčev V, Konstantinović M, Filipović D, Bočvarski V and Marinković B, , *J. Phys .B: At. Mol. Opt. Phys.* **26**, 1005 (1993)
4. Marinković B, Pejčev V, Filipović D and Vušković L, *J. Phys .B: At. Mol. Opt. Phys.* **24**, 1817 (1991)
5. Panajotović R, Šević D, Pejčev V, Filipović D M, Marinković B P, *Int. J. Mass Spectrom.* **233**, 253 (2004)
6. Milisavljević S, Šević D, Pejčev V, Filipović D M, and Marinković B P, *J. Phys .B: At. Mol. Opt. Phys.* **37**, 3571 (2004)
7. Williams W and Trajmar S, *J. Phys. B: At. Mol.Phys.* **11** 2021 (1978)
8. Leep D and Gallagher A, *Phys. Rev.* A **13** 148 (1976)
9. Brunger J, Riley J L, Scholten R E and Teubner P J, *J. Phys. B: At. Mol. Opt. Phys.* **21** 1639 (1988)
10. Brown D O, Crowe A, Fursa D V, Bray I and Bartschat K, *J. Phys. B: At. Mol. Opt. Phys.* **38** 4123 (2005)
11. Brown D O, Cvejanović and Crowe A, *J. Phys. B: At. Mol. Opt. Phys.* **36** 3411 (2003)
12. Fabrikant I I, *J. Phys. B: At. Mol. Phys.* **13** 603 (1980)
13. Mitroy J and McCarthy I E, *J. Phys. B: At. Mol. Opt. Phys.* **22** 641 (1989)
14. McCarhy I E, Ratnavelu K and Zhou Y, *J. Phys. B: At. Mol. Opt. Phys.* **22** 2597 (1989)
15. Meneses G D, Pagan C B and Machado L E, *Phys. Rev.* A **41** 4740 (1990)
16. Clark R E H, Csanak G and Abdallah J, *Phys. Rev.* A **44** 2874 (1991)
17. Kaur S, Srivastava R, McEachran R P and Stauffer A D, *J. Phys. B: At. Mol. Opt. Phys.* **30** 1027 (1997)
18. Fursa D V, and Bray I, *Phys. Rev.* A **63** 032708 (2001)
19. Brinkmann R T and Trajmar S, *J. Phys. E: Sci. Instrum.* **14** 245 (1981)
20. Liljeby L, Lindgard A, Mannervik S, Veje E, Jelenković B, *Phys. Scripta* **21** 805 (1980)
21. Avdonina N B, Felfli Z and Msezane A, *J. Phys. B: At. Mol. Opt. Phys.* **30** 2591 (1997)
22. Srivastava R and Stauffer A D, (private communication)
23. Felfli Z and Msezane A Z, *Proc. XX Int. Conf. on the Physics of Electronic and Atomic Collisions (Vienna)* Eds. F. Aumayr, G. Betz and H. P. Winter, p MO 105 (1997)
24. Predojević B, Šević D, Pejčev V, Marinković B P and Filipović D M, *J. Phys .B: At. Mol. Opt. Phys.* **36** 2371 (2003)

Quantum and Classical Description of H Atom Under Magnetic Field and Quadrupole Trap Potential

J. Mahecha* and J. P. Salas†

*Institute of Physics, University of Antioquia, AA 1226, Medellín, Colombia.
LPMC, Institute of Physics, University Paul Verlaine, 1 Bv Arago, 57078 Metz Cedex 3, France
†Area of Applied Physics, University of La Rioja, C/Madre de Dios 51, 26006, Logroño, Spain

Abstract. A discussion regarding the energy levels spectrum of quantum systems whose classical analogous has states of chaotic motion is presented. The chaotic dynamics of the classical underlying system has its manifestation in the wave functions (in the form of "scars") and in the energy levels (in the form of "statistical repulsion" of the energy levels). The above mentioned signatures are named "quantum chaos". A typical study of quantum chaos requires finding accurate energy eigenvalues of highly excited states, to calculate the nearest neighbors spacing between levels, to perform the "unfolding" of the spectrum in order to separate the fluctuations, and finally to find the probability distribution of the unfolded spectrum. This is exemplified by the hydrogen atom under uniform magnetic field and a quadrupole electric field.

Keywords: Atoms in electromagnetic traps, Nonlinear dynamics and chaos, Quantum chaos.
PACS: 32.80.Pj; 33.55.Be; 05.45.-a; 05.45.Mt; 33.80.Ps; 39.10.+j; 42.50.Vk

INTRODUCTION

We numerically study the relation between the classical dynamics and the quantum spectrum of the hydrogen atom under the combined effect of a uniform magnetic field and a quadrupolar electric field (HAUMQE). The classical dynamics exhibits a phase space whose regular and chaotic regions can be put in correspondence with the parameters of the Hamiltonian; this work has been done in [1]. Quantum analysis, when the two fields can be considered as a small perturbation to the free hydrogen atom, was performed in [2], there the connection between the exact integrable cases found in [1] and the quantum spectrum when the energy levels can be labeled by a principal quantum number was shown. Here we continue this study in a nonpertubative, covering the whole regular and chaotic classical dynamics, and calculating thousands of energy levels and their statistical properties. The purpose of this work is to investigate the set of quantum energy levels of a hydrogen atom in the presence of uniform magnetic and quadrupolar electric fields. The classical analog of this system has a rich dynamics [1] which it is expected to manifest itself into the statistical properties of the energy levels of the quantum system. We pretend to describe a novel system which exemplifies the existing link between integrability/regularity of the classical dynamics and clustering/repulsion of the quantum energy levels. Furthermore, we consider the scaling parameter of magnetic field γ as a quantized variable, with a set of discrete values whose statistical properties are related to the underlying regular/chaotic/mixed classical dynamics.

CP876, *The Physics of Ionized Gases: 23rd Summer School and International Symposium*,
edited by L. Hadžievski, B. P. Marinković, and N. S. Simonović

Chaotic classical systems present exponential sensitivity to initial conditions, feature which depends on the nonlinearity of the equations of motion, but their quantum analogs apparently do not require special concepts to be described with quantum mechanics. Bound quantum systems in all cases have a well defined set of energy levels. The answer to the question of whether quantum systems can exhibit sensitivity to initial conditions is not related to the linearity of the Schrödinger equation. We look for the "traces" of classical chaos in the quantum world and we focus the problem from the Haake's [3] point of view: Having lost the classical distinction between regular and chaotic motion when turning in quantum mechanics, quantum chaos can be considered as a set of criteria that allows one to recognize two types of quantum dynamics. The most important criteria are the "scars" of the wave functions and the statistical properties of the energy levels spacing.

This last criterion is based on the fundamental concept of level repulsion. We note that this fact is a generic feature in the quantum world as nonintegrability is in classical systems. Hence, the statistical properties of the energy levels allow one to classify quantum systems in two groups. On the one hand, integrable/regular classical systems correspond to quantum systems whose quantum signature is that energy levels tend to clustering. On the other hand, nonintegrable classical systems (with phase space dominated by chaos), correspond to quantum systems whose quantum signature is that energy levels show resistance to crossing.

Random Matrix Theory (RMT) is used to quantify the level of clustering/repulsion. The behavior of a classically chaotic system is reflected in a universal statistical fluctuation of the energy levels of its quantum analog. Variation of parameters of Hamiltonian can induce changes into the classical dynamics and also into the statistical properties of quantum energy levels. Due to "symmetry $\Longleftrightarrow$ integrability" and "lost of symmetry $\Longleftrightarrow$ non-integrability", the difference is that energy levels of integrable system can be labeled by a complete set of good quantum numbers, fact that is not possible in the case of non-integrable systems.

Random spectrum of the energy levels of a quantum system whose classical analog is in a regular state of motion is called a *Poisson* spectrum. In the non integrable case, the spectrum is of *Wigner* type.

Now we will describe the systematic procedures to characterize the Level Statistics. Let's assume that we have a set of energy levels $E_1 \leq E_2 \leq ...E_M$. The first thing that we can do is to count them. This can be done with the counting function $N(E) = \sum_{j=1}^{M} \theta(E - E_j)$ where $\theta(x)$ is the Heaviside step function. The staircase function $N(E)$ counts the number of levels with energy up to E. It fluctuates around a mean function $\overline{N}(E)$. Function $\overline{N}(E)$ can be obtained by using semiclassical arguments, or by fitting $N(E)$. The derivative of $N(E)$ defines the levels density $\rho(E) = \sum_{j=1}^{M} \delta(E - E_j)$, and derivative of mean counting function $\overline{N}(E)$ define the mean levels density $\overline{\rho}(E) = d\overline{N}(E)/dE$.

A key point of level statistics is the so called Unfolded Spectrum. Fluctuations of the spectrum are *local* (dependent on $\rho(E)$) or *systematic* (dependent on $\overline{\rho}(E)$). Systematic fluctuations must to be removed. In order to compare two spectra according to their "degree of level clustering", they must be expressed in the same "currency". It is necessary to *unfold* the spectrum in order to have a new normalized spectrum where only local fluctuations are present.

In order to find the NNS distribution we define the energy level spacing $s_i = E_{i+1} - E_i$. We define the *mean level spacing* $\bar{s}(E)$ by $\bar{s}(E) = 1/\bar{\rho}(E)$. The *unfolding* consists in eliminating the energy dependence in the energy level spacing by defining the normalized level spacing $x_i = s_i/\bar{s}(E) = s_i\bar{\rho}(E)$. Next, the spacing are normalized to a unit mean spacing.

NNS of a *Poisson* spectrum follows a distribution $P(x) = e^{-x}$. High probability at zero spacing express level clustering. The NNS of a *GUE* spectrum follows a distribution $P_U(x) = (32/\pi^2)x^2 e^{-4x^2/\pi}$. The NNS of a *GOE* spectrum follows a distribution $P_W(x) = (\pi/2)xe^{-\pi x^2/4}$. High probability at nonzero spacing express level repulsion.

Most classical Hamiltonians have mixed regular-chaotic phase space. Then, spectrum of most quantum Hamiltonians has NNS distribution which is a mixing of Poisson and Wigner. In order to describe the transition from integrability to nonintegrability, Brody proposed a distribution which for $v = 0$ reduces to Poisson distribution, and for $v = 1$ the Wigner distribution for the GOE is obtained [4].

See details in the literature [5, 6] where this approach was applied to the problem of hydrogen atom in a magnetic field.

This introduction deal to fundamental concepts of Quantum Chaos, Random Matrices and Level Statistics. A brief review regarding the classical dynamics of Hydrogen atom in a constant magnetic field and a quadrupolar electric field is presented in Section "Classical dynamics of HAUMQE". The main results of the paper, involving the study of quantum energy levels, obtained by a nonperturbative approach, and the study of their statistical properties are presented in Section "Energy levels of HAUMQE". A discussion of our results is given in Section "Conclusions".

CLASSICAL DYNAMICS OF HAUMQE

Now we consider Hamiltonian of hydrogen atom in the presence of uniform magnetic and quadrupolar electric fields. In cylindrical coordinates it has the form [1],

$$H = \frac{1}{2}\left(p_\rho^2 + p_z^2\right) + \frac{p_\phi^2}{2\rho^2} - \frac{1}{(\rho^2+z^2)^{1/2}} - \gamma p_\phi + \frac{\gamma^2}{2}\rho^2 + \frac{\omega_z^2}{2}\left(z^2 - \frac{\rho^2}{2}\right). \quad (1)$$

We use atomic units and put infinite the nuclear mass. p_ϕ is the z component of canonical angular momentum. The Larmor frequency $\gamma = B/(2B_0)$ is the scaled magnetic field, $B_0 \sim 2.35 \cdot 10^5\,T$, The electric quadrupolar field induces axial oscillations at frequency ω_z. This frequency is expressed in units of $\omega_0 = me^4/(4\pi\varepsilon_0)^2/\hbar^3 \sim 4.13 \cdot 10^{16}s^{-1}$. Hamiltonian of free hydrogen atom is separable both in the classical and quantum regimes. The interaction with the fields mixes the coordinates in a no separable way.

Next, rotating coordinates system with angular velocity γ are introduced to eliminate the paramagnetic term. Then, coordinates and momenta are scaled according to $\boldsymbol{r}' = \gamma^{2/3}\boldsymbol{r}$, $\boldsymbol{p}' = \gamma^{-1/3}\boldsymbol{p}$. Dynamics depends only on the parameters λ and ε, after dropping the primes, equation (1) becomes

$$\gamma^{-2/3}H = \frac{1}{2}(p_\rho^2 + p_z^2) - \frac{1}{(\rho^2+z^2)^{1/2}} + \frac{1}{2}\left[\left(1 - \frac{\lambda^2}{2}\right)\rho^2 + \lambda^2 z^2\right], \quad (2)$$

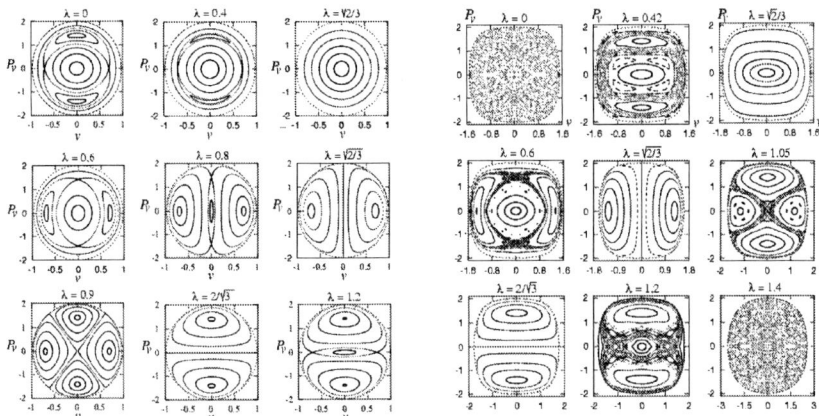

FIGURE 1. (a) Weak field case. Evolution of the surfaces of section ($u = 0$, $p_u \geq 0$) as a function of λ for $\varepsilon = -2$. The three oyster bifurcation are observed. (b) Strong field case. Evolution of the surfaces of section ($u = 0$, $p_u \geq 0$) as a function of λ for $\varepsilon = -0.2$. The three chaos-order transitions are observed.

we define $\lambda = \omega_z/\gamma$, which describes the ratio between the intensities of the fields and we put $p_\phi = 0$. Last two terms represent the external potential, which confines the electron along the vertical coordinate, and also along the horizontal coordinates if $1 - \lambda^2/2 > 0$. See [1]. For $\lambda < 2^{1/2}$ the effective potential shows only a minimum at $(0,0)$, while for $\lambda > 2^{1/2}$, the potential energy surface shows a saddle point at $(2^{1/3}/(\lambda^2 - 2)^{1/3}, 0)$, through which atom can ionize.

Now, by fixing the value of λ and the value of the scaled energy ε, and by simultaneously varying the energy E and the magnetic field γ, we can explore different regions of the quantum spectrum with the same classical dynamics.

Singularity in (2) can be removed by means of the Levi-Civita regularization. This procedure is done in two steps. First, semi parabolic coordinates (u, v) are introduced, $z = (u^2 - v^2)/2$, $\rho = u\,v$, then, the momenta p_u, p_v are defined with respect to the coordinates dependent scaled time $d\tau = dt/(u^2 + v^2)$. Finally, Hamiltonian takes the form [5],

$$K = 2 = \frac{1}{2}(p_u^2 + p_v^2) - \varepsilon(u^2 + v^2) + \frac{1}{2}\left(1 - \frac{\lambda^2}{2}\right)u^2 v^2(u^2 + v^2) + \frac{1}{8}\lambda^2(u^2 - v^2)^2(u^2 + v^2).$$

(3)

The 2 describes the Coulomb interaction. For negative scaled energy $\varepsilon < 0$, Hamiltonian (3) represents two harmonic oscillators with the same frequency $(-2\varepsilon)^{1/2}$ perturbed by two six-degree polynomial terms describing the diamagnetic and the electric interactions.

We define the Poincaré surface of section projecting the phase space on the $u = 0$ plane with $p_u \geq 0$. In order to study the evolution of the structure of the phase space, several surfaces of section were generated by keeping the scaled energy ε constant while varying the parameter λ.

For ε large, system is nearly integrable, due to at $\gamma \to 0$ we have the free hydrogen atom, and the external fields are weak enough to be considered as perturbations. In Fig. 1(a) is shown the evolution of the surfaces of section for $\varepsilon = -2$ and λ in the interval $[0, 1.2]$. Note that for these parameter values system is still close to the integrable limit $\varepsilon \to -\infty$, and all orbits are regular and confined on adiabatic invariant torii. We observed in Fig. 1(a) that system suffers three consecutive oyster bifurcations at the integrable values $\lambda = (\sqrt{2}/3, \sqrt{2/3}, 2/\sqrt{3})$ [1]. It is worth noting that at each of these λ values a degenerate curve of fixed points appears in the surface of section. In many case degeneracy is the fingerprint of integrability [7].

Let's see the strong field regime. It occurs if energy ε is low enough. Fig. 1(b) shows a gallery of surfaces of section for $\varepsilon = -0.2$ and increasing values of λ. The sequence begins with $\lambda = 0$ and global chaos completely dominates the dynamics. As λ varies in the interval $[0, 1.4]$, we observe three chaos-order transitions. As λ approaches each of the three integrable values of λ, the stochastic motion gradually disappears in such a way that at the corresponding integrable λ value the expected regularity is reached. On the other hand, when λ moves away from each integrable value, the size of regions of stochastic motion grows again. These order-chaos transitions were illustrated by computing the fraction of the phase space where the trajectories are chaotic [2], and it confirms the presence of these three consecutive chaos-order transitions.

ENERGY LEVELS OF HAUMQE

Hamiltonian of a pair of isotropic 2D oscillators has energy levels given by $E_{uv} = 2\omega[n_u + n_v + (|m_u| + |m_v|)/2 + 1]$, where n_u, $n_u - m_u$, n_v, and $n_v - m_v$ are non-negative integers. Free 3D Hydrogen atom is dynamically equivalent to two non-interacting isotropic 2D harmonic oscillators. Those oscillators have equal and opposite angular momentum. $m_H = m_u \equiv m$. The principal quantum number of atom is related to oscillators' quantum numbers, $n_H = n_u + n_v + |m| + 1$. Hydrogen energy levels are obtained from $E_{uv} = 2$, $\omega = (-2E)^{1/2}$, or $2 = 2\omega n_H$. Result is $E = -1/(2n_H^2)$. In order to find the quantum spectrum of HAUMQE, Schrödinger equation must be solved. The no scaled form is,

$$\left[T - E(u^2 + v^2) - 2 + \frac{\gamma^2}{2}u^2v^2(u^2 + v^2) + \frac{1}{8}\omega_z^2(u^4 + v^4 - 4u^2v^2)(u^2 + v^2) \right] \psi = 0, \quad (4)$$

where $2T = -\nabla_u^2 - \nabla_v^2$. Hamiltonian in (4) corresponds to two 2D harmonic oscillators perturbed by a sixth-degree term in the coordinates u and v.

For weak fields system is regular and a perturbative approach is appropriate [2]. From a qualitative point of view, this assumption holds when the perturbations strengths γ and λ are smaller than the energy spacing between consecutive hydrogenic manifolds, i.e., $\gamma^2 n^7 \ll 1$ and $w_z^2 n^7 \ll 1$. Energy shifts for the $n = 20$ manifold and $\gamma = 1/(10n^{7/2}) \sim 2.8 \times 10^{-6}$ in the range $\lambda \in [0, 1.3]$ are shown [2], they were obtained by using first-order degenerate perturbation theory. We observe three zones of clustering (strong degeneration) near the classical integrable cases for λ, away from these values splitting is bigger. See Fig. 2(a).

For a given n-manifold, the eigenstates of (1) for $p_\phi = m = 0$ can be expressed as a function of the pure hydrogenic basis by using the usual expansion over the orbital quantum number l. The energy levels $E_{n,l} = -1/2n^2 + \Delta E_{n,l}$ are the eigenvalues of the secular matrix, being $\Delta E_{n,l}$ the energy shifts due to the fields.

Both parities energy-level shifts (in atomic units) for the $n = 20$ manifold and $\gamma = 2.8 \times 10^6$ in the range $\lambda = [0, 1.3]$ are shown in Fig. 2(a). We observe in this figure that the quantum spectrum presents three zones of accumulations (strong degeneration) around the $\lambda's$ for which the classical system is integrable. As we note in the Introduction, the quantum mark of integrability is that energy levels tends to clustering, that is to say, degeneracy increases.

In the strong field regime, perturbative approach is no longer valid. We observe that the classical system presents three transitions from irregular to regular motion when λ passes through the values for which system is integrable. See Fig. 1(b). Moreover, because of the scaling property, the dynamics does not depend on E and γ independently, but on their combination $\varepsilon = \gamma^{-2/3}E$. The usual way to solve the Schrödinger equation corresponding to (1) is to use, for fixed values of γ and λ, a Sturmian basis expansion. However, a much more powerful method takes advantage from the fact of the $O(4)$ symmetry of the field-free hydrogen atom [8, 10].

Note that in the quantum case, the scaling invariance appears as an effective field dependent Planck's constant $\gamma^{1/3}\hbar$ [9]. If we assume that $\varepsilon < 0$, we are dealing with two two-dimensional harmonics oscillators of equal frequency perturbed by two six-order terms. In the above generalized eigenvalue problem, rather than to find the energy levels for given values of γ and λ, we determine the values of $\gamma^{2/3}$ where an energy level exists for a given classical scaled energy ε. These values $\gamma_i^{2/3}$ give the energy levels by using the scaling property $E_i = \gamma_i^{2/3}\varepsilon$. In other words, we obtain the intersections γ_i of the curves $E(\gamma_i)$ with the curve $E = \gamma^{2/3}\varepsilon$ [8].

Scaled and regularized Schrödinger equation is transformed in a matrix equation by expanding solution into the basis of eigenstates of H_0, the free hydrogen atom, given by $\{|n_u, n_v\rangle\}$. This basis is dependent on frequency ω, a free parameter which can be used like a variational parameter. Matrix equation is, in scaled form,

$$\frac{\varepsilon}{\omega^2}A + \frac{2}{\omega}I + \frac{4}{\omega^4}B + \frac{\lambda^2}{\omega^4}(C - 3B) = \gamma^{2/3}\left(h_0 - \frac{1}{2}A\right),\qquad(5)$$

where A, I, B, h_0 and C are, respectively, matrices of $\omega(u^2 + v^2)$, 1, $(-1/8)\omega^3 u^2 v^2(u^2 + v^2)$, H_0/ω, and $(-1/8)\omega^3(u^6 + v^6)$. Those matrices are independent of ε, ω and γ. There not exists a unique global value of the free parameter ω for which all energy levels attain simultaneously a minimum, See Fig. 2(b) and Fig. 2(c). By systematically changing ω, we were able to find very well converged energy levels.

The level statistics of H atom in $\mathbf{B}$ (spectrum of $\gamma^{2/3}$) is obtained by applying the general approach presented in the Introduction. For a fixed scaled energy ε, we can study the statistical properties of spectrum obtained from the scaled Hamiltonian. Spectrum depends only on ε. In this way we follow the evolution of the spectrum when the

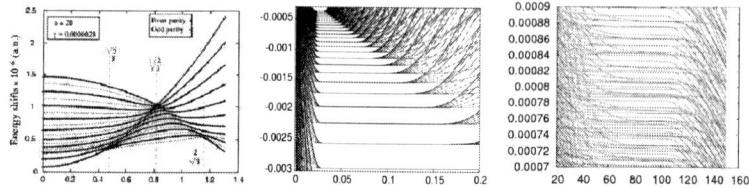

FIGURE 2. (a) Energy-level shifts of the $n = 20$ manifold for $\gamma = 2.8 \cdot 10^{-6}$ as a function of λ. At the λ's corresponding to integrable cases, a clustering occurs but there are no mixings with levels of other manifolds. (b) Eigenvalues as function of the basis parameter ω. Perturbative limit. Note that levels are grouped in manifolds, that a value of ω occurs for which there is a minimum, and that there is no a global value of ω. The case $\lambda = 0$ and $\gamma = 1.6 \cdot 10^{-6}$ is considered and the eigenvalues are ε, (c) Eigenvalues as function of the basis parameter ω. Nonperturbative case. Note that levels are not grouped in manifolds, that a value of ω occurs for which there is a plateau, and that there is no a global value of ω. The case $\lambda = 2^{1/2}/3$ and $\varepsilon = -0.2$ is considered and the eigenvalues are $\gamma^{2/3}$.

corresponding classical dynamics changes from regular to chaotic. We observe that the NNS distribution shows a clear transition from Poisson to Wigner.

When the eigenvalue problem (5) is solved, the spectrum of $\gamma^{2/3}$ along the classical curves $E = \gamma^{2/3}\varepsilon$ is clearly useful for a comparison between the quantum mechanical level statistics and the phase space structure. In this sense, by fixing the scaled energy $\varepsilon = -0.2$, we calculated the evolution of the statistical properties of the spectrum as a function of λ. This calculation allows us to illustrate the quantum consequences of the chaos-order transitions illustrated in Fig. 1(b).

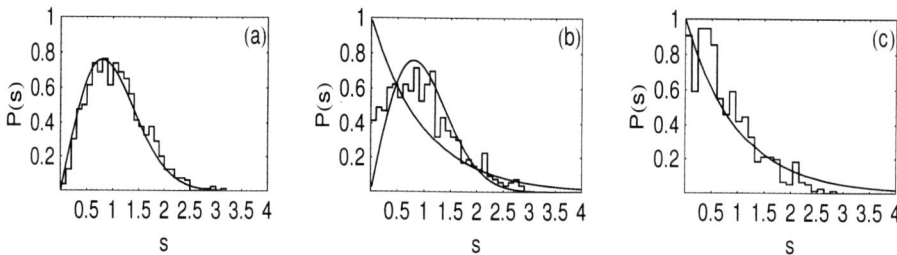

FIGURE 3. NSS histogram for energy $\varepsilon = -0.2$ and different relative strengths of the fields. (a) $\lambda = 0$. (b) $\lambda = 0.42$. (c) $\lambda = 2^{1/2}/3$.

In Fig. 3 are depicted the NNS distributions obtained for $\varepsilon = -0.2$ and $\lambda = (0, 0.42, \sqrt{2}/3)$. These figures show a clear evolution form a Wigner distribution for $\lambda = 0$ to a Poisson distribution for $\lambda = \sqrt{2}/3$. The evolution of the NNS distributions corresponds to the first chaos-order transition in the classical counterpart. This fact demonstrates that a chaos-order transition in the classical system has deep consequences in the level statistics.

An analogous study can be done where for a given value of γ the NNLS of the scaled energy is obtained. The convergence of the eigenvalues has been checked by variation of the basis size. We have used matrices of dimension 2070 and 2025, in order to obtain the eigenvalues corresponding to even and odd parities respectively. Diagonalization has been done separately for eigenstates of parity even and odd, and the obtained spectra of converged scaled NNS were joined in order to analyze the NNLS.

We have solved the scaled matrix equation for different values of the parameters. For $\varepsilon = -0.2$ we followed the evolution of system for λ less than the value of the integrable case $\lambda = 2^{1/2}/3$. See Fig. 1(a) and Fig. 2(a). For $\lambda = 0$ we have the quadratic Zeeman effect, this for this energy has a dynamics dominated by global chaos [1]. For $\lambda = 0.42$ phase space has structure with chaotic and regular regions.

CONCLUSIONS

Our numerical study of the hydrogen atom in the presence of magnetic and quadrupolar electric fields, shows that: (a) In the low field regime, the mark of integrability in classical mechanics is the presence of degeneracy while in the corresponding quantum system, the mark of integrability is the energy levels clustering. (b) In the high field regime, classical chaos-order transitions are associated with significant changes in the level statistics according to the predicted behavior of the random matrix theory. We have shown that the spectrum of eigenvalues $\gamma^{2/3}$ of the scaled problem of HAMQUE has signatures of the underlying classical dynamics. In the regular regime, the NNS distribution is in agreement with the Poisson model. In the chaotic regime the NNS distribution is in agreement with the Wigner model. In the mixed regime an intermediate distribution, like the Brody one, could be used in order to fit the NNS distribution numerically found in Fig. 3(b). A more complete study, in progress, consists of systematically changing λ into the interval $[0, 2^{1/2}]$ for a conveniently chosen set of values of ε.

ACKNOWLEDGMENTS

JM thanks the Université Paul Verlaine - Metz UPVM for the awarding of a visiting fellowship to the Laboratoire de Physique Moléculaire et des Collisions LPMC of the UPVM.

REFERENCES

1. M. Iñarrea, J. P. Salas, V. Lanchares. Phys. Rev. E. **66**, 056614 (2002).
2. M. Iñarrea, J. P. Salas. Eur. Phys. J. D. **27**, 3 (2003).
3. F. Haake. *Quantum Signatures of Chaos*. Springer, Berlin (2001).
4. T. A. Brody. Lett. Nuovo Cimento 7. 482 (1973).
5. H. Friedrich. *Theoretical Atomic Physics*. Springer, Berlin (1990).
6. D. Delande, J. C. Gay. Phys. Rev. Lett. **57**, 2006 (1986).
7. D. Farrelly and T. Uzer, Celest. Mech. Dyn. Astron. **61**, 71 (1995).
8. D. Wintgen and H. Friedrich, Phys. Rev. Lett. **35**, 1464 (1987).
9. H. Friedrich, D. Wintgen. Phys. Rep. **183**, 37 (1989).
10. D. Delande, *Chaos in Atomic and Molecular Physics* in *Chaos and Quantum Physics*, M.-J. Giannoni, A. Voros, J. Zinn-Justin (eds.). Elsevier Science Publishers, Amsterdam (1991).

Electron Transport Coefficients and Scattering Cross Sections in CH₄, HBr and in Mixtures of He and Xe

Olivera M. Šašić

Institute of Physics, POB 68, 11080 Belgrade, Serbia,
also at The Faculty of Transport and Traffic Engineering, Belgrade, Serbia

Abstract. We have applied a standard swarm procedure in order to obtain electron scattering cross sections and transport coefficients that provide a data base for plasma modeling. In case of CH₄ the dissociative excitation cross sections from binary collision experiments were renormalized by fitting the measured excitation coefficients with our calculations. In case of HBr we have produced a complete set of cross sections based on available data from the literature, with some extrapolations. We have also tested the cross sections in He-Xe mixtures and the application of Blanc's law and common mean energy procedure in calculating drift velocities in by comparison with recent measurements. Finally, a well tested Monte Carlo code was used in wide range of both DC and RF electric and magnetic fields in order to calculate a number of transport coefficients in case of CH₄ and HBr.

Keywords: swarm procedure, plasma modeling, methane, hydro bromide, mixtures, Blanc's law, transport coefficients.
PACS: 52.20.Fs 51.50.+v 52.25.Fi

INTRODUCTION

In recent years we have seen an extensive use of low temperature plasmas in numerous applications such as plasma display panels [1], plasma etching devices [2], electron pumped lasers, fabrication of optoelectronic devices and many more. At the same time many of the gases which are used in the industry are major pollutants of the atmosphere and contribute a great deal to the global worming [3]. Various theoretical techniques have been developed for the simulation and modeling of plasma devices. All of these techniques have the same goal – a better understanding of complex kinetic phenomena which leads to optimization and further development of technological applications. In achieving this goal all of them have the same obstacle – a constant lack of reliable and complete sets of electron collision cross section data and transport parameters, such as drift velocities, characteristic energies etc. This is the problem that we are dealing with in this paper.

The results presented here, can be divided into three parts. The first part is related to the review the existing electron impact cross sections and transport coefficients in helium, xenon and their mixtures with the view of possible improvements of existing data and application of laws for mixtures. The second part is an attempt to produce a

CP876, The Physics of Ionized Gases: 23ʳᵈ Summer School and International Symposium,
edited by L. Hadžievski, B. P. Marinković, and N. S. Simonović
© 2006 American Institute of Physics 978-0-7354-0377-2/06/$23.00

more complete and detailed set of cross sections in methane. Also a set of calculated transport coefficients is obtained under various conditions in situations interesting from the point of view of applications. The third part represents a significant extrapolation of the existing cross section data for hydro bromide to produce a complete set. This set has been used to calculate a wide range of transport coefficients.

We have used our Monte Carlo code [4], for the calculation of transport and rate coefficients both in case of DC and RF electric and magnetic fields. In Monte Carlo (MC) simulation we followed the spatiotemporal evolution of each electron through time steps in order to represent correctly the motion of electrons in external fields. The moment of collision was determined from the integral collision probability. The time steps were determined by the minimum of three constants: mean collisional time, cyclotron period (in case of constant crossed electric and magnetic fields) and period of time depended fields. Our simulations were performed with 5×10^5 initial electrons. The gas number density was 3.54×10^{22} m^{-3} which corresponded to the gas pressure of 1 Torr (133.3 Pa), at 273 K. Initial electron energy distribution was a Maxwellian with mean energy of 1 eV. All scattering events were assumed to be isotropic.

Since duration of a single simulation could be the critical point in some cases, in the swarm procedure (that is, the iterative process of adjusting the cross sections until the agreement of calculated and measured transport coefficients was achieved) we have also used a two term code ELENDIF [5]

RESULTS AND DISCUSSION

Mixtures of Helium and Xenon

The basic set of cross sections was adopted from the Sigmalib data base of the two term (TT) code itself [5], and it was supplemented by Hayashi's data [6, 7] and extrapolated towards higher energies. For He, the momentum transfer cross section coincides with that of Crompton et al. [8, 9], and in case of Xe, the data were mainly based on the results of Frost and Phelps [10] and Hunter et al. [11]. As a test of the cross section set we calculated drift velocities in pure gases by using both TT and MC codes. Calculated drift velocities were shown to be in a very good agreement both in case of He, and in case of Xe, both with old [11-15] published data and with recent measurements of de Urquijo et al. [16].

We calculated drift velocities in 1, 5, 10, 20, 50 and 70% mixtures of Xe in He by using TT and MC codes and three different techniques for predicting data in mixtures. The first one was the application of the well known Blanc's law [17]. In this case the data for pure gases are to be taken at the same value of the reduced electric field (E/N). The second one was the so-called common mean energy procedure (CME) [18]. In this case drift velocities were found from:

$$1 = \sum_{\alpha=1}^{n} x_\alpha \frac{(E/N)_\alpha}{(E/N)_{mix}} \left(\frac{W_{mix}}{W_\alpha} \right)^{\pm 1} \tag{1}$$

where n is the number of components in the mixtures and drift velocities for pure gases W_α and for the mixture W_{mix} are taken at the same mean energy. Correct theoretical basis of that procedure, was presented by Jovanović *et al.* [19], who showed that the equation with exponent +1 could be obtained from the momentum balance, while the form with exponent –1 could be obtained from the energy balance.

Mutual grater similarity of distribution functions in mixtures than in pure gases and pure gases and mixtures, gave us the idea to combine the values of drift velocities in two mixtures by applying standard Blanc's law in order to obtain drift velocity in the third mixture, performing the scaling by adequate concentrations. That was the third technique in calculating drift velocities.

Figure1 shows the results obtained for 50% mixture.

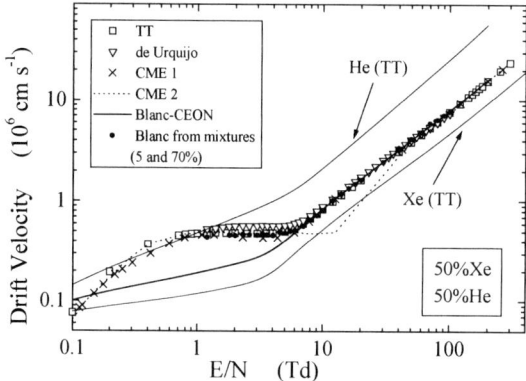

FIGURE 1. .Comparison of measured and calculated drift velocities in the 50%Xe + 50% He mixture. CME1 and CME2 in the legend denote the calculation of drift velocities with the positive and negative exponent, respectively, in Eq.(1). The drift velocities for the pure gases are shown with separately marked solid lines.

There is a visible discrepancy of the prediction by using the standard Blanc's law while the CME1 procedure (Eq.1 with positive exponent) gives an excellent agreement with the experimental and calculated mixture data. CME2 (Eq.1 with negative exponent) procedure departs from the correct values, although still indicating an existence of the negative differential conductivity (NDC).

Dissociative Excitation Cross Sections in Methane

The basic set of cross sections in methane was adopted from Hayashi [20]. As the chosen set of cross sections does not contain any information on dissociative excitation we used the data of Tsurubuchi and coworkers [21] and de Heer and coworkers [22, 23] as initial guesses. Namely, the most intensive in CH spectra is the AX molecular band ($A^2\Delta$-$X^2\Pi$), while the BX band ($B^2\Sigma$-$X^2\Pi$) is somewhat weaker.

It is also possible to notice the emission of hydrogen lines (H_α and H_β) of Ballmer's series which also appear in the process of dissociative excitation. Since the data for BX band do not exist in the literature; we have used the same energy dependence as in the case of AX band and shifted the cross section curve towards the threshold for that process. We have also made appropriate extrapolations of the data towards higher energies. The initial set was tested by making comparisons with the available experimental transport data for drift velocities and characteristic energies.

We fitted the calculated excitation coefficients to recent experimental data [24, 25]. The *E/N* dependence of calculated excitation coefficients was shown to be in agreement with the experimental data. However, the magnitude of cross sections had to be modified. As can be seen from Fig. 3, a very good fit of the experimental data may be obtained when the cross section for H_α line is normalized by 0.7 and for H_β line by 0.72. At the same time the initial cross sections for AX molecular band had to be multiplied by 1.98, and by 0.12 to fit the excitation of the BX band.

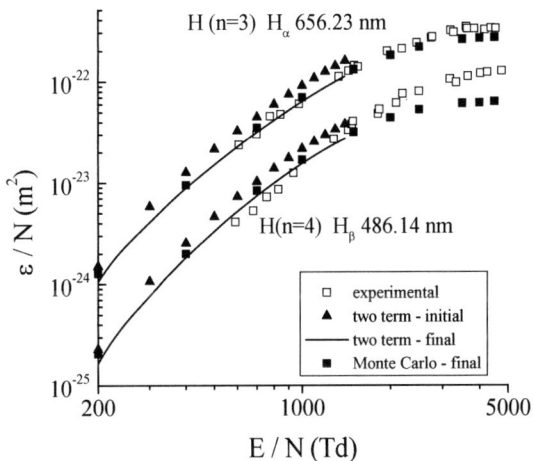

FIGURE 3. Electron excitation coefficients for H (n=3) and H (n=4) levels. TT calculations were performed with the initial cross section set (solid triangles) and with rescaled cross sections (solid lines), and the MC simulations were performed only with rescaled cross sections (solid squares)

Figure 4 shows the final, complete set of cross sections for electrons in methane. We used this set for calculation of transport coefficients in case of DC and RF electric and magnetic fields.

We also made calculations of rate coefficients for ionization and individual inelastic processes [26] for a wide range of DC and RF fields.

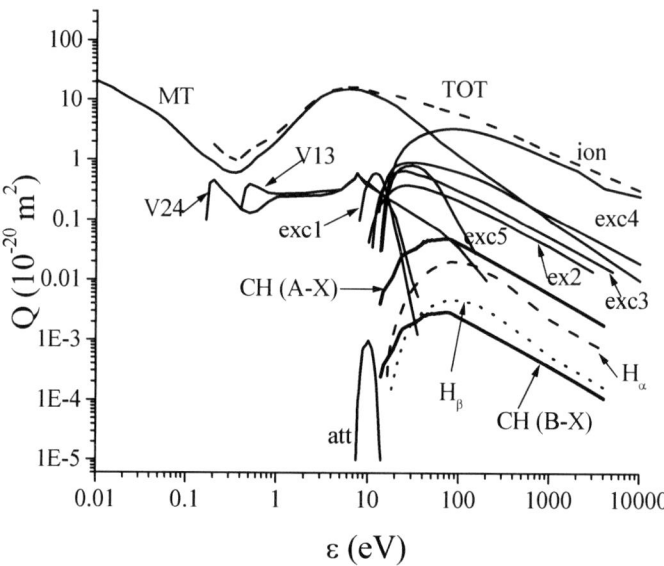

FIGURE 4. Final set of cross sections for electrons in methane. MT-denotes momentum transfer, TOT-total, V-vibrational excitation, exc -electronic excitation, att –attachment, ion- ionization and dissociative excitation cross sections are marked by their fragments and their excited state.

Cross Section Set for Electrons in HBr

In order to obtain the first complete set of cross sections for HBr, we compiled the best available data from the literature. Starting from theoretical calculations of Rescigno [27] for elastic momentum transfer cross section, given in a very narrow energy interval, we made the extrapolation of these results towards higher and lower energies. We calculated rotational excitation cross sections by using the well known Takayanagi formulae [28]. In the region of low electron mean energies, the dominant feature in the inelastic energy losses are the vibrational excitations and dissociative attachment. The situation with these processes in the literature is much better than with other processes. We recommend the most recent results of groups of Horáček and Allan [29]. Dissociative excitation cross sections that we recommend are the results of theoretical calculations of Rescigno [27], and we made some extrapolations of these results towards higher energies. We also introduce a fictitious excitation cross section that has similar energy dependence to the one that exists for HCl molecule according to Hayashi [30]. This cross section takes the contribution of all missing inelastic processes and it is presented with reduction by a factor of 10 in the Figure. At last, we recommend ionization cross section which is the result of semi empirical calculation of Ali and Kim [31]. The cross section set is shown in Fig. 5.

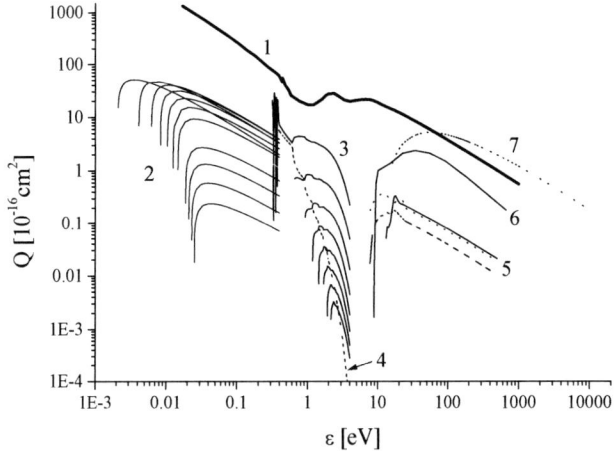

FIGURE 5. Complete set of cross sections for electron in HBr: momentum transfer (1), rotational excitation (2), vibrational excitation (3), dissociative attachment (4), electronic excitation (5), effective excitation and dissociation (6) and ionization (7).

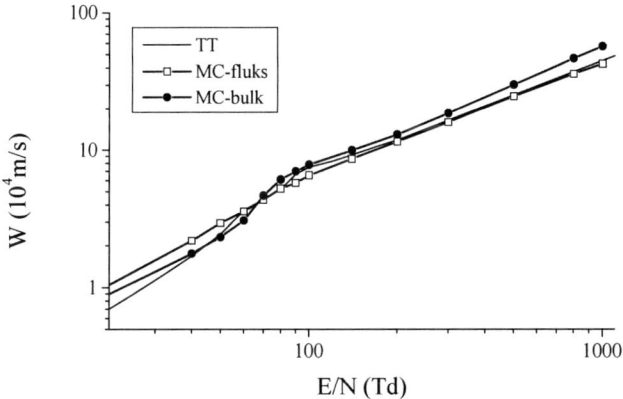

FIGURE 6. Drift velocity for electrons in HBr in DC electric fields. Flux and bulk in the legend denote drift velocities defined in the real space and in the velocity space, respectively.

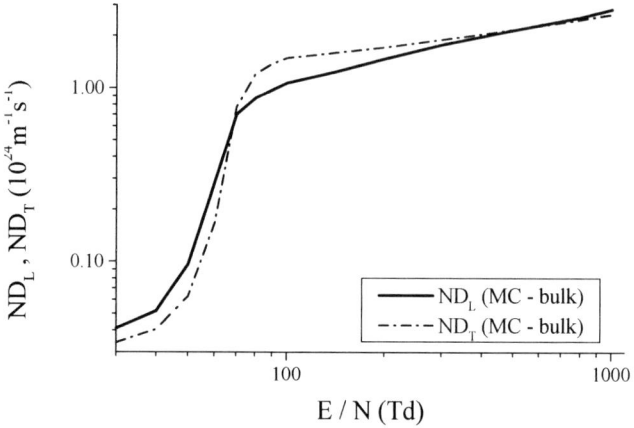

FIGURE 7. Longitudinal (ND$_L$) and transversal (ND$_T$) diffusion coefficients for electrons in HBr in DC electric fields.

This cross section set was used as an input parameter in both TT and MC code in order to calculate transport coefficients for electrons under various conditions. In case of DC electric field the most significant conclusions that can be stressed are that despite the fact that one could expect the occurrence of the NDC (due to the shape of the cross sections) it does not exist in pure HBr (Fig 6.), and that diffusion is only slightly anisotropic (Fig. 7). We also did calculations for dc *ExB* and time dependent *ExB* fields.

CONCLUSION

In this work we have applied a standard swarm procedure in order to revise, or to improve the existing cross section data, and to construct the complete set that hasn't been available in the literature in case of three different systems interesting from the point of view of applications. In case of He – Xe mixtures, the available cross section data for these gases (with some extrapolations that were made) were tested against recent measurements of de Urquijo. The accuracy of calculated drift velocities appears to be within the error bars of the measured values. Laws for mixtures have also been tested. It was shown that the best results could be obtained with the common mean energy procedure derived from the momentum balance equation as well as with the third mixture procedure

In case of methane, the most detailed set of cross sections available in the literature was supplemented with the data for the formation of exited radicals CH[*] and H[*]. The present data could be of use for normalization of fast neutral excitation cross sections and for modeling of kinetics of DC and RF collisional plasmas that are used for plasma technologies.

Finally, we have compiled a set of cross sections for HBr from the literature and calculated a number of transport coefficients and rate coefficients for individual processes under the various conditions interesting from the application point of view. The characteristics of the cross sections are very high magnitude of the momentum transfer cross section and the pronounced vibrational excitation and dissociative attachment cross sections in the region of Ramsauer – Townsend minimum, which lead to the situation quite close to the constant collision frequency cross section and, consequently, to rather uneventful E/N dependences of the transport data.

ACKNOWLEDGMENTS

Author is grateful to Dr. Z. Lj. Petrović for guidance, constant interest, support and useful suggestion and ideas. This work was supported partially by MNZŽS 141025 project

REFERENCES

1. G. J. M. Hagelaar, G. M. W. Kroesen and M. H. Klein, *J. Appl. Phys* **88**, 2240 (2000).
2. F. Tochikubo, Z. Lj. Petrović, S. Kakuta, N. Nakano and T. Makabe, *J. Appl. Phys.Part I* **33**,4271 (1994)
3. E. Nisbet, *Nature* **347**, 23 (1990)
4. Z. M. Raspopović, S. Sakadžić, S. Bzenić and Z. Lj. Petrović, *IEEE Trans. Plasma Sci.* **27**, 1241 (1999)
5. W. L. Morgan and B. M. Penetrante, *Comput. Phys. Commun.* **58**, 127 (1990)
6. M. Hayashi, IPP-AM-19, Nagoya Institute of Technology, Nagoya, (1980)
7. M. Hayashi, *J. Phys. D* **16**, 581 (1983)
8. R. W. Crompton, M. T. Elford, and A. G. Robertson, *Aust. J. Phys.* **23**, 667 (1970)
9. H. B. Milloy and R. W. Crompton, *Phys. rev. A* **15**, 1847 (1977)
10. L. S. Frost and A. V. Phelps, *Phys. Rev.* **136** A1538 (1964)
11. S. R. Hunter, J. G. Carter and L. G. Christophorou, *Phys. Rev. A* **38** 5539 (1988)
12. R. W. Crompton, M. T. Elford and R. L. Jory, *Aust. J. Phys.* **20** 369 (1967)
13. J. Dutton, *J. Phys. Chem. ref. Data* **4** 577 (1975)
14. J. L. Pack, R. E. Voshall and A. V. Phelps, *Phys. Rev.***127** 2084 (1962
15. K. H. Wagner, *Z. Phys.* **178** 64 (1964)
16. O. Šašić, J. Jovanović, Z. Lj. Petrović, J. de Urquijo, J. R. Castrejón-Pita, J. L. Hernández-Ávila and E. Basurto, *Phys. Rev. E* **71** 046408 (2005)
17. A. Blanc, *J. Phys* **7** 825 (1908)
18. R. V. Chiflikyan, *Phys. Plasmas* **2** 3902 (1995)
19. J. V. Jovanović, S. B. Vrhovac and Z. Lj. Petrović, *Eur. Phys. J. D* **28** 91 (2004)
20. M. Hayashi (personal communication from Y. Nakamura)
21. K. Motohashi, H. Soshi and S. Tsurubuchi, XIX ICPEAC **1** 422 (1995)
22. J. E. M. Aarts, C. I. M. Beenakker and F. J. de Heer, *Physica* **53** 32 (1971)
23. D. A. Vroom, and F. J. de Heer, *J. Chem. Phys.* **50** 573 (1969)
24. O. Šašić, G. Malović, A. Strinić, Ž. Nikitović and Z. Lj. Petrović, New Journal of Physics 6 **74** (2004)
25. G. N. Malović, "Measurements and Analysis of Excitation Coefficients in Noble Gases and in Methane",Ph.D. Thesis, University of Belgrade (1999).
26. Ž. Nikitović, O. Šašić, Z. Lj. Petrović, G. Malović, A. Strinić, S. Dujko, Z. M. Raspopović and M. Radmilović-Radjenović, *Material Science Forum* **453-454** (2004)
27. T. N. Rescigno, *J. Chem. Phys* **104** 125 (1996)
28. K. Takayanagi, *J. Phys. soc. Japan* **21** 507 (1966)
29. M. Čížek, J. Horáček, A. Sergenton, D. Popović, M. Allan, W. Domcke, T. Leininger and F. Gadea, *Phys. Rev. A* **63** 062710,1 (2001)
30. M. Hayashi (personal communication, 1992)
31. M. Ali, Y. Kim (personal communication, 2004)

Amino Acid Formation by Electron Irradiation

D. Cáceres[1,2], M. Bertin[1], A. Lafosse[1], A. Domaracka[1,3], D. Pliszka[1], R. Azria[1].

[1] *Laboratoire des Collisions Atomiques et Moléculaires, CNRS-Université Paris-Sud (UMR 8625, FR LUMAT), Bât. 351, Université Paris Sud, F-91405 Orsay Cedex, France.*

[2] *Departamento de Física. Escuela Politécnica Superior. Universidad Carlos III de Madrid. Av. Universidad, 30. 28911 Leganés (Madrid), Spain.*

[3] *Atomic Physics Group*
Gdansk University of Technology, Gdansk, Poland.

Abstract. The origin of amino acid formation in the early years is still not well known. Muñoz Caro and collaborators [1] showed the generation of amino acid by UV irradiation of a mixture of organic molecules reproducing the interstellar media. Although the photochemistry is one of the mechanism which would be active in the early years, it is also important to study other mechanisms such as the electrochemistry or electron induced chemistry which would act as well as a tool for the amino acid formation. The aim of this work is to demonstrate the amino acid formation by low energy electron irradiation of condensed organic molecules. In particular, our purpose is the formation of glycine by low energy electron irradiation of a film composed of a mixture of condensed NH_3 and CH_3COOD. Prior to the study of amino acid formation it is important to know about the effect of the interaction between electrons and organic molecules, so the first step in this work is the study of the electron irradiation of NH_3 and CH_3COOD films condensed on a passivated substrate (hydrogenated diamond) at a temperature of about 40 K. The next step involves the electron irradiation of the condensed NH_3 and CH_3COOD mixture on the same substrate. In this case, the reactions induced by electron irradiation would permit the formation of glycine in the zwitterionic form. Induced modifications on the NH_3 and CH_3COOD molecules, as well as the formation of glycine will be monitored by High Resolution Electron Energy Loss Spectroscopy (HREELS).

Keywords: Amino acids, glycine, induced reactivity, HREELS.

PACS: 34.80.Gs; 34.80.Ht; 68.47.Pe; 79.20.Uv.

CP876, *The Physics of Ionized Gases: 23rd Summer School and International Symposium,*
edited by L. Hadžievski, B. P. Marinković, and N. S. Simonović
© 2006 American Institute of Physics 978-0-7354-0377-2/06/$23.00

INTRODUCTION

The search for the origin of life is one of the fundamental questions for the humanity. The experiment by Miller [2] in 1953 opened the field of research in the origin of the amino acid generation, which was followed by many works.

The research is centered in trying to explain the amino acid formation from prebiotic organic molecules such as H_2O, CO, CO_2, NH_3 or CH_3COOH. This explanation is essential in the searching for the origin of life because amino acids are the main blocks of the proteins, while proteins are the main blocks of the DNA molecule or the molecule of life as it is sometimes cited.

The search for amino acid formation leaves other question that remains without answer about the extraterrestrial or terrestrial origin of life. The extraterrestrial generation of amino acids is based in their presence, together with prebiotic molecules, in meteorites or comets as was indicated by Oró in the sixties [3] and confirmed in more recent studies [4,5]. When these bodies impacted the Earth in the early years, they bought the generated molecules to the planet.

On the other hand, the formation of prebiotic molecules as well as amino acid is supposed to have occurred in the primitive reducing Earth atmosphere [2,6,7] or in the deep-sea hydrothermal vents where superheated water generates abundant biological activity [6-9].

In order to answer these questions some experiments have been conducted simulating the interstellar medium or the primitive atmosphere of the Earth and trying to generate amino acid by glow discharge [2,10], UV-irradiation [1,11-13], energetic charged particles [14], laser irradiation [15] or XUV synchrotron radiation [16].

Secondary electrons with low energies are found in the interstellar media which are able to induce reactions in the prebiotic molecules. The aim of this work is the study of the glycine generation by low energy electron irradiation of an organic film formed of a mixture of ammonia and acetic acid (CH_3COOD/NH_3). Results will be probed by High Resolution Electron Energy Loss Spectroscopy (HREELS).

EXPERIMENTAL SET-UP

Experiments were performed in an Ultra-High Vacuum (UHV) system with a base pressure below 5×10^{-11} Torr.

Figure 1 shows the experimental set-up composed of three different vacuum chambers. The main one is equipped with the Electron Stimulated Desorption (ESD) technique and a helium cryostat which allows us to cool down our samples to about 25 K. The sample is attached to the bottom part of a transfer rod to descent it to the lower chamber were the HREELS is located. The spectrometer is an IB500 by OMICRON equipped with a double monochromator and a single analyzer. Spectra were obtained in the specular geometry with an incident angle of 55° from the surface and an overall resolution of about 7 meV, measured as the full width at half maximum (FWHM).

Another smaller chamber is equipped with Auger Electron Spectroscopy (AES), Low Energy Electron Diffraction (LEED) and different facilities for the substrate preparation.

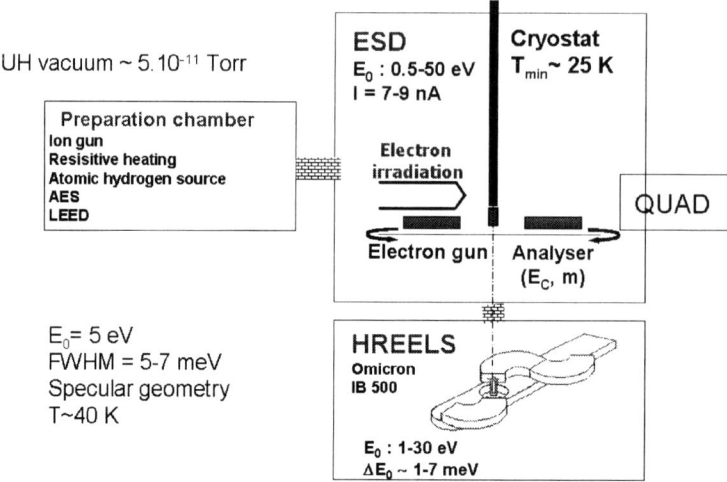

FIGURE 1. Experimental set-up.

Molecular films composed of CH_3COOD, NH_3 and a mixture of both molecules were deposited on hydrogenated diamond at a temperature of 40 K. The substrate was selected for its non-catalytic behavior and details of its preparation were explained elsewhere [17].

Organic compounds (CH_3COOD and NH_3) were dosed into the vacuum chamber from different sources with a typical pressure of 2×10^{-8} Torr.

Finally, organic films were irradiated with low energy electrons (0.5-20 eV) at different doses.

PROCEDURE

The aim of the study is the amino acid generation by electron irradiation of molecular films composed of a mixture (1:1) of NH_3 and CH_3COOD condensed at 40 K on hydrogenated diamond. The selection of this passive substrate is very important when studying the induced reactivity by electron irradiation. You must be sure that the observed reactions are due to the electron irradiation and not to a catalytic effect from the substrate.

Before studying the composed film we started by condensing, separately, NH_3 and CH_3COOD pure films and studying them before electron irradiation to be sure that we are able to condense these molecules and to know about their main vibrations. Then, we will study the effect of electron irradiation on these pure films.

Finally, we will study the composed NH_3/CH_3COOD films before electron irradiation to be sure that no reaction occurs between the two molecules. Then we will irradiate the condensed mixture with different electron doses and incident energies and will check the possible induced reactions in this system by means of HREELS.

CONCLUSIONS

The experimental set-up described in this work will allow us to prepare the hydrogenated surface of our substrate and then to condense organic molecules with the help of the helium cryostat.

Electron irradiation of our samples will induce reactions and modifications which would give as a final result the glycine formation from the NH_3/CH_3COOD mixture condensed films.

HREELS is a good technique to determine the presence of the new species after electron irradiation.

ACKNOWLEDGEMENTS

This work has been partly performed within the EPIC EU network, "Electron and Positron Induced Chemistry", Framework V, 2002-05. D. Cáceres thanks EPIC network for its support during his stay at LCAM. A. Domaracka was supported by a Marie Curie Fellowship of the European Community program Improving Human Potential and the Socio-economic knowledge Base under number HPMT-CT-2001-00358.

REFERENCES

1. G.M. Muñoz Caro et al. Nature 416 (2002) 403-406.
2. S.L. Miller. Science 117 (1953) 528-529.
3. J. Oró. Nature 190 (1961) 389-390.
4. J.R. Cronin, "Organic molecules on the early Earth. Clues from the origin of the Solar System: meteorites", in The molecular origins of life: assembling pieces of the puzzle, ed. A. Brack, Cambridge University Press, Cambridge, 1998, pp. 119-146.
5. A. H. Delsemme, "Nature and history of the organic compounds in comets: an astrophysical view", in Comets in the post Halley era, ed. R.L. Newburn Jr., M. Neugebauer and J. Rahe, Kluwer, Dordrecht, 1991, pp. 377-428.
6. L.E. Orgel. Trents in biochemistry science. 23 (1998) 491-495.
7. L.E. Orgel. Crit. Rev. in Biochem. Mol. 39 (2004) 99-123.
8. G. Wächtershäuser. Microbiol. Rev. 52 (1988) 452-484.
9. A. Brack. Adv. Space Res. 24 (1999) 417-433.
10. S. Miyakawa, A.B. Sawaoka, K. Ushio and K. Kobayahi. J. Appl. Phys. 85 (1999) 6853-6857.
11. H. Mita, N. Shirakura, H. Yokoyama, S. Nomoto and A. Shimoyama. Adv. Spa. Res. 33 (2004) 1282-1288.
12. M. Morita, Y. Harada, K. Iseki, S. Izumi and A. Hiraya. Ana. Sci. 21 (2005) 1085-1090.
13. J. Takahashi, H. Masuda, T. Kaneko, K. Kobayashi, T. Saito and T. Hosokawa. J. Appl. Phys. 98 (2005) 024907.

14. K. Kobayashi, T. Kaneko, T. Saito and T. Oshima. Origins Life Evol. Biosphere. 28 (1998) 155-165.
15. S. Civis et al. Chem. Phys. Lett. 386 (2004) 169-173.
16. J. Takahashi, T. Hosokawa, H. Masuda, T. Kaneko, K. Kobayashi, T. Saito and Y. Utsumi. Appl. Phys. Lett. 74 (1999) 877-879.
17. A. Hoffmann, A. Laikhtmann, S. Ustaze, M. Hadj-Hamou, M.N. Hedhili, J.P. Guillotin, Y. Le Coat, D. Teillet-Billy, R. Azria and M. Tronc. Phys. Rev. B. 63 (2001) 045401.

Gas Phase Dissociative electron attachment study to L-Valine

Š. Matejčík, J. Kočíšek, D. Kubala, M. Stano* and O. Ingolfsson†

*Department of experimental physics, Comenius University, Mlynska dolina F2, 84248 Bratislava, Slovakia
†University of Iceland, Science Institute, Dunhaga 3, 107 Reykjavik, Iceland

Abstract. Using a crossed electron/molecule beams apparatus the dissociative electron attachment (DEA) to gas phase amino acid L-valine has been studied. The DEA to valine at low electron energies exhibits several common features with DEA to other amino acids (glycine, alanine) and to simple organic acids. Like in the case of previously studied amino acids glycine and alanine no stable parent anion was detected. The transient negative ions is formed at low electron energies via electron capture to the unoccupied π^* orbital of the COOH group, the majority of the products is associated with the COOH group. The dominant product of the DEA reaction is $(\text{Val-H})^-$ (m/Z=116) observed at two resonances of 1.2 eV and 5.3 eV. Other fragment anions (m/Z=100,72,56,26 and 17) are formed via the low energy resonances (below 5 eV) and core excited resonances at about 5.0 and 9.0 eV.

Keywords: electron, dissociative, attachment, amino acid, valine
PACS: 34.80.Ht Dissociation and dissociative attachment by electron impact

INTRODUCTION

The interaction of the ionizing radiation with the living organism besides the direct damages is responsible for the production of secondary reactive species along the track of the radiation. The secondary species may undergo subsequent reactions with the medium. To the most abundant secondary species belongs the electrons with kinetic energies below 20 eV [1] with a yield of 5×10^4 per MeV [2] deposited. The dissociative electron attachment (DEA) is a reaction between the electrons and molecules. This reaction results in formation of negative ions and radicals in the gas (and aggregate) phase and may play important role in the radical formation in the cell. In many molecular systems already low energy electrons (0 eV) are able to dissociate the molecules and the reactions may be very efficient.

In this paper we present an experimental study of dissociative electron attachment (DEA) to the amino acid L-Valine (Valine, Val - $(\text{CH}_3)_2\text{CHCH(NH}_2)\text{COOH}$, Figure 1) in the gas phase. This study is a contribution to the better understanding of the mechanism of the negative ion and radical formation from this amino acid and amino acids general. The amino acids are the building blocks of the peptides and the the proteins. The studies of DEA to amino acids may also improve our insight into the radical formation and the fragmentation of the peptides and proteins, moreover, present study may give better insight on the processes of radical formation in the crystalline, polycrystalline valine.

The DEA to a molecule in the gas and condensed phase is a resonant process which proceeds via two steps. The first step is the formation of the transient negative ion (TNI)

CP876, *The Physics of Ionized Gases: 23ʳᵈ Summer School and International Symposium*, edited by L. Hadžievski, B. P. Marinković, and N. S. Simonović

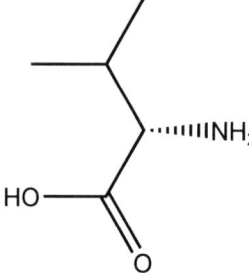

FIGURE 1. L-valine

$(Val^-)^{\#}$ which may subsequently (in the second step) dissociate into negative ions and neutral fragments or radicals. The resonant character of the DEA reactions is represented by the form of the cross sections. The cross section for particular DEA reaction has a form of a more or less narrow peak (resonance), which is in contrast to the monotonic form of the cross sections for the positive ion formation by electron impact. The position and width of the DEA cross section are characteristic for the molecule and given ion and depends on the electronic structure of the molecule. Using the crossed electron/molecule beams technique combined with the mass analysis of the negative ions we were able to measure the ion yields for the formation of particular negative ions.

Recently several studies on the DEA to amino acids (glycine and alanine) [3, 4, 5] have been published. These molecules together with the valine belong to the aliphatic amino acids. These studies revealed some characteristic dissociation patterns in the DEA to aliphatic amino acids. The dominant DEA channel in the molecules was the $(M-H)^-$ ion, where M denotes the parent molecule. Also additional reaction channels were associated with the COOH group ($(M-OH)^-$, $(M-COOH)^-$, $COOH^-$ nd OH^-). According to these studies a low energy π^* (COOH) resonance peaking at about 1.8 eV have been found for several fragment ions. The Electron Transient Spectroscopy (ETS) study concerning formic acid and glycine and other amino acids [6] shows that a π^* (COOH) resonance is typical for the molecules containing the COOH (carboxyl) functional group. The DEA studies of glycine and alanine [3, 4, 5] showed that besides the π^* resonance additional resonances at higher electron energies (around 6 and 10 eV) exist. These resonances were tentatively assigned to the lowest excited states of the molecules. Present work is according to our knowledge the first study on DEA to the gas phase valine and one of the aims of the study is to compare present results with the early alanine and glycine one.

EXPERIMENTAL SETUP

The experiment has been carried out using a crossed electron/molecule beams apparatus at Comenius University Bratislava. The apparatus has been described in detail in [7, 8] therefore we give only a brief description. Schematic view of the apparatus is shown in the Figure 2. The electron beam is formed by Trochoidal Electron Monochromator

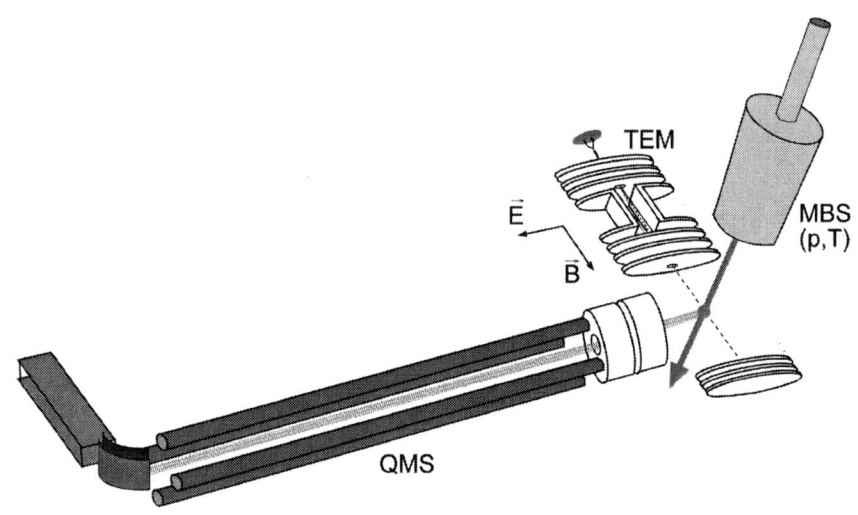

FIGURE 2. A schematic view of the apparatus

(TEM)[9]. The electron energy resolution of the electron beam in present experiment was was set to 140 meV. The resolution of the electron monochromator has been kept relatively low in order to increase the electron current and to achieve sufficient ion yield. The calibration of the electron energy scale and the estimation of the electron energy resolution were performed through the measurement of the non dissociative electron attachment to SF6. This anion exhibits a narrow s-wave resonance at 0 eV. The apparent width of the resonance was used as a measure for the electron energy resolution of the electron beam and its position to define the zero energy point of the energy scale.

The molecular beam was formed in a heatable effusive molecular beam source (EMBS). The solid valine sample was placed directly into the stainless steel container of the EMBS. The valine is at room temperature in solid state and the vapor pressure is not sufficient for this type of experiment. For this reason was the EMBS resistively heated to temperatures of 140°C in order to increase the vapor pressure on value sufficient for the experiment (1Pa). The beam of the molecules is formed by molecular flow (effusion) of the gas trough a narrow channel (0.5mm diameter and 4 mm long) and an external aperture. The TEM and EMBS are located in a vacuum chamber pumped by a turbomolecular pump with a background pressure of 10^{-7} Pa.

The negative ions formed at the intersection of the electron and the molecular beams are extracted by from the reaction volume by a weak electric field (1 Vcm^{-1}) and by the ion optic into the quadrupole mass spectrometer (QMS). Using this setup we are able to measure either the negative mass spectra of the molecules at different electron energies, or the ion yield of mass selected negative ions as a function of the electron energy.

TABLE 1. Negative ions formed via DEA to valine and the peak positions of the resonances

Reaction	m/Z	Peak energies (eV)		
$(Val-H)^-$	116	1.2	5.3	
$(Val-OH)^-$	100	1.6	5.0	7.5
$(Val-COOH)^-$	72			8
$C_3H_6N^-, C_3H_4O^-$	56			9
$COOH^-$	45	0.2, 1.4, 2.5	5.6	7.9
CN^-	26	1.8	5.8	8.3
OH^-	17	0.2	5.3	8.4

RESULTS AND DISCUSSION

Using the experimental setup described in previous section we were able to detect 7 different negative ions formed via DEA to the gas phase valine (m/Z=100,72,56,26 and 17). We assume that also the H^- ion is formed, however, in present experiment we were not able to detect this ion due to the fact that the mass range of the quadrupole mass spectrometer is limited, starting with the mass 2. The survey of the detected negative ions and the resonances observed (positions of the peaks of the resonances) are listed in the Table 1.

The molecular ion $(Val^-)^{\#}$ is not present in the list. We were not able to detect this anion. The absence of the molecular ion is most probably due to the fact that the life time of the molecular ion is to short to be detected in present experiment. In order to detect metastable ions in present experiment the life time of the ions has to be more then 10^{-5}s. The metastable, or transient negative ion $(Val^-)^{\#}$ decays in the reaction volume or in the ion optics via electron detachment or via dissociation into the observed negative fragment ions.

The negative ions may be formed via DEA following reactions:

$$e + Val \rightleftharpoons (Val^-)^{\#} \rightarrow (Val-H)^- + H \quad (1)$$
$$\rightarrow (Val-OH)^- + OH \quad (2)$$
$$\rightarrow (Val-COOH)^- + COOH \quad (3)$$
$$\rightarrow (CH_3)_2CHCH^- + fragments \quad (4)$$
$$\rightarrow C_3H_6N^- + fragments \quad (5)$$
$$\rightarrow CN^- + fragments \quad (6)$$
$$\rightarrow OH^- + fragments \quad (7)$$

The ion yields for the ions formed via reactions (1) - (7) are presented in the Figures 3 and 4. We were not able to estimate the cross section for the DEA to valine. Beside other problems of the estimation of the absolute magnitudes of the cross sections (transmission of the mass spectrometer, discrimination in the reaction chamber, the detection efficiency), the main problem in present experiment was related to the measurement of the pressure of the valine in the reaction volume. Nevertheless, the intensities of the ions

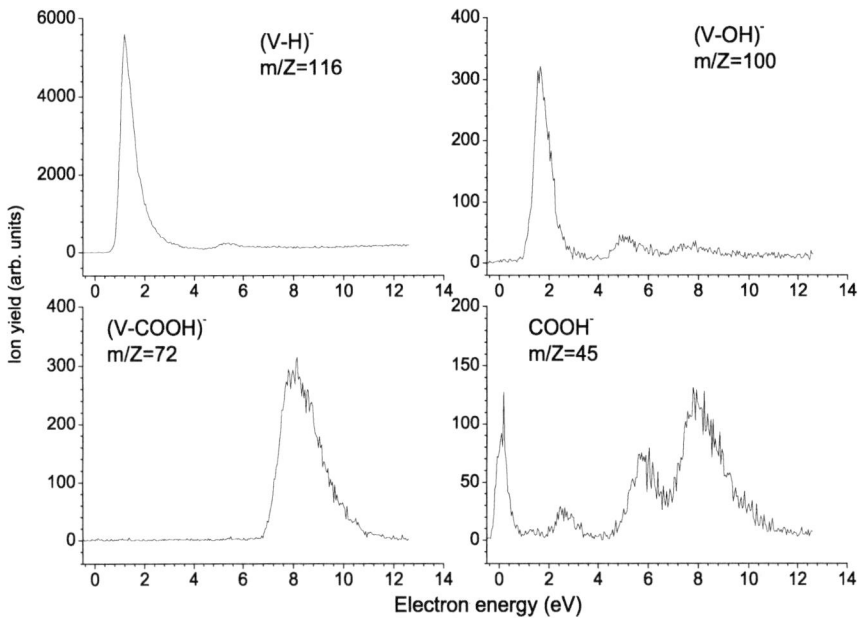

FIGURE 3. The ion yields for the products of the DEA to valine a) (Val-H)⁻ – m/Z=116, b) (Val-OH)⁻ – m/Z=100 c) (Val-COOH)⁻ – m/Z=72 d) COOH⁻ – m/Z=45

in the Figures 3 and 4 give insight on the relative magnitudes of the DEA cross sections for particular fragment ions, because the measurements were performed under identical conditions and the signal was collected for the same time period. On the basis of the ion yields we may resume that the strongest reaction channel for DEA to valine at low electron energies (below 5 eV) is the (Val-H)⁻ ion. Concerning the magnitude of the absolute cross sections for DEA to valine, on the basis of our experiences with other molecules and the ion intensities measured with present experimental setup the DEA to valine belongs to less or moderate efficient DEA reactions.

After close inspection of the DEA reaction channels (1) - (7) we see that majority of the negative ions formed via DEA to valine is associated with the COOH group. These ions are formed via cleavage of COOH group or its part. Moreover, many of these ions are formed via simple cleavage, via single bond dissociation. Similar observation have been done also for glycine and alanine [3, 5]. This fact is most probably related to the chemical properties of the COOH group and its fragments (the electro-negativity of the group and of the fragments). The important role of the π^* resonance in the low energy interactions of the electrons has been pointed out already by Aflatooni et al. [6].

The main product of the DEA reaction to valine at low electron energies (below 5 eV) is the negative ion with the m/Z=116. This is a close shell negative ion formed via

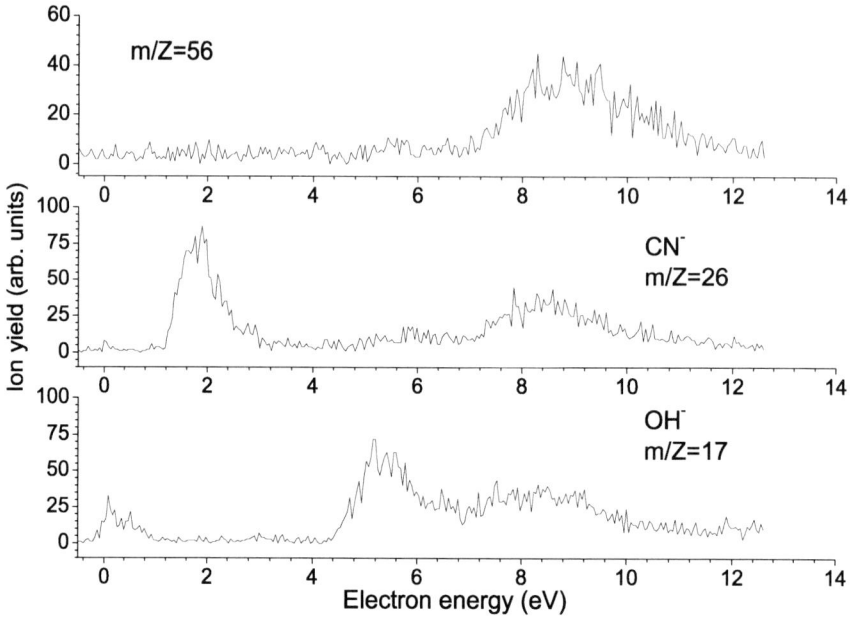

FIGURE 4. The ion yields for the products of the DEA to Valine a) m/Z=56, b) CN⁻ – m/Z=100 c) OH⁻ – m/Z=17 d) COOH⁻ – m/z=45

hydrogen cleavage from the the COOH group (equation 1). The ion yield for the (Val-H)⁻ is presented in the Figure 3. We associate the strong 1.2 eV resonance with the π^* resonance of the -COOH group which is typical for the organic acids and amino acids [6]. The 5 eV resonance has for this ion only weak intensity and the 9 eV resonance is not present at all in the ion yield of the (Val-H)⁻ ion.

The negative ion with m/Z=100 we assign to the (Val-OH)⁻ which is formed via the reaction 2. The ion yield in the Figure 3 exhibit 3 resonances. The π^* resonance at approximately 1.6 eV and two core excited resonances at 5.0 eV and 7.5 eV. The ion with m/Z=72 we associate with the cleavage of the COOH group from TNI (Val-COOH)⁻. This reaction occurs only at the highest resonance. A complementary negative ion is the ion with m/Z=45 which is formed via several resonances. The radical COOH has an electron affinity of 3.498 eV ([10] and for this reason is the COOH⁻ preferred over the (Val-COOH)⁻ channel. The last ion associated with COOH group is the ion OH⁻ (m/Z=17) 4. The electron affinity of OH radical is according to[10] 1.828 eV. The OH⁻ is complementary to the (Val-OH)⁻ ion. The (Val-OH)⁻ ion is in the low electron energy range much stronger then the OH⁻ ion, which indicate that the electron affinity of (Val-OH) radical is larger than the electron affinity of OH radical.

The ion CN⁻ (m/Z=26) and the ion with m/Z=56 are formed via different mechanism.

The formation of these ions could not be explained by single bond cleavage and and also are not associated with COOH group. The formation of these ion is associated with complex rearrangement reactions. The ion with m/Z=26 we easily identify as CN^-. The CN^- is known to have strong electron affinity (EA=3.869(eV) [10]. Nevertheless, it is surprising to see the formation of the ion already via the low energy resonance, because several bonds have to be broken in order to form CN^-. The energy necessary for this process has to be gained from the electron affinity of the CN radical and from formation of new bonds in the products. However, in present experiment we are not able to get any informations about the neutral products. While we assume the structure of the ion with m/Z=26, the nature of the ion with m/Z=56 is still not clear. We suggest that it could be an ion with the formula $C_3H_6N^-$, or $C_3H_4O^-$. In order to find the nature of this ion and also to identify other ions more rigorously we plan to apply quantum chemistry methods in order to calculate reaction enthalpies for DEA reactions (1-7) to valine and electron affinities of neutral precursors.

CONCLUSIONS

The DEA to gas phase L-valine leads to the formation of at least 7 fragment anions. Like in the case of many biological important molecules (formic acid, acetic acid, thymine, cytosine, uracil) the most abundant anion is the closed shell anion ($(M-H)^-$). In the case of DEA to L-valine formation of $(Val-H)^-$ anion is dominant in the electron energy range below 5 eV. The appearance of the negative ions in this energy range results from the decay of the π^* resonance of the COOH group. The majority of the negative ions formed via DEA is associated with COOH group ($(Val-OH)^-$, $(Val-COOH)^-$, $COOH^-$, OH^-). Additional DEA channels are at low electron energies also associated with the π^* resonance and at high electron energies with the core excited resonances at about 5 and 9 eV. In the electron energy range above 5 eV the OH^- and $COOH^-$ are most abundant products of DEA reaction. For all reaction channels with exception of the $(Val-H)^-$ we were not able to identify unambiguous the neutral fragments of the DEA reactions.

The comparison of the DEA to valine with the previously studied DEA studies to alifatic amino acids glycine and alanine [3, 4, 5] indicates that there exists very common features. The DEA at low electron energies is dominated by the COOH group and also the fragmentation patterns are very similar. The DEA studies to the alifatic amino acids indicate that at the beginning of the complex reaction chains responsible for the formation of the radicals in peptides and proteins we should also consider the products formed via DEA reactions.

ACKNOWLEDGMENTS

This work has been supported by the Slovak Research and development agency, project Nr. APVT-20-007504 and by the ESF program EIPAM.

REFERENCES

1. International Comission on Radiation Units and Measurements, *ICRU Report 31* ICRU, Washington, DC, (1979)
2. T. Cobut, Y. Fongillo, J. P. Patau, T. Goulet, M. J. Fraser and J. P. Jay-Gerin, *Radiat, Phys. Chem.* **51**, 229 (1998)
3. S. Gohlke, A. Rosa, E. Illenberger, F. Brüning, M. Huels, *J. Chem. Phys.* **116**, 10164 (2002)
4. S. Ptasinska, S. Denifl, A. Abedi, P. Scheier, T. D. Märk, *Anal. Bioanl. Chem.* **377**, 1115 (2003)
5. S. Ptasinska, S. Denifl, P. Scheier, T.D. Märk, S. Matejcik *Chem. Phzs. Lett.* **403**, 107 (2005)
6. K. Aflatooni, B. Hitt, G. A. Gallup, P. D. Burrow, *J. Chem. Phys.* **115**, 6489 (2001)
7. M.Stano, S.Matejcik,J.D.Skalny,T.D.Mark,*J.Phys.B:At.Mol. Opt.Phys.* **36**, 261 (2003)
8. S. Matejcik, V. Foltin, M. Stano, J.D. Skalny, *Int. J. Mass Spect.* **223-224**, 9 (2003)
9. A. Stamatovic, G. Schulz,*Rev. Sci. Instr.* **39**, 1752 (1968)
10. Handbook of Chemistry and Physics, CRC Press LCC 2005

High resolution cross section measurements using a trap based positron (electron) beam

J. P. Marler[*] and C. M. Surko[†]

[*]Lawrence University, Appleton, WI, 54911
[†]University of California, San Diego 92093

Abstract. The advent of the trap based beam has made possible the highest resolution cross section measurements of positrons interacting with atoms and molecules to date. The strong magnetic field needed for the trap required new methods of making cross section measurements in such a field. We first describe absolute, integrated inelastic cross section measurements. However, these techniques can also be applied to electron scattering and in fact provide some advantages for integrated electron-impact cross section measurements. Additionally the ability to do both in the same apparatus minimizes systematic effects in comparative measurements. This paper reports on the results on the first of these measurements, vibrational excitation in CF_4.

Keywords: cross section, scattering, positron
PACS: 34.85.+x,34.50.Gb,34.80.Gs

INTRODUCTION

Positron interactions with matter play important roles in many physical processes of interest. Examples include the origin of astrophysical sources of annihilation radiation [1], the use of positrons in medicine (e.g., positron emission tomography); the characterization of materials [2]; and the formation of antihydrogen [3, 4], which is the simplest form of stable, neutral antimatter. While the interactions of positrons with atomic targets have been studied for decades [5, 6, 7], many fundamental questions remain open [8, 9].

This area is much less advanced, as compared, for example, with the study of electron scattering processes, particularly at low energies. The reason for this is twofold. From an experimental viewpoint, positrons are much less common than electrons, and consequently techniques for using them to study scattering are more difficult and less well developed. One solution, using a trap-based beam, has provided the highest energy resolution to date. From a theoretical viewpoint, positron interactions with atoms and molecules provide additional challenges with respect to calculations. In particular, the exchange interaction is absent, and a new process, the formation of positronium, Ps (i.e., the "atom" which consists of an electron and a positron), is believed to play an important role, either as an open or closed channel. It is hoped that benchmark experimental study of positronium formation will help to illuminate and encourage further theoretical study of this process.

Another important contribution to this field will be comparative cross section measurements of the same processes by electron and positron impact. To minimize systematic differences, new modifications to the detection scheme of the current apparatus have allowed for electron impact cross section measurements.

CP876, *The Physics of Ionized Gases: 23rd Summer School and International Symposium*,
edited by L. Hadžievski, B. P. Marinković, and N. S. Simonović

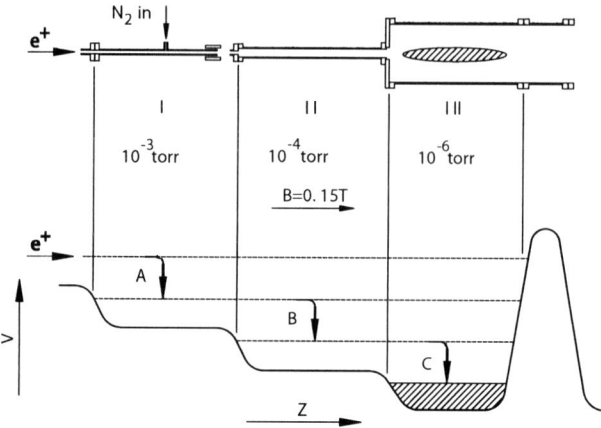

FIGURE 1. Schematic diagram of the three stage buffer gas trap (top) and the electric potential in the three regions (below). In the final stage where the plasma is stored there is the lowest gas pressure to minimize loss.

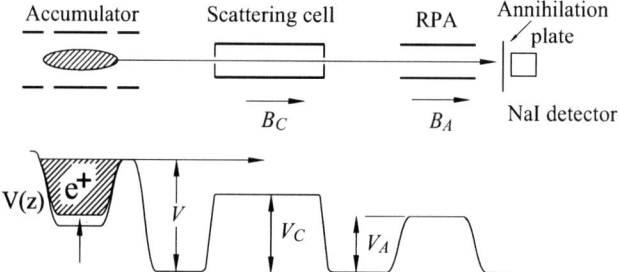

FIGURE 2. Schematic diagram of the experimental apparatus. Note that the magnetic field in the scattering cell and analyser can be varied independently.

EXPERIMENTAL TECHNIQUES

The design and operation of the trap has been described in more detail elsewhere [10]. Briefly, positrons emitted from a radioactive ^{22}Na source are initially cooled by interaction with the frozen ($\sim$7 K) neon walls nearby. These slow (eV's of energy) positrons are magnetically guided into the three stage buffer gas trap (see Fig. 1). The positrons lose energy via inelastic collisions with the N_2 buffer gas. Adding a small (relative to the N_2) amount of CF_4 in the third stage decreases the necessary cooling time. The positrons cool to 25 meV in 0.1 s.

Figure 2 is a schematic diagram of the scattering cell and retarding potential annalyser (RPA). Bunches of cooled positrons are pulsed out of the trap and magnetically guided through the rest of the apparatus. The magnetic field in the trap is quite sizeable ($\sim$0.1 T) and therefore it would be a real challenge to extract the positrons from the magnetic field to use traditional electrostatic scattering techniques. New techniques were developed to make the measurements in the magnetic field and several advantages over electrostatic techniques have been noted and exploited.

A more in depth discussion of the procedure is given in Ref.[11]. Basically the energy of the positrons in the magnetic field can be divided into energy in the parallel direction (i.e. along the magnetic field lines) and energy in directions perpendicular to the magnetic field lines (i.e. the energy in the cyclotron motion around the magnetic field lines). The positrons are accelerated out of the trap with significant parallel energy but maintain their relatively small ($\sim$25 meV FWHM) perpendicular energy. The beam is guided through the gas cell, and analyser and directed into an annihilation plate at the end of the vacuum chamber. The annihilation gamma rays are then detected with a NaI crystal. It is important to note that the retarding potential analyser only measures the parallel energy distribution of the positron beam. The ratio of the final parallel energy to its initial energy gives information about the scattering angle for elastic collisions taking place in the gas cell.

The cross section measurements presented here were done using a technique that relies on the fact that the positron orbits are strongly magnetized [11, 12]. For the experiments described here, the magnetic field in the scattering region, B_S, and in the analyzing region, B_A, can be adjusted independently. When $B_S \gg B_A$, the invariance of the quantity, $\xi = E_\perp/B$ (where $E_\perp$ is the energy in the particles cyclotron motion) allows us to measure the approximate *total* positron energy using the RPA [12, 11]. In practice, we lower the magnetic field in the analyzing area thereby lowering the component in the perpendicular direction and transferring that energy into the parallel component. This in essence makes the parallel component approximately equal to the final total energy of the particles. The ratio of particles losing energy corresponding losing exactly the energy of certain inelastic processes to the initial beam strength determines the cross section for that process. The gas cell has been specially designed with small entrance and exit apertures so that the effective path length of the collisions. This allows us to make absolute cross section measurements with no fitted parameters.

Adaptations for making electron measurements

Secondary electrons are emitted in the source region. If the source is biased negatively, the electrons (and not the positrons) are directed towards the trap guided by the magnetic field lines. The electrons can be cooled and trapped in the same three stage buffer gas trap simply by reversing the potential on the electrodes.

The detection scheme was altered to allow for electron measurements. The previous annihilation gamma ray detector was replaced with a charged particle diode internal to the vacuum chamber [13]. Since it does not depend on annihilation gammas, the charged particle detector works with particles of either sign.

IONIZATION RESULTS

Using the above techniques results in positron impact vibrational excitation [14], electronic excitation [15], positronium formation, and ionization [16, 17] have been made

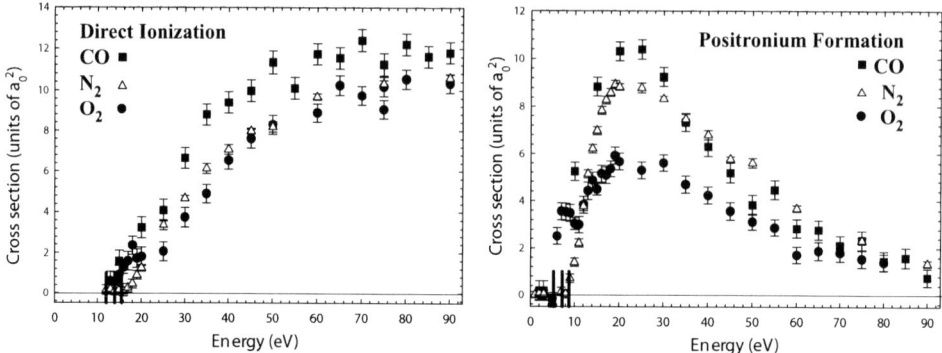

FIGURE 3. Integral cross sections for the direct ionization and positronium formation of CO, N_2 and O_2 respectively. Vertical bars mark the positions of the thresholds for O_2, CO and N_2 respectively.[17]

in a variety of atoms and molecules. (Differential cross sections and total cross sections have also been made in the same apparatus, see Ref. [11] for more details.)

I chose to highlight the most recent results in ionization as it may be of particular interest to this community. Note that in positron scattering there are multiple processes which result in ionization. Direct ionization is the analogue to ionization by electrons and has the same threshold. However, positronium formation is also possible. In this process the positron ionizes the atom or molecule but remains bound to the ionized electron as a positronium atom. The binding energy of the positronium atom is 6.8 eV (1/2 that of the hydrogen atom) and therefore the threshold for this process is 6.8 eV less than the electron ionization threshold. A final possibility is the direct annihilation of the positron on one of the electrons of the atom or molecule. This process is also under investigation [18] but the cross sections for direct annihilation are orders of magnitude smaller than either direct ionization or positronium formation at the energies studied and can be ignored in these measurements.

Figure 3 shows a comparison of the direct ionization and positronium formation cross sections for N_2, CO and O_2. For a more comprehensive examination of these results see Ref [17]. The isoelectronic molecules N_2 and CO have similar positronium formation and ionization cross sections as might be expected. However, the positronium formation cross section for O_2 is qualitatively different near threshold from those for N_2 and CO. The characteristic shape of the near-threshold feature in O_2 has qualitative similarities to a feature observed previously in the total ionization cross section for O_2 in this region of energies [19]. This feature is not fully understood and warrants further examination.

ELECTRON AND POSITRON VIBRATIONAL CROSS SECTIONS

Study of electron-CF_4 interactions is important in many plasma-assisted material-processing applications as well as in space and atmospheric sciences [26]. We report here the first direct integral measurements of this electron-impact cross section. There have been previous, systematic comparisons of total and differential elastic cross sec-

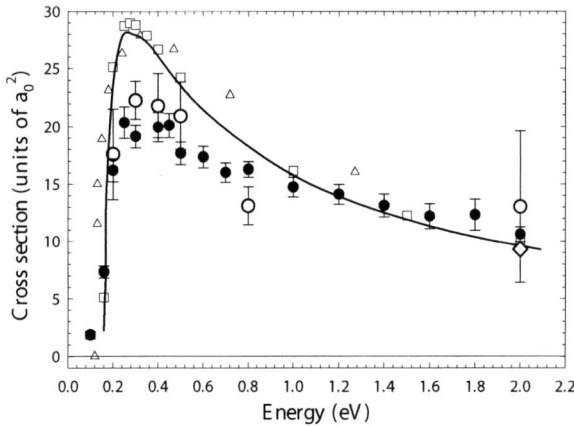

FIGURE 4. Integral electron-impact vibrational excitation of the v_3 vibrational mode of CF_4: (○) current data [13] and (△) results of [20] scaled by a factor of 0.7 as suggested in Ref. [21]. Shown for comparison (●) are the current positron-impact results [13]. Also shown are (—) the results of an analytic, Born-dipole approximation calculation for the cross section [22], using infrared measurements to fix the dipole strength, M_n^d; and (□) [23] the Born model, fixing M_n^d using electron differential scattering cross section measurements. Shown by the (◇) symbol is the result of a recent local interaction potential calculation for electron impact [24].

tions for electrons and positrons [27, 28, 29, 25]. There have also been reported positron cross section measurements for the sum of excitation to the three vibrational modes in CO_2 [25]. The experiments reported here, on the other hand, are capable of sufficiently high energy resolution to measure state-resolved, integral inelastic cross sections, and to perform these state-selective measurements for both electrons and positrons in the same apparatus in order to minimize systematic differences.

Figure 4 shows the result for the vibration excitation of the v_3 mode in CF_4[13]. The positron cross section is the largest positron vibrational cross section seen to date. This may explain on how adding CF_4 in the the third stage of the trap is effective in reducing the cooling time. Also noticeable is how similar the positron and electron impact cross sections are to each other.

A comparison of the Born Dipole Model with measured cross sections for both positron and electron impact excitation of the v_3 vibrational mode of CF_4,CO, CH_4 and CO_2 has recently been presented in Ref. [30]. For the targets studied thus far that have nonzero transition dipole moments, the *shapes* of the Born-dipole cross sections are in fair agreement with the measurements. For example, shown in Fig. 5 is a comparison of the Born model with data for the v_3 mode of CO_2 [14]. In this case, the Born dipole prediction accounts for about 60% of the measured value and has a shape virtually identical to that measured.

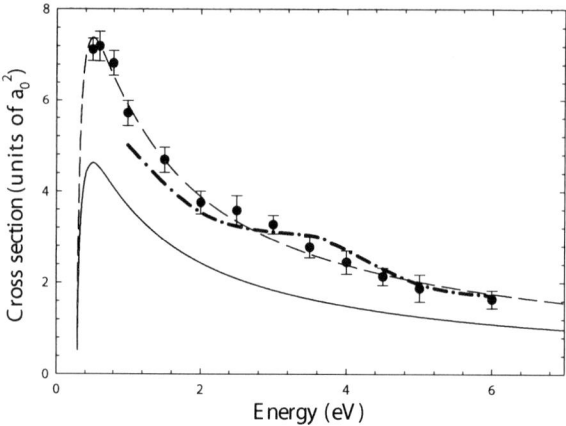

FIGURE 5. Comparison of (•) the integral positron-impact cross section for vibrational excitation of the v_3 mode of CO_2 [14] with the predictions of the Born dipole model. Shown are (—) the results of an analytic, Born-dipole calculation, (— —) the Born-dipole calculation scaled by a factor of 1.6, and (— · —) a calculation for positron-impact using a close coupling and continuum multiple scattering approach [25]. This comparison indicates that the predictions of the long-range dipole-coupling model can account for both the shape and much of the magnitude of the observed cross section.

OUTLOOK

The advent of the buffer gas trap has allowed for high resolution measurements of positron and electron cross section measurements. However, to date the resolution is limited to ~ 25 meV. This limits near threshold measurements and low energy closely spaced inelastic cross sections (e.g. rotational excitation). Current work towards obtaining even finer energy resolution includes the construction of a 5 T superconducting Penning-Malmberg trap [31]. This field is strong enough to allow the positrons to cool by cyclotron radiation and therefore eliminates the need for buffer gas cooling. Cryogenic electrodes (~10K) in the trap will allow for lower energy beams.

We also note of great interest to this group is the measurement of direct annihilation cross sections. These measurements also currently use the buffer gas trap as a source of low energy positron pulses and could also benefit by an improved (i.e. lower energy with a more narrow energy spread) beam. Current progress on these results are described in Refs. [18, 32]

ACKNOWLEDGMENTS

This work is supported by the National Science Foundation.

REFERENCES

1. N. Guessoum, R. Ramaty, and R. E. Lingenfelter, *Astrophysical Journal* **378**, 170 (1991).

2. L. D. Hulett, Jr., D. L. Donohue, J. Xu, T. A. Lewis, S. A. McLuckey, and G. L. Glish, *Chemical Physics Letters* **216**, 236–40 (1993).
3. M. A. et al., *Nature* **419**, 456 – 459 (2002).
4. G. G. et al., *Physical Review Letters* **89**, 213401 (2002).
5. T. C. Griffith, and G. R. Heyland, *Physics Reports* **39**, 169 (1978).
6. W. E. Kauppila, and T. S. Stein, *Advances in Atomic, Molecular, and Optical Physics* **26**, 1 – 49 (1990).
7. M. Charlton, and J. Humberston, *Positron Physics*, Cambridge University Press, New York, 2001.
8. C. M. Surko, and F. A. Gianturco, editors, *New Directions in Antimatter Chemistry and Physics*, Kluwer Academic Publishers, Dordrecht, 2001.
9. C. M. Surko, G. F. Gribakin, and S. J. Buckman, *Journal of Physics B* **38**, R57–126 (2005).
10. S. J. Gilbert, C. Kurz, R. G. Greaves, and C. M. Surko, *Applied Physics Letters* **70**, 1944–1946 (1997).
11. J. P. Sullivan, S. J. Gilbert, J. P. Marler, R. G. Greaves, S. J. Buckman, and C. M. Surko, *Physical Review A* **66**, 042708 (2002).
12. S. J. Gilbert, R. G. Greaves, and C. M. Surko, *Physical Review Letters* **82**, 5032–5035 (1999).
13. J. P. Marler, and C. M. Surko, *Physical Review A* **72**, 062702 (2005).
14. J. P. Sullivan, S. J. Gilbert, and C. M. Surko, *Physical Review Letters* **86**, 1494 (2001).
15. J. P. Sullivan, J. P. Marler, S. J. Gilbert, S. J. Buckman, and C. M. Surko, *Physical Review Letters* **87**, 073201 (2001).
16. J. P. Marler, J. P. Sullivan, and C. M. Surko, *Physical Review A* **71**, 022701 (2005).
17. J. P. Marler, and C. M. Surko, *Physical Review A* **72**, 062713 (2005).
18. L. D. Barnes, S. J. Gilbert, and C. M. Surko, *Physical Review A* **67**, 032706 (2003).
19. G. Laricchia, J. Moxom, and M. Charlton, *Physical Review Letters* **70**, 3229–3230 (1993).
20. M. Hayashi, "Electron collision cross sections for molecules determined from beam and swarm data," in *Swarm Studies and Inelastic Electron-Molecule Collisions*, edited by L. C. P. et al., Springer, Berlin, 1987, pp. 167–87.
21. L. E. Kline, and T. V. Congedo, *Bull. Am. Phys. Soc.* **34**, 325 (1989).
22. A. Mann, and F. Linder, *Journal of Physics B* **25**, 545–556 (1992).
23. R. A. Bonham, *Japanese Journal of Applied Physics* **33**, 4157–4164 (1994).
24. S. Irrera, and F. A. Ginaturco, *New Journal of Physics* **7**, 1 (2005).
25. M. Kimura, M. Takekawa, Y. Itikawa, H. Takaki, and O. Sueoka, *Physical Review Letters* **80**, 3936–3939 (1998).
26. L. G. Christophorou, J. K. Olthoff, and M. V. V. S. Rao, *Journal of Physical and Chemical Reference Data* **25**, 1341–1388 (1996).
27. E. Surdutovich, W. E. Kauppila, C. K. Kwan, E. G. Miller, S. P. Prikh, K. A. Price, and T. S. Stein, *Nuclear Instruments and Methods B* **221**, 97–99 (2004).
28. S. Zhou, H. Li, W. E. Kauppila, C. K. Kwan, and T. S. Stein, *Physical Review A* **55**, 361–368 (1997).
29. C. Makochekanwa, M. Kimura, and O. Sueoka, *Physical Review A* **70**, 022702 (2004).
30. J. P. Marler, G. F. Gribakin, and C. M. Surko, *Nuclear Instruments and Methods B* **247**, 87–91 (2005).
31. J. R. Danielson, P. Schmidt, J. P. Sullivan, and C. M. Surko, "A Cryogenic, High-field Trap for Large Positron Plasmas and Cold Beams," in *Non-Neutral Plasma Physics V*, edited by M. Schauer, T. Mitchell, and R. Nebel, AIP, New York, 2003, pp. 149–161.
32. L. D. Barnes, J. A. Young, and C. M. Surko, *Physical Review A* **74**, 012706 (2006).

Electron Excitation Coefficients in Helium, Neon, Oxygen and Methane at High E/N

Željka D. Nikitović

Institute of Physics, P.O.B. 68, 11080 Belgrade, Serbia

Abstract. Swarm analysis is performed by comparing experimental and calculated transport coefficients. Comparisons are repeated until a satisfactory agreement is achieved after modifications of the cross sections. We have made an analysis of our excitation coefficient data for neon and methane by using detailed Monte Carlo simulation scheme. In this work we also present experimental electron excitation coefficients for other gases: helium, neon and oxygen. We used a drift tube technique to measure the absolute emission intensities in low current self sustained Townsend type discharges.

Keywords: Townsend discharges, helium, neon, oxygen, methane, Monte Carlo code
PACS: 52.20.-j, 52.20.Hv, 52.25.Ya, 52.80.Dy

INTRODUCTION

Understanding the kinetics of excited rare gases is of practical importance for modeling and optimizing kinetics process in gas discharges devices, such as plasma displays, light sources, excimer lasers, ion lasers, ion thrusters, particle detectors and in microwave afterglows. Methane is the most popular molecular additive in gas mixtures for drift chambers used as track detectors for high energy particles. Discharges in methane are used for deposition of diamond like films. Energy dependent cross sections for electron excitation of atomic levels of rare gases may be of interest for modeling and diagnostics of a great number of different discharges and gas discharge devices.

We used a drift tube technique for measuring the absolute emission intensities in low current self sustained Townsend type discharges. The excitation coefficients were determined from the measurements of the optical signal at the anode after correction for detector quantum efficiency. The data were obtained between moderate E/N values where electrons are in equilibrium and very high E/N values where electrons may not be in equilibrium with the field and where heavy particles may contribute to the excitation.

EXPERIMENTAL SETUP AND PROCEDURE

The experimental setup has been described in greater detail [1-3].

CP876, *The Physics of Ionized Gases: 23rd Summer School and International Symposium,*
edited by L. Hadžievski, B. P. Marinković, and N. S. Simonović
© 2006 American Institute of Physics 978-0-7354-0377-2/06/$23.00

The drift tube had two parallel plane electrodes of 79 mm in diameter separated by 14.7 mm, placed inside a closely fitting quartz tube. The cathode was made of stainless steel, while the anode was made of vacuum-grade sintered graphite in order to minimize the effect of backscattered electrons from the anode.

Vacuum chamber (stainless steel) is evacuated by a turbomolecular pump typically to 3×10^{-6} Torr (4×10^{-4} Pa) and then it is filled by research grade purity gas. Gas pressure is measured by a capacitance manometer. The discharge current was kept below 2 µA, in order not to perturb the homogeneous electric field between the electrodes [4].

The light emitted from the discharge was spectrally selected by Jobin Yvon M25 monochromator with a spectral resolution of 0.7 nm and recorded by a cooled Hamamatsu R928 photomultiplier tube operating in a single photon counting regime. The monochromator was positioned at the wavelength of the maximum of the line and the line was integrated by opening the exit slit. The entire detection system was placed on a platform and moved by a stepper motor, which was controlled by a computer.

The absolute calibration of the quantum efficiency of the detection system was done using a standard tungsten lamp. Calibration of the system for the wavelength region 330-830 nm was done with the lamp operating at two different temperatures. The light intensity was measured with and without an interference filter and with different neutral density filters to reduce the signal. Standard tungsten lamp enabled us continuous control of optical calibration and reproducibility of detection system response.

System was recalibrated frequently by making measurements in Ar at 100 Td, 500 Td and 1000 Td and by measuring the signal from a standard wolfram lamp. In addition, several times a day measurements were repeated for a selected line of the gas that was studied.

In Townsend discharges where electron impact excitation dominates over other processes populating the upper level and where electrons are in equilibrium with the electric field, the electron excitation coefficient ε_m/N of the level m is given by:

$$\frac{\varepsilon_m}{N} = \frac{S_a}{j_e} \frac{e}{\frac{\Omega}{4\pi}\Delta x Q(\lambda)N} \frac{A_m}{A_{mn}}(1 + \frac{N}{A_m}k_q) \qquad (1)$$

where S_a is the emission signal at the anode, j_e is the electron current density at the anode, e is the electron charge, Ω is the effective solid angle of the detector, Δx is the width of the entrance slit of the monochromator, $Q(\lambda)$ is the quantum efficiency of the detector, A_{mn} is the Einstein coefficient of the observed transition m-n, A_m is the summed Einstein coefficient for the level m, and k_q is the rate coefficient for the collisional quenching of the state m.

EXPERIMENTAL RESULTS AND DISCUSSION

In this section we shall show examples of results that were presented in specialized publications and examples of the still unpublished work.

In Figure 1. we show the spatial distribution of 504.774 nm He I emission obtained at two different E/N [5]. The positions of the electrodes are shown by the vertical dashed lines. The spatial scans of lines show an exponential increase of the measured intensity as we scan from the cathode to the anode.

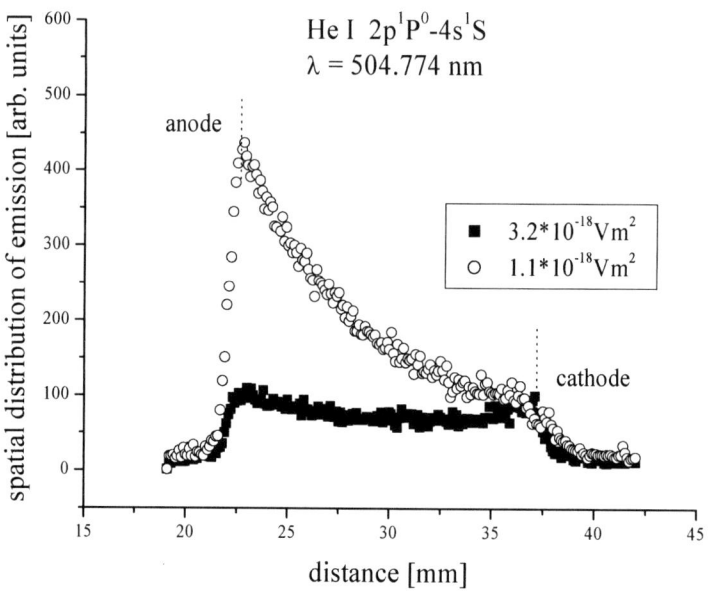

FIGURE 1. Spatial distribution of He I (λ=504.774 nm) at two different E/N.

As we operate in the self sustained (albeit low current) we first determine the breakdown voltage. In case of the it was difficult to achieve very high E/N (low pressure) because of a very sudden increase in the voltage in the left hand side of the Paschen curve. Our limitation was in the voltage which could be applied to the high voltage feed through. We could not approach what is presumably S shaped Paschen curve duo to limitation of voltage.

We have measured spatial profiles for several transitions of helium. These profiles may be normalized to the excitation coefficients and be used to obtain the cross-section data with the aid of some numerical techniques [6]. In particular, such data may be used to identify heavy particle excitation [7] that is recognized by its growth towards the cathode. From the slope of the curves of the spatial distribution of emission we determined the ionization coefficients. The spatial distribution of emission is extrapolated and normalized to give the excitation coefficients at the anode.

In Figure 2. the excitation and emission coefficients for He I (λ=388.865 nm) transition are shown as a function of E/N. As E/N values increases the number of electrons which have sufficient energy to excite the state increases rapidly. With further increase of E/N, the average electron energy approaches the energy of the maximum of the cross-section and the value of the excitation coefficients levels off. For even higher E/N, a decrease of the excitation coefficients is observable, which is in agreement whit the shape of the cross-sections.

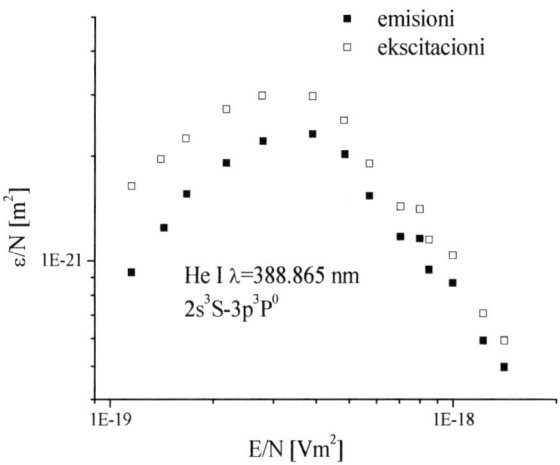

FIGURE 2. Emission and excitation coefficients for He I (λ=388.865 nm).

In Figure 3. we show experimental data for both absolute emission and excitation coefficients for the level O I $2p^33s-2p^3(4S^0)3p$ (leading to λ=777.4 nm transition), as a function of E/N. The E/N dependence follows the expected rapid increase, almost constant values at around 2 kTd as the average electron energy approaches the energy of the maximum of the cross-section leading to the expected decrease at even higher mean energies.

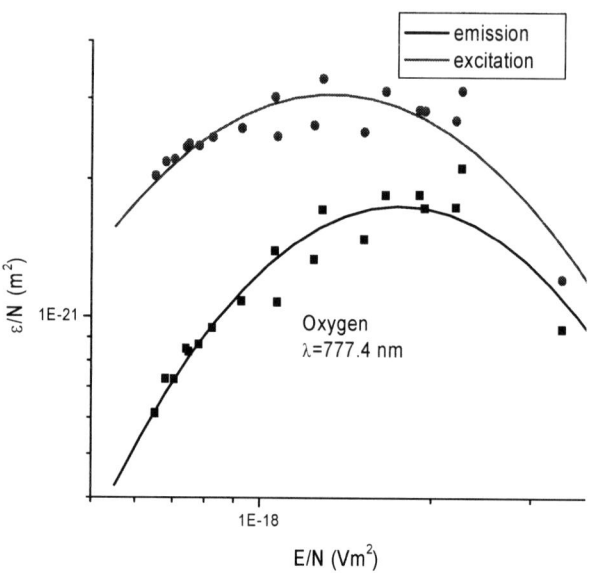

FIGURE 3. Emission and excitation coefficients for O I (λ=777.4 nm).

MONTE CARLO SIMULATION AND DISCUSSION

Swarm analysis was performed by comparing experimental data and the data calculated by a Monte Carlo simulation (MCS) procedure. The Monte Carlo code is based on our null-collision code developed in particular for our steady state Townsend (SST) conditions [8-10]. The validity of the code was shown in several previous publications [8, 9].

In the simulation we have followed the kinetics of electrons first, followed by ions and finally fast neutrals. A large number of initial electrons (between 10.000 and 1.000.000, but typically 100.000) were released one after the other from the cathode until a satisfactory overall statistics was achieved. The basic set of cross sections for electron scattering was obtained from Hayashi [11] and Puech and Mizzi [12].

Having completed the calculation for the "direct" electrons, sampled properties of electrons hitting the anode were used to release reflected electrons and secondary electrons induced by the primary electrons. The properties of the backscattered electrons were taken numerically from the relevant literature including realistic energy distributions, energy dependence of the reflection coefficient, secondary electron yield and angular distribution (all the relevant references were cited and details of the model of reflection were given by [13]). Calculations of the spatially resolved emission

136

profile and of the excitation coefficients were made. The spatial profiles were put on the absolute scale there by determining the excitation coefficient at the anode.

In addition to the electrons we have followed the kinetics of the ions and the fast neutrals produced in charge transfer collisions of the ions and the parent gas atoms. The procedure was similar to that used for argon [9]. The basic cross sections for neon ions [14] and fast neutrals [15] were used.

Monte Carlo calculations were performed for low current (diffuse) Townsend discharges in neon and methane.

The basic set of cross sections for electron scattering for neon was obtained from Hayashi [11] and Puech and Mizzi [12].

In Fig. 4. we show comparison of the electron excitation coefficients obtained from the experimental data and MCS [8] for the $2p_1$ ($\lambda = 585.25$ nm) level of neon. These data were obtained under conditions when electrons dominate. We were able to achieve a reasonable agreement of the calculated excitation coefficients with the experimental data with further adjustments to the initially selected cross sections that were based on the available cross section data. However further adjustments are needed to take the full advantage of the available experimental data.

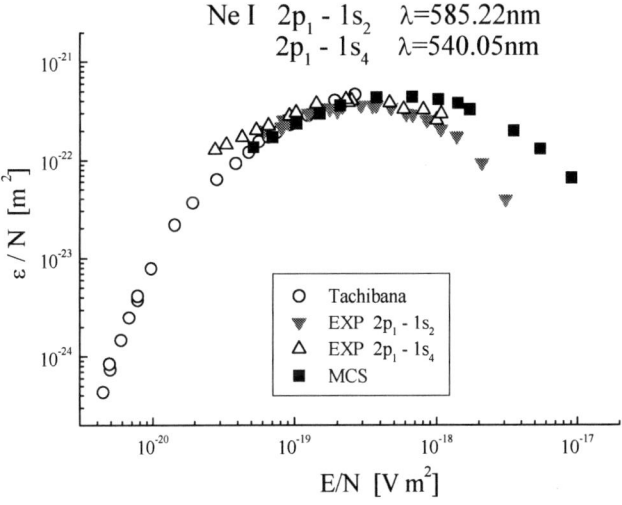

FIGURE 4. Excitation coefficients for 2p1 Ne I ($\lambda = 585.25$nm).

The basic set of electron scattering data was taken from Hayashi [16] for methane and it was tested through comparisons with experimental drift velocities, characteristic energies and ionization coefficients [17]. The dissociative excitation cross sections were taken from Hayashi and renormalized to fit the measured excitation coefficients [17].

In Fig. 5. we show spatial profiles of the excitation coefficients for the methane CH(A$^2\Delta$ –X$^2\Pi$) molecular band obtained at different reduced electric field (E/N = 1.5 kTd, 2.5kTd, 5kTd and 10kTd) and comparing with experimental data and the Monte Carlo simulations. Those fits were obtained by significantly modifying the available cross sections for excitation and the best cross sections for excitation of these levels are given in the next slide. The calculations were based on the heavy particle cross sections of Petrović and Phelps [18]. The excellent agreement between predicted emission profiles and their absolute magnitude inducates that the collisional scheme and the cross sections that were produced by Petrović and Phelps [18] are good.

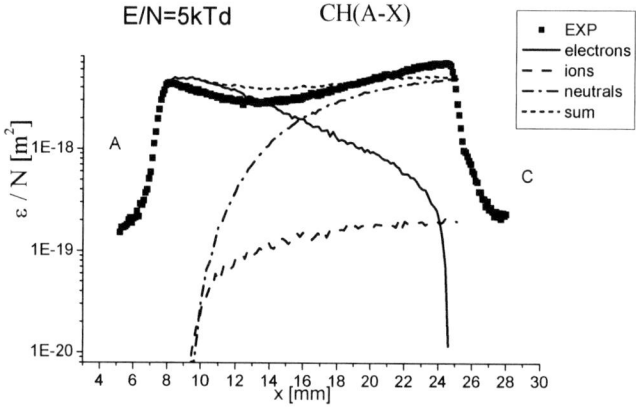

FIGURE 5. Spatial profiles of excitation coefficients for CH (A$^2\Delta$ –X$^2\Pi$).

CONCLUSION

In this work we presented new measurements of excitation coefficients for helium, neon, oxygen and methane for a large number of atomic lines.

In addition, we have used MCS to calculate the spatial profiles of emission and the values of excitation coefficients for neon and methane. This was done for an assumed cross section until a good fit of the experimental data was achieved.

Using MCS and compared experimental data with results of simulations we determined cross sections for ten atomic Ne I levels and for one molecular band in methane.

ACKNOWLEDGMENTS

Work at the Institute of Physics is supported by the MNZŽS, under grants 141025.

Author is grateful to Z. Lj. Petrović, G. Malović, A. Strinić, V. Stojanović and O. Šašić for colaboration on the related projects.

REFERENCES

1. Z. M. Jelenak., Z. B. Velikić, J. V. Božin, Z. Lj. Petrović and B. M. Jelenković, *Phys. Rev. E* **47** pp. 3566-3573 (1993).
2. G. N. Malović, J. V. Božin, B. M. Jelenković and Z. Lj. Petrović, *Nucl. Instrum. and Methods B* **129** pp. 317-322 (1997).
3. G. N. Malović, A. I. Strinić, Z. Lj. Petrović, J. V. Božin and S. S. Manola, *Eur.Phys. J.* D **10** pp.147-151(2000).
4. S. Živanov, J. Živković, I. Stefanović, S. Vrhovac and Z.Lj. Petrović, *Eur.Phys.J.* AP 11, pp.59-69 (2000).
5. Željka Nikitović, Aleksandra Strinić, Vladimir Šamara, Gordana Malović and Zoran Petrović *Acta Chimica Slovenica, vol.52,* pp.463-466 (2005).
6. V. D. Stojanović and Z. Lj. Petrović, *J. Phys. D 31,* pp. 834-846 (1998).
7. B. M. Jelenković and A. V. Phelps, *Phys. Rev. A 36,* pp.5310-5326 (1987).
8. V. D. Stojanović and Z. Lj. Petrović, *J. Phys. D: Appl. Phys.* 31, pp. 834-846 (1998).
9. Z. Lj. Petrović and V. D. Stojanović, *J.Vac.Sci. Technol* A 16 (1), pp. 329-336 (1998).
10. S. Sakadžić and Z. Lj. Petrović, 2000. Proc. 20th SPIG 2000 Zlatibor (Ed.s Z. Lj.Petrović, M. M. Kuraica, N. Bibić and G. Malović), pp. 127-130 (2000).
11. M. Hayashi (private communication) (2001).
12. V. Puech and S. Mizzi, *J. Phys. D: Appl. Phys.* 24, pp. 1974-1985 (1991).
13. S. B. Vrhovac, V. D. Stojanović, B. M. Jelenković and Z. Lj. Petrović, *J.Appl. Phys.* 90 (12), pp. 5871-5877 (2001).
14. J. V. Jovanović, S. B. Vrhovac and Z. Lj Petrović, *Eur. Phys.* J. D 21, pp. 335-342 (2002).
15. A. V. Phelps (private communication) (2003).
16. M. Hayashi (private communication) (2003).
17. O. Šašić, G. N. Malović, A. I. Strinić, Ž. Nikitović and Z. Lj. Petrović, *New Journal of Physics* **6** 74, pp. 1-11 (2004).
18. Z. Lj. Petrović and A. V.Phelps (1992) unpublished.

139

SECTION 2

PARTICLE AND LASER BEAM INTERACTION WITH SOLIDS

Invited Lectures
Topical Invited Lectures
Progress Reports

Interaction of ultra-short laser pulses with clusters: short-time dynamics of a nano-plasma

Cornelia Deiss*, Nina Rohringer*,† and Joachim Burgdörfer*

*Institute for Theoretical Physics, Vienna University of Technology, A-1040 Vienna, Austria, EU
†Argonne National Laboratory, Argonne, IL 60439, USA

Abstract. We study the dynamics of the interaction of short infrared laser pulses with large rare-gas clusters. Special attention is given to the microscopic atomic collision dynamics in order to explain the efficient heating of the quasi-free electrons in the nano-plasma formed during the interaction. In the framework of our mean-field classical transport simulation we are able to explain the emission of characteristic x-rays at moderate laser intensities ($I \sim 10^{15} W cm^{-2}$) where the ponderomotive energy of the electrons is by far to low to allow for the creation of inner-shell vacancies. We identify large-angle elastic electron-ion scattering as an important heating mechanism at moderate laser intensities.

INTRODUCTION

The study of the interaction of intense short and ultra-short laser pulses with clusters has received much attention during the last decade [1]. KeV x-ray photon emission from rare-gas clusters unite the advantages of solid and gaseous targets: like solids they can provide large yields, yet they are relatively debris-free, just like gas targets.

In a simple picture, the dynamics during the laser-cluster interaction can be summarized as follows (Fig. 1): the atoms of the cluster are first ionized by the incident laser pulse (inner ionization) and a cold "nano-plasma" of solid density is formed. The quasi-free electrons take part in a collective oscillation driven by the laser field and moreover interact with the field of the surrounding particles. Electron-impact ionization of cluster ions produces additional quasi-free electrons and inner-shell vacancies which are at the origin of x-ray radiation. As a fraction of the electrons leaves the cluster (outer ionization), a net positive charge is left behind and the cluster begins to expand before disintegrating completely in a Coulomb explosion.

From the experimental point of view, the spectroscopy of the emitted ions [2] and electrons [3] gives information on the system a few microseconds after the femtosecond laser-pulse and the cluster disintegration. X-ray spectroscopy [4], on the other hand, allows measurements on a much shorter time-scale, down to a few femtoseconds. The inner-shell vacancies in argon responsible for the 3.1 keV characteristic x-ray radiation can not be explained by field ionization for laser pulses with a peak intensity less than $I \lesssim 10^{21} W cm^{-2}$. Consequently, the origin of these vacancies must be impact ionization by energetic electrons. The x-ray emission thus probes the high energy tail of the electron energy distribution, thereby providing valuable insight into the electronic dynamics which are the key to a detailed theoretical understanding of laser-cluster interaction. Furthermore, high-resolution x-ray spectroscopy determines the charge-state

CP876, *The Physics of Ionized Gases: 23ʳᵈ Summer School and International Symposium*,
edited by L. Hadžievski, B. P. Marinković, and N. S. Simonović
© 2006 American Institute of Physics 978-0-7354-0377-2/06/$23.00

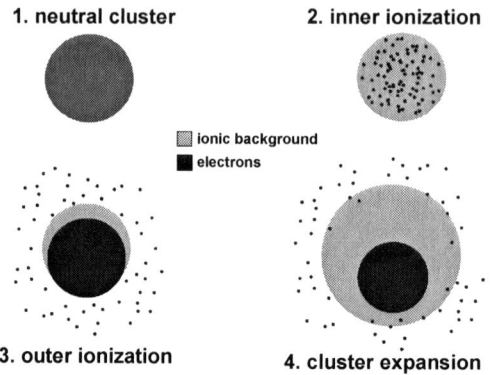

FIGURE 1. Stages of cluster dynamics (see text)

distribution of ions emitting the x-ray radiation. Recent experiments [5, 6] found an unexpectedly low laser intensity threshold for x-ray production. When irradiating large clusters with $N > 10000$ argon atoms with infrared ($\lambda = 800nm$) laser pulses of duration $\tau = 60fs$ at FWHM, characteristic x-ray radiation could be measured for laser peak intensities as low as $I_{th} \simeq 2.2 \cdot 10^{15}$ Wcm^{-2}. At this intensity, the ponderomotive energy $U_P = F^2/(4\omega^2)$ (atomic units are used unless otherwise stated) associated with the oscillatory motion of a free electron in a laser field with field strength F and frequency ω is $U_P \simeq 130$eV. This value is more than one order of magnitude below the binding energy $E_K \simeq 3.1$keV of K-shell electrons in argon. Moreover, the analysis of the high-resolution x-ray spectra show that the mean charge state of the emitting ions can be as high as 13+ when irradiating argon clusters with pulses of an intensity of $I = 3 \cdot 10^{16}$ Wcm^{-2}. Both these observations raise questions as to the additional heating mechanisms at play in a cluster environment, which allow the electrons to be effectively accelerated well beyond the ponderomotive energy.

After a decade of experimental and theoretical studies, the mechanisms causing the efficient heating of electrons in a cluster environment are still a matter of debate [7]. In the nano-plasma model [8] fast electrons are produced by inverse bremsstrahlung. The energy absorption is greatly enhanced when the plasma frequency of the electrons in the cluster matches the laser frequency. It has, however, been argued [6, 9] that this resonance should be strongly damped and that the high energy contribution of the electronic energy distribution should not be sufficient to explain the creation of inner-shell vacancies. Molecular dynamics simulations are limited to clusters of about 1000 atoms [10, 11, 12], and a scaling of the results to larger cluster sizes is difficult. Clusters of $\sim 10^4$ atoms have been simulated using a microscopic particle in cell (MPIC) code [13]. However only predictions for the energy distribution of the emitted electrons for a laser intensity $I = 8 \cdot 10^{15} Wcm^{-2}$ were made. This makes it difficult to draw conclusions as to the energy distribution of the electrons inside the cluster at lower laser intensities. Moreover, the predicted maximum ionic charge state of Ar^{6+} differs significantly from

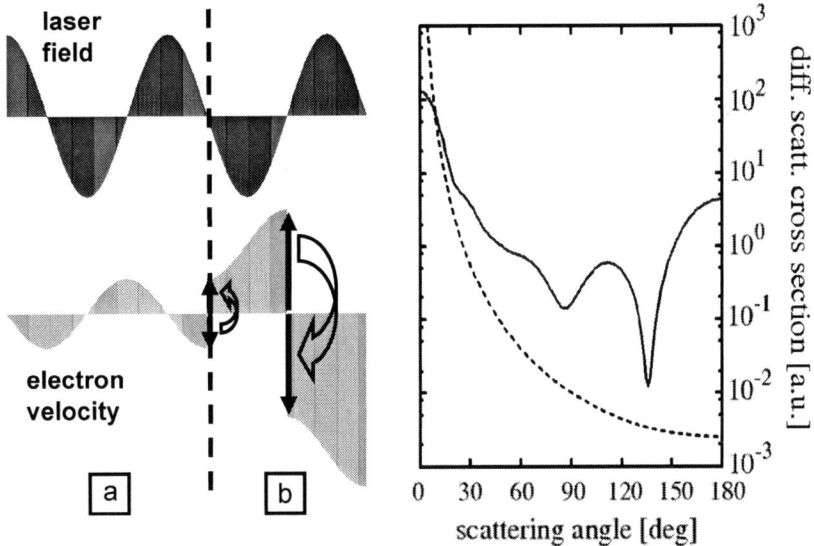

FIGURE 2. Left: Schematic time evolution of the velocity for a subensemble of electrons in a laser field.
Right: Differential cross-section $d\sigma_e/d\theta$ for elastic scattering of an electron with kinetic energy $E = 10$a.u. at Ar^{2+} (solid line). The Rutherford cross-section is also displayed (dashed line).

the measured mean ionic charge state of Ar^{13+} where the L-shell population is efficiently depleted.

In the next section we will explore elastic electron-ion scattering as an efficient electronic heating mechanism and discuss its appropriate theoretical description. In order to study the heating of the electrons quantitatively, we have developed a classical trajectory Monte Carlo (CTMC) simulation describing the laser-cluster interaction, the details of which will be presented in the third section. The last section will contain simulation results concerning the electronic dynamics and the resulting x-ray yield, which allow to quantify the efficiency of heating by elastic electron-ion scattering.

MICROSCOPIC DYNAMICS

Elastic backscattering of an electron at the core potential of an atom or ion can flip the velocity vector of the electron, and thus allows, with a non-negligible probability, to remain synchronized with the alternating laser field vector during the subsequent half-cycle. It is then possible for the electron to gain additional kinetic energy instead of losing energy and momentum it has gained during the previous laser half-cycle (Fig.

2). A small sub-ensemble of electrons can thus be rapidly heated to high kinetic energies well beyond the maximum quiver energy $E_P = 2U_P$. This heating mechanism by repeated back-scattering resembles the Fermi-shuttle acceleration [14, 15] and is also related to the lucky-electron model proposed for IR photoemission from metallic surfaces [16]. It can also be viewed as a classical realization of inverse bremsstrahlung. The estimate of this heating mechanism depends critically on the accurate theoretical description of the elastic scattering process. The scattering into backward angles ($\theta \gtrsim 90°$), which is of particular importance for this mechanism, is determined by the short-ranged non-Coulombic contributions to the ionic core potential. As these contributions are neglected in frequently used descriptions of ions in terms of softened Coulomb potentials, the effect of backscattering is left unaccounted for in many simulations [11, 12, 13]. In the present work we calculate the differential scattering cross-sections for the electrons $d\sigma_e/d\theta$ by partial wave analysis of parametrized Hartree-Fock potentials [17, 18]. The differential cross-section is typically dominated by few low-order partial waves giving rise to generalized Ramsauer-Townsend minima [19] and diffraction oscillations [20] (Fig. 2). To account for solid-state effects a muffin-tin potential is employed in the interstitial region [18], which, however, has effectively little influence on the scattering into large angles. For moderate electron energies ($\lesssim$ keV), the cross-section for backward angles exceeds the Rutherford cross-section (pure Coulomb case) by several orders of magnitude for all charge states (Fig. 2).

NUMERICAL MODEL

Due to the large size of the clusters ($N > 10000$ atoms), a full ab-initio simulation seems still impractical. We therefore opt for a simplified theoretical description of the dynamics of the laser-cluster interaction [21]. It consists of the separation of electronic and ionic dynamics valid for short pulse durations. Furthermore, the many-electron system is treated as an open effective mean-field one electron system, in which many-particle effects are included via stochastic processes. Finally, we employ a classical-trajectory Monte Carlo (CTMC) test particle discretization with a typical representation fraction of $\alpha \simeq 0.1$, thereby drastically reducing the number of test particles to be followed.

Ionic Dynamics

Because of their large inertia, the dynamics of the cluster ions proceeds on the time scale of the laser pulse. We therefore describe the ions of mass M and mean charge state $\langle q(t) \rangle$ as a uniform spherical positive background charge with a time dependent radius $R(t)$. As some quasi-free electrons leave the cluster (outer ionization), the cluster of radius $R(t)$ is charged and acquires the positive cluster charge $Q(R,t)$. The monopole field resulting from this charging drives the cluster expansion. For a realistic estimate of the time evolution of the cluster radius $R(t)$, we take into account that the ionic charge $\langle q(t) \rangle$ is screened by the surrounding quasi-free electrons. We determine $R(t)$ from the equation of motion of an ion with the screened charge state $\langle q_{scr}(t) \rangle$, situated at the

146

surface of the cluster ($r = R(t)$):

$$M\frac{\mathrm{d}^2 R(t)}{\mathrm{d}t^2} = \frac{\langle q_{scr}(t)\rangle Q(R,t)}{R(t)^2}.$$ (1)

To estimate the screened charge state $\langle q_{scr}(t)\rangle$ of the ions on the cluster surface, we reduce $\langle q(t)\rangle$ corresponding to the number of quasi-free electrons per ion:

$$\langle q_{scr}(t)\rangle = \langle q(t)\rangle - n_e(R,t)/\rho(t),$$ (2)

where $n_e(R,t)$ is the local electronic number density on the surface of the cluster and $\rho(t)$ is the ionic number density.

Electronic Dynamics

To describe the electronic dynamics, we employ a test-particle discretization, i.e. we solve the equations of motion only for a representative fraction of the ensemble of particles. The representation fraction α is limited by computational capabilities. The test-particle ensemble of the $i = 1, \ldots, N_{\text{test}}(t) = \alpha N_e(t)$ electrons obeys the following Langevin equation:

$$\ddot{\mathbf{r}}_i = -\mathbf{F}_{\text{L}}(t) - \mathbf{F}_{\text{mean}}(\mathbf{r}_i, t) + \mathbf{F}_{\text{stoc}}(\mathbf{r}_i, \dot{\mathbf{r}}_i, t).$$ (3)

This approach can also be seen as a generalization of classical transport theory (CTT) [22] for a dynamical system open to both particle number variation $N_{test}(t)$ and energy exchange with the many-particle reservoir (ions and electrons), as well as with the laser field. As the cluster is much smaller than the wavelength of the laser pulse ($R(0) \ll \lambda$), the laser can be described as a uniform time-dependent external electric field linearly polarized in the z-direction:

$$\mathbf{F}_{\text{L}}(t) = F_0 \hat{\mathbf{z}} \sin(\omega t) \sin^2\left(\frac{\pi t}{2\tau}\right).$$ (4)

Electron-electron and electron-ion interactions are taken into account by the time-dependent mean field $\mathbf{F}_{\text{mean}}(\mathbf{r}, t)$, which depends on the positions of all test particles. The mean field is approximated by the monopole and dipole contributions of its multipole expansion. The monopole term is given by:

$$\mathbf{F}^0_{\text{mean}}(\mathbf{r}, t) = Q(r,t)\frac{\mathbf{r}}{r^3},$$ (5)

where $Q(r,t)$ stands for the instantaneous charge contained in the sphere of radius r resulting from the possible imbalance between the number of electrons and the ionic background charge. $Q(r,t)$ is evaluated by discretizing a sphere of radius $2R(t)$ including the cluster of radius $R(t)$ and the surrounding simulation volume up to $2R(t)$ in 20 concentric spherical shells of equal distance. The dipole contribution to the mean field arises from the laser induced shift of the electron sphere with respect to the ion sphere.

147

The dipole field inside the cluster ($r < R(t)$ can therefore be estimated as the field of a polarized sphere:

$$\mathbf{F}_{\text{mean}}^{(1)}(\mathbf{r},t) = -\frac{\mathbf{p}(t)}{R(t)^3}, \tag{6}$$

where the dipole moment $\mathbf{p}(t)$ is determined by the position of all the test particles within the sphere:

$$\mathbf{p}(t) \simeq -\frac{1}{\alpha} \sum_{r_i < R(t)} \mathbf{r}_i. \tag{7}$$

Outside the cluster, $\mathbf{F}_{\text{mean}}^{(1)}$ is assumed to be simply the field of a central dipole $\mathbf{p}(t)$. Momentum changes due to collision processes are taken into account by the stochastic forces $\mathbf{F}_{\text{stoc}}(\mathbf{r}_i, \dot{\mathbf{r}}_i, t)$. For example, elastic electron-ion scattering is controlled by the probability of each electron to scatter elastically during the time step Δt:

$$P_e = \sigma_e(\langle q \rangle, E)\rho(t)\dot{r}_i \Delta t, \tag{8}$$

which is determined by the energy (E) and charge state (q) dependent total elastic scattering cross-section σ_e, the ionic density $\rho(t)$ and the velocity of the electron. If the electron scatters, the scattering angle θ is determined randomly according to the differential cross-section $d\sigma_e/d\theta$. The change in momentum associated with this scattering of angle θ is one contribution to the stochastic force $\mathbf{F}_{\text{stoc}}(\mathbf{r}_i, \dot{\mathbf{r}}_i, t)$. Further contributions are, for example, electron-impact ionization of the outer-shell electrons and K-shell ionization. These events are not only marked by a change in momentum, but also by a change in the number of test particles $N_{\text{test}}(t)$ i.e. of ionized electrons, as well as in the mean number of bound electrons in the respective shells $N_M(t)$, $N_L(t)$ and $N_K(t)$.

Ionization mechanisms

The binding energies E_L and E_M of electrons in the L-shell and M-shell of an argon ion are estimated by the following empirical formulas (in atomic units):

$$E_M(N_L, N_M) = 12.47 - 0.89N_L - 0.62N_M \tag{9}$$
$$E_L(N_L, N_M) = 33.3 - 2.25N_L - 0.79N_M \tag{10}$$

where N_L and N_M stand for the number of electrons in the L-shell and M-shell respectively. These formulas are approximations for argon ions with 2 electrons in the K-shell and are based on spectroscopic data of binding energies [23, 24]. The binding energy E_K of the K-shell electrons is assumed to be only dependent on the charge state q of the ion [23]:

$$E_K(q) = \begin{cases} 117.8 + 1.3q & \text{for } q \leq 8 \\ 108.6 + 2.4q & \text{for } 8 < q \leq 16 \end{cases} \tag{11}$$

Field ionization takes place when the combined field of the laser field and dipole field inside the cluster exceeds the threshold field for over-barrier ionization

$$F_{OBI} = \frac{W(N_L, N_M)^2}{4(q+1)},$$ (12)

which is determined by the ionization potential W of an argon ion with charge q, two electrons in the K-shell, N_L electrons in the L-shell and N_M electrons in the M-shell. In the event of field ionization, all ions are ionized at once, i.e. αN new test particles with zero velocity are distributed uniformly in the cluster. Additional quasi-free electrons can be created by electron-impact ionization which is treated as stochastic process. The cross-section for impact ionization by an electron with kinetic energy E is evaluated from the Lotz formula [25] (the step functions $\Theta(x)$ assure that the corresponding energy thresholds are exceeded):

$$\sigma_{ei} = 2.17 \left(N_M \frac{\ln(E/E_M)}{EE_M} \Theta(E - E_M) + N_L \frac{\ln(E/E_L)}{EE_L} \Theta(E - E_L) \right).$$ (13)

Furthermore, the contribution of two-step ionization (electron-impact excitation from L-shell to M-shell followed by an impact ionization) is roughly estimated by assuming the cross-section for impact excitation to be:

$$\sigma_{L \leftarrow M} = 2.17 N_L \frac{\ln(E/(E_L - E_M))}{E(E_L - E_M)} \Theta(E - (E_L - E_M)).$$ (14)

The Auger decay of the L-shell vacancies is incorporated in the simulation by creating new test particles with the rate $1/\tau_A \approx 5.7 \cdot 10^{-3}$a.u. [26]. It has been proposed [27] that, in the case of a cluster ion, ionization may be enhanced by the proximity of the surrounding highly charged ions. The additional Coulomb potential of its neighbor lowers the Coulomb barrier of the ion. We estimate the lowering of the ionization threshold due to the superposition of the two Coulomb potentials to be:

$$\Delta W = -2 \frac{\langle q_{scr}(t) \rangle}{d_{ii}/2},$$ (15)

where $d_{ii} = \rho^{1/3}$ is the mean ion-ion distance.

SIMULATION RESULTS

Figure 3 presents typical simulation results obtained when solving Eq. (3) for a cluster with $N = 2.8 \cdot 10^5$ argon atoms irradiated by an infrared ($\lambda = 800$nm) laser pulse of pulse duration $t = 60$fs with a peak intensity of $I = 1.9 \cdot 10^{16}$Wcm^{-2}. The cluster has initially a solid density of $\rho(t = 0) = 2.66 \cdot 10^{22}$cm^{-3} and the initial cluster radius is $R(0) = 258$a.u.. The first $N_{test}(t_1)$ test particles are released with zero velocity and random positions in the cluster as soon as the laser field reaches the threshold for the over barrier ionization of the neutral argon atoms ($F_{OBI}(t_1) \simeq 0.08$a.u.). These particles provide the

149

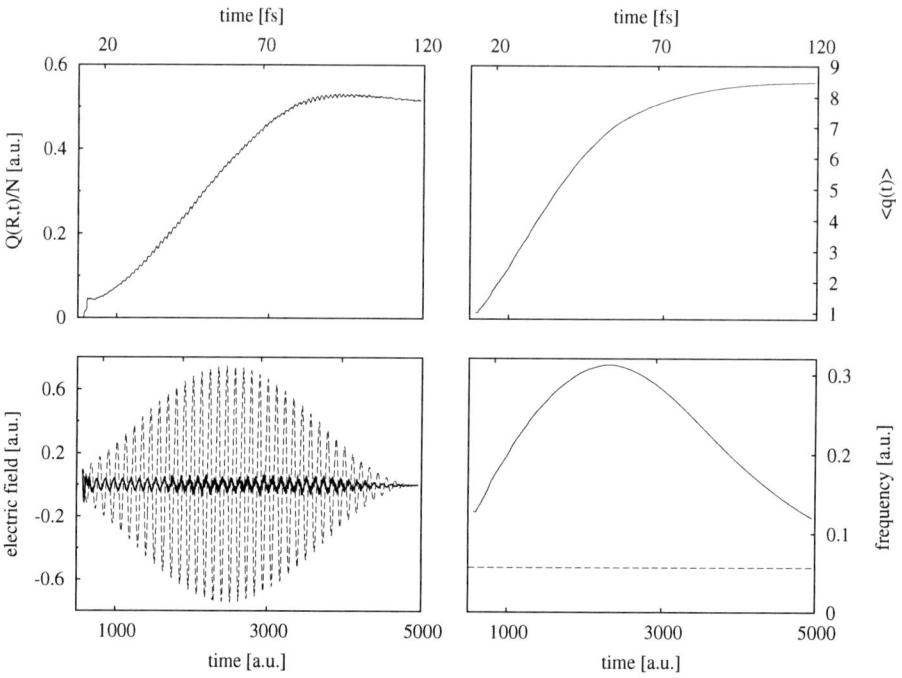

FIGURE 3. Top left: total cluster charge per atom. Top right: mean ionic charge state. Bottom left: effective field inside the cluster (solid) and laser field (dashed). Bottom right: plasma frequency (solid) and laser frequency (dashed). All quantities are shown as a function of time.

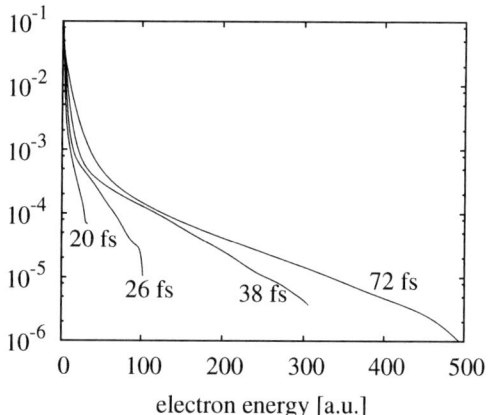

FIGURE 4. Kinetic energy distribution of the electrons at different times during the laser pulse

initial conditions for the propagation of Eq. (3). In the present case the representation fraction is chosen to be $\alpha = 0.05$. The number of test particles increases rapidly due to the efficient ionization of the cluster. The secondary ionization mechanisms discussed in the previous section (excitation, Auger decay and ion proximity) form only a minor contribution to the ionization process, which is dominated by electron impact ionization of the outermost atomic shell. At the conclusion of the laser pulse, a mean ionic charge state of Ar^{8+} is reached. Even though this is significantly higher than the maximum charge state of Ar^{6+} found in Ref. [13], the efficient depletion of the L-shell associated with the mean charge state of Ar^{13+} found experimentally [6] can not yet be explained. The total cluster charge resulting from electrons leaving the cluster results in an increase of the cluster radius by a factor 2 during the laser pulse. Furthermore, the simulations show that the charge is concentrated on the cluster surface, the interior of the cluster being well shielded by the quasi-free electrons. Inside the cluster the effective field acting on the electrons is therefore the sum of the laser field and the dipole field:

$$\mathbf{F}_{\text{eff}}(t) = \mathbf{F}_{\text{L}}(t) + \mathbf{F}_{\text{mean}}^{(1)}(t) \tag{16}$$

Due to the polarization of the cluster, the effective field inside the cluster is strongly reduced compared to the laser field. This can also be understood in terms of the plasma frequency ω_P. After the first ionization event ($t = t_1$), the plasma frequency is $\omega_P^2(t_1) = N_{\text{test}}(t_1)/(\alpha R(t_1)^3) \approx 5\omega^2$. Due to the efficient ionization, ω_P first increases rapidly, before diminishing in the second half of the pulse as the cluster begins to expand. Following Ref. [8], the mean field can be estimated as

$$\mathbf{F}_{\text{eff}}(t) \approx \Re \left\{ \int_{\omega - \Delta\omega}^{\omega + \Delta\omega} \tilde{\mathbf{F}}_{\text{L}}(\omega') \left(1 - \frac{\omega_P^2}{\omega_P^2 - \omega'^2 - i\omega'\gamma} \right) e^{i\omega't} d\omega' \right\}, \tag{17}$$

where $\Delta\omega$ is the Fourier width associated with the temporal profile of the pulse (4) and γ stands for the damping constant due to scattering events. As the plasma frequency exceeds the laser frequency ω during the duration of the laser pulse, the effective field is smaller than the laser field (Fig. 3). It is worth noting that no plasma resonance $\omega_P = \omega$ is encountered for this range of parameters. Figure 4 displays the time evolution of the electronic kinetic energy distribution. During the entire pulse the average kinetic energy of the electrons is well below the ponderomotive energy associated with the laser pulse of $U_P = 41$ a.u.. The high energy tail of the energy distribution reaches, however, quickly energies well above the K-shell binding energy of $E_K \approx 114$ a.u.. One notes, moreover, that the time-scale of the laser-cluster interaction is too short for the electrons to reach a thermal equilibrium, and that the energy distribution can therefore not be described by a Maxwell-Boltzmann distribution.

The x-ray yield per cluster is determined by the number of K-shell vacancies created, corrected by the mean fluorescence yield $\eta = 0.12$ [26]. Furthermore, the spatial Gaussian intensity profile of the laser beam [6] has to be taken into account to obtain the absolute x-ray yield. The simulation results are compared to the experimental results in Fig. 5. In order to estimate the efficiency of heating by elastic electron-ion scattering, the simulation carried out with the parametrized Hartree-Fock potential describing the ionic potential is compared to an otherwise identical simulation where a pure Coulomb

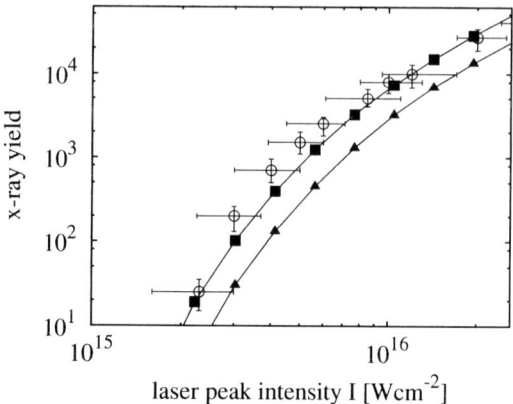

FIGURE 5. Absolute x-ray yield as a function of the laser peak intensity. Experimental results [6] (○), simulation results obtained with a Coulomb potential for the description of the ions (▲) and obtained with the realistic ionic potential (see text) (■). (Lines to guide the eye).)

potential describes the elastic electron-ion interaction. In both cases a small fraction of electrons gains sufficient energy to produce K-shell vacancies for $I > 2 \cdot 10^{15} \mathrm{Wcm}^{-2}$. This can be associated with the heating by the monopole field of the charged-up cluster. However, including the realistic scattering potential and thereby allowing for elastic backscattering events, drastically increases the x-ray yield by a factor 3 compared to the pure Coulomb case. This effect is the most pronounced at low intensities, i.e. close to the threshold. As the laser intensity increases, the cross-section for backscattering decreases due to the higher energy of the electrons.

In summary, we have developed a simple mean-field approach to the complex dynamics of laser-cluster interaction. We find x-ray yields in surprisingly good quantitative agreement with the experimental results. We have shown that the microscopic atomic dynamics play an important role and have identified elastic electron-ion large-angle scattering as a heating mechanism efficient for moderate laser intensities ($I = 10^{15} - 3 \cdot 10^{16} \mathrm{Wcm}^{-2}$) and large clusters ($N > 10^4$ atoms). Work has still to be done to account for the high ionic charge states observed in the experiments (up to Ar^{16+}), so far not explained by any theoretical approach. Furthermore, an improvement of the description of the ionic dynamics is planned in order to study the effect of an inhomogeneous cluster expansion and explore the dependency of the x-ray yield on cluster size, pulse length, and laser wavelength.

ACKNOWLEDGMENTS

This work is supported by FWF SFB-16 (Austria). Illuminating discussions with Thomas Brabec are gratefully acknowledged.

REFERENCES

1. V.P. Krainov and M.B. Smirnov, *Phys. Rep.* **370**, 237 (2002).
2. M. Lezius, S. Dobosz, D. Normand and M. Schmidt, *J. Phys. B* **30** L251 (1997).
3. V. Kumarappan, M. Krishnamurthy, and D. Mathur, *Phys. Rev. A* **67**, 043204 (2003)
4. S. Dobosz, M. Lezius, M. Schmidt, P. Meynadier, M. Perdrix, D. Normand, J.-P. Rozet, and D. Vernhet, *Phys. Rev. A* **56**, R2526 (1997)
5. E. Lamour, C. Prigent, J.P. Rozet and D. Vernhet, *Nucl. Instr. and Meth. Phys. Res. B* **235**, 408 (2005).
6. C. Prigent, Ph.D. thesis, Université Pierre et Marie Curie, Paris VI, 2004.
7. U. Saalmann, Ch. Siedschlag, and J.M. Rost, *J. Phys. B* **39** R39 (2006).
8. Ditmire T., Donnelly T., Rubenchik A.M. et al., *Phys. Rev. A* **53**, 3379-3402 (1996).
9. F. Megi, M. Belkacem, M.A. Bouchene, E. Suraud and G. Zwicknagel, *J. Phys. B* **36**, 273 (2003).
10. Rose Petruck C., Schafer K.J., Wilson K.R. et al., *Phys. Rev. A* **55** (1997) 1182-1190.
11. I. Last and J. Jortner, *Phys. Rev. A* **60**, 2215 (1999).
12. U. Saalmann and J.M. Rost, *Phys. Rev. Lett.* **91**, 223401 (2003).
13. C. Jungreuthmayer, M. Geissler, J. Zanghellini and T. Brabec, *Phys. Rev. Lett.* **92**, 133401 (2004).
14. E. Fermi, *Phys. Rev.* **75**, 1169 (1949).
15. J. Burgdörfer, J. Wang and R.H. Ritchie, *Phys. Scr.* **44**, 391 (1991).
16. F. Pisani et al., *Phys. Rev. Lett.* **87**, 187403 (2001).
17. P.P. Szydlik and A.E.S. Green, *Phys. Rev. A* **9**, 1885 (1974).
18. F. Salvat and R. Mayol, *Comp. Phys. Comm.* **74**, 358 (1993).
19. N. Mott and H. Massey, *Theory of Atomic Collisions*, Oxford University Press, New York, 1965.
20. J. Burgdörfer, C.O. Reinhold, J. Sternberg and J. Wang, *Phys. Rev. A* **51**, 1248 (1995).
21. C. Deiss et al., *Phys. Rev. Lett.* **96**, 013203 (2006).
22. J. Burgdörfer and J. Gibbons, *Phys. Rev. A* **42**, 1206 (1990).
23. W. Lotz, *J. Opt. Soc. Am.* **58**, 915 (1968).
24. Y. Ralchenko et al., NIST Atomic Spectra Database (version 3.0.3), http://physics.nist.gov/asd3 (2005).
25. W. Lotz, *Z. Phys.* **216**, 241 (1968).
26. C.P. Bhalla, *Phys. Rev. A* **8**, 2877 (1973).
27. Ch. Siedschlag and J.M. Rost, *Phys. Rev. A* **67**, 013404 (2003).

Shiny quartz: luminescence in ion-implanted and epitaxially recrystallizing α-quartz

Klaus-Peter Lieb[1], Pratap K. Sahoo[1,2] and Juhani Keinonen[3]

[1] *II. Physikalisches Institut, Universität Göttingen,*
Friedrich-Hund-Platz 1, D-37077 Göttingen, Germany
[2] *Present address: Instituut voor Kern- en Stralingsfysica,*
K. U. Leuven, B-3001 Leuven, Belgium
[3] *Accelerator Laboratory, University of Helsinki,*
P. O. Box 43, FI-00014 Helsinki, Finland

Abstract. Ion implantation is a promising route to doping quartz with selected luminescent impurity atoms and to grow photoactive nanoparticles of them. The present work review some aspects of cathodoluminescence (CL) and epitaxy during or after Na, Rb, Cs, Ge, Ba, and Rb/Ge double implantation in crystalline α-quartz, under the conditions of dynamic, chemical or laser epitaxy. Different classes of CL bands will be discussed: intrinsic bands, which are associated with specific defects in the silica matrix, and very intense blue or violet ion-specific bands, which are related to the presence of particular implanted ions and the degree of epitaxial regrowth of the matrix. The case of double Rb/Ge implantation not only appears to be the most promising one with respect to combine a high CL output <u>and</u> full chemical epitaxy, but it also provides important information about the role of alkali ions and/or substitutional Ge nanoclusters for the luminescence in quartz and silica.

Keywords: Quartz; Ion Implantation, Luminescence, Epitaxy.

PACS: 71.55.-I; 78.60.Hk; 78.70.-g; 81.10.-h.

1. INTRODUCTION

Silicon dioxide in its amorphous and crystalline forms and doped with group-IV or rare-earth atoms or nanoclusters is an important material for optoelectronics, photonic and microelectronic device fabrication [1-4]. Doping α-quartz with ions which substitute Si in the crystalline SiO_2 matrix may be a challenging step in tailoring novel optical properties suitable for quantum semiconductor devices [5]. Quartz amorphizes easily under the impact of low fluences of heavy ions [6]. For that reason, defect production and annihilation by energetic ion bombardment and thermal annealing is an essential technological issue for understanding the defect dynamics responsible for the optical properties of SiO_2. Attempts at dynamic, chemical or laser-induced solid phase epitaxial growth of such amorphized layers have been made in order to remove any radiation damage induced during implantation <u>and</u> to achieve high quantum efficiency of the visible cathodoluminescence (CL) light output [7-24].

In this work, we discuss CL spectroscopy in correlation with full or partial epitaxy of α-quartz achieved during or after Ge, Ba, alkali (Na, Rb, Cs), or double Rb/Ge-ion implantation [14-21]. When combining the microstructural results on dynamic or

CP876, *The Physics of Ionized Gases: 23rd Summer School and International Symposium,*
edited by L. Hadžievski, B. P. Marinković, and N. S. Simonović
© 2006 American Institute of Physics 978-0-7354-0377-2/06/$23.00

chemical epitaxy with the corresponding luminescence data, one may draw conclusions concerning the origin of the CL bands and their relation to the surrounding matrix (crystalline, amorphous, nanoparticles) and the type of photoactive defects. For instance, it is an interesting question whether CL emission is related to the excitation of single atoms, molecular-type structures or nanoclusters. We try to distinguish the ion-specific CL-emission bands, which are associated with the particular ion species implanted, from those intrinsic bands, which are related to defect structures of the matrix due to any ion and/or electron bombardment. In all the cases considered, the ion-specific defect centres were found to emit blue-violet CL bands in the 2.9-3.7 eV range. Furthermore, we discuss recent results of high-resolution transmission electron microscopy (TEM) after Ge and Sn implantation [25,26] and Rutherford back-scattering channeling spectroscopy (RBS-C) after Ge/Rb implantation [21], which provided first insight into the structure of the luminescent Ge nanoparticles responsible for the blue/violet bands.

2. EPITAXY OF THE MATRIX AND STRUCTURES OF PHOTOACTIVE DEFECTS

While Ge implantation at elevated sample temperatures leads only to incomplete dynamic epitaxy [16,17], nearly full dynamic epitaxy was found during Ba implantation [14,15]. Irradiation of α-quartz with alkali ions followed by annealing in air or oxygen, a process called chemical epitaxy, is known to produce full planar recrystallization of the amorphous region [9-13]. After alkali ion implantation and excimer laser annealing, only partial epitaxy has been found so far [22]. Table 1 summarizes the processing parameters in the various systems studied, while details of the experimental methods are given in the respective publications [14-24].

TABLE 1: Ion implantation and processing parameters.

Ions	Energy (keV)	$\Phi\,(10^{15}/cm^2)$	T_{impl} (K)	Processing after implantation	Ref.
Ge	120	≤ 10	300-1220	None	[16,17]
Ba	175	1	300-1170	None	[14,15]
Cs	250	28	80	Excimer laser, 308 nm, 55 ns	[22]
Rb	175	25	80	Excimer laser, 308 nm, 55 ns	[22]
Rb	175	25	80	Anneal. in air/$^{18}O_2$, 673-1223 K	[12,18,23]
Na	50	0.1 - 100	80	Anneal. in $^{18}O_2$, 673-1173 K	[10,13,19]
Ge/Rb	Rb: 175	20	80	Anneal. in air/$^{18}O_2$, 1173 K	[20,21]
	Ge: 120	0.1-10	300		

Information on the microstructure of implanted ions in the quartz or silica matrix and the re-crystallization process itself has been gained by RBS-C and high-resolution TEM. It has been observed that Ge-ion implantation does not provide full dynamic epitaxy of the quartz matrix, not even at an implantation temperature as high as 1220 K [16,17]. The situation is much more favourable for dynamic epitaxy during Ba-ion implantation [14,15]. In both cases the RBS spectra did not reveal any changes in the Ge or Ba depth profiles or any loss of implanted ions from the matrix. This finding ruled out long-range diffusion, but not local clustering to form nanoparticles.

155

It is interesting to compare these results with those obtained by Lopes *et al.* [25,26] from high-resolution TEM data after Ge and Sn implantation of 180-nm thick SiO_2 layers on single-crystal Si substrates at room temperature and subsequent thermal annealing in nitrogen. In the case of Ge, the authors observed a coarsening of single-crystal Ge nanoparticles: at 1173 K the mean diameter had increased from 2.2 nm to 5.6 nm. About 20% of the implanted Ge remained distributed in the matrix around the single-crystal nanoparticles. A similar coarsening was also observed in the Sn-implanted layers during annealing at 973 – 1173 K. Annealing in vacuum resulted in a more homogeneous size distribution, with a mean diameter of 2.5 nm. The larger size distribution was explained as due to the oxidation of Sn during the annealing in nitrogen. The photo-activity was related to the structure of the interface between the nanoclusters and the SiO_2 matrix and to the location of single implanted atoms at the interface.

Still another mechanism seems to pertain to <u>chemical</u> epitaxy after alkali-ion implantation (Na [10,13], Rb [12], Cs [9,10]). In these cases, the ions were implanted at room or liquid-nitrogen temperature and the samples were annealed in air or oxygen up to about 1200 K. This treatment led to full epitaxy at about 1120 K, accompanied by out-diffusion of the alkali ions and oxygen exchange between the matrix and the annealing gas. A model has been applied [10], in which the alkali ions are bound to dangling oxygen ions. At about 1000 K, this bond breaks, the alkali ions diffuse to the surface and complete epitaxy of the matrix occurs.

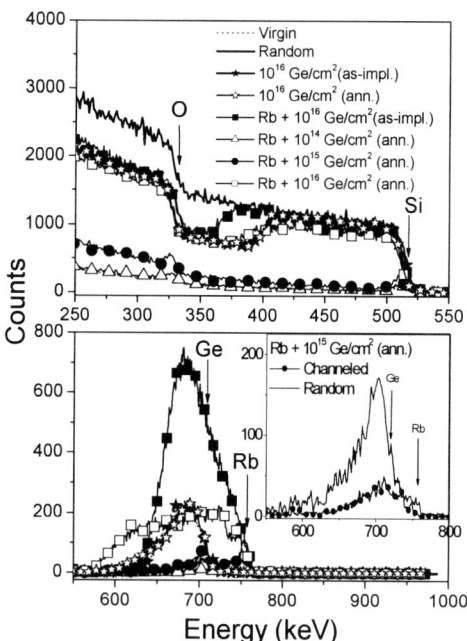

FIG. 1. RBS-C spectra measured after Ge or double Ge//Rb implantation in α-quartz and annealing in air at 1170 K. The upper part illustrates the degree of epitaxy of the matrix, while the lower part shows the signals of Rb and Ge in random and channeling conditions (from [21]).

Technologically, the case of chemical epitaxy after double Rb/Ge-ion implantation appears to be the most interesting one, because complete chemical epitaxy and a high CL output were achieved after thermal annealing at 1120 K in air [20,21]. The corresponding RBS-C spectra displayed in Fig. 1 clearly proved epitaxy for Ge-fluences up to $1 \times 10^{15}/cm^2$, but not at the highest fluence of 1×10^{16} Ge-ions/cm^2. The overlapping Ge and Rb signals in the RBS spectra (see Fig. 1, bottom) exhibit very interesting features: clearly, Rb diffused out during the annealing process and the Ge signal decreased, due to the fact that up to 80% of Ge substituted Si atoms in the re-grown quartz matrix and thus escaped detection under channelling conditions (see insert of Fig. 1, bottom). This finding confirms the TEM data mentioned before [25]: at low fluences, Ge probably forms substitutional nanoclusters (crystalline Ge quantum dots), in addition to non-clustered solute atoms in the re-crystallized matrix and non-crystalline nano-particles for larger Ge-ion fluences (about $10^{16}/cm^2$).

3. CATHODOLUMINESCENCE

3.1 General shape of the CL spectra

The CL measurements were performed using a 5-keV electron beam, whose current (power density) was maintained at 2 μA (1 W/cm^2). The sample temperature was varied between 12 K and room temperature (RT), using a closed-cycle helium refrigerator. The luminescence light was detected by means of a Hamamatsu R928 photomultiplier after focusing it into a Czerny-Turner spectrograph (Jobin Yvon 1000M). The CL spectra were collected in the wavelength range from 200 to 900 nm at a speed of 1 nm/s, with a grating of 1200 lines/mm. They were deconvoluted with up to six Gaussian-shaped sub-bands and the CL emission intensities of all the sub-bands were corrected for the instrument response function, which was determined using a standard halogen calibration lamp. As some of the bands are rather broad (up to 0.45 eV FWHM) compared to peak distances, it was necessary to iterate the fitting procedure in order to fix as many parameters as possible. For details, see [15,17]. The energies of all the bands fitted in all the systems studied by our group are listed in Table 2.

3.2 Intrinsic CL bands

Intrinsic CL bands in SiO$_2$ are those arising from photoactive defects in the quartz or silica matrix after any ion or electron irradiation and/or thermal treatment. These defects are therefore not related to the specific ions implanted. In general, their intensities vary only weakly with the annealing conditions and ion fluences. Fig. 2 illustrates the CL spectra taken at RT for samples of pure α-quartz, fused silica, amorphous SiO$_2$ grown by oxidation of Si, and α-quartz amorphized by 800-keV Ba-ion implantation [23]. Concerning the latter spectrum, we emphasize that the Ba-ion range was much larger than the CL depth sensitivity of some 300 nm. In all these samples, the ion-specific blue/violet bands in the 2.9 - 3.7 eV range were absent. The spectra from amorphous silica (i. e. fused silica, a-SiO$_2$ on Si, quartz amorphized by Ba-ions) exhibited mainly intrinsic bands at 2.79 and 4.30 eV, while crystalline quartz showed very weak bands at 2.0 and 2.4 eV. It must be concluded from this comparison that two classes of intrinsic defect centres exist in these matrices, those correlated with the electron bombardment of quartz during the CL measurements themselves (2.0 and 2.4 eV), and those in the amorphous SiO_2 matrix (2.79 and 4.3 eV).

TABLE 2. Summary of the CL emission bands observed at RT after Ge- or Ba-ion implantation at a sample temperature of 300 – 1220 K, and after Na, Rb or double Rb/Ge implantation and annealing in air up to 1170 K.

CL band (eV)	Wavelength (nm)	Identification	Ref.
Virgin α-quartz, Ba: 2.00(2)	620, red	Non-bridging oxygen-hole centre (NBOH)	[14,15]
Ge, Na, Rb, Rb/Ge: 2.40(2)	511, green	Oxygen-vacancy interstitial pairs	[17-21,23]
Ba: 2.42(2)	510, green		[14,15]
Ge, Rb/Ge: 2.72(1)	455, blue	Oxygen-deficiency centre (ODC)	[15-20]
Ba, Na, Rb: 2.79(4)	442, blue		
Ge, Rb/Ge: 2.95(2)	420, violet	Ge- and/or Rb-related defects	[16,17,20,21]
Na, Rb, Cs, Ge, Rb/Ge: 3.25(2)	382, violet	Charge compensated alkali-ion Centre; Neutral oxygen vacancy	[17-21,22,23]
Ba: 3.40(5)	363, violet	Ba-related defect centre	[14,15]
Rb/Ge: 3.53(3)	350, violet	Ge- or Rb-related defect centre	[20,21]
Na, Rb, Cs: 3.65(4)	338, violet	Rb-related defect centre	[18,19,23]
Ba: 4.2(2)	294, UV	E´-centre	[14,15]
Na, Rb, Ge, Rb/Ge: 4.30(5)	288, UV		

On the basis of mainly electron and laser irradiations, various defect configurations and luminescence mechanisms have been proposed for the intrinsic bands, e.g. Refs. [27-35], and will be discussed below.

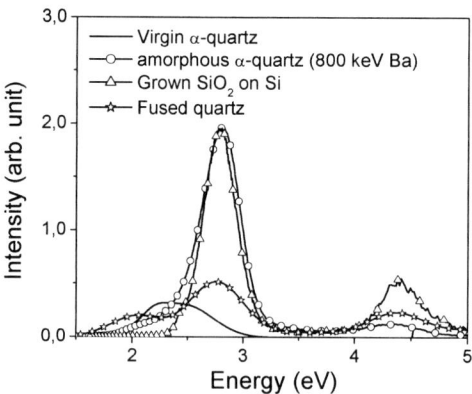

FIG. 2. CL spectra taken at room temperature for α-quartz and various forms of silica (fused silica, oxidized Si, and α-quartz irradiated with 800-keV Ba ions); from [23].

3.3 Ion-specific CL bands

Fig. 3 illustrates the decomposition of the CL spectra obtained after Ge or double Ge/Rb-ion implantation and annealing in vacuum, air or $^{18}O_2$-gas. One can distinguish between two classes of sub-bands: the mostly weakly populated intrinsic bands at 2.40, 2.79 and 4.30 eV, and the strong blue/violet ion-specific bands in the range 2.9 – 3.7 eV, and possibly also at 4.30 eV. In the particular case of double Rb/Ge implantation and chemical epitaxy, the blue/violet bands are located at 2.95, 3.25 and 3.53 eV. As mentioned before, the various blue/violet bands in the 3.0 – 3.7 eV range clearly dominate the CL spectra after irradiation/annealing around 1200 K (see Table 2). One notes that the energies of these CL bands differ for some ions, but agree for others (Na, Rb, Cs).

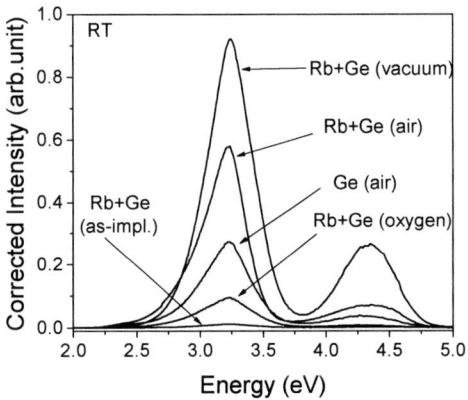

FIG. 3. CL-spectra taken at room temperature after double Rb/Ge-ion implantation into α- quartz and annealing for 1 h at 1170 K in air, vacuo or $^{18}O_2$-gas. From [20].

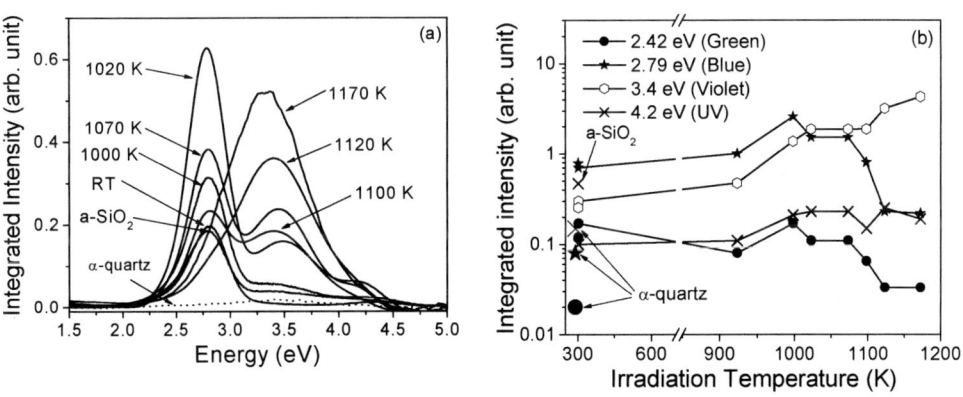

FIG. 4. CL spectra, and (b) deduced integrated intensities obtained at room temperature after implanting Ba ions in quartz at the temperatures indicated. From [15].

159

Fig. 4a illustrates the CL spectra taken at room temperature, while Fig. 4b shows the integrated intensities of the bands in the case of dynamic epitaxy of Ba-irradiated quartz [15]. It is interesting to note that the intrinsic bands at 2.42 and 2.79 eV bands decrease in intensity during the epitaxy, which occurs mainly at 900-1100 K, because the thickness of the amorphous layer decreases. On the other hand, the intensity of the ion-specific band at 3.4 eV steadily increases with the sample temperature.

In the following, we shall discuss to what extent these bands may be associated with the particular ion species used. Note that the various conditions of implantation and annealing provide different local environments for the implanted ions, i.e. formation of nanoparticles, out-diffusion of alkali ions and solid-phase chemical epitaxy, crystalline and/or amorphous SiO_2 for Ba and/or Ge. For that reason, one expects correlations to exist between the CL intensity and the ion fluence or the fraction of ions retained in the samples. Other important quantities are the dependence of the CL intensities on the implantation, annealing and/or CL temperatures, and results of time-differential CL measurements.

We start the discussion with the 3.25-eV and 3.65-eV bands (see Table 2). No explanation has been given so far concerning the nature of the 3.65-eV source. Fig. 5 illustrates the correlations of the 3.65-eV intensity with the fractions of Rb-ions retained in the matrix during chemical epitaxy in air or $^{18}O_2$-gas [18]. Clearly the 3.65-eV intensity was reduced to the extent alkali ions diffused to (and evaporated at) the surface; the intensity of the 3.65-eV band decreased by two orders of magnitude. In the case of Cs-implantation and subsequent laser annealing, only about 50% of Cs ions left the sample and, correspondingly, the 3.65-eV intensity decreased only by one order of magnitude [22]. Similar results as for Rb have also been found for chemical epitaxy after Na implantation [19]. Fig. 6 shows the temperature dependence of the 3.25 and 3.65-eV intensities during chemical epitaxy after Rb and Na ion implantation. T_X and T_D mark the temperatures, where half the implanted alkali atoms have out-diffused and half the thickness of the amorphized zone has recrystallized

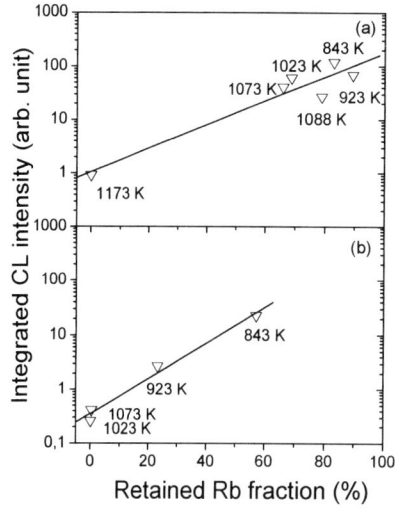

FIG. 5. Correlation between the intensity of the 3.65-eV CL band and the retained Rb fraction, after 175-keV Rb-ion implantation and chemical epitaxy in air (a) or ^{18}O-gas (b), at the temperatures indicated. From [18].

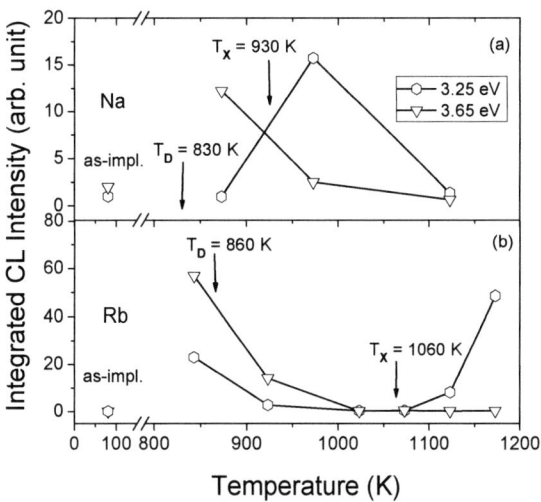

FIG. 6. Variation of the intensities of the 3.25-eV and 3.65-eV CL bands during chemical epitaxy after Na and Rb ion implantation in α-quartz and annealing in ^{18}O-gas (from [23]).

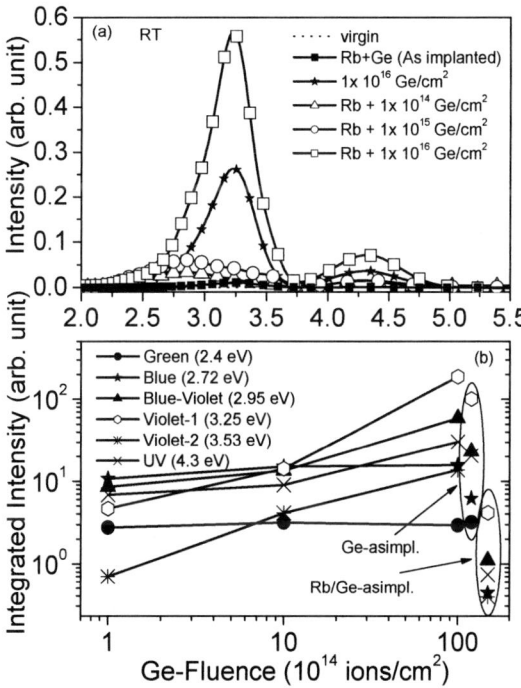

FIG. 7. CL spectra (a) and intensities (b) measured after double Rb/Ge implantation and annealing in air at 1170 K. Note the strong dependence of the CL light on the implanted Ge fluence (from [21]).

161

Fig. 7 displays the variation of the CL spectra (a) and intensities (b) for the case of Ge/Rb double implantation and chemical epitaxy in air at 1170 K. For comparison, the intensities immediately after Ge and/or Ge/Rb implantation are also shown. One notes a strong dependence of the 2.95, 3.25, 3.53 and 4.30 eV intensities on the Ge fluence, while the 2.40 and 2.79 eV intrinsic bands have rather constant intensities. This reflects the fact that the Rb or Ge implantation alone produces enough damage to amorphize the CL-sensitive volume. On the other hand, the ion-specific blue/violet bands of course depend on the Ge content [20,21].

4. DISCUSSION

4.1 Intrinsic CL bands

Ion implantation introduces damage in SiO_2 due to the deposited energy and the implanted atoms. A high-energy irradiation, governed by electronic energy loss, induces the formation of transient defects, the so-called self-trapped excitons (*STEs*), which leads to the formation of stable point defects such as *E'* centers by non-radiative recombination of the *STEs* (see Refs. [24,36] and references therein). Low-energy irradiation deposits energy via atomic collisions and removes atoms from their lattice sites (in α-quartz) or local SiO_2 structures. During implantation and/or annealing, the defect structures are modified and the implanted atoms can stay as isolated atoms, form molecules or nanoparticles. The luminescent defects may be localized in the irradiation-induced damage, in dispersed implanted atoms or clusters, or at the cluster-matrix interface.

In crystalline SiO_2 the bond lengths and the Si-O-Si bond angles connecting the silicon dioxide tetrahedra are well defined [37] (See Fig. 8). There is a broader range of slightly different atom sites in the random amorphous structure of fused SiO_2 [1,38] or ion-irradiated quartz. Although the basic tetrahedral configuration is maintained in amorphous SiO_2, the local lattice structure is less dense and more easily deformed and there is more local disorder due to slightly different bond lengths and bond angles.

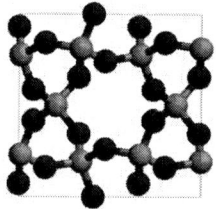

FIG. 8. Visualization of the structure of crystalline α-quartz

Particular defect configurations and luminescence mechanisms have been discussed in the literature [27-35]: interstitial O_2 molecules, *E'* centers (paramagnetic, positively charged oxygen vacancies, ≡Si•Si≡, or neutral dangling Si bonds, ≡Si•), non-bridging oxygen-hole centres (*NBOHC*; dangling oxygen bonds, O≡Si–O•), three-coordinated silicon with a trapped electron (≡Si•) or with other associated precursors, peroxy radicals (*POR*, ≡Si–O– O•) or oxygen-deficient centres (*ODC* = neutral oxygen vacancies, ≡Si–Si≡). The *STEs* and *E'* centers are by far the best-studied defects because they are easily accessible to electron spin resonance. Since diamagnetic defects can be potential precursors for photon- or ion-induced paramagnetic defects, the characterization of diamagnetic defects has been of importance. Fig. 9 illustrates the various defect structures classified according to dangling-bond-type defects and atomistic defects [29,30].

Dangling-bond-type defects

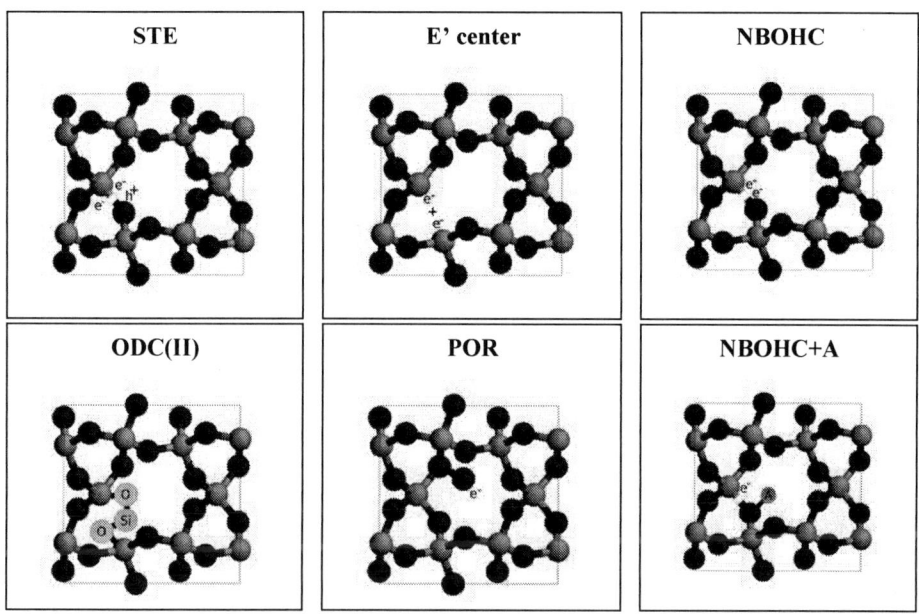

Atomistic defects

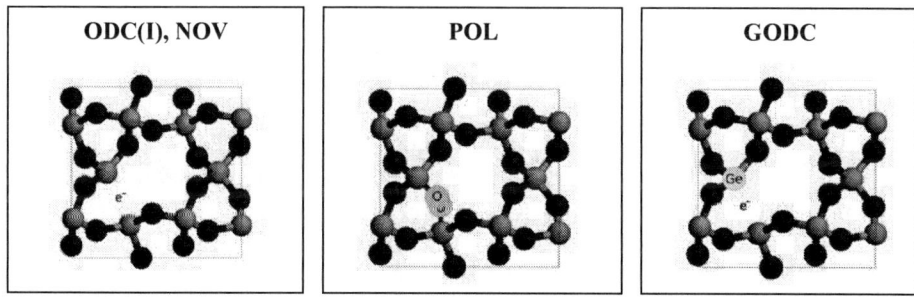

Fig. 9: Illustration of photoactive defects in α-quartz. Note that bond lengths or bond angles have not been changed when a defect has been added in the defect-free α-quartz. From [29,30]. The various abbreviations denote: STE = self-trapped exciton, NBOHC = Non-bridging oxygen hole centre, ODC = Oxygen deficiency centre, POR = peroxy radicals, GODC = Ge oxygen deficiency centre.

163

The 2.4-eV green band was connected either to oxygen vacancy interstitial pairs $(V_O;O_2)$ [34] or irradiation-induced self-trapped excitons within the a-SiO_2 outgrowth at the top of the sample containing a large amount of peroxy linkages (Si-O-O-Si) [29]. The blue and UV bands at 2.79 and 4.30 eV were related to *ODC*-type defects [29,32].

In our CL experiment with virgin α-quartz, the 2.40-eV CL band was observed as the only line and was present with a constant intensity in all the annealed samples (see Fig. 2). We connect it to vacancy-interstitial pairs $(V_O;O_2)$ [28] induced by the CL electron bombardment. The 2.79 and 4.30-eV bands observed in the cases of fused quartz, SiO_2 grown on Si, and quartz amotphized by Ba-ion irradiation (see Fig. 2), were present in all strongly damaged SiO_2 structures and did not anneal out in the temperature region used. The strong peak at 2.79 eV in the Ba-irradiated a-quartz and in SiO_2 grown on Si is in accordance with a very strong 2.75-eV peak dominating the PL spectrum in a heavy-ion-irradiation experiment of a-quartz [36] and with the statement that the radiative recombination of *STE*s can produce 2.7 eV PL radiation [39].

4.2 Ion-specific CL bands

We observed two groups of ion-implantation induced defects in quartz (see Table 2): the 3.65-eV line related to the alkali-ion implanted matrix and the 2.95, 3.25, and 3.53-eV triplet related to the Ge- or Rb-implanted and Rb/Ge-doubly-implanted matrix (see Figs. 3, 6a and [25,26]).

We suggest that the 3.65-eV sub-band is due to the defect center $O_3\equiv Si-O-A$, i.e. an atomic configuration where the alkali ion (A) is connected to a dangling oxygen bond of the non-bridging oxygen-hole center ($O\equiv Si-O\bullet$). This configuration is produced in the recrystallization process, where the effect of implanted alkali-ions and migrating oxygen (O^*) is described as

$$O_3\equiv Si-O-Si\equiv O_3 + 2A + O^*$$
$$\rightarrow O_3\equiv Si-O-A \quad A-O^*-Si\equiv O_3$$
$$\rightarrow O_3\equiv Si-O^{(*)}-Si\equiv O_3 + A_2O.$$

O^* denotes external ^{18}O and $O^{(*)}$ denotes ^{18}O or ^{16}O. For the chemical epitaxy of Li-, Na-, Rb-, and Cs-implanted α-quartz and details of the process, see Refs. [9-12,20]. Note that as the A–O bond is energetically weaker ($\approx$ 3 eV) than the Si–O bond (4.57 eV), annealing first disrupts the alkali-oxygen bond leading to alkali outdiffusion and then the silicon-oxygen bond leading to the *E'* center ($O\equiv Si\bullet$).

The 3.25-eV band has been tentatively associated either with the positively charged alkali- (or hydrogen) compensated $[A^{3+}/M^+]$ centre, where M^+ denotes a H^+ or alkali ion [29], or with impurities incorporated during growth into nanoparticles [40]. The strong 3.25-eV PL peak has been observed in the case of nanocluster experiments.

In the past, implantation of group-IV atoms has been employed to produce photo-luminescent nanoparticles. Investigations on nanoparticles of 1–10 nm diameter of Si, Ge, and Sn in SiO_2 structures have shown that the emission mechanism is a transition located at the nanoparticle-SiO_2 matrix interface [25,26]. Violet or blue PL observed in SiO_x layers has been reported to be due to radiative centers identified as neutral oxygen vacancies (*NOV*s) [41-44]. The optical transition of the electrons in the centers from the ground (singlet) to an excited (triplet) state is the origin of the PL [45]. In particular, *NOV* structures such as oxygen-deficient Ge centers (*GODC*s) ($\equiv$Ge–Si$\equiv$) have been obtained by implanting Ge ions into SiO_2 [25]. The *GODC*s and Sn-implanted nanoclusters ($\equiv$Sn–Si$\equiv$) in SiO_2 [26] emit the violet PL at 3.25 eV, while Si-ion implantation leads to the formation of Si *NOV*s ($\equiv$Si–Si$\equiv$) connected to the emission of the blue PL at 2.79 eV [39]. The 3.25-eV line from SiO_x nanoparticles has been associated with Si–Si vibrations [46].

We interpret the 3.25-eV line observed in the SPEG to be also connected to the formation of the NOV structure $\equiv$Si–Si$\equiv$ [47]. During chemical epitaxy, the matrix network is modified in such a way that *E'* centers (O$\equiv$Si$\bullet$) are produced and join to form the NOV structure.

Annealing the Na-implanted samples at temperatures clearly above T_X, led to the dissociation of the *NOV*s. For the Rb-implanted samples, such temperatures were not reached. In Ge-implanted α-quartz, a saturated increase in the 3.25-eV CL intensity was observed when only about 40 % of the full SPEG was achieved and the content of implanted Ge atoms stayed constant [14]. In doubly implanted Rb/Ge samples, again no loss of Ge, but out-diffusion of Rb and full SPEG led to a high CL intensity of the 3.25-eV line [17]. The PL intensity of the single-crystalline Ge and Sn nanoparticles, revealed in high-resolution TEM, was related to *NOV*s around the nanoparticles [25,26]. We conclude that the formation of the *NOV* structures between group-IV atoms indicated by the intensive 3.25-eV light indicates a strongly damaged SiO_2 matrix.

The molecule-like luminescence centers in $\equiv$Si–Si$\equiv$, $\equiv$Ge–Ge$\equiv$, $\equiv$Ge–Si$\equiv$, $\equiv$Sn–Sn$\equiv$, and $\equiv$Si–Sn$\equiv$ structures have a three-level energy system, in which blue-violet emission (≈ 3.2 eV) and UV emission (≈ 4.3 eV) are due to a triplet-to-singlet ($T_1 \rightarrow S_0$) and singlet-to-singlet ($S_1 \rightarrow S_0$) transition, respectively. If the 3.25-eV line is related to the triplet-to-singlet transition, it should have a long lifetime. Our time-differential CL data measured in quartz at RT after double Rb/Ge implantation and chemical epitaxy at 1170 K [21] gave lifetimes of $\tau = 5.7$ µs and 4.5 µs for the 3.25- and 2.95-eV bands, respectively (see Fig. 10).

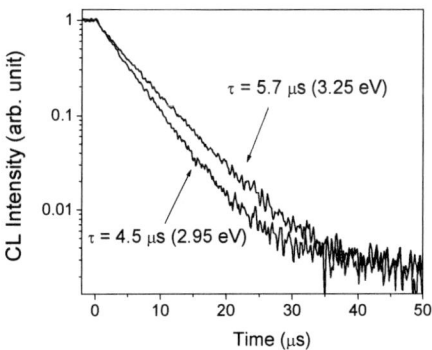

FIG. 10. Time dependence of the CL intensities of the 2.95- and 3.25-eV bands taken at room temperature and giving the values of the time constants τ indicated (from [21]).

The values are typical of Ge-related defects, which remain after complete out-diffusion of Rb. According to Skuja *et al.* [29], the Ge-related *ODC* centers have luminescence lifetimes of 110 µs for the $T_1 \rightarrow S_0$ transition (≈ 3.2 eV) and about 6 ns for the $S_1 \rightarrow S_0$ transition (≈ 4.3 eV). We attribute the observed few-µs lifetimes to $T_1 \rightarrow S_0$ emission in Ge-related *ODC* centres in re-crystallized quartz and interpret this finding to indicate the formation of Ge-Ge and Ge-Si defect centers uniformly spread across the crystallized layer [21]. A reason for the difference between the measured value and the long lifetime typical of spin-forbidden triplet-to-singlet transitions is probably the inhomogeneous strain field at the Ge-nanocrystal/matrix interface, which makes the forbidden transition partially allowed, and shortens the lifetime.

5. CONCLUSIONS AND OUTLOOK

The present experiments were first successful attempts to monitor cathodo-luminescence during chemical, dynamic and laser epitaxy of α-quartz after heavy-ion implantation. For Na, Rb and Cs ions, complete epitaxy was achieved during thermal annealing in air or oxygen, but only partial epitaxy during excimer laser annealing. Nearly complete dynamic epitaxy was achieved during Ba irradiation, but only little dynamic epitaxy (but strong CL emission) during Ge implantation.

Evidently, the evolution of the CL spectra, which picture various intrinsic and ion-specific photoactive centres, strongly depends on the ion species and the implantation and/or annealing conditions. The main finding of this work is the identification of several strong violet bands in the 2.9-3.7 eV range, which dominate the CL spectra around 1120 K implantation or annealing temperature. Most important is the fact that the high CL efficiency of Ge and the tendency of alkali ions for chemical epitaxy can be combined with each other during chemical epitaxy after Rb/Ge double implantation. We also have identified, for the first time, 3.65-eV luminescence from a defect connected to the implanted alkali ions and related to the regrowth of the ion-implantation-amorphized lattice structure. We have been able to prove that two types of intrinsic CL bands exist: the 2.4-eV band due to electron bombardment of α-quartz, and the 2.79- and 4.3-eV bands associated with amorphous silica. Finally, for the first time, a comparison of the CL luminescence efficiency for the various ions and types of epitaxy has been developed [47].

This survey is only a first step in interpreting the photoactive centers associated with implanted ions in quartz [30,48]. In particular, detailed knowledge of the critical ion fluences, up to which substitutional implantation <u>and</u> chemical epitaxy with alkali ions occur, is needed. Furthermore, the roles of the 3.25 and 4.30 eV bands and their defect anatomy require further investigations. Clearly, time-differential and temperature-dependent CL and PL data (see the example displayed in Fig. 10) and the results obtained by means of other methods such as TEM, EPR, NMR, PL, and absorption spectroscopy are useful to systematize the present findings and to learn more about the nature of the optically active centers and the luminescence mechanism. Finally, the role of substitutional nano-clusters in the luminescence process of quartz as seen in the case of Ge remains another hot topic.

ACKNOWLEDGMENTS

Most of the CL experiments described here have been carried out in Göttingen in collaboration with Drs. S. Dhar and S. Gasiorek, The authors wish to thank these colleagues for their excellent work. They are further indebted to Prof. H. Hofsäß, D. Purschke, Dr. U. Vetter (University of Göttingen), Dr. K. Arstila, Dr. T. Savajaara (University of Helsinki), and Prof. V. N. Kulkarni (IIT Kanpur) for their cooperation at various stages of this work, which was supported by Deutsche Forschungs-gemeinschaft, DAAD and the Academy of Finland.

REFERENCES

[1]   *Structure and Imperfections in Amorphous and Crystalline Silicon Dioxide*, edited by Devine, R. A. B., Duraud, J.-P., and Dooryhee, E. (John Wiley & Sons, 2000).
[2]   Rebohle, L., *et al.,* Appl. Phys. Lett. **77,** 969 (2000).
[3]   Wang, Y. Q., *et al.,* Appl. Phys. Lett. **81,** 4174 (2002).
[4]   Kabashin, A. V., and Meunier, M., Appl. Phys. Lett. **82,** 1619 (2003).
[5]   Fang, Y. C., *et al.,* Nanotechnology **15,** 494 (2004).
[6]   Harbsmeier, F., and Bolse, W., J. Appl. Phys. **83,** 4049 (1998).
[7]   Devaud, G., *et al.,* J. Non-Cryst. Solids **134,** 129 (1991).
[8]   Dhar, S., Bolse, W., and Lieb, K. P., J. Appl. Phys. **85,** 3120 (1999).
[9]   Roccaforte, F., Bolse, W., and Lieb, K. P., Appl. Phys. Lett. **73,** 134 (1998).
[10]   Roccaforte, F., Bolse, W., and Lieb, K. P., J. Appl. Phys. **89,** 3611 (2001).
[11]   Gustafsson, M., *et al.,* Phys. Rev. B **61,** 3327 (2000).
[12]   Gąsiorek, S., *et al.,* J. Appl. Phys. **95,** 4705 (2004).
[13]   Dhar, S., *et al.,* Surf. Coat. Technol. **158/159,** 436 (2002).
[14]   Dhar, S., *et al.,* Appl. Phys. Lett. **85,** 1341 (2004).
[15]   Dhar, S., *et al.,* J. Appl. Phys. **97,** 014910 (2005).
[16]   Sahoo, P. K., *et al.,* Nucl. Instr. Meth. B **216,** 324 (2004).
[17]   Sahoo, P. K.,, *et al.,* J. Appl. Phys. **96,** 1392 (2004).
[18]   Gąsiorek, S., *et al.,* Appl. Phys. B **84,** 357 (2006).
[19]   Gąsiorek, S., *et al.,* J. Non-Crystal. Solids, in press.
[20]   Sahoo, P. K., Gąsiorek, S., and Lieb, K. P., Nucl. Instr. Meth. B **240,** 188 (2005).
[21]   Sahoo, P. K., *et al.,* Appl. Phys. Lett. **87,** 021105 (2005).
[22]   Gąsiorek, S., *et al.,* Appl. Surf. Sci. **247,** 396 (2005); Sahoo, P. K., *et al.,* Nucl. Instr. Meth. B, online.
[23]   Keinonen, J., Gąsiorek, S., Sahoo, P. K., Dhar, S., and Lieb, K. P., Appl. Phys. Lett. **88,** 261102 (2006).
[24]   Lieb, K. P., in *Encyclopedia on Nanoscience and Nanotechnology*, H. S. Nalwa, Ed., American Scientific Publishers (2004) vol. **3**, pp. 233-251.
[25]   Lopes, J. M. J.. *et al.,* J. Appl. Phys. **94,** 6059 (2003).
[26]   Lopes, J. M. J., *et al.,* Appl. Phys. Lett. **86,** 023101 (2005).
[27]   Tohmon, K., *et al.,* Phys. Rev. Lett. **62,** 1388 (1989).
[28]   Stevens-Kalceff, M. A., and Philips, M. R., Phys. Rev. B**52,** 3122 (1995); Stevens-Kalceff, M. A., Phys. Rev. B **57,** 5674 (1998).
[29]   Skuja, L., Hirano, M., Hosono, H., and Kajihara, K, phys. stat. sol. (c) **2,** 15 (2005); Skuja, L. N., J. Non-Cryst. Solids **239,** 16 (1998); Skuja, L. N., Güttler, B., Schiel, D., and Silin, A. R., Phys. Rev. B **58,** 4296 (1998).
[30]   Lieb, K. P., and Keinonen, J., Contemp. Phys., submitted.
[31]   Stevens-Kalceff, M. A., Phys. Rev. Lett. **88,** 3137 (2000).
[32]   Fitting, H. -J., Barfels, T., Trukhin, A. N., and Schmidt, B., J. Non-Cryst. Solids **279,** 51 (2001).
[33]   Fitting, H. -J., *et al.,* J. Non-Cryst. Solids **303,** 218 (2002).
[34]   Yoshikawa, M., *et al.,* J. Appl. Phys. **92,** 7143 (2002).
[35]   Stevens-Kalceff, M. A., and Wong, J., J. Appl. Phys. **97,** 113519 (2005).
[36]   Costantini, J. M., *et al.,* J. Appl. Phys. **88,** 1339 (2000).
[37]   Griscom, D. L., Rev. Solid State Sci. **4,** 565 (1990).
[38]   Song, K. S., and Williams, R. T., *Self-Trapped Excitons* (Springer, Berlin, 1992).
[39]   Siu, G. G., Wu, X. L., Gu, Y., and Bao, X. M., Appl. Phys. Lett. **74,** 1812 (1999).

[40] Götze, J., and Zimmerle, W., *Quartz and silica as guide to provenance in sediments and sedimentary rocks*, Sediment. Geology, **21** (E. Schweizerbart' sche Verlagsbuchhdl., Nägele & Obermiller, Stuttgart 2000).

[41] Awazu, K., and Kawazoe, H., J. Appl. Phys. **68,** 3584 (1990).

[42] Skuja, L. N., Silin, A. R., and Boganov, A. G., J. Non-Cryst. Solids **63,** 431 (1984).

[43] Munekuni, S., *et al.*, J. Appl. Phys. **68,** 1212 (1990).

[44] Bae, H. S., *et al.* J. Appl. Phys. **91,** 4078 (2002).

[45] Kuzuu, N., Matsumoto, Y., and Murahara, M., Phys. Rev. B **48,** 6952 (1993).

[46] Geohegan, D. B., *et al.*, Appl. Phys. Lett. **73,** 438 (1998).

[47] Sahoo, P. K., Gąsiorek, S., Dhar, S., and Lieb, K. P., Nucl. Instr. Meth. B (2006) in press.

[48] Lieb, K. P., *et al.*, Nucl. Instr. Meth. B **244**, 272 (2006); Physica B (2006) in press.

168

Role of Plasma in Femtosecond Laser Pulse Propagation

Vladimir Mezentsev*, Mykhaylo Dubov*,
Jovana S. Petrovic*, Ian Bennion*,
Jürgen Dreher† and Rainer Grauer†

*Photonics Research Group, Aston University, Birmingham B4 7ET, United Kingdom
†Theoretische Physik I, Ruhr-Universität Bochum, D-44780 Bochum, Germany

Abstract. This paper describes physics of nonlinear ultra-short laser pulse propagation affected by plasma created by the pulse itself. Major applications are also discussed. Nonlinear propagation of the femtosecond laser pulses in gaseous and solid transparent dielectric media is a fundamental physical phenomenon in a wide range of important applications such as laser lidars, laser micro-machining (ablation) and microfabrication etc. These applications require very high intensity of the laser field, typically 10^{13}–10^{15} TW/cm2. Such high intensity leads to significant ionisation and creation of electron-ion or electron-hole plasma. The presence of plasma results in significant multiphoton and plasma absorption and plasma defocusing. Consequently, the propagation effects appear extremely complex and result from competitive counteraction of the above listed effects and Kerr effect, diffraction and dispersion. The theoretical models used for consistent description of laser-plasma interaction during femtosecond laser pulse propagation are derived and discussed. It turns out that the strongly nonlinear effects such self-focusing followed by the pulse splitting are essential. These phenomena feature extremely complex dynamics of both the electromagnetic field and plasma density with different spatio-temporal structures evolving at the same time. Some numerical approaches capable to handle all these complications are also discussed.

Keywords: Self-focusing, femtosecond inscription, adaptive mesh refinement
PACS: 42.65.Re Ultrafast processes; optical pulse generation and pulse compression 42.65.Sf Dynamics of nonlinear optical systems; optical instabilities, optical chaos and complexity, and optical spatio-temporal dynamics 52.38-r Laser-plasma interactions-in plasma physics

INTRODUCTION

The main objective of this paper is to outline the role of plasma created during the course of propagation of the intense femtosecond (fs) laser pulse. To be more specific, the applications of this phenomenon in fs microfabrication used as an example. Direct inscription of the complex microstructures in refractive materials by means of intense femtosecond radiation is one of the novel enabling technologies in modern photonics. This technology implies that pre-focused femtosecond light pulses produce phase transitions and create domains with a modified refractive index. Nonlinear propagation of femtosecond intense laser pulses in dielectrics exhibits a wide range of fascinating phenomena including conical emission[1], X-waves[2], self-reconstruction[3], self-healing[2] light filaments plasma breakdown[4]. Proof of principle experiments of the potential of fs inscription for microfabrication of photonic structures was demonstrated almost a decade ago[5]. It has now become a very promising method of microfabrication[6, 7, 8]. Other interesting applications aim to achieve the longest possible self-guiding filaments to de-

CP876, *The Physics of Ionized Gases: 23rd Summer School and International Symposium*,
edited by L. Hadžievski, B. P. Marinković, and N. S. Simonović
© 2006 American Institute of Physics 978-0-7354-0377-2/06/$23.00

liver the energy through the bulk of material[9].

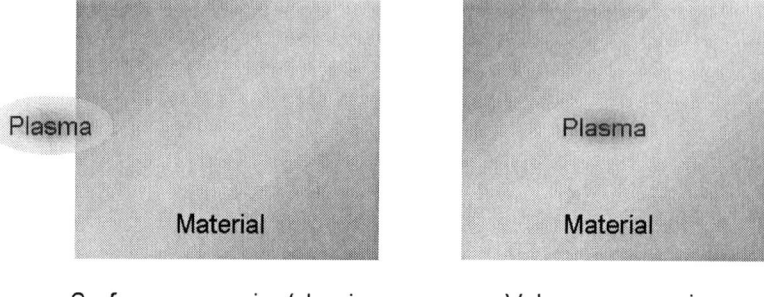

Surface processing/shaping   Volume processing

FIGURE 1. Plasma assisted material processing is usually associated with surface treated with plasma (left). We aim internal processing in the bulk of the material with the plasma created by the high power laser pulse

Experimental setup for volume plasma processing is shown in Fig. 2. Intense laser pulse is focused under the surface of the planar sample typically made of fused silica.

Self-focusing of the intense laser pulse is a key physical phenomenon leading to a multi-photon ionization at its final stage. In fact the very formation of plasma filaments limits the catastrophic damage due to defocusing and multi-photon absorption. Eventually, the thermalization and recombination of the plasma filament leads to the modification of medium and a distributed profile of refractive index is produced. The dynamics of the light-induced plasma filaments is extremely complex and defined by many factors. It is an extremely fast process evolving at the very fine spatial scales.

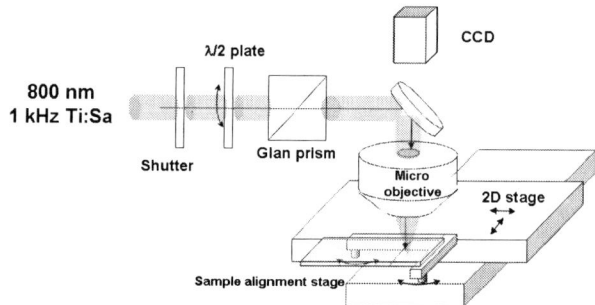

FIGURE 2. Experimental setup for implementation of plasma assisted material processing in the bulk of transparent dielectric

In this paper, we demonstrate and adaptive numerical approach to the detailed study of the evolution of plasma filaments and the role of pulse and media parameters on the shape of resulting filaments. We consider a geometry in which an incident Gaussian input field is focused through a lens into the planar silica sample sample.

PLASMA PARAMETERS

This section describes a set of typical parameters for plasma generated by the femtosecond pulse. We consider a couple of limiting factors which define the range of plasma concentration and temperature. One important limitation can be defined from the matching condition between the laser frequency f_e and plasma frequency $f_p = \omega_p/2\pi = (2\pi)^{-1}(\rho e^2/\varepsilon_0 m_e)^{1/2}$ where ρ is an electron concentration. The condition $f_e = f_p$ defines the critical (breakdown) density of plasma $\rho_{BD} = \varepsilon_0 m_e e^{-2}\omega^2$. This is a resonant condition for conversion of electromagnetic wave into plasma waves. It gives a breakdown electron density for a given laser frequency. Typical region of operation parameters is shown in Fig. 3

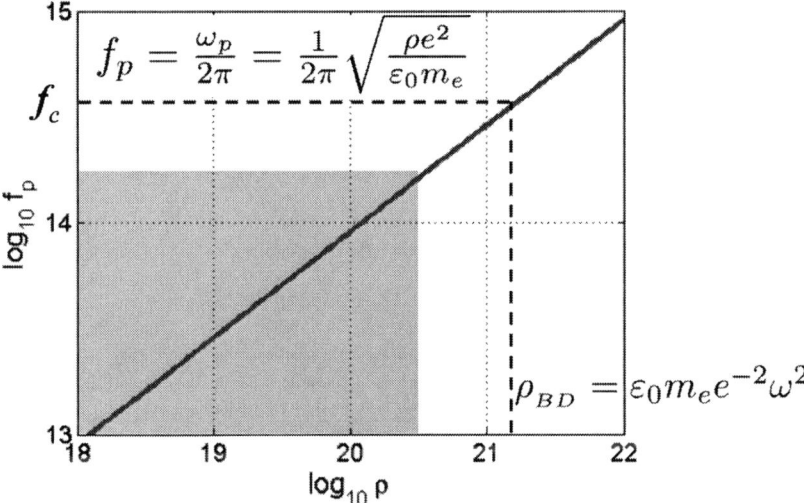

FIGURE 3. Matching of plasma frequency and laser frequency resulting in breakdown condition.

The plasma temperature can be estimated from the analysis of the electron oscillations in high frequency electromagnetic field. Then average kinetic energy can be expressed as

$$K = \left\langle \frac{e^2 \mathscr{E}(t)^2}{4m_e\omega^2} \right\rangle_t = 9.3 \times 10^{-14} I\lambda^2 ,$$

where e and m_e are electron charge and mass respectively, $\mathscr{E}$ is an envelope amplitude of the electric field and I is its intensity, omega is a carrier angular frequency of the electromagnetic wave, and λ is its wavelength. The numerical expression in the right hand side gives a value of the kinetic value in eV provided that intensity is given in W/cm^2 and the wavelength is given in microns. Typical values of electron temperatures versus laser intensity are summarized in Table 1.

Both plasma concentration and temperature characterize a femtosecond laser produced plasma as shown in Fig. 4 in comparison with other types of plasmas.

TABLE 1. Average electron energy versus light intensity for typical focusing conditions

Intensity, W/cm^2	10^{13}	10^{14}	10^{15}
Electron energy, eV	0.01	1	100

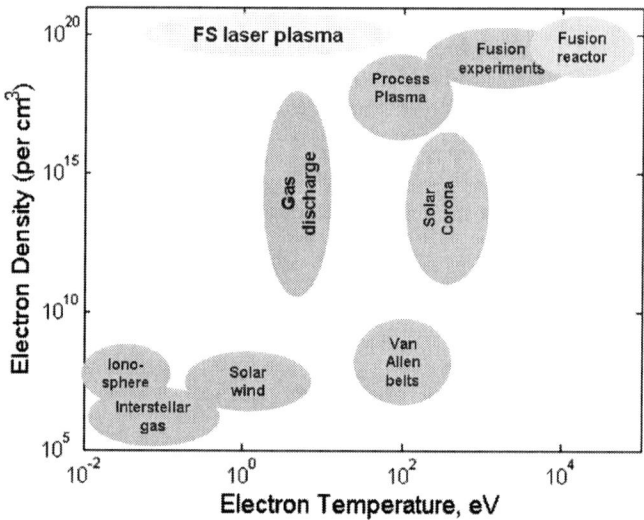

FIGURE 4. Plasma parameters typical for femtosecond laser pulse propagation

THEORETICAL MODEL

Equations

This section describes a theoretical model used for femtosecond pulse propagation in dielectrics. Electromagnetic wave is described by a set of Maxwell equations

$$\nabla \times \mathbf{E} = -\frac{\partial B}{\partial t}$$

$$\nabla \times H = \frac{\partial D}{\partial t} + J$$

$$D = \varepsilon E \; ; \quad B = \mu H \, ,$$

where J is an electron current density.

Description of plasma is based on relaxation dynamics of the electrons driven by the electromagnetic wave. The major source of plasma in strong electromagnetic field is

172

multi-photon and avalanche ionization to be included in the continuity equation.

$$\frac{d\mathbf{v_e}}{dt} = -\tau_c^{-1}\mathbf{v_e} - \frac{e}{m_e}e\mathbf{E}$$

$$\mathbf{J} = -e\rho\mathbf{v_e}$$

$$\frac{d\rho}{dt} = ionization\ sources\ ,$$

where τ_c is the shortest collision time. Envelope approximation can be used to describe quasi-monochromatic paraxial evolution

$$\mathbf{E}(\mathbf{r}_\perp,z,t) = \hat{\mathbf{y}}\mathcal{E}(\mathbf{r}_\perp,z,t)\exp[i(kz-\omega t)]$$

$$\frac{\partial}{\partial z} = ik + \frac{\partial}{\partial z}\ ;\ \frac{\partial}{\partial t} = -i\omega + \frac{\partial}{\partial t}$$

Finally, Kerr nonlinearity must be taken into account for strong laser field $n = \sqrt{\varepsilon} = n_0 + n_2|\mathcal{E}|^2$. Such approach was originally suggested by Feit and Fleck [14] and later developed into fairly complex models, see e.g. Refs.[15, 16, 17, 9, 18, 19, 20, 21] For the purposes of this paper, a simplified model is used, essentially similar to that described by Feng et al.[10]:

$$i\mathcal{E}_z + \frac{1}{2k}\Delta_\perp\mathcal{E} - \frac{k''}{2}\frac{\partial^2\mathcal{E}}{\partial t^2} + k_0 n_2|\mathcal{E}|^2\mathcal{E} = -\frac{i\sigma}{2}(1+i\omega\tau)\rho\mathcal{E} - i\frac{\beta^{(K)}}{2}|\mathcal{E}|^{2(K-1)}\mathcal{E} \quad (1)$$

$$\frac{\partial\rho}{\partial t} = \frac{1}{n_b^2}\frac{\sigma_{bs}}{E_g}\rho|\mathcal{E}|^2 + \frac{\beta^{(K)}}{K\hbar\omega}|\mathcal{E}|^{2K} \quad (2)$$

The terms on the left-hand side of Eq.(1) describe effects of beam diffraction, group velocity dispersion (GVD), and Kerr nonlinearity. The latter is responsible for a catastrophic self-focusing which is limited by the effects described by terms on the right-hand side of Eq.(1), namely plasma absorption and multi-photon absorption. In Eq.(1) the laser beam propagation along the z axis is assumed and this equation is essentially a reduced paraxial approximation of the wave equation for the complex electric field envelope $\mathcal{E}$ with a carrier frequency ω in the moving frame of coordinates. Here $k = n_b k_0 = n_b\omega/c$ is the propagation vector, $k'' = \partial^2 k(\omega)/\partial\omega^2$ is the GVD parameter, $n_b(\omega)$ is a linear refractive index of the bulk medium, n_2 is the nonlinear coefficient describing nonlinear self-modulation (Kerr effect) such that $n_2|\mathcal{E}|^2$ is a nonlinear contribution to the refractive index, σ_{bs} is the cross section for inverse Bremsstrallung, τ is the electron relaxation time, E_g is the ionization energy, and the quantity $\beta^{(K)}$ controls the K–photon absorption. Equation (2) implements the Drude model for electron-hole plasma in the bulk of silica and describes the evolution of the electron density ρ. The first term on the right-hand side is responsible for the avalanche impact ionization and the second term — for the ionization resulting from MPA. Equation (2) is suitable for description of the sub–picosecond laser pulses when plasma diffusion is negligible. Here, the wave equation describing the evolution of the focused optical beam in the form of NLSE (left-hand terms in Eq.(1)) which is extended to include plasma generation, pulse-Ũ plasma interaction, and MPA (terms on the right-hand of Eq.(1). Group velocity dispersion included in Eq.(1) has been shown to lead to pulse splitting and to arrest the collapse [22, 23, 24, 25, 26, 13].

Equations Eqs(1,2) were used for numerical modeling of the plasma formation using adaptive mesh approach described below. We use a simplified version of the widely accepted model of the nonlinear propagation of the laser pulse as it was formulated in Ref.[10]. It is essentially a nonlinear Schrödinger equation (NLSE) coupled with an equation describing the plasma generation. This basic model describes effects of self-focusing, multiphoton absorption (MPA), and group velocity dispersion (GVD). Pure NLSE is a generic model which ultimately appears in consistent description of envelope amplitude of the nonlinear wave packets. It is widely used to describe fs light propagation[11] and has been extensively studied. One of the most striking features of the NLSE is catastrophic self-focusing or beam collapse which means a formation of a singularity in finite propagation length[12]. Beam collapse happens in the framework stationary NLSE if the pulse power exceeds a certain critical value. Formally, the on-axis intensity achieves an infinitely high value at some critical propagation length. However, in the extended physical models which account for dispersion and nonlinear absorption a formation of singularity is arrested[13]. The nonlinear evolution of the collapsing beam with the presence of the arresting effects is extremely rich. Mathematically it poses a stiff multidimensional evolution problem. Straightforward numerical modeling of such problems is a very difficult challenge due to the multiscale nature of underlying physical phenomena. In this paper, we address an intrinsic stiffness of the mathematical problem by introducing a hierarchy of the adaptively refined grids which are dynamically adjusted for proper resolution of the fine spatial structures and temporal features of the beam. We report on the development of a portable computational framework for the parallel, mesh-adaptive solution of system of a 3D parabolic wave equation for envelope amplitude of electromagnetic field coupled with the rate equation for plasma density. Local mesh refinement is realized by the recursive bisection of grid blocks along each spatial and temporal dimensions. Implemented numerical schemes include standard finite-difference and spectral methods. Non-adaptive solver has also been implemented for back-to-back accuracy tests and performance profiling. Parallel execution is achieved through a configurable hybrid of POSIX-multi-threading and MPI-distribution with dynamic load balancing.

Physical parameters

In all our simulations the Gaussian initial condition:

$$\mathscr{E}(z=0,r,t) = \sqrt{\frac{2P_{in}}{\pi r_0^2}} \exp\left(-\frac{r^2}{r_0^2} - \frac{ikr^2}{2f} - \frac{t^2}{t_p^2}\right), \tag{3}$$

where r_0 is the waist of the incident beam, t_p defines the conventionally defined pulsewidth $t_{FWHM} = \sqrt{2\ln 2}t_p \approx 1.177t_p$, and f is a focal length of the objective lens.

For all our simulations a fixed single values of $r_0 = 2.5$ mm and $t_p = 60$ fs was used. This fixed pulsewidth corresponds to the critical energy of 116 nJ for a critical power

174

$P_{cr} = \lambda_0^2/2\pi n_b n_2 \approx 2.3$ MW in fused silica with $n_b = 1.453$ being the linear refraction index and $n_2 = 3.2 \times 10^{-16}$ cm^2/W the nonlinear refraction index. Critical power is proven to be a crucial parameter to determine the evolution of the collapsing beam. We assume the laser wavelength λ_0 to be 800 nm and the focusing lens to have $f = 40$ mm.

The other parameters for fused silica, used in simulations are described below. GVD coefficient $k'' = 361$ fs^2/cm, inverse Bremsstrahlung cross section $\sigma_{bs} = 2.78 \times 10^{-18}$ cm^2. Multiphoton absorption coefficient can be expressed as $\beta^{(K)} \hbar \omega \sigma_K \rho_{at}$, whith $\rho_{at} = 2.1 \times 10^{22}$ atoms/cm^3 being a material concentration and $\sigma_K = 1.3 \times 10-55$ cm^{2K}/W^K/s. We assume five–photon ionization with $K = 5$ and $E_g = 7.6$ eV in fused silica.

The quation (2) can be expressed as

$$\frac{\partial}{\partial t}\frac{\rho}{\rho_{BD}} = \frac{1}{n_b^2}\frac{\sigma_{bs}}{E_g}\frac{\rho}{\rho_{BD}}|\mathcal{E}|^2 + \left(\frac{|\mathcal{E}|^2}{I_{MPA}}\right)^K \qquad (4)$$

where

$$I_{MPA} = \left(\frac{K\hbar\omega\rho_{BD}}{\beta^{(K)}}\right)^{1/K} \qquad (5)$$

is defined to be an MPA threshold as the plasma ionization rate becomes very steep when the intensity I exceeds I_{MPA}. $\rho_{BD} = 1.7 \times 10^{21}$ is a plasma breakdown density. Being important physical thresholds, both I_{MPA} and ρ_{BD} were used in a very useful normalization of physical variables to dimensionless ones for performing simulations. The choice of normalization is irrelevant to discuss as all the parameters and fields throughout this paper are produced with their physical values.

ADAPTIVE NUMERICAL METHOD

The principle of adaptive mesh refinement is rather simple. Starting with one grid of given resolution (in most of our 3D configurations we currently chose 512×64×64 mesh points) called master grid, the partial differential equations (1-2) are solved with a scheme summarized below. After a certain number of steps along propagation axis z, it is checked whether the local numerical resolution is still sufficient on the entire grid. If it is detected that finer grid is locally needed, a refinement is carried out and a child grid is created using the interface with a parent grid as a new boundary. In order to prepare for it, the points where the error of discretization exceeds a given value are marked on the grid. In addition to these grid points, adjacent ones are included. These marked points of insufficient numerical resolution have to be covered with rectangular grids of finer resolution as efficiently as possible. Our algorithm for this purpose is very similar to the one used by Berger and Colella[27], and it was described in detail by Friedel et al.[28]. On the child grids, the spatial discretization length and the time step are reduced by a certain refinement factor. The new grids are filled with data obtained by interpolation from the preceding parent (coarser) level. The integration advances on both the parent and the child levels until the local resolution becomes insufficient again. The rebuilding of the grid hierarchy starting with the current level and proceeding on all subsequent levels begins when the above-mentioned threshold for the error is locally exceeded, e.g., if the domains of high intensity have left the finer grids, or if local gradients have developed, such that the prescribed accuracy is not guaranteed. The points of insufficient numerical resolution are collected on all grids of each level. On the basis of the resulting list of these points, new grids are generated. After assuring that the newly generated grids are properly embedded in their parent grids, interpolated data are filled in. If data existed on grids of the same level before the regridding, these are substituted to the interpolated data from the parents grids.

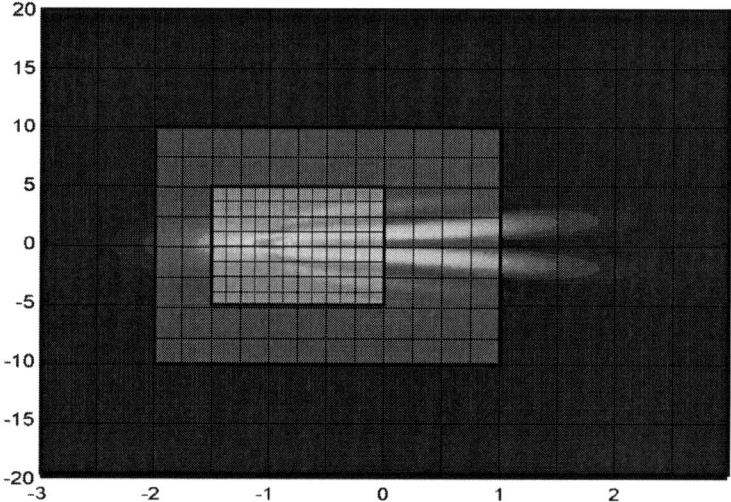

FIGURE 5. Principle of adaptive mesh refinement. Dynamically created child meshes adaptively adjusted to form a hierarchy of meshes for resolving finest details (refining). Fine meshes are removed when fine pattern disappears (coarsening).

Solution of the model described in Introduction in the outlined framework requires selecting an appropriate numerical integration scheme. It turned out that the optimal performance can be achieved by using a different scheme on the refined levels than on the base level, when taking into account that arbitrary mesh sizes on the refined levels occurred. On the base level, we applied an operator–splitting method which is second–order accurate in z. Radial diffraction term in the Laplacian operator of Eq.(1) is integrated by means of Crank–Nicholson scheme with zero boundary condition at the maximum radius. Both the angular term of the transverse diffraction operator and the dispersion operator are diagonal in Fourier space thus Fast Fourier Transform was utilized for numerical integration of these terms. It is worth noting that neither periodic or zero global boundary conditions both in transverse space and time do not impose a significant restriction on the problem considered, because the localized wave-packets vanish at the boundaries during the entire process of focusing and defocusing after the pulse passes the focal domain. On the refined meshes we apply a semi-implicit scheme of Crank-Nicholson type, which was used by, e.g., Pietsch et al.[29] and also utilized in a previous modelling on NLSE with normal dispersion[13].

It is important to comment on the refinement criterion. We calculate the discretized equations based on the actual grid spacing and twice the grid spacing. When the difference exceeds a given threshold, those mesh points are marked under-resolved and are subject to refinement. The threshold value conditioning the refinement was determined such that a sufficient resolution was guaranteed during the evolution along the propagation axis z. The length of the integration step Δz was dynamically adapted to ensure that at all times the i) Courant–Friedrich–Levy condition was met to enforce the numerical stability and the iterative method converged at a prescribed minimal rate and ii) the maximal relative increment of both the amplitude and the nonlinear phase was always kept less than a prescribed limit, usually 1%.

The implementation of the adaptive mesh refinement strategy described above is done in C++. We make use of a non-adaptive solver described above for handling the master grid. This solver was also used for benchmarking and back-to-back accuracy testing of adaptive solver. Handling of the data structures is separated from the problem under consideration. Therefore, it is relatively easy to use the code for other types of problems including various generalizations of the system Eqs.(1-2). Since on each grid the advance along z and the Helmholtz-type equation can be solved independently and the number of grids supersedes the number of processors available, parallelization is highly efficient.

NONLINEAR DYNAMICS

There are two distinct typical setups related to the focusing geometry. If the goal is to produce the longest possible filament, then the loose focusing and small spot size are required [31, 9, 2]. However in the microfabrication context the opposite goal is usually desirable. The focused spot is often required to be as small as possible and the absorbed energy is needed to be within a narrow window between the thresholds of inscription and damage. This is an extremely difficult challenge because of the huge difference in spatial and temporal scales of the incident laser beam and fine features of light and plasma patterns in the vicinity of the focal point. The adaptive procedure described above allows for accurate treatment of multiscale evolution which results in stationary (in the framework of the model considered) distribution of plasma. The mechanisms of eventual plasma recombination and subsequent relaxation of the medium are extremely complex. For example, the latter can be described as a sophisticated 3D thermo-elasto-plastic processes[32]. The purpose of the present work is to find an accurate spatial distribution of plasma needed for such or similar subsequent analysis. Typical asymptotic plasma density profiles for different initial pulse energies are shown energies are shown in Figure 7. Density plots of plasma concentrations are shown in one transverse and one propagation coordinates. It shows that subcritical evolution ($P_{in} < P_{cr}$)leads to a smooth plasma cloud without visible fine structure. This regime is probably the most attractive from the microfabrication viewpoint because it is characterized by the smooth subcritical evolution of the peak intensity. Larger energies lead to development of the pronounced periodic fringes which result from the relaxation oscillations of a collapsing beam after the collapse is arrested by the multiphoton absorption.

ρ/ρ_{BD} I/I_{MPA}

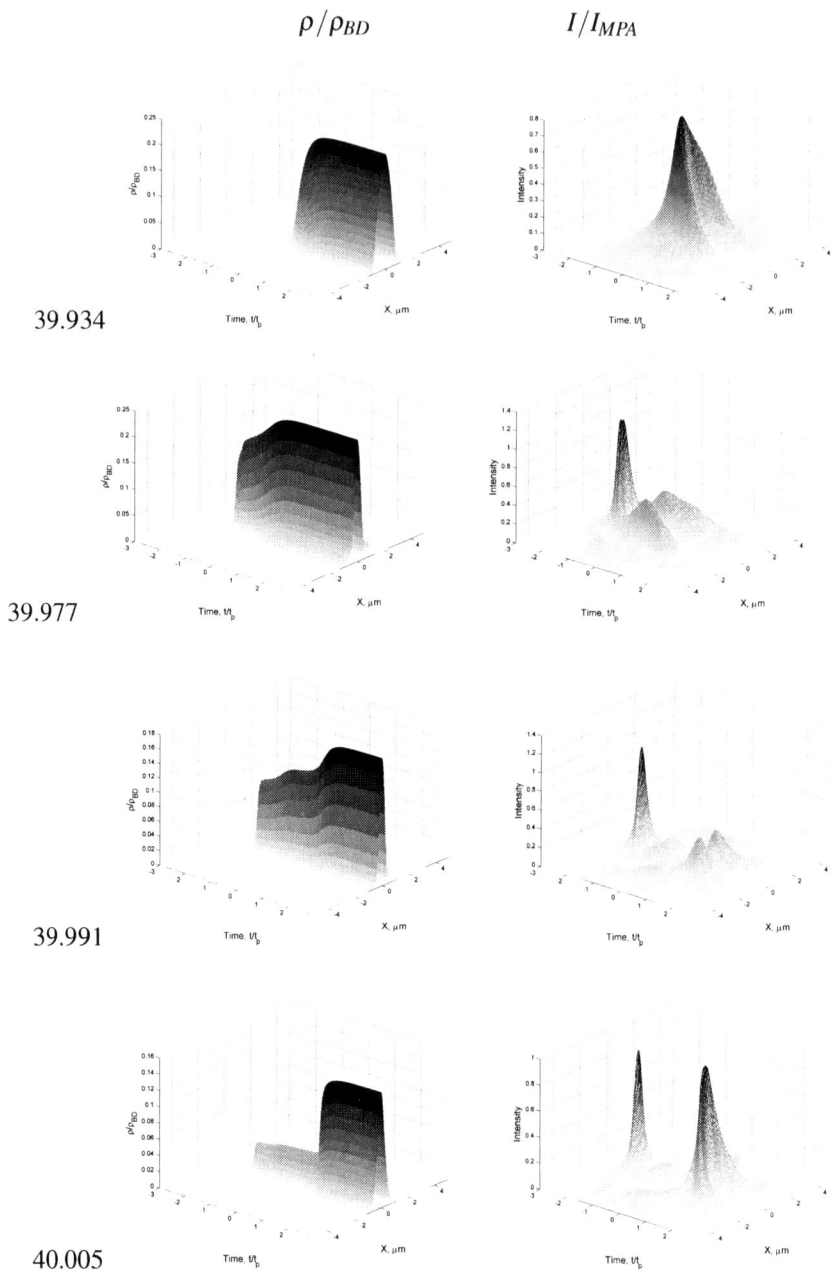

FIGURE 6. Dynamics of the intensity and the plasma density at different positions along axis z.

Nontrivial complex light and plasma dynamics and formation of interesting light patterns are illustrated by Figure 6. Snapshots of the beam intensity profile in transverse space and time are presented at different points along the propagation axis in the vicinity of the focal point for supercritical case $P_{in} > P_{cr}$. Originally, an intensity profile forms a distinct crescent in $x - t$ plane. It is then evolved so that the shoulders of this crescent split from the front of the pulse and form a pair of satellite pulses which rise being fed by a contracting beam. This pair is eventually coalesce to form a secondary crescent pulse following the remains of the front pulse. Then this scenario repeats for the newborn crescent pulse which even has the same amplitude just under the MPA threshold. A cascade of such crescents result in periodic fringes of plasma density until finally significant fraction of the original pulse energy is absorbed.

The most important result of the intense laser propagation is creation of the cloud of plasma due to ionization. The distribution of electrons can be found as

$$\rho(\mathbf{r},z) = \int_{-\infty}^{\infty} \left[\frac{1}{n_b^2} \frac{\sigma}{E_g} \rho |\mathscr{E}(\mathbf{r},z,t)|^2 + \frac{\beta^{(K)}}{K\hbar\omega} |\mathscr{E}(\mathbf{r},z,t)|^{2K} \right] dt \ .$$

Typical stationary plasma profiles are shown in Fig. 7.

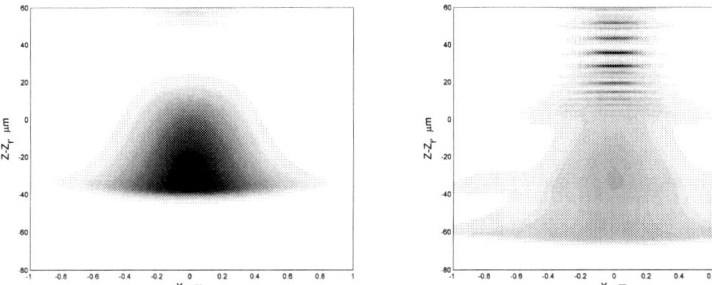

FIGURE 7. Contours of the plasma density at infinite time after electric field is vanished for different energies: subcritical (a) and supercritical (b).

CONCLUSION

It is shown that the dynamics of creation of plasma filaments is a crucial phenomenon in propagation of femtosecond laser pulses in dielectrics. Detailed adaptive numerical modeling of fine dynamics in the vicinity of the focal point reveals an extremely complex behavior of the focused light coupled to generated plasma.

ACKNOWLEDGMENTS

We acknowledge technical support of technical staff running Cray XD1 computing cluster facility in School of Engineering and Applied Science at Aston Univercity.

REFERENCES

1. O. G. Kosareva, V. P. Kandidov, A. Brodeur, C. Y. Chien, and S. L. Chin, *Opt. Lett.* **22**, 1332–1334 (1997).
2. M. Kolesik, E. M. Wright, and J. Moloney, *Phys. Rev. Lett.* **92**, 253901 (2004).
3. A. Dubietis, E. Kucinskas, G. Tamosauskas, E. Gaizauskas, M. A. Porras, and P. D. Trapani, *Opt. Lett.* **29**, 2893–2895 (2004).
4. C. W. Carr, M. D. Feit, A. M. Rubenchik, P. D. Mange, S. O. Kucheyev, M. D. Shirk, H. B. Radousky, and S. G. Demos, *Opt. Lett.* **30**, 661–663 (2005).
5. K. M. Davis, K. Miura, N. Sugimoto, and K. Hirao, *Opt. Letters* **21**, 1729–1731 (1996).
6. C. B. Schaffer, A. Brodeur, J. F. Garcìa, and E. Mazur, *Opt. Lett.* **26**, 93–95 (2001).
7. A. H. Nejadmalayeri, P. R. Herman, J. Burghoff, M. Will, S. Nolte, and A. Tuennermann, *Opt. Lett.* **30**, 964–966 (2005).
8. A. M. Kowalevicz, V. Sharma, E. P. Ippen, J. G. Fujimoto, and K. Minoshima, *Opt. Lett.* **30**, 1060–1062 (2005).
9. S. Tzortzakis, L. Sudrie, M. Franco, B. Prade, A. Mysyrowicz, A. Couairon, and L. Bergé, *Phys. Rev. Lett.* **21**, 213902 (2001).
10. Q. Feng, J. V. Moloney, A. C. Newell, E. M. Wright, K. Cook, P. K. Kennedy, D. X. Hammer, B. A. Rockwell, and C. R. Thompson, *IEEE Journal of Quantum Electronics* **33**, 127–137 (1997).
11. J. H. Marburger, *Prog. Quantum Electron.* **4**, 35Ű–110 (1975).
12. V. P. S.N. Vlasov, and V. Talanov, *Izv. Vuzov, Radiofizica* **14**, 1353–1364 (1971).
13. K. Germaschewski, R. Grauer, L. Bergé, V. K. Mezentsev, and J. J. Rasmussen, *Physica D* **151**, 175–198 (2001).
14. M. D. Feit, and J. A. Fleck, *Appl. Phys. Lett.* **24**, 169–Ű172 (1974).
15. T. Brabec, and F. Krausz, *Phys. Rev. Lett.* **78**, 3282Ű–3285 (1997).
16. J. K. Ranka, and A. L. Gaeta, *Opt. Lett.* **23**, 534–536 (1998).
17. A. L. Gaeta, *Phys. Rev. Lett.* **84**, 3582–3585 (2000).
18. M. Kolesik, J. V. Moloney, and M. Mlejnek, *Phys. Rev. Lett.* **89**, 283902 (2002).
19. S. Tzortzakis, L. Bergé, M. Franco, B. Prade, A. Mysyrowicz, and A. Couairon, *J. Opt. Soc. Am. B* **19**, 1117–1129 (2002).
20. M. Kolesik, G. Katona, J. Moloney, and E.M.Wright, *Phys. Rev. Lett.* **91**, 043905 (2003).
21. R. Nuter, S. Skupin, and L. Bergé, *Opt. Lett* **30**, 917–919 (2005).
22. N. A. Zharova, A. G. Litvak, T. A. Petrova, A. M. Sergeev, and A. D. Yunakovskii, *PisŠma Zh. Eksp. Teor. Fiz. (JETP Lett.)* **44**, 12–15 (1986).
23. P. Chernev, and V. Petrov, *Opt. Lett.* **17**, 172–Ű174 (1992).
24. J. E. Rothenberg, *Opt. Lett.* **17**, 583–Ű585 (1992).
25. G. G. Luther, J. V. Moloney, A. C. Newell, and E. M. Wright, *Opt. Lett.* **19**, 862–864 (1994).
26. G. G. Luther, A. C. Newell, and J. V. Moloney, *Physica D* **74**, 59–73 (1994).
27. M. J. Berger, and P. Colella, *J. Comp. Phys.* **82**, 64–84 (1989).
28. H. Friedel, R. Grauer, and C. Marliani, *J. Comp. Phys.* **134**, 190–198 (1997).
29. H. Pietsch, E. Laedke, and K.-H. Spatschek, *Phys. Rev. E* **47**, 1977–1995 (1993).
30. W. Hackbusch, *Iterative Solution of Large Sparse Systems of Equations*, Springer, New York, 1994.
31. A. Dubietis, G. Tamoauskas, I. Diomin, and A. Varanaviiuset, *Opt. Lett* **28**, 1269–1271 (2003).
32. X. Zhang, X. Xu, and A. Rubenchik, *Appl. Phys. A* **79**, 945Ű–948 (2004).

Ultra-shallow Junction Formation in SOI using Vacancy Engineering

R.M. Gwilliam, N.E.B Cowern[1], B. Colombeau[2], B. Sealy, A.J. Smith

Ion Beam Centre, University of Surrey, Guildford, Surrey, GU2 7XH, UK
[1]Advanced Technology Institute, University of Surrey, Guildford, Surrey, GU2 7XH, UK
[2]Chartered Semiconductor Manufacturing Ltd, 60 Woodlands, Industrial Park D, Street 2, Singapore 738406

Abstract. Forming highly conducting, ultra-shallow boron doped layers, is well known to be a challenge for future CMOS devices. This paper reviews a technique known as vacancy engineering, which is a co-implant process that has been proven to be efficient in reducing anomalous effects, such as transient enhanced diffusion and dopant clustering. Due to relatively low improvement factors, vacancy engineering has never been implemented as an industrial process. However, recent advancements demonstrate that by optimizing the implant, substrate and anneal parameters it is possible to produce low resistive, p-type layers with a high degree of thermal stability which rival the more preferred techniques used today.

Keywords: vacancy engineering, SOI, source/drain extensions, boron
PACS: 61.72.-y, 61.72.Cc, 61.72.Ji, 61.72.Ss, 66.30.Jt, 66.30.Lw

INTRODUCTION

For the last 40 years a natural demand for faster, more complex, and therefore, more functional electronic systems, has been the fundamental driving force behind the miniaturization of the Complementary Metal Oxide Semiconductor (CMOS) transistors. As the devices evolve, the physical requirements become ever increasingly more difficult to achieve. In fact, as the 45nm technology node approaches, the actual requirements push the fundamental limits of the starting substrate [1].

A key component within the device architecture is known as the Source/Drain Extension (SDE), which is essential for reducing problems categorized as the 'short channel effects'. The specification of the SDEs are defined in terms of junction depth (X_j) and sheet resistance (Rs), where today, the junction has to be formed by a 'diffusionless' process with the level of electrical active dopant being well above the solubility limit. However, the most popular p-type dopant, boron, is susceptible to process-induced phenomena known as Transient Enhanced Diffusion (TED) and Boron Interstitial Clustering (BICs) which hinder the junction depth and sheet resistance, respectively [2,3].

It is well known that a supersaturation of interstitial defects remaining after the implantation process is the underlying cause of such detrimental effects. Therefore, to

CP876, The Physics of Ionized Gases: 23rd Summer School and International Symposium,
edited by L. Hadžievski, B. P. Marinković, and N. S. Simonović
© 2006 American Institute of Physics 978-0-7354-0377-2/06/$23.00

continue with the current rate of scaling, these dopant-defect interactions need to be inhibited.

The vacancy engineering technique uses a co-implant process to generate an excess of vacancy defects within the proximity of the boron implant, in an attempt to 'mop up' the excess interstitials through an interstitial-vacancy annihilation mechanism, and therefore, reduce their unfavorable effects. The underlying principle of this technique relies on the inherent properties of the ion implantation itself. Choosing the right implant species (typically silicon) and energy, it is possible to transfer enough momentum from the impinging ion to the host lattice to cause a spatial separation of the Frenkel pair population [4,5], resulting in a net excess of vacancies close to the surface and a corresponding net excess of silicon interstitials around the projected range of the implant.

This paper reviews the current progress associated with vacancy engineering as a viable alternative for producing highly stable, low resistive, ultra-shallow junctions for future CMOS devices.

Historical Review

The history of vacancy engineering can be traced back to 1980 when Winterbon et al.[6] observed an increase in vacancy concentration at the surface when high energy implants were simulated using the Net Recoil Density (NRD) algorithm, which is simply expressed as:

$$C_D(x) = Si_I(x) - Si_V(x) \qquad (1)$$

Where $C_D(x)$ is the defect distribution calculated from $Si_I(x)$ and $Si_V(x)$, representing the generated silicon interstitial and vacancy concentrations respectively. Therefore, if $C_D(x)$ is negative, the defect distribution is determined to be vacancy rich and conversely interstitial rich when $C_D(x)$ is positive. The average separation can be approximately described by Rp-ΔRp, where Rp is the projected range and ΔRp is the range straggling of the incident ion [7]. Eleven years after Winterbon's initial theory, Holland et al. [8] proved experimentally the presence of voids (agglomeration of vacancies) in the near surface region after a high energy (1.25MeV) silicon implant via TEM analysis.

The influence of a vacancy engineering process on a boron doping implant was not examined until the early 90s by Raineri et al. [9]. Using a 1MeV Silicon implant followed by a 10keV boron implant, it was clearly demonstrated that during a subsequent anneal TED was retarded. However, the actual cause of such an effect was under some debate. It was speculated that the damage created by the high energy silicon implant acted as a gettering layer, trapping the excess silicon interstitials generated by the boron implant. This theory was also supported by Saito et al. [10] who found that a phosphorus co-implant had a similar effect in reducing boron enhanced diffusion. It was not until 1997, when Roth et al. [11] used a Silicon On Insulator (SOI) substrate to investigate the effect of a high energy co-implant, that the effect of excess vacancies was really considered as the probable cause for the reduction in boron TED.

Roth et al. were the first group to experimentally investigate the origin of the diffusion reduction by actually physically decoupling the two excess defect regions (interstitial and vacancies) via what was thought as a diffusion barrier. It was suggested that the Buried Oxide (BOX) layered structure in SOI would restrict the excess silicon interstitials from diffusing back into the silicon top layer and annihilating the generated vacancies, providing the majority of the silicon interstitials were positioned beneath the BOX. Using a 2MeV silicon implant energy, the majority of the excess interstitials were positioned well below the BOX and a successful reduction in boron TED was still achieved. However, it was not possible to unambiguously determine the vacancy effect as the boron implant itself was a source of silicon interstitials due to the implant damage.

In the late 90s Venezia et al. [12] proved conclusively that vacancies were behind the reduction of boron TED, using a SOI structure in the same manner as Roth. However, this study used epitaxially grown boron doped marker layers instead of a boron implant to detect the effect of a 1MeV silicon vacancy generating implant. Then by controllably introducing damage into the top silicon layer, it was possible to distinctly show that the excess vacancies efficiently 'moped up' the problematic interstitials.

Vacancy engineering has been shown by a range of groups to reduce anomalous diffusion, see references [9,11,13,14,15,16,17,18] and references within. The research dedicated to investigating the effects on electrical activation is significantly less. As the focus at that time was more on diffusion, the experiments tended to use low boron doses combined with high temperature anneals, where the doping levels were well below the solid solubility limit. Thus, many opportunities to observe improved activation were missed.

Shao et al. [7] were one of the first groups to perform electrical measurements on vacancy engineered substrates using a 500keV silicon implant combined with a 2keV boron implant. Even though an improvement in Rs was observed at low temperatures (400°C and 600°C after 10s), the resulting Rs was still ~2500 ohms/sq, much greater than required by the International Roadmap for Semiconductors (ITRS). A more detailed study was performed by Kalyanaraman et al. [19] a year later. The active fraction of a 40keV boron implant with a dose of 2×10^{14} cm^{-2} was studied over a wide range of anneal temperatures. It was clearly shown that a distinct improvement in dopant activation is achieved showing an improvement factor of ~2.4x. Even the studies on boron activation discussed above, typically used low boron doses and/or relatively high boron ion energies. It was not until recently that Shao et al. [20] investigated the effects of vacancy engineering on an ultra-low energy (500eV) boron implant with a relatively high dose (10^{15} cm^{-2}). This was the first instance of vacancy engineering implants (MeV silicon) having the ability to actually raise boron activation above solid solubility. However, the observed enhancements were not solely from a vacancy engineering implant, since a combination of pre-amorphisation and vacancy engineering was used. Therefore, the vacancy engineering technique always showed promise but due to the low improvement factors compared to rival processes, and uneconomical implantation steps, the full potential of such a technique was never realized.

OPTIMISATION OF THE VACANCY ENGINEERING TECHNIQUE

Traditionally, to generate an excess of vacancies, the silicon ion energies used have been in the order of 1 to 2MeV. There have even been accounts of ion energies of up to 100MeV [21], in an attempt to position the interstitial rich region deep into the substrate so that during subsequent annealing, the excess interstitials do not interact with a doping implant or vacancies within the surface layer, as observed Eaglesham et al. [13].

This section presents simulations and experiments to demonstrate how it is possible to reduce the co-implant dose and energy, making this technique more appealing as an industrially relevant process with advantages which rival the currently preferred techniques.

Using Monte-Carlo simulations [22] and the NRD algorithm, it is possible to estimate the vacancy distributions for a range of silicon implants. Figure 1 illustrates the defect distributions for a 100keV (a), 500keV (b) and 1MeV (c) co-implants for a fixed dose.

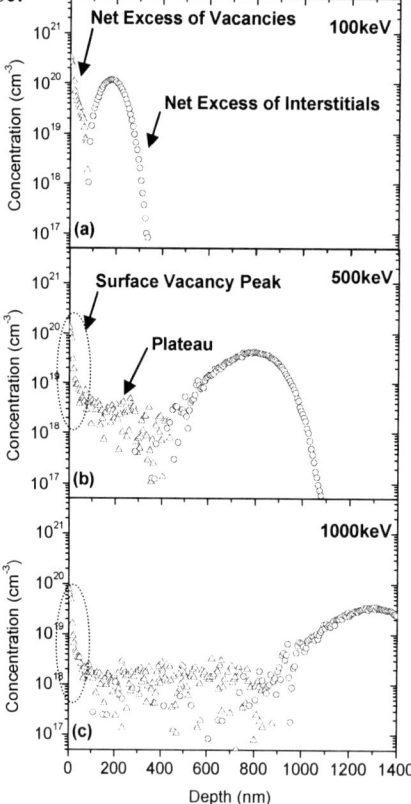

FIGURE 1. Estimated defect distributions calculated using Monte Carlo simulations and the NRD algorithm for 100, 500 and 1000keV silicon implants, at a fixed dose. The net vacancy and interstitial regions are represented by the open triangles and circles, respectively.

It is possible to see that as the ion energy is increased, the vacancy concentration (open triangles) decreases and a plateau forms to the point where the interstitial region (open circles) dominates. This trend suggests that that as the co-implant energy decreases the vacancy generation per ion increases. This indicates that for optimal vacancy generation, the ion energy should be reduced, contrary to the majority of studies presented in the scientific literature. However, as the ion energy is reduced, the probability of the excess interstitials interacting with the dopant/vacancies in the surface layer increases. Therefore, using a SOI structure similar to Roth, it is possible to restrict this interaction and optimize this process, where the minimum co-implant energy is dictated by the dimensions of the SOI structure [23].

To prove this experimentally, two vacancy generating implants were chosen. The first of these, a high, 1MeV silicon energy was chosen as this has been typically used in vacancy engineering experiments, and therefore known to generate an excess of vacancies. The second, a much lower, 300keV ion energy was chosen as this is the minimum ion energy to position the majority of the excess silicon interstitials beneath the BOX of a SOI substrate comprising of a 110/200nm structure, which is the configuration used for this experiment.

The specific silicon doses were tailored to theoretically generate the same aerial density of vacancies for the 1MeV and 300keV silicon co-implants. Due to the nuclear collision cross section increasing as the incident ion energy decreases, the 300keV co-implant has roughly a 2.5x reduction in dose to generate the same number of vacancies within the first 100nm surface region. To illustrate the similarities of not only the total number of vacancies but also the vacancy distributions (simulations), figure 2 presents the vacancy profiles for the two highest dose co-implants with respect to a 2keV boron implant (SIMS measurement) which was used as a detector to asses the success of the co-implants [24]. Throughout this study the co-implants were always performed prior to the doping implant, as it has been shown that due to boron clustering at low temperatures or even dynamically during the implant procedure, performing the vacancy generating implant after the boron implant, degrades the efficiency of the vacancies due to boron already being captured within stable clusters [25].

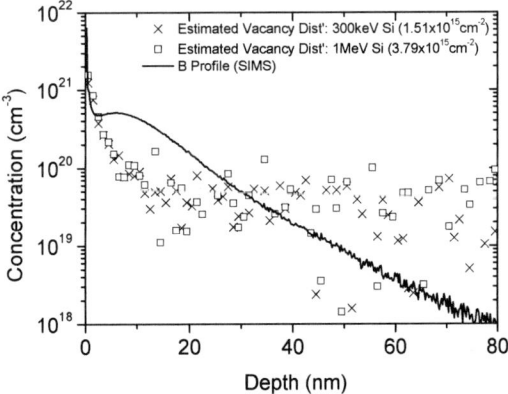

FIGURE 2. simulated vacancy distributions for the high dose 300keV (crosses) and 1MeV (open squares) silicon implants. A 2keV boron SIMS profile is shown to indicate the relation of the boron distribution to the estimated vacancy distributions.

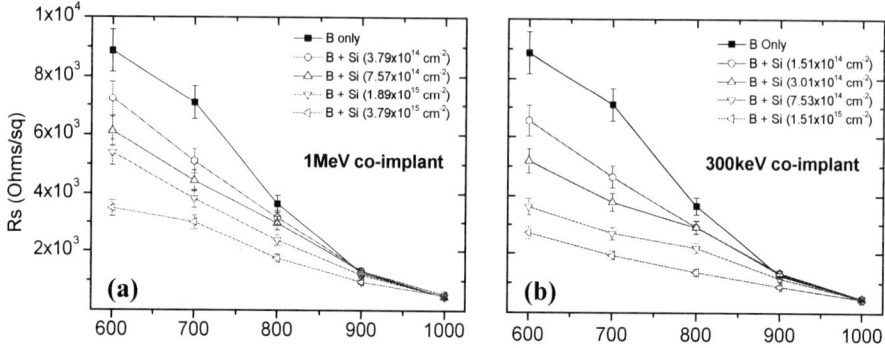

FIGURE 3. Van der Pauw Rs measurements of a 2keV boron implant with a 1MeV (a) or 300keV (b) co-implant as a function of anneal temperature and silicon dose. For comparison a boron reference curve is presented as the solid squares.

Using a 10s isochronal annealing scheme in the temperature range of 600°C to 1000°C, the resulting boron layers were analyzed via Van der Pauw Rs measurements (figure3). Comparing the 1MeV (a) and 300keV (b) at low temperatures, one can see that as the co-implant dose increases, the Rs improvement over the reference boron curve (solid squares) increases. Then as the anneal temperature is increased, the effect of the excess vacancies diminishes until no observable effect is seen at 1000°C. This was shown to be due to two phenomena: firstly, the natural dissolution of BICs at high temperatures and secondly, the inherent increase in boron solubility, causing all the curves to merge at high temperatures [24]. Comparing the silicon doses for both the 1MeV and 300keV co-implants as they are incremented from low to high doses, the degree of Rs improvement is similar, indicating that the mechanisms behind the boron improvements are the same. As it is known, in the 1MeV co-implant case this is due to a generation of excess vacancies, and it is possible to conclude that the improvements seen with the 300keV co-implant are also due to an excess of vacancies. This proves that a lower energy, lower dose, silicon co-implant can produce an excess of vacancies which has same levels of improvement in Rs at low temperatures compared to the traditional high energy co-implants. However, the level of Rs reduction is still not comparable to techniques such as pre-amorphization. Therefore, the improvement factors need to be increased, requiring further optimization of not only the co-implant but also the boron doping implant.

Creating Highly Active Ultra-shallow Junctions

In an attempt to increase the performance of the vacancy engineering technique, it is crucial to also optimize the boron doping implant. The majority of the published research within this area uses high energy co-implants with relatively high energy boron implants, 2keV and above. Therefore, the peak of the boron implant is normally located within the plateau region of the vacancy distribution shown in figure 1, resulting in a large fraction of the boron implant above the vacancy concentration. Therefore, locally, there is still a significant amount of silicon interstitials available to form BICs, assuming the plus 1 model [26].

186

The estimated vacancy distributions illustrated within figures 1 and 2 clearly show a surface vacancy distribution (0-20nm) which does not change significantly with ion energy. It is thought that this surface vacancy distribution is the result of a combination of effects arising from sputtering and primary recoils [27]. This rapidly decaying vacancy distribution from the entry surface has been neglected in the scientific literature due to the high boron implant energies. Therefore, the following experiment studies the effect of a vacancy engineering implant designed to push the boundaries of such a process, using an ultra-low boron implant which shows a much superior fit to the vacancy distribution [28].

Using a thinner SOI structure (55nm/145nm) it was possible to use a lower energy co-implant (160keV) than previously used, and as the Rs reduction is a function of silicon dose, the co-implant doses were designed to approach the amophization threshold ($4 \times 10^{14} cm^{-2}$. $8 \times 10^{14} cm^{-2}$ and $1.1 \times 10^{15} cm^{-2}$). The ultra-shallow boron implant was designed to be at 500eV, and to a dose of $10^{15} cm^{-2}$. Figure 4 illustrates a schematic of the experimental layout, indicating the vacancy (open squares) and interstitial (open triangles) rich regions with respect to the SOI structure (solid lines) and boron implant (dotted line) distribution.

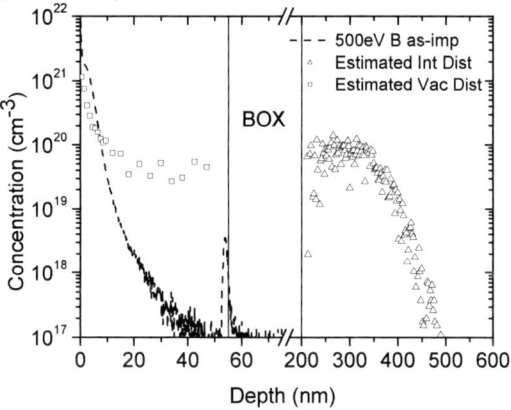

FIGURE 4. Schematic the experimental layout, illustrating the SOI structure, the distribution of excess vacancies (□) and interstitials (Δ) resulting from the highest-dose 160keV silicon co-implant (estimated from Monte Carlo simulations), and the 500eV B implant profile (measured by SIMS).

To evaluate the level of damage generated by the highest dose co-implant ($1.1 \times 10^{15} cm^{-2}$), TEM analyses (figure 5) were used before (a) and after a 700°C/10s anneal (b). In the as-implanted state, the TEM analysis illustrates that the SOI structure has sustained a high level of damage, resulting in the formation of a buried amorphous layer which extends from beneath the BOX into the silicon over-layer, whilst leaving a 30 nm crystalline surface region. On annealing, a reverse solid phase epitaxial process occurs, where the remaining crystalline surface layer acts as a seed to re-grow the amorphous layer down towards the BOX, and the underlying silicon acts as a seed to re-grow the amorphous layer up towards the BOX [29,30]. It is interesting to note that no interstitial defect band remains within the active device layer, and the only interstitial band formed is actually beneath the BOX in the structure of an End of Range (EOR) defect band, physically isolated from the top silicon layer.

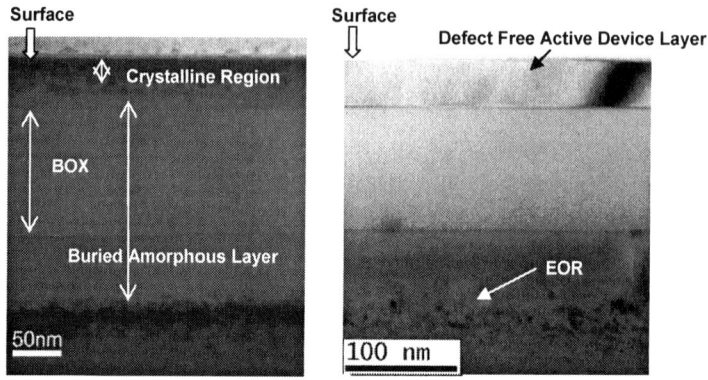

FIGURE 5. XTEM analysis of the SOI structure with a 160keV silicon co-implant, to a dose of 1.1×10^{15} cm^{-2} before (a) and after (b) annealing at 700°C for 10s.

The electrical properties of the resulting boron layers were examined as a function of isochronal annealing (10s) and co-implant dose via Rs and Hall analyses, presented within figures 6 and 7, respectively.

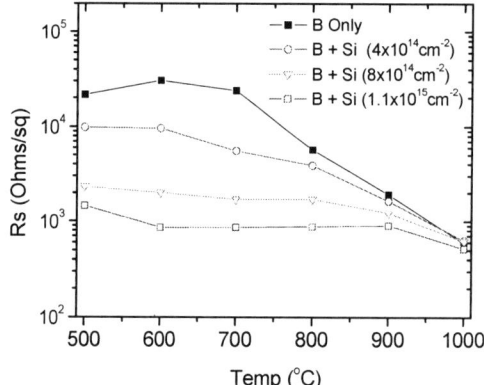

FIGURE 6. Van der Pauw Rs comparison of boron doped layers with and without a 160keV silicon co-implant in SOI as a function of 10s isochronal annealing, and co-implant dose. The silicon implant has been performed to a range of doses: 4×10^{14} cm^{-2} (open circles), 8×10^{15} cm^{-2} (open triangles) and 1.1×10^{15} cm^{-2} (open squares). For comparison the boron curve without a co-implant is represented by the solid squares.

In terms of Rs, the boron trends as a function of anneal temperature and co-implant dose are similar to the previous section, showing distinct improvements at low temperatures which increase with co-implant dose, with all the curves merging at high temperatures. At 700°C the Rs with the highest dose co-implant (open squares) is roughly 30 times lower than the reference boron curve (solid squares), resulting with a boron layer which has an average Rs of ~870 Ohms/sq over the temperature range of 600°C – 900°C, clearly illustrating a high degree of thermal stability.

To gain a greater understanding of the Rs curves, the corresponding substitutional fraction of the boron atoms (Ns) is presented in figure 7 using the same notation as figure 6. Without the co-implant, the boron only curve shows a distinct deactivation in the 500°C to 700°C temperature range, with a subsequent re-activation when the temperature is increased above 700°C. This deactivation/re-activation can be attributed to the formation and dissolution kinetics of the boron clusters [31,29]. By adding the lowest dose co-implant, it is possible to see that the level of deactivation is reduced. Furthermore, when the highest silicon dose is used, no significant deactivation is observed. Thus, illustrating that the excess vacancies inhibit the formation of BICs, and therefore, enable boron to become electrically active at low temperatures, and remain active at higher temperatures.

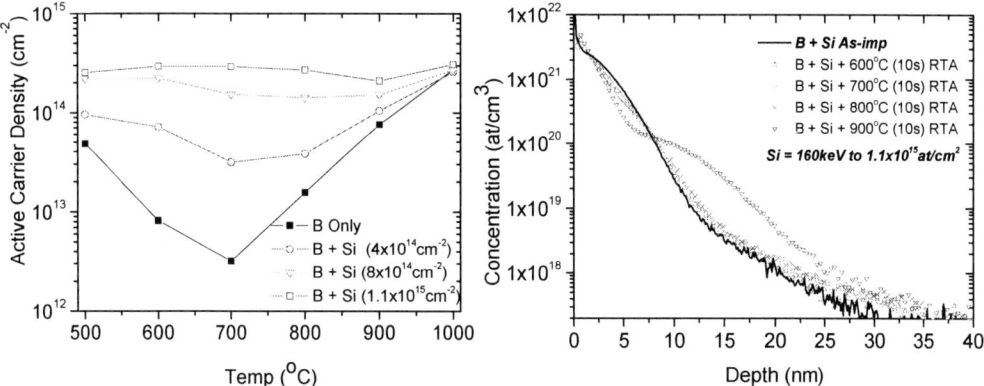

FIGURE 7. Ns measurements of boron doped layer with and without a 160keV co-implant using the same notation as figure 6.

FIGURE 8. SIMS analyses of boron doped layers with a 160keV silicon co-implant, to a dose of 1.1×10^{15} cm^{-2}, as a function of 10s isochronal annealing in the temperature range of 600°C to 900°C.

So far only sheet electrical measurements have been presented, which is only one component of the junction requirements. Therefore, to finish this study the effect of the highest dose co-implant in terms of boron diffusion is examined via SIMS analyses (figure 8), which are presented as a function of anneal temperature, from 600°C-900°C. Without any thermal processing (solid line), the metallurgical junction depth taken at 3×10^{18}cm^{-3} corresponds to ~15nm. Therefore, in an as-implanted state, the junction depth is only just shallow enough to fulfill the requirements. This presents a challenge as any thermal processing implemented to achieve a high level of electrical activation has to be ascertained with minimal diffusion. However, it is clearly seen that with the vacancy engineering co-implant (symbols), the junction remains close to as-implanted distribution up to an anneal temperature of 800°C, and only a significant amount of diffusion is seen when the temperature is increased to 900°C. This provides a large process integration window for ease of implementation in industrial manufacturing.

SUMMARY

An alternative method for creating ultra-shallow p-type junctions, known as vacancy engineering, has been reviewed to date. It has been shown that by understanding the vacancy generation through simulations, it is possible to optimize the implant parameters and subsequent thermal processes, in a manner which results in the ability to be able to engineer a concentration of vacancies to efficiently improve the electrical and diffusion properties of a boron doped layer. Therefore, resulting in a highly stable and virtually defect free shallow junction. It is speculated that with further optimization it is possible to increase the improvement factors even further, making this technique more appealing for industrial application.

REFERENCES

1. ITRS http://public.itrs.net, 27 (2004).
2. A.Michel, W.Rausch, P.Ronsheim, and R.Kastl , Appl.Phys.Lett. 50, 416 (1987).
3. N.Cowern, K.Janssen, and H.Jos , J.Appl.Phys. 68, 6191 (1990).
4. O.Holland, J.Budai, and B.Nielsen , Mat.Sci.Eng. A253, 240 (1998).
5. B.Nielsen, O.Holland, T.Leung, and K.Lynn , J.Appl.Phys. 74, 1636 (1993).
6. K.Winterbon , Rad.Eff. 46, 181 (1980).
7. L.Shao, X.Wang, J.Liu, J.Bennett, L.Larsen, and W.Chu , J.Appl.Phys. 92, 4307 (2002).
8. O.W.Holland and C.W.White , Nucl.Inst.Meth.Phys.Res. B59/60, 353 (1991).
9. V.Raineri, R.Schreutelkamp, F.Saris, K.Janssen, and R.Kaim , Appl.Phys.Lett. 58, 4, 922 (1991).
10. S.Saito, K.Hamada, and A.Mineji , Nucl.Inst.Meth.Phys.Res. B 120, 37 (1996).
11. E.Roth, O.Holland, V.Venezia, and B.Nielsen , J.Elec.Mat. 26, 11, 1349 (1997).
12. V.Venezia, T.Haynes, A.Agarwal, L.Pelaz, H.-J Gossmann, D.Jacobson, and D.Eaglesham , Appl.Phys.Lett. 74, 9, 1299 (1999).
13. D.Eaglesham, T.Haynes, H.-J Gossmann, D.Jacobson, and P.Stolk , Appl.Phys.Lett. 70, 24, 3281 (1997).
14. Lin Shao, X.lu, X.Wang, I.Rusakova, J.Liu, and W.Chu , Appl.Phys.Lett. 78, 16, 2321 (2001).
15. W.Chu, J.Liu, Lin Shao, X.Wang, and P.Ling , Nucl.Inst.Meth.Phys.Res. B 190, 34 (2002).
16. Lin Shao, J.Liu, X.Wang, H.Chen, P.Thomson, and W.Chu , Nucl.Inst.Meth.Phys.Res. B 206, 413 (2003).
17. Lin Shao, P.Thomson, P.van der Heide, S.Patel, Q.Chen, X.Wang, H.Chen, J.Liu, and W.Chu , Extended Abstracts, 59 (2003).
18. A.Nejim and B.Sealy , Semicond.Sci.Tech. 18, 839 (2003).
19. R.Kalyanaraman, V.Venezia, L.Pelaz, T.Haynes, H.-J Gossmann, and C.Rafferty , Appl.Phys.Lett. 82, 2, 215 (2003).
20. Lin Shao, J.Zhang, J.Chen, D.Tang, P.Thomson, S.Patel, X.Wang, H.Chen, J.Liu, and W.Chu , Appl.Phys.Lett. 84, 3325 (2004).
21. V.Skuratov, N.Dinu, I.Antonova, and V.Obodnikov , Sur.Eng.Sur.Ins.Vac.Tec. 63, 571 (2001).
22. R.Webb , Monte Carlo Simulator of Ions in Solids, Cascade, IBC, University of Surrey (2005).
23. A.J.Smith, B.Colombeau, R.Gwilliam, E.Collart, N.Cowern, and B.Sealy , Mat.Res.Soc.Symp.Proc. 810, C3.8.1 (2004).
24. A.J.Smith, B.Colombeau, N.Bennett, R.Gwilliam, N.Cowern, and B.Sealy , Mat.Res.Soc.Symp.Proc. 864, E7.1 (2005).
25. R.Gwilliam, N.Cowern, B.Colombeau, B.Sealy, and A.Smith , CAARI 2006 Proceedings: Fortworth (2006).
26. M.Giles , J.Electrochem.Soc. 138, 1160 (1991).
27. N.Cowern, A.J.Smith, B.Colombeau, R.Gwilliam, B.Sealy, and E.Collart , IEDM Tech.Digest 2005 (IEE, Piscataway, NJ), 39.1.1 (2005).
28. A.J.Smith, N.Cowern, R.Gwilliam, B.Sealy, B.Colombeau, E.Collart, S.Gennaro, D.Giubertoni, M.Bersani, and M.Barozzi , Appl.Phys.Lett. 88, 082112 (2006).
29. A.Smith, B.Colombeau, R.Gwilliam, N.Cowern, B.Sealy, M.Milosavljevic, E.Collart, S.Gennaro, M.Bersani, and M.Barozzi , Mat.Sci.Eng.B. 124-125, 210 (2005).
30. M.El-Ghor, O.Holland, C.White, and S.Pennycook , J.Mater.Res 5, 352 (1990).
31. S.Mirabella, E.Bruno, F.Priolo, D.De Salvador, E.Napolitani, A.Drigo, and A.Carnera , Appl.Phys.Lett. 83, 680 (2003).

Unusual Application Of Ion Beam Analysis For The Study Of Surface Layers On Materials Relevant To Cultural Heritage

F. Mathis[1,2,3]; J. Salomon[1]; P. Trocellier[2]; M. Aucouturier[1]

1 Centre de Recherche et de Restauration des Musées de France CNRS UMR 171 Palais du Louvre – Porte des Lions 75001 Paris, France
2 Service de Recherche en Métallurgie Physique CEA Saclay 91191 Gif sur Yvette Cedex, France
3 Centre Européen d'Archéométrie Université de Liège Sart Tilman B15 4000 Liège, Belgique

Abstract. Recently a new thematic of research – intentional patinas on antic copper-base objects – lead the AGLAE (Accélérateur Grand Louvre pour l'Analyse Elémentaire) team of the C2RMF (Centre de Recherche et de Restauration des Musées de France) to improve its methods of analyzing thin surface layers both in their elemental composition and in-depth elemental distribution. A new beam extraction set-up containing a particle detector has been developed in order to use a 6 MeV alpha beam both in PIXE and RBS mode and to monitor precisely the ion dose received by the sample. Both RBS and ionization cross sections were assessed in order to make sure that the analysis can be quantitative. This set up allows great progresses in the understanding of both nature and structure of this very particular oxide layer obtained in the antiquity by chemical treatment on copper alloys, containing gold and/or silver and presenting very interesting properties of color and stability.

Besides the non destructive properties of the IBA in external beam mode, this method of analyzing allows the study of samples in interaction with its environment. This was used to study the high temperature oxidation of Cu-Sn alloys using a furnace developed in order to heat a sample and analyze it in RBS mode at the same time. This new way of studying the growth of oxide layers permits to understand the oxidation mechanism of this system and to propose an experimental model for the identification of oxide layers due to an exposition to a high temperature, model needed for a long time by curators in charge of the study and the conservation of archaeological bronzes.

Keywords: IBA, PIXE, RBS, External beam mode, Archaeometry, Bronze, Non destructive Analysis
PACS: 29.27.Ac, 29.30.Ep, 81.65.Mq, 82.80.Yc

INTRODUCTION

The study of artifacts relevant to cultural heritage is submitted to many constraints due to the fragility, the rarity, and the complexity of such materials. IBA (Ion Beam Analysis) is therefore a well-adapted method for analyzing this type of material thanks to the precision of the results, the localization of the analysis, especially in micro-beam mode, and the non destructivity in external beam mode. The most commonly used IBA method in cultural heritage artifacts analysis is the Particle Induced X-ray Emission (PIXE) method which provides the elemental composition of the object with

CP876, The Physics of Ionized Gases: 23rd Summer School and International Symposium,
edited by L. Hadžievski, B. P. Marinković, and N. S. Simonović

a precision sufficient to allow trace analyses and for example provenance study. A well-known specificity of the IBA by conventional users is the access to the elemental profile allowed by the use of Rutherford Backscattering Spectroscopy (RBS). This method is to our knowledge less developed by external beam users, and this is a limit to the comprehension of layered objects. This is particularly due to the difficulty of the extraction at the atmosphere of a $^4He^{2+}$ beam which is one of the most powerful beams for RBS. The C2RMF (Centre de Recherche et de Restauration des Musées de France) IBA facility AGLAE (Accélérateur Grand Louvre d'Analyse Elémentaire)[1,2] is specially devoted to the analysis of museum artefacts and has increased the possibilities of extracted beam, the major method used in our laboratory. AGLAE is one of the only IBA facility which uses in routine a 3 MeV $^4He^{2+}$ beam extracted in atmosphere with sufficiently good characteristics to make RBS experiments [3,4]. The particularity of this set-up permits to develop new fields of research in the study of thin layers on metal archaeological objects.

Recently a new thematic of research – the study of intentional patinas on antic copper-base objects – lead the AGLAE team to improve its methods of analyzing thin layers both in their elemental composition and in-depth distribution. A new beam extraction set-up, containing a particle detector has been developed in order to use a 6 MeV alpha beam both in PIXE and RBS mode and to monitor precisely the current sent to the sample. Both RBS and ionization cross sections were verified to be sure of the quantization of these analyzes. This set up has allowed great progresses in the comprehension of the nature and the structure of a very particular oxide layer with very interesting properties of color and stability produced during the antiquity by chemical treatment on copper alloys containing gold and/or silver.

Besides the non destructive properties of IBA in external beam mode, this way of analysis permits also to study a sample in interaction with its environment. We use this particularity to study high temperature oxidation of Cu-Sn alloys using a furnace developed in order to heat a sample and allow analyses by RBS at the same time. This new way to study the growth of oxide layers leads us to understand the oxidation mechanism of this system and propose an experimental model for the identification of oxide layers due to an exposition to a high temperature, model needed since a long time by curators in charge of the study and the conservation of archaeological bronzes.

THE 6 MEV $^4HE^{2+}$ EXTERNAL BEAM: USE AND APPLICATION TO THE STUDY OF ANTIC BLACK INTENTIONAL PATINAS

Several research works have been recently published about the existence of a known intentional black patination of copper alloys containing small amounts of gold and sometimes silver, on Egyptian, Minoan, Roman and Japanese objects [5-9]. These black coloured alloys, called "black bronze", "black copper", "Corinthian bronze", *shakudo*, etc., were used either for bulk objects, subsequently inlaid with other alloys, or for the inlays themselves applied on copper-base objects (Fig. 1). Patinas are layers generally oxidised, from 1 to 20 µm thick containing mostly the same metallic elements as the substrate. The complexity of the study (poor knowledge of the techniques used in the antique world, important corrosion phenomena) implies that, one has to get as much information as possible from analyses.

For a good understanding of the structure of this type of superficial layers, one needs to determine the quantitative elemental composition of the patina to distinguish the elements contained in the substrate from those originating from the chemical treatment or from the burying environment. The use of a 3 MeV proton beam for PIXE and a fortiori the use of XRF (X-ray Fluorescence) as in the published analyses of this type of objects does not permit the quantitative analysis, because it penetrates too deep into the material, which causes a resulting influence of the substrate on the results. One needs also the elemental depth distribution in order to evaluate the in-depth homogeneity of the layer.

FIGURE 1. Example of black patinated antic objects.
Left: roman Inkpot of Vaison-La-Romaine, Louvre DAGER; the himation of Venus in patinated.
Right: Egyptian Feminine Statuette, Louvre DAE; all the dress is patinated.

This precision in the results is needed because the objects are very precious, no sampling is allowed that limits the allowed analyses, and also because the properties of this oxide layer are not understood. Thanks to previously published analyses it is known that the patina layer is constituted of copper oxide cuprite, Cu_2O, which is normally red. The black coloration must be due to other elements present in the layer and it is important to know the nature, the quantity and the distribution of these elements.

The quality of the RBS results is function of the beam used, the mass discrimination depending of the mass of the incident particle: an alpha beam is therefore more valuable than a proton beam. Considering the possible heterogeneous nature of the objects, PIXE and RBS analyses have to be done at the same time on the same point. The use of a unique beam is also better considering the time constraints (objects are brought to the laboratory for a very short time).

These reasons made us change our previous experimental protocol and choose one beam having a lower penetration in the material than the 3MeV proton beam, while keeping good ionisation cross sections for PIXE and a good mass resolution for RBS. The use of a 6 MeV ^{4}He beam in normal incidence (the most convenient when analysing objects of complex shape) seems to meet all these constraints.

The previous experimental set-up presented some drawbacks which could be critical when using this new type of beam for simultaneous PIXE–RBS measurements, and this lead us to design a new extraction system including an annular particle detector [10] able to permit the PIXE measurement without changing the precedent set-up, do the RBS measurement and also monitor the particles current; the latter is necessary to interpret the RBS spectra and is one of the major difficulties in external beam mode. We also compared the ionization cross sections of the 6 MeV alpha beam [11] with the 3 MeV proton cross sections and the theoretical cross sections present in the GUPIX package [12].

The New Experimental Beam Extraction System

The new device must respect several constraints:

• a beam extracting nozzle containing an annular particle detector (20 mm diameter) easy to fix and not prone to mechanical displacement;

• a system for current monitoring with at least the same reliability as the previous one;

• a set-up usable for all types of beams and experiments, implying keeping the configuration of the other detectors.

To meet these constraints and for mechanical reasons, it was impossible to use a complete commercial detector; thus it was decided to insert only the detector crystal without its housing in the nozzle. Figure 2 shows the scheme of the new device. The way of setting the crystal and making the electrical contacts was directly inspired by the housing of the commercial detectors.

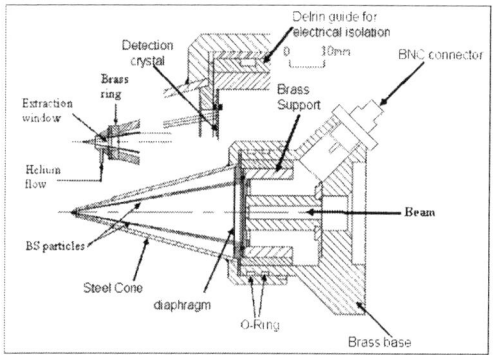

FIGURE 2. Scheme of the new beam extraction system

The new set-up is no longer compatible with the previous beam monitoring based on the measurement of X-ray emission by the exit Si_3N_4 window. The chosen solution is to deflect the beam periodically (about 10% of the measuring time) on a reference (brass ring) located behind the exit window and to send the RBS signal to another detection channel. The deflection of the beam is produced by the magnetic deflector of the micro beam, driven by a program in Labview® code. The signal of the detector is sent after amplification to an ADC (Analog to Digital Converter) and a SCA (Single Channel Analyser) which are gated in synchronization with the beam movement.

The PIXE Cross-Sections

The cross section measurements were carried out at the micro-spot beam line of the AGLAE accelerator, mainly designed to perform measurements with an external beam. The targets were thin layers of pure elements chosen with respect to a variety of Z numbers. The samples can be considered as thin targets: the energy loss is always negligible and the self-absorption of the induced X-rays has not to be considered.

The experimentally calculated X-ray yields are shown in Figure 3. For the $^4He^{2+}$ beams of different energies, the yields are shifted parallel in height. The yield of the proton beam shows another slope, since the amount of yielded X-rays does not only depend on the energy of the ion but also on the Z number of the ion. Therefore, the yield from the 3 MeV proton beam crosses the 6 MeV $^4He^{2+}$ X-ray yield at a Z number of 26, showing the fact that α-PIXE is favourable for low Z elements. For the yields of L-lines, the α-PIXE gives approximately the same yield as the proton beam with a slight advantage for the 6 MeV $^4He^{2+}$ at Z numbers around 50 (Sn) and less significantly lower yields for Au and Pb.

The experimental data follows in general the behaviour predicted by the theory, but all experimental yield ratios are slightly less than the theoretical calculated ratios given in the GUPIX package. However all discrepancies show a lower yield ratio for the experimental data. This difference might come from a systematic experimental error (e.g. due to incorrect sample positioning or charge measuring) or from a slight inaccuracy in the theoretical α-PIXE cross sections for those lines.

The background amount in the spectra is another major criterion for a good limit of detection. Two PIXE spectra from 3 MeV protons and 6 MeV ^4He are compared in Figure 3. The latter shows a lower background in the area around 3–6 KeV. This is due to the reduction by a factor of two of the maximum energy of the electron bremsstrahlung for 6 MeV $^4He^{2+}$, since 6 MeV $^4He^{2+}$ has an E/amu-value of only 1.5 MeV/amu. The higher background for α-PIXE in below 2.5 keV is compensated by the much higher yields for elemental lines in this area. Both provide a better peak-to-noise ratio and thus a better sensitivity of α-PIXE for elements having usable X-ray lines in this region of the spectra.

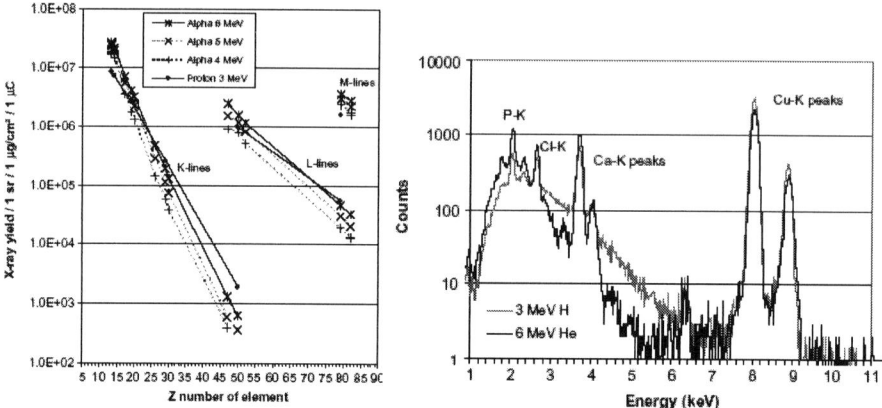

FIGURE 3. <u>Left</u>: measured X-ray yields per 1 sr, per 1 $\mu g/cm^2$ target material and per 1 μC particle fluence. <u>Right</u>: PIXE spectra of a 3 MeV proton beam and a 6 MeV $^4He^{2+}$ beam. The target is made of copper deposited on Mylar backing containing some additional elements. Both spectra are done under the same particle fluence of 0.6 μC

The results of the use of PIXE-RBS with 6 MeV alpha beam on antic black patinated artifacts

This study is a main part of the first author's PhD thesis [13] and the detailed results can be found in it. For this study nine objects were analyzed, four Egyptian artifacts dated from the middle kingdom to the third intermediary period (1800-600 BC) and five Roman artifacts dated from the 1st century to the 4th century AD. As can be seen on Figure 4, which is representative of a typical *"black bronze"* all patinas are made of copper oxide containing gold and/or silver. They can contain also others elements in small amounts like tin (when it is present in the base alloy), arsenic and sulfur. The precious elements added are present in all the layers and are certainly responsible of this particular color. The dispersion of the results is very important, and no correlation could be performed between the composition and the production period. The Egyptian alloys used to be patinated are very different from each other as if the techniques were not precisely codified. The use of tin in the alloy is almost systematic in opposition to the Roman's alloys which are only copper-based alloys without tin. The Egyptian alloys contain also more gold than silver contrarily to the Roman ones. The presence of sulfur detected in the majority of the patinas seems to come from the chemical treatment used to make the patina. The patinas are generally thick (>10µm).

The use of 6MeV $^4He^{2+}$ and this new set-uplead to make the only study on a large corpus of objects analyzed with the same experimental protocol, which allow a precise comparison of the results. One of the great improvement of this protocol in terms of non destructive analyses is the differentiation between the analysis of the bulk metal and that of the patina, differentiation which is needed in order to compare different objects and to understand the realization of the patina. This realization depends on two criteria: the metal used and the recipe of the chemical treatment.

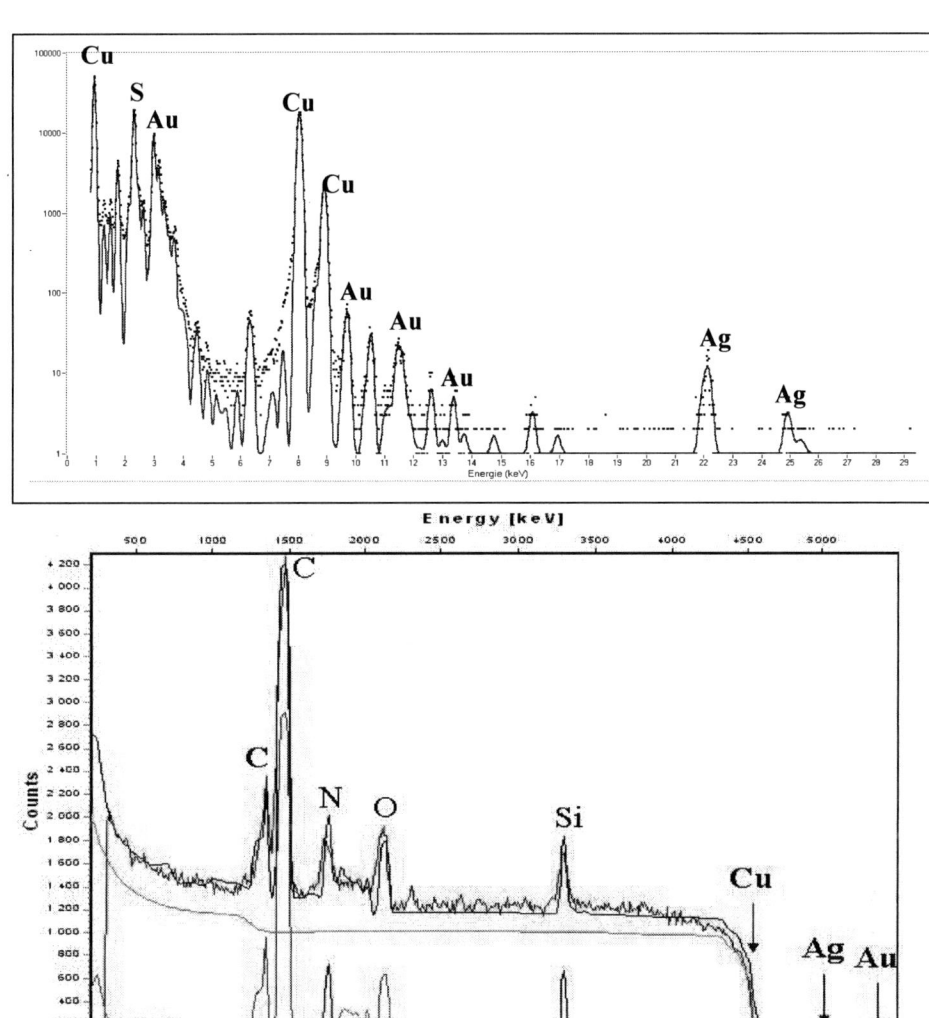

FIGURE 4. Typical PIXE and RBS spectra of a black bronze patina with a 6 MeV alpha beam. On the RBS spectrum, elements in black are to be attributed to the experimental set up while elements in blue are from the sample

THE OPEN AIR FURNACE FOR THE RBS STUDY OF SAMPLE EVOLUTION WITH TEMPERATURE: APPLICATION TO THE STUDY OF THE OXIDATION OF Cu-Sn ALLOYS

Besides the non destructive properties of the IBA in external beam mode, this way of analysis gives also the great opportunity to study a sample by IBA in interaction with its environment and to use the advantages of these techniques to a new domain of application.

An open air furnace especially developed to permit the study of the heated sample by RBS in extracted beam mode has been designed by the AGLAE team. This furnace has been previously used to study the high temperature oxidation of galena [14] (to understand recipes of antic cosmetics) and the formation of gilded decoration on glass [15]. It was here improved and applied to study the kinetics of the high temperature oxidation of Cu-Sn alloys [16].

The development of the open-air furnace in view of studying high-temperature oxidation by real time RBS imposed certain constraints:

• the sample surface must be in contact with the atmosphere for undergoing oxidation;

• the sample surface must remain directly accessible to the beam, and the beam exit window must be at most 3 mm from the surface to reduce beam straggling in air;

• the heating system should reach 650–700 °C with variations of less than a few degrees;

• the beam exit window and the detectors must be shielded from the heat source.

The system that has been developed meets all the above constraints (Fig. 6).

This system has been used to measure in function of time the nature and the thickness of the oxide layers growing at the sample surface and to determine a law of growth of these layers. The RBS has the great advantage on other techniques generally used in kinetics study (e.g. thermo-gravimetry) to separate the different layers and to offer then the possibility to determine individual growth kinetics.

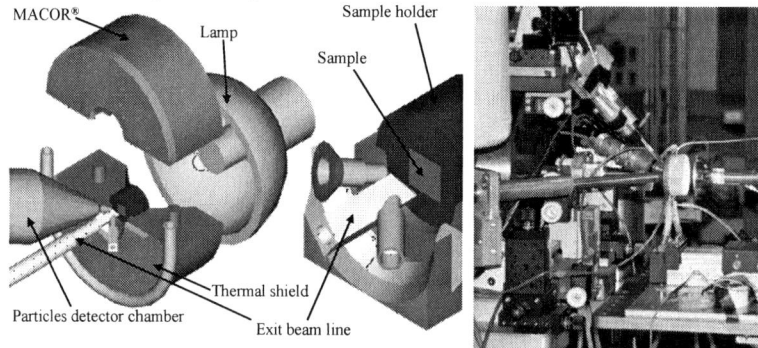

FIGURE 5. Sketch of the furnace and image of the set-up in function

Figure 7 presents different spectra obtained on a sample in function of the time and the parabolic law which was determined.

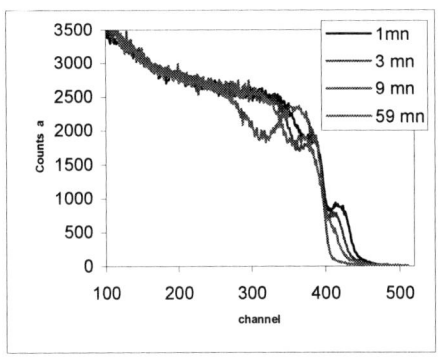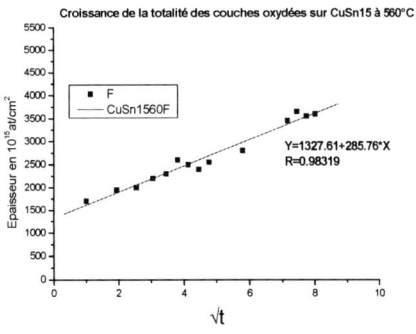

FIGURE 6. RBS Spectra and kinetic law obtained on a Cu 10 wt % tin heated at 560°C during 90 minutes. The graph at the right presents the thickness of the layer in function of the square root of the time illustrating a parabolic law of growth.

This study brings totally new results on the oxidation mechanism of this system. Tin addition in copper produces a significant decrease in the bronze oxide growth rate. This can be explained by the effect of tin oxide on the migration of the elements in the scale. The role of tin on the inhibition of oxidation rate is relatively different than that of classical alloying elements in oxidation protection. Tin does not create a protective continuous layer at any amount in the alloy. Copper-tin alloys develop a mixed layer of Cu_2O and SnO_2 which plays a similar role: the layer is almost totally protective at temperature under 400°C with kinetics near those known in the case of passivating layers with a non-parabolic growth law, indicating that the protection mechanism inhibits the diffusion of the species; it is partially protective at higher temperature, with growth rates considerably decreased comparing to copper oxidation, but keeping a parabolic law of growth, indicating that diffusion is the major growth mechanism like in copper oxidation.

Two critical parameters are revealed:

- A critical temperature around 400°C under which the formation of a passivating layer is observed and over which the parabolic law of growth known for copper oxidation is operating, but with a dramatically decreased growth rate as compared to copper oxidation.
- A critical tin concentration around 7 wt % under which the growth rate decrease is relatively moderate and over which the inhibition is maximum.

CONCLUSION

The technical developments in terms of extraction beam lead to increase the possibility of use of the IBA to different application fields. Here is presented the contribution it has given in a classical material science problematic like the high temperature oxidation study of a given alloy. It has been proved that IBA are perfectly adapted to the study of the materials surface even when it is in evolution.

The continuous improvement of the experimental set-up and the particular adaptability of IBA in terms of nature and energy of the ions used and therefore in terms of analysed thickness and results accuracy brings also great advantages for the research in complex fields like the study of antic artefacts and the comprehension of fabrication techniques.

A well adapted protocol allows making significant progress in the study of a particular and until recently not well-known decoration technique used by antic craftsmen. This may have important consequences in the field of the technical history especially in the evaluation of the knowledge of antic artisans in chemistry and material science.

ACKNOWLEDGMENTS

The authors wish to acknowledge here the contribution of the members of the technical team of AGLAE, Laurent Pichon and Brice Moignard who are responsible of all the technical realization developed in this paper, and also the contribution of Stefan Röhrs post-doctorate in the AGLAE team for his measurement and interpretation of $^4He^{2+}$ PIXE cross-sections.

REFERENCES

1. Amsel G., Menu M., Moulin J. and Salomon J., *Nucl. Instr. and Meth. B* **45**, 296 (1990).
2. Calligaro T., Dran J.-C., Ioannidou E., Moignard B., Pichon L. and Salomon J., *Nucl. Instr. and Meth. B* **161-163**, 328-333 (2000).
3. Ioannidou E., Bourgarit D., Calligaro T., Dran J.-C., Dubus M., Salomon J. and Walter P., *Nucl. Instr. and Meth. B* **161-163**, 730-736 (2000).
4. Dran J.-C., Salomon J., Calligaro T. and Walter P., *Nucl. Instr. and Meth. B* **219-220**, 7-15 (2004).
5. Craddock P. and Giumlia-Mair A., "Hsmn-Km, Corinthian bronze, shakudo: black-patinated bronze in the ancient world" in *Metal Plating and Patination,* edited by S. La Niece and P. Craddock, London: Butterworth-Heinemann Ltd, 1993, pp. 101-127.
6. Giumlia-Mair A. and Craddock P., *Antike Welt* **22**, (1993).
7. Giumlia-Mair A., *Antike Welt* **27**, 3136321 (1996).
8. Giumlia-Mair A. and Quirke S., *Revue d'Egyptologie* **48**, (1997).
9. Delange E., Meyohas M.-E. and Aucouturier, *Journal of cultural heritage* **6**, 99-113 (2005).
10. Mathis F., Moignard B., Pichon L., Dubreuil O. and Salomon J., *Nucl. Instr.and Meth.B* **240**, 532-538 (2005).
11. Röhrs S., Calligaro T., Mathis F., Ortega-Feliu I., Salomon J., Walter P., *Nucl. Instr.and Meth B.* **249**, 604-607 (2006).
12. Hopman T.L., Nejedly Z., J.A. Maxwell J.A., Campbell J.L., *Nucl. Instr.and Meth.B* **189**, 138 (2002).
13. Mathis F., *Croissance Et Proprietes Des Couches D'oxydation Et Des Patines A La Surface D'alliages Cuivreux D'interet Archeologique Ou Artistique*, PhD Thesis Université Paris Sud XI, http://tel.ccsd.cnrs.fr/tel-00011255, 2005.
14. Martinetto P., Dran J. C., Moignard B., Salomon J. and Walter P., *Nucl. Instr.and Meth B* **181**, 703-706 (2001).
15. Deram V. *Décor doré à base d'organométalliques sur verre. Formation et durabilité* PhD Thesis, Ecole des Mines de Paris, 2005
16. Mathis F., Salomon J., Moignard B., Pichon L., Aucouturier M. and Dran J. -C., *Nucl. Instr.and Meth B* **226**, 147-152 (2004).

Chemical Sputtering of Fusion Plasma-Facing Carbon Surfaces

P. S. Krstić[1], S. J. Stuart[2], and C. O. Reinhold[1]

[1]Oak Ridge National Laboratory, Physics Division, P.O. Box 2008, Oak Ridge, TN 37831-6372 USA
[2]Clemson University, Department of Chemistry, Clemson, SC 29634 USA

Abstract. We perform molecular dynamics simulations of the chemical sputtering of deuterated amorphous carbon surfaces irradiated by low energy deuterium atoms and molecules (≤ 30 eV/D). Particular attention was paid to the proper preparation of the surfaces, as well as to the internal (rovibrational) state of impinging molecules. Sputtered hydrocarbons are analyzed with respect to their mass, kinetic energy and angular distribution. The sputtering yields are in good agreement with recent experimental results.

Keywords: chemical sputtering, low energy, impacting molecules, vibrational excitation, kinetic energy spectra, angular spectra
PACS: 34.50.Dy, 52.40.Hf, 68.03.Hj, 61.43.Bn, 61.80.x

1. INTRODUCTION

Interactions of the plasma-facing walls (C, Be, W) with the surrounding edge-plasma particles is one of the most important, yet poorly understood issues for construction of ITER and other future magnetically confined plasma fusion reactors. Due to its thermal and mechanical properties, carbon is the most likely constituent of the divertor tiles, which experience the largest particle fluxes. Chemical sputtering, the most complex of the plasma-surface interactions, is a process where bombardment by ions, atoms or molecules induces a chemical reaction which produces a particle that is weakly bound to the surface and hence ejected or easily desorbed into the gas phase. The actual nature of these chemical processes is not well understood. For carbon surfaces, it is hypothesized that incident ions break bonds within the collision cascade which are then passivated by the abundant flux of atomic hydrogen from the hydrogen fusion plasma environment. This leads to the formation of hydrocarbon molecules underneath the surface, which desorb from the surface. These are further transported within the plasma, possibly dissociated and ionized and redeposited on the plasma-facing materials. Besides eroding the tiles, the sputtered hydrocarbons pollute the plasma, and can be a source of tritium deposition and retention at the surfaces.

The typical energies in the plasma surrounding the divertor tiles is a few eV to tens of eV, though they can reach keVs in Edge-Localized Mode (ELM) plasma bursts. Depending on the amount of plasma detachment, particle fluxes are expected to reach 10^{25} m^{-2}s^{-1}. With approximately the same concentrations of ions and neutrals at the divertor, the most abundant constituents are deuterium and tritium, both ions and

CP876, *The Physics of Ionized Gases: 23rd Summer School and International Symposium*,
edited by L. Hadžievski, B. P. Marinković, and N. S. Simonović

neutrals in atomic and molecular forms, possibly excited, which irradiate the surface of the carbon tiles. In addition, other particles present in the plasma, like inert gases (injected for diagnostic purposes), various metals (W, Be, Fe), oxygen, carbon and hydrocarbons may impinge upon the surface as well. Under continuous bombardment, one may expect amorphization of an initially crystalline surface, for which reason we focus our attention on amorphous target surfaces. Amorphous carbon is a mixture of sp, sp^2 and sp^3 hybridized carbon atoms, with "dangling" bonds that can be easily saturated by hydrogen (deuterium, tritium) atoms from the plasma environment. This leads to the process of hydrogenation (deuteration, tritiation) of carbon with H/C ratios reaching values of ~0.4 in the bulk [1,2], although the local hydrogen content at the top surface layers can reach significantly higher levels.

A description of our molecular dynamics (MD) approach, preparation and thermostatting of the surfaces, as well as an analysis of the target surfaces, is given in Section 2. Our results for the spectra of ejected particles, including their type, kinetic energy and angular distributions, and influence of the vibrational excitation of D_2 on the sputtering yield are presented in Section 3. Finally, our conclusions are given in Section 4.

2. PREPARATION OF TARGET SURFACES

The yields of the sputtered particles are strongly dependent on the details of the potential in which the particles move, on the microstructure of the target, and on the type and internal state of the impacting particle. Simulations were performed with the reactive empirical bond order (REBO) potential [3]. This potential has been widely and successfully used for a number of different carbon-based and hydrocarbon systems. As a member of the classical bond order family of potentials (Tersoff–Brenner potentials [4]) it provides a good, empirical description of the covalent bond for nonpolar systems, without incurring the enormous expense of a quantum mechanical calculation. More advanced bond order potentials are available that include better descriptions of the torsional and van der Waals interactions (AIREBO [5]), but these enhancements are not expected to contribute significantly for covalent network solids such as a-C:D.

In all of our simulations the impacting particles (either a D atom or D_2 molecule) were introduced at a distance of more than 20 Å from the initial position of the surface with a translational kinetic energy between 7.5 and 30 eV/D directed normal to the surface. The placement of the center of mass of the incoming particle was chosen uniformly at random within a plane parallel to the surface. For molecular D_2 impacts, the initial angular orientation was also chosen randomly with uniform distribution. The molecular projectile was given a rotational kinetic energy corresponding to a temperature of 1000 K, with the direction of the angular momentum vector chosen randomly from the unit sphere. In order to prevent net motion of the center of mass of the system upon repeated impacts, the atoms in the bottom 2 Å layer of the substrate furthest from the impact surface were held rigid in the direction of impact (orthogonal to the surface). These atoms were always more than 25 Å below the impact surface.

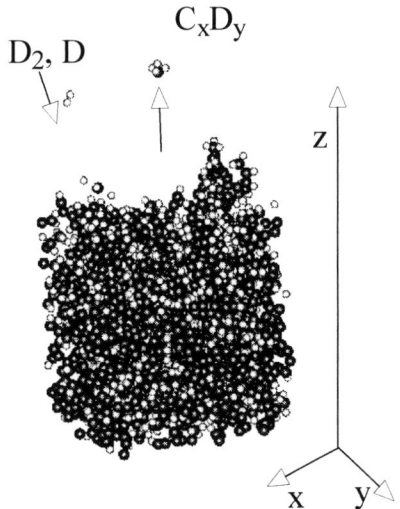

FIGURE 1. Geometry of the simulation cell.

We performed two types of bombardment simulations. The first one was devoted to the surface preparation of bombarded amorphous carbon (a:C) while the second one is used to calculate the sputtering yields. The surface preparation was done in several phases. The initial substrate cell, the "virgin surface", a cube of 26.5 Å in each direction, consisted of a hydrogenated amorphous carbon sample with an initial density of 2.0 g/cm^3 and a D/C ratio of 0.4 (700 deuterium atoms and 1750 carbon atoms), prepared by heating at 10^4 K and subsequent quenching to room temperature (300 K), followed by removal of the periodic boundary conditions in the z direction (perpendicular to the surface, see Fig. 1). The final hydrogenated amorphous carbon surfaces, which were used for the sputtering studies, are obtained by D and D$_2$ bombardment of the "virgin" surface at a flux of 1.4×10^{28} D m^{-2} s^{-1}, or one impact every 2 ps on the surface, with freely evolving dynamics in the first ps, and with a Langevin thermosat applied (at 300 K) in the second ps, in order to remove the excess thermal energy that resulted from the impact. While a typical experiment has a time scale measured in minutes, and sample dimensions that may reach thicknesses of µm to mm, the typical time scale of a MD sputtering simulation is measured in ps, and the cell thickness in nm [6-8]. As a consequence, the energy deposition in a small simulation cell at rates too fast for diffusion of heat requires some kind of forced cooling. The surface is bombarded repeatedly in this manner, progressively accumulating more damage. A total fluence of up to of 2.8×10^{20} D m^{-2} was accumulated for the a-C:D surface for each set of initial conditions for the impact projectile (i.e., energy, type and internal state of particle). The effect of the projectile impacts is to gradually increase the D content at the interface, while progressively etching away the carbon. The carbon depletion and deuterium enrichment proceeds up to fluences of $\sim1\times10^{20}$ D m^{-2}, at which point the D/C ratio in the outermost 10 Å of the surface reaches a value of approximately 1. The kinetically formed steady state surface is thus quite different from the equilibrium bulk material [6], with a D/C ratio substantially enhanced from the bulk value of 0.4 to a value of 1 at the interface. This modified surface generates different sputtering products than does the initial surface. The higher D/C ratio in the modified surface produces heavier hydrocarbon products, while the equilibrium bulk surface generates small hydrocarbon radicals or C atoms. We selected six surfaces for each set of the impact particle initial conditions in the "steady state" range of fluences, and all subsequent sputtering simulations at a given energy and with a given impacting particle were performed by averaging the results across these six surfaces. Considering all studied cases (at various impact energies, with various impact particles and their internal states), more than 300 target surfaces were created.

The second type of simulation was designed to calculate the sputtering yield of a particular surface. In order to least influence the physical conditions during a sputtering event, we perform a 5 ps MD simulation, bombarding a particular surface with one D or D_2 projectile without any forced cooling during the projectile cascade. In order to obtain a statistically meaningful estimate of the sputtering yields (which are typically less than 1%), 2000 statistically independent impacts are performed on the same surface, resetting the surface after each 5ps interval. In these statistically independent impacts, the position, orientation, and angular momentum direction are chosen randomly. These simulations evolve for 5 ps in the absence of any thermostat, and ejected molecules are removed once they are more than 10 Å from the surface. For each initial choice of the impinging particles (energy, kind and internal state of the particle), this procedure was repeated with six surfaces, differing in accumulated fluence during the phase of surface preparation. Being statistically independent, the sputtering data count of a particular kind, n, from all six surfaces were combined, producing a yield n/N, for a total of N impinging particles, with an absolute standard error of the mean of $\sqrt{n}/N$. All classical molecular dynamics simulations were performed with the velocity Verlet integrator and a time-step of 0.2 fs.

3. SPUTTERED SPECTRA

The spectrum of sputtered hydrocarbon molecules is a quantity of substantial interest, both in beam–surface experiments and in fusion applications. This product distribution is in turn closely related to the hybridization state of the carbon atoms in the surface. Methyl radicals and methane molecules will result primarily from surface R–CD_3 moieties; saturated molecular products will in general result from cleavage at sp^3-hybridized carbon atoms. The content of sp^2, sp and sp^3 components in the initial, "virgin" surface changes upon bombardment. The sp^2 content decreases slightly as the surface is modified by collisions, with higher energy impacts causing greater changes in hybridization. Simultaneously, the sp^3 content increases, while the sp content stays mainly constant. We show the yields of various hydrocarbons in Fig. 2, as a function of the impact energy of vibrationally excited D_2 (only species with yields greater than 0.001 are displayed). Noteworthy trends include the increasing prevalence of acetylene and CD ejecta with increasing energy (note that CD may be a byproduct of the breakup of energetic acetylene). There is a corresponding decrease of C_2D_n species that does not quite rise to the level of statistical significance. Lastly, methyl radical sputtering yields exhibit a distinct peak at 15 eV, which correlates with the enhanced concentration of terminal $-CD_3$ moieties in the surface bombarded at 15 eV, and decrease slightly for higher impact energies. We also show in Fig 2 recent experiments for methane sputtering [9,10], which agree approximately with our yields for methyl radical. Similar agreement was found with other experimental results [11] (not shown in the figure). While experiments [9-12] indicate mainly saturated methane production, the simulations predict a preponderance of the radical methyl, CD_3, with little methane. The mechanism of passivation of methyl into methane, missing in the simulation, has been a mystery so far.

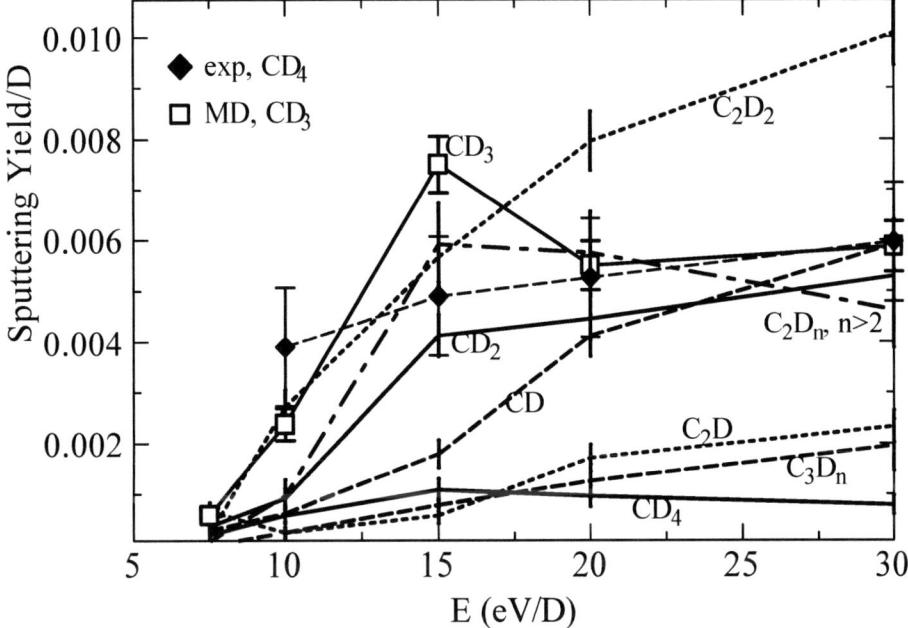

FIGURE 2. Sputtered hydrocarbons as a function of impact projectile energy of excited D_2. The experimental data are from Ref. [9].

We have also studied the energetic and angular spectra of sputtered particles. This information, much needed in the modeling of the fusion plasma–surface interactions has not been available for sputtering of hydrogenated carbon, the main problem being the small number of sputtered particles and thus, poor statistics in the energetic and angular bins of the spectra. We have found that the translational energy spectrum of sputtered particles, E_s, takes the form of a Boltzman profile, $\sim\exp(-E_s/E_{s0})$ for both partial and total hydrocarbon yields. The temperatures associated with the mean energy E_{s0} are near 5,000 K.

Fig. 3(a) shows that the translational energy averaged over all sputtered particles is a weakly increasing function of impact energy, except at the lowest energies in the considered range. The error bars represent the standard error of the mean. A similar dependence (not shown) is present also for various individual components of the particle spectra, such as CD_3, C_2D_2, CD and others.

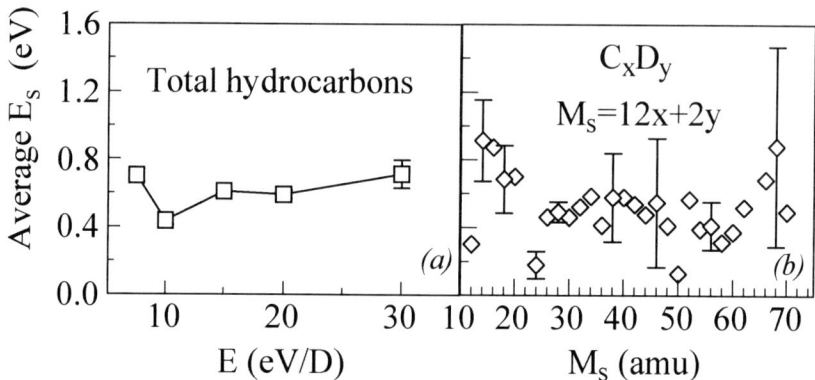

FIGURE 3. (a) Mean translational kinetic energy of sputtered hydrocarbons, as a function of impact projectile energy. (b) Mean translational kinetic energies of the sputtered particles, averaged over the impact energies, as functions of their masses (in amu).

Fig 3(b) shows that lightest sputtered particles are the most energetic. Similar high energies are also seen also for the heaviest particles, although this might be an artifact of a low particle counts for the most complex hydrocarbons. The error bars here represent the standard deviation defining the dispersion of E_s with impact projectile energy. We note that the dependence of the kinetic energy, averaged over the impact E, is a weak function of the sputtered particle mass. The energies of all sputtered hydrocarbons stay in the range of 0.2 to 1 eV, most of them clustering around 0.5 eV, indicating kinetic desorption. The mechanism for this elevation of the sputtered energies is not clear in this moment.

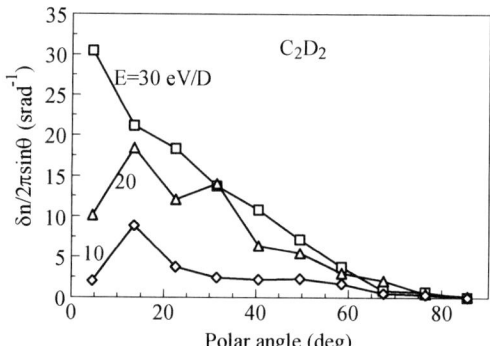

FIGURE 4. Angular distributions of sputtered acetylene for various impact energies of D_2, normalized per unit solid angle.

As expected for normal incidence, the velocity distribution of ejecta is uniform in the x-y plane. Fig. 4 displays the polar angular distribution of acetylene per unit solid angle, for various energies of the impact projectile. These were obtained from the binned counts of the sputtered particles in 10-degree intervals, scaled by $2\pi \sin\theta$, where θ is the polar angle (between the z-axis, Fig. 1, and the ejected particle velocity vector).

We find a significant dependence of the sputtering yield on vibrational excitation of the molecular projectile, especially at low impact energies, as illustrated in Fig. 5 for the yields of methyl+methane, acetylene and total carbon. We associate this change with the energy needed to dissociate the molecule, which takes the highest value in the ground state, and vanishes in the dissociation limit [13]. With further increase of vibrational energy (or internuclear separation R), the correlated energy deposition from two particles disappears and the yields tend (within statistical uncertainty) to the yields for D impact.

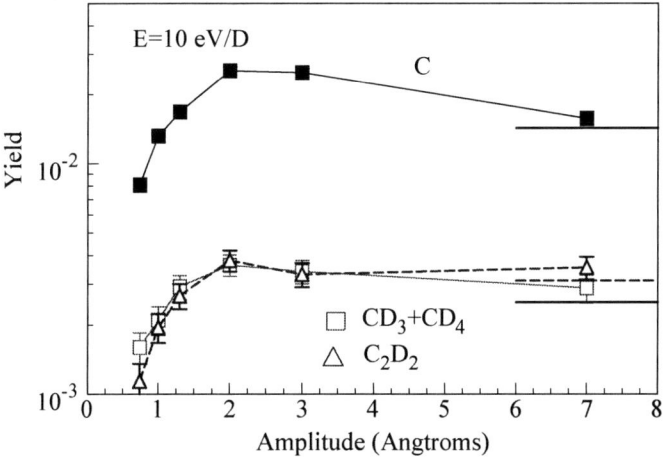

FIGURE 5. Sputtering yields for various "bond" lengths (amplitudes of atomic vibrations) R of the impacting D_2.; $R=0.74$ Å for the ground vibrational state; $R>1.7$ Å is a dissociative state. The horizontal lines are the calculated sputtering yields of carbon, acetylene and methyl+methane by D impact.

4. CONCLUSIONS

We have shown that after careful preparation of the target carbon deuterated surface, obtained upon bombardment of a virgin surface, calculated methane and acetylene yields are found to be in good quantitative agreement with recent experimental results. The kinetic energy spectra of the sputtered particles are weakly dependent on impact energy and ejected particle mass and have a mean value around 0.5 eV. The only preferential direction of the angular distributions of the sputtered particles is in the direction orthogonal to the surface. Finally, we find that the sputtered yields at low impact energies (< 15 eV/D) are strongly dependent on the vibrational excitation state of the impact molecule, increasing with the level of excitation.

ACKNOWLEDGMENTS

We acknowledge support by the Office of Fusion Energy Sciences (PSK) and the Office of Basic Energy Sciences (COR) of the U.S. DoE under contract No. DE-AC05-00OR22725 with UT-Battelle, LLC, and partial support through SciDAC. SJS acknowledges support by the DoE (DEFG0201ER45889), NSF (CHE0239448) and a DOD MURI administered by the ARO. This research was performed in large part using 512-1024 processors at the MPP2 in the MSCF located at PNL, through DoE INCITE project SC18392. We are grateful to Fred W. Meyer for fruitful discussions and making available experimental data prior to publication.

REFERENCES

1. J. Roth, B. M. U. Scherzer, R. S. Blewer, D. K. Brice, S. T. Pieraux and W. R. Wampler, J. Nucl. Mater. **93&94**, 601 (1980).
2. B. L. Doyle, W. R. Wampler and D. K. Brice, J. Nucl. Mater. **103&104**, 513 (1981).
3. D. W. Brenner, O. A. Shenderova, J. A. Harrison, S. J. Stuart, B. Ni and S. A. Sinnott, J. Phys.: Condens. Matter **14**, 783 (2002).
4. J. Tersoff, Phys. Rev. B **37**, 6991 (1988).
5. S. J. Stuart, A. B. Tutein, and J. A. Harrison., J. Chem. Phys. **112**, 6472 (2000).
6. E. Salonen, K. Nordlund, J. Keinonen, and C. H. Wu, Europhys. Lett, **52**, 504 (2000).
7. E. Salonen, K. Nordlund, J. Keinonen, and C. H. Wu, Phys. Rev. B **63**, 195415 (2001).
8. J. Marian, L. A. Zepeda-Ruiz, G. H. Gilmer, E. M. Bringa, and T. Rognlien, Phys. Scr. **T124**, 65 (2006).
9. L. I. Vergara, F. W. Meyer, H. F. Krause, P. Träskelin, K. Nordlund, and E. Salonen, J. Nucl. Mater., in press (2006) .
10. F. W. Meyer, L. I. Vergara, H. F. Krause, Phys. Scripta **T124**, 44 (2006).
11. G. M. Wright, A. A. Haasz, J. W. Davis, and R. G. Macaulay-Newcombe, J. Nucl. Mater., **337-339**, 74 (2000).
12. E. Vietzke, J. Nucl. Matls. **290-293**, 158 (2001).
13. P. S. Krstić, and R. K. Janev, Phys. Rev. A**67**, 022708 (2003).

Ion Beam Mixing at Crystalline and Amorphous Fe/Si Interfaces

V. Milinović[1], K. Zhang[1], N. Bibić[1,2], K. P. Lieb[1], M. Milosavljević[2] and P. K. Sahoo[1,3]

[1] II. Physikalisches Institut, Universität Göttingen, 37077 Göttingen, Germany
[2] VINCA Institute of Nuclear Sciences, Belgrade, Serbia-Montenegro
[3] Now at K. U. Leuven, B-3001 Leuven, Belgium

Abstract. Ion beam irradiation of a-Si/Fe/c-Si trilayers with 350-MeV Au ions and of Fe/a-Si bilayers with 250-keV Xe ions were carried out in order to measure the interface mixing rates and microstructure, phase formation, and magnetic polarization in the regimes of electronic and nuclear stopping. For Fe/a-Si and nuclear stopping, an enhancement of the interface mixing rate of 1.75 ± 0.15 was observed relative to Fe/c-Si. For electronic stopping, the enhancement is 3.21 ± 0.34. A plausible explanation of this enhancement lies in the much smaller thermal conductivity in a-Si relative to c-Si, which prolongates the relaxation phase of the ion-induced thermal spikes.

Keywords: Ion beam mixing; swift heavy ions; silicides.
PACS: 61.80.-x; 61.80.Jh; 68.55.-a.

1. INTRODUCTION

Motivated by the prospect of producing semiconducting ß-FeSi$_2$ films, which are potential candidates for infrared detecting and emitting devices, ion-beam mixing in the iron/silicon system has attracted considerable attention in the past decade [1-6]. Single-phase ß-FeSi$_2$ films were indeed produced under appropriate conditions of ion mass and fluence and substrate temperature [5,6], when irradiating Fe/Si bilayers with ions at energies of several hundred keV. The model of local spike mixing [7] was found to reproduce the experimental interface mixing rates in this regime quite well [5,6]. Recently, the mixing studies of Fe/Si bilayers were extended to chemically active ions (nitrogen) [8], highly charged ions [6,9], and swift heavy ions with more than 100 MeV, i.e. in the regime of electronic stopping [10-15].

In the present work, we investigated the effect of amorphous Si layers on the mixing process. Either Si wafers, which were pre-amorphized by low-energy Ar-ion bombardment, or thin amorphous Si layers were used, on which the Fe films were deposited. For both nuclear and electronic stopping, the use of a-Si was found to enhance the mixing rates by a factor of 2-3, relative to those in the Fe/c-Si system. Besides measuring the mixing rates (by means of Rutherford backscattering spectroscopy, RBS), we also followed the evolution of phase formation (by means of X-ray diffraction, XRD, and high-resolution transmission electron spectroscopy, TEM), surface roughness (scanning electron microscopy, SEM), and magnetic

CP876, *The Physics of Ionized Gases: 23rd Summer School and International Symposium*,
edited by L. Hadžievski, B. P. Marinković, and N. S. Simonović

properties (magneto-optical Kerr effect, MOKE) as function of the ion fluence. The present article refers to the results on mixing and the phases formed, while other properties are discussed in Refs. [8,9,16].

2. EXPERIMENTS

The 350-MeV Au^{26+} ion beam provided by the ISL/HMI facility was directed towards tri-layers of a-Si(12 nm)/Fe/c-Si(100), which were prepared by electron evaporation and cooled to 87 K during the ion irradiations. The ion beam entered at normal incidence with fluences of up to $1.1 \times 10^{15}/cm^2$ and covered the 8×8 mm^2 implantation spots homogeneously by means of an XY-sweeping system. According to SRIM calculations [17], the mean ion range was some 30 μm and therefore the Au ions were stopped neither in the a-Si top layer nor in the intermediate Fe layer. The nuclear and electronic stopping powers were calculated as $S_n = 90$ eV/nm and $S_e = 51$ keV/nm in Fe, and $S_n = 23$ eV/nm and $S_e = 20$ keV/nm in the top a-Si film and in the Si-substrate near the Fe/c-Si interface, respectively [17]. Hence the energy loss at both interfaces was almost completely due to electronic stopping.

In the second set of experiments, Fe(32 nm)/Si bilayers were irradiated at room temperature with 250-keV Xe ions, either singly charged or in the 17+ charge state. Crystalline Si wafers or those pre-amorphized with 2×10^{17} Ar^+ ions/cm^2 at 1.0 keV were used. The projected Ar-ion range in Si was about 3.5 nm [17]. Ion beams of Xe^+ and Xe^{17+} were provided by the Göttingen ion implanter IONAS [18] and the VINCA facility [19], respectively. The deposited energy density at the Fe/Si interface was about 2.8 keV/nm and almost exclusively due to nuclear stopping.

The Fe and Si depth profiles were measured by means of Rutherford backscattering spectros-copy (RBS) using the 0.9 MeV $^4He^{++}$ beam of IONAS and two Si surface barrier detectors placed at 165° to the beam. The spectra were de-convoluted with the software WIN-DF [20]. For further experimental details concerning the XRD, TEM, SEM and MOKE analyses, see Refs. [9,21,22].

3. RESULTS

3.1 Electronic stopping

Fig. 1 illustrates the XRD spectra taken for a-Si/Fe/c-Si trilayers in glancing angle incidence (3°) as a function of the Au-ion fluence Φ. One notes the strong (311) reflex from the c-Si wafer and several peaks due to the polycrystalline Fe layer. Both types of peaks decrease in intensity with increasing ion fluence Φ; the (311) peak due to amorphization of the Si wafer and the Fe peaks due to reactions of the intermediate Fe layer with Si at both interfaces. No indication of crystalline iron-silicide phase formation was observed. This finding is in agreement with the Mössbauer effect and XRD data collected in various swift heavy ion irradiation experiments of Fe/Si bi- and multilayers [10,11,14,15], but in contrast to the results of Assmann et al. [23], who - on the basis of CEMS - claimed that crystalline α-$FeSi_2$ was formed.

The measured backscattering spectra and deduced concentration profiles and mixing rates are illustrated in Figs. 2 and 3. The data show very different mixing speeds at the two interfaces: the front a-Si/Fe mixes with a very high average mixing rate of $\Delta\sigma_a^2/\Phi = 184(20)$ nm^4, where $\Delta\sigma^2$ denotes the ion-induced change of the interface variance and 2σ the width, at which the Fe or Si depth profiles change from 16% to 84% of the pure materials. At a fluence of 3.4×10^{14} Au-ions/cm^2, the front layer of the sample was fully intermixed, while the (second) Fe/c-Si just started to intermix, with a much lower mixing rate of $\Delta\sigma_c^2/\Phi = 60(7)$ nm^4 (see Fig. 2b). This value is in good agreement with the one measured for Fe/Si and Ni/Si bilayers and 350-MeV Au ions, $\Delta\sigma_c^2/\Phi(\text{Fe/Si}) = 55(5)$ nm^4 [15] and $\Delta\sigma_c^2/\Phi(\text{Ni/Si}) = 64(4)$ nm^4 [15]. For Fe/Si bilayers and 350-MeV Au ions, we thus arrive at the following ratio of mixing rates: $\Delta\sigma_a^2/\Delta\sigma_c^2 = 3.21(34)$.

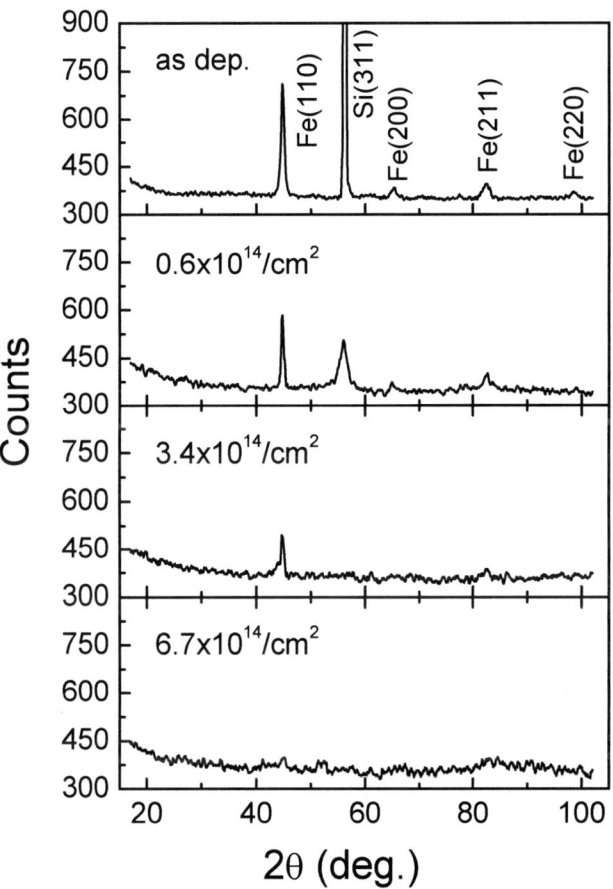

FIG. 1. Grazing incidence XRD spectra of a-Si/Fe/c-Si trilayers measured as a function of the fluence of the 350-MeV Au beam.

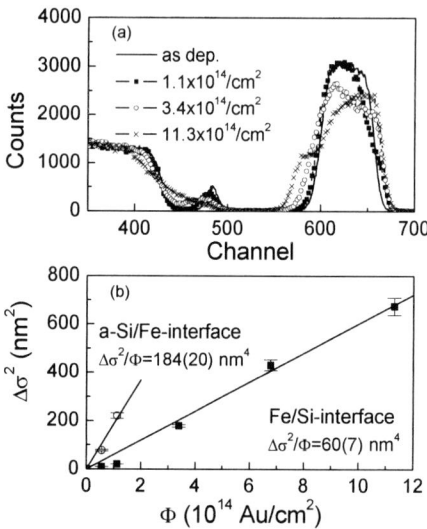

FIG. 2. a) RBS data of a-Si/Fe/c-Si trilayers taken with a 0.9 MeV He[++] beam at 165°.
b) Fluence dependence of the variance $\Delta\sigma^2$ at the a-Si/Fe and Fe/c-Si interfaces, giving the mixing rates $\Delta\sigma^2/\Phi$ indicated.

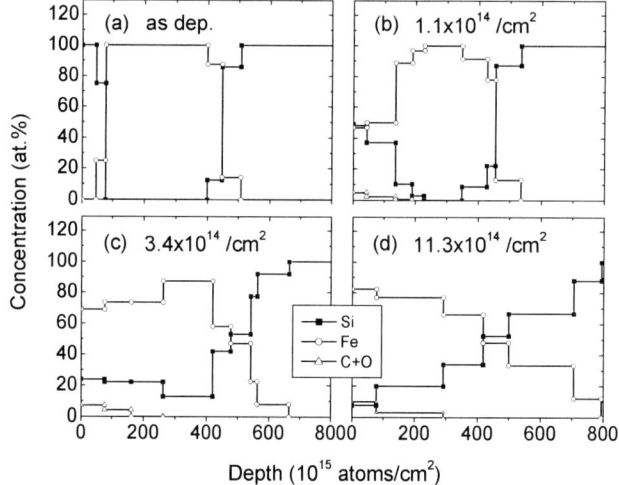

FIG. 3. Fe and Si concentration profiles resulting from the de-convoluted RBS spectra shown in FIG. 2a by means of the program WIN-DF.

3.2 Nuclear stopping

The concentration profiles within the Fe/a-Si bilayers irradiated with 250-keV Xe-ions are illustrated in Fig. 4, which also shows those of the implanted Xe ions. As the Xe-ion range was chosen to match the top Fe film thickness, one notes non-Gaussian Xe implantation profiles peaked at the interface, which are typical of the presence of end-of-range spikes and possibly Xe precipitates near the interface [7,25]. The mixing rates obtained at the Fe/a-Si and Fe/c-Si for Xe^+ and Xe^{17+} ions are summarized in Table 1. One notes that the values do not depend on the ion charge, but clearly on the microstructure of the Si substrate. Averaging the values for the two charge states, we arrive at the ratio $\Delta\sigma_a^2/\Delta\sigma_c^2 = 1.85(15)$ and therefore again at an enhanced interdiffusion at the a-Si interface.

4. DISCUSSION AND CONCLUSIONS

The experimental mixing rates in the case of 250-keV Xe ions were modeled with the ballistic and thermal (local and global) spike approaches, details of which have been described in Refs. [6,7,9,26]. The results presented in Table 1 clearly demonstrate that local thermal spikes reproduce the Fe/c-Si mixing rate quite well, while the ballistic approach underestimates it by one order of magnitude. However, none of the models in their present form accounts for the enhanced mixing rate in the Fe/a-Si system.

Concerning mixing with swift Au ions, Wang et al. [27] and Srivastava et al. [13] recently calculated Fe/Si interface mixing by swift heavy ions by considering "instantaneous" energy transfer from the electronic system to track-like spikes. These calculations demonstrate that a molten track is formed during some 1.2 ps on both sides of the interface and that atomic diffusion and possibly silicide phase formation occurs mainly during the spike and its relaxation. This approach is expected to account also for the mixing rates measured in the present work.

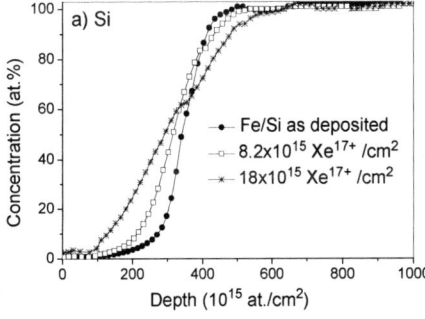 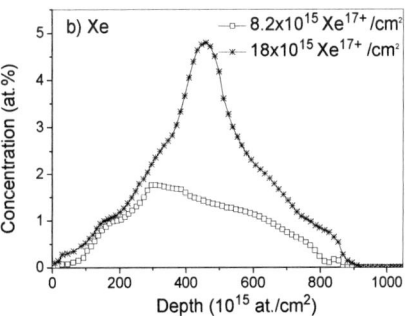

FIG. 4. Si and Xe depth profiles deduced from the RBS spectra taken for Fe(32 nm)/a-Si bilayers after irradiation with 250-keV Xe^+ ions to the fluences indicated.

TABLE 1. Measured and calculated mixing rates at Fe/Si interfaces.

Sample	Ion	Mixing rate (nm⁴)	Ref.
Experiment			
a-Si(12 nm)/	350 MeV Au^{26+}	184(20) @ a-Si/Fe	Present work
Fe(45 nm)/c-Si		60(7) @ Fe/c-Si	
Fe(65 nm)/c-Si	350 MeV Au^{26+}	55(5)	[15]
Fe(32 nm)/c-Si	250 keV Xe^{1+}	4.8(3)	Present work
Fe(32 nm)/c-Si	250 keV Xe^{17+}	4.5(5)	"
Fe(32 nm)/a-Si	250 keV Xe^{1+}	8.1(7)	"
Fe(32 nm)/a-Si	250 keV Xe^{17+}	9.3(6)	"
Theory			
Ballistic	250 keV Xe	0.50	[9]
Local spikes	250 keV Xe	4.0	[9]
Global spikes	250 keV Xe	3.2	[9]

Clearly, interface mixing for both nuclear and electronic stopping is more effective at the Fe/a-Si interface than at the Fe/c-Si interface. A possible reason for this effect may be the much smaller heat conduction in a-Si compared to that in c-Si, which leads to smaller cooling rates and longer atomic diffusion across the a-Si/Fe interface in the spike phase. It has been argued that energy deposition to the electronic system via laser and heavy-ion irradiation (in the regime of electronic stopping) bear some similarities. In their work on pulsed-laser hydriding of a-Si and c-Si, Schwickert *et al.* [28] simulated the temperature evolution in the molten zone generated by excimer laser pulses. The authors argued that the much smaller cooling rates, leading to smaller amounts of hydrogen stored in a-Si, are due to the longer preservation of a molten Si layer. Concerning the similarities existing between stopping of swift heavy ions and laser irradiations, with both processes coupling energetically first to the electronic system, which then rapidly couples to the lattice, this kind of reasoning may well apply to the present mixing experiments.

ACKNOWLEDGMENTS

The authors gratefully acknowledge the help of D. Purschke with the ion irradiations and RBS measurements in Göttingen, of Dr. S. Klaumünzer with the swift Au-ion irradiation at ISL/HMI in Berlin, and of M. Siljegović with the Xe^{17+} implantations at VINCA. This project has been funded by Deutsche Forschungsgemeinschaft and the Ministry of Science and Environmental Protection of the Republic of Serbia (Project No. 1960).

REFERENCES

[1]   M. C. Boast and J. E. Mahan, J. Appl. Phys. **58**, 2696 (1985).
[2]   D. N. Leong, M. A. Harry, K. J. Reeson, and K. Homewood, Nature (London) **387**, 686 (1997).
[3]   I. Dézsi, Cs. Fetzer, M. Kiss, H. Pattyn, A. Vantomme, and G. Langouche, Appl. Phys. Lett. **76**, 1917 (2000).
[4]   V. E. Borisenko, *Semiconducting Silicides*, Springer, Berlin-Heidelberg-New York, 2000.
[5]   M. Milosavljević, S. Dhar, P. Schaaf, N. Bibić, Y.-L- Huang, M. Seibt, and K. P. Lieb, J. Appl. Phys. **90**, 4474 (2001).
[6]   S. Dhar, P. Schaaf, N. Bibić, E. Hooker, M. Milosavjević, and K. P. Lieb, Appl. Phys. A **76**, 773 (2003).
[7]   W. Bolse, Mat. Sci. Eng. **R12**, 40 (1994); Nucl. Instr. Meth. B **148**, 83 (1999).
[8]   V. Milinović, N. Bibić, S Dhar, M. Siljegović, P. Schaaf, and K. P. Lieb, Appl. Phys. A **79**, 2093 (2004).
[9]   V. Milinović, N. Bibić, K. P. Lieb, M. Milosavljević, and F. Schrempel, in preparation.

214

[10] C. Dufour, P. Bauer, G. Marchal, J. Grilhé, C. Jaouen, J. Pacaud, and J. C. Jousset, Europhys. Lett. **21**, 671 (1993).

[11] Ph. Bauer, C. Dufour, C. Jaouen, G. Marchal, J. Pacaud, J. Grilhé, and J. C. Jousset, J. Appl. Phys. **81**, 116 (1997).

[12] S. K. Srivastava, S. Ghosh, A. Gupta, V. Ganesan, W. Assmann, S. Kruijer, and D. K. Avasthi, Hyp. Int. **133**, 53 (2001).

[13] S. K. Srivastava, D. K. Avasthi, W. Assmann, Z. G. Wang, H. Kucal, E. Jacquet, H. D. Carstanjen, and M. Toulemonde, Phys. Rev. B **71**, 193405 (2005).

[14] K. P. Lieb, K. Zhang, V. Milinovic, P. K. Sahoo, and S. Klaumünzer, Nucl. Instr. Meth. B **245**, 121 (2006).

[15] V. Milinović, K. P. Lieb, P. K. Sahoo, P. Schaaf, K. Zhang, S. Klaumünzer, and M. Weisheit, Appl. Surf. Sci. **252**, 5339 (2006).

[16] K. Zhang, K. P. Lieb, V. Milinović, P. K. Sahoo, and S. Klaumünzer, J. Appl. Phys., submitted.

[17] http://www.srim.org/

[18] M. Uhrmacher, K. Pampus, F. J. Bergmeister, D. Purschke, and K. P. Lieb, Nucl. Instr. Meth. B **9**, 234 (1985).

[19] A. Dobrosavljević, M. Milosavljević, N. Bibić, and A. Efremov, Rev. Sci. Instrum. **71**, 786 (2000).

[20] N. Barradas, C. Jeynes, and R. P. Webb, Appl. Phys. Lett. **71**, 291 (1997).

[21] G. A. Müller, doctoral thesis, Göttingen (2004); G. A. Müller, E. Carpene, R. Gupta, P. Schaaf, K. Zhang, and K. P. Lieb, Eur. Phys. J. B **48**, 449 (2005).

[22] V. Milinović, doctoral thesis, Göttingen (2005).

[23] W. Assmann, M. Dobler, D. K. Avasthi, S. Kruijer, H. D. Mieskes, and H. Nolte, Nucl. Instr. Meth. B **146**, 271 (1998).

[24] K. Zhang, K. P. Lieb, V. Milinović, M. Uhrmacher, and S. Klaumünzer, Nucl. Instr. Meth. B, online.

[25] W. Bolse and T. Weber, Nucl. Instr. Meth. B **85**, 188 (1994).

[26] S. Dhar, M. Milosavjević, N. Bibić, and K. P. Lieb, Phys. Rev. B **65**, 024109 (2002).

[27] Z. G. Wang, C. Dufour, S. Euphrasie, and M. Toulemonde, Nucl. Instr. Meth. B **209**, 194 (2003).

[28] M. Schwickert, E. Carpene, K. P. Lieb, M. Uhrmacher, P. Schaaf, and H. Gibhardt, Phys. Scripta T **108**, 113 (2004).

Plasma Assisted Femtosecond Laser Inscription in Dielectrics

Jovana S. Petrović, Vladimir Mezentsev, Mykhaylo Dubov,
Amós Martínez, and Ian Bennion

Photonics Research Group, Aston University, B4 7ET, Birmingham, UK
petrovij@aston.ac.uk

Abstract. Principles of the femtosecond fabrication of the optoelectronic components in glass are explained and illustrated by examples of the in-bulk writing. The results of the experimental investigation of the dependence of the induced index change on the pulse energy and the numerical modelling of the corresponding laser–glass interaction are presented. The distribution of the plasma density is simulated that may bridge the gap between the models of the pulse propagation and the induced permanent refractive index change.

Keywords: femtosecond laser, micromachining, plasma
PACS: 52.50.Jm, 78.20.Bh

INTRODUCTION

Over the last decade the interaction of the femtosecond laser pulses with different materials has been studied in regard to the in-bulk fabrication of optoelectronic devices [1, 2, 3]. Fabrication of various planar and 3D devices such as waveguides [4, 5], couplers [6], wavelength multiplexers [7], optical storage [8], and in-fibre devices such as fibre Bragg [9, 10] and long period gratings [11] has been demonstrated. The major advantages of the femtosecond micromachining are the possibility of the inscription into the bulk of materials, versatility of the geometries that can be inscribed, stability of the induced index change, and the possibility to modify different materials such as pure silica, doped silica, polymers [12], semiconductors, as well as living tissues [13]. By adjusting the pulse energy, pulse duration and focusing conditions along with the translation speed of the sample, various devices can be fabricated, for instance vectorial bend sensors [14, 15] and a fibre laser [16]. However, due to the incomplete knowledge of the physical processes involved in the femtosecond inscription the full exploitation of this technique has not been achieved yet.

During the fabrication of aforementioned devices femtosecond pulse is focused inside the material, its intensity in and near the focal point being high enough to cause the nonlinear absorption of the pulse. The pulse energy deposited in the material causes a permanent structural change which results in a change in the refractive index. Whereas the propagation of the ultra-short pulse has been well understood [17, 18, 19], the mechanisms leading to the modification of the material are still under investigation. The widely adopted model suggests that an intense femtosecond pulse causes multiphoton ionisation of electrons that are further heated by the rest of the pulse generating a hot electron plasma [3]. The mechanisms of the energy transfer from plasma to the

CP876, *The Physics of Ionized Gases: 23ʳᵈ Summer School and International Symposium*,
edited by L. Hadžievski, B. P. Marinković, and N. S. Simonović

ion lattice and the subsequent structural changes in material are highly material dependent. Models implying thermo-plastic stresses [20], colour centre formation [21], shockwaves [22], microexplosions [8] have been considered. The first two mechanisms lead to the smooth index change which is needed for the fabrication of waveguides. The microexplosions and the shockwaves explain the formation of the complex index profiles often containing voids. In [20] the model is proposed that calculates the index change starting from the thermal distribution of the ionised electrons.

In this paper the principles of the femtosecond fabrication are reviewed and illustrated by the examples of the devices fabricated in our research group. The results of the experimental investigation of the permanent refractive index change and the results of numerical modelling of the spatial distribution of plasma density, which is required as an input to the model in [20], are reported. Qualitative agreement between these results indicates the analogy between the plasma distribution and the induced index change.

FABRICATION METHOD

A significant impetus to the femtosecond micromachining was the development of the Kerr mode locking and the chirped pulse amplification (CPA) that enabled bringing the solid state femtosecond laser system to the optical table. A standard set-up is shown in Fig. 1. The source of the 120 fs pulses at 800 nm with the peak power of 10 GW

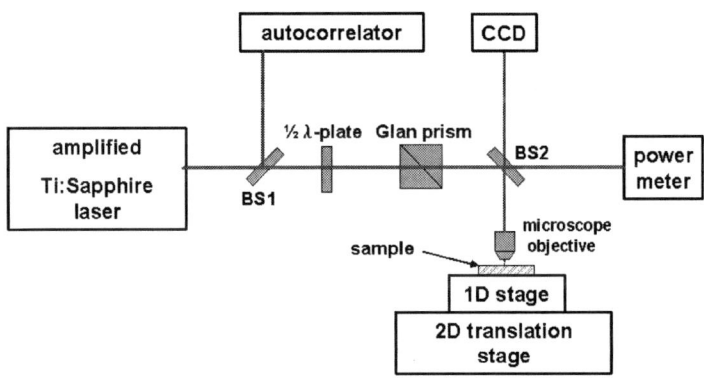

FIGURE 1. Setup for the femtosecond inscription.

is an amplified Ti:Sapphire laser. The half wavelength plate and Glan prism are used for adjusting the pulse power. The pulse is focused to the sample by a microscope objective or a lens and the sample is moved by a precision translation stage. This is a generic set-up for several scenarios of inscription that emerge from different focusing conditions, pulse powers and pulse repetition rates, which is explained below.

Loose focusing performed by the lens enables smooth index change and can be used with the sample translated both transversally and longitudinally (along the beam). Tight focusing appeared as a solution to the problem of creating an index change in the bulk of the material without allowing for the beam collapse or the surface ablation and is nowadays the prevailing method of inscription. In practise, the tight focusing is achieved by the microscope objectives with high NA, resulting in either smooth index change or large structural changes such as voids with complex index profiles. However, due to the short working distance of these objectives the longitudinal translation of the sample is severely limited.

In the case of a slow pulse repetition rate (e.g. 1 KHz) all the processes are finished before the next pulse arrives to the interaction region. When the repetition rate is high (e.g. 80 MHz) thermal diffusion ($\tau \sim 1\mu s$) cannot be completed before the arrival of the next pulse and the effects of the subsequent pulses are accumulated. In the experiments and simulations described in this paper tightly focused pulses with the slow repetition rate were used.

DEVICES INSCRIBED BY FS-LASER

Planar and 3D devices

Fabrication of the waveguides was one of the first targets of the femtosecond micro-machining. It requires smooth index change of the order 10^{-3} which can be achieved by moderate pulse powers and the transversal translation of the sample with the speed adjusted so that the structures inscribed by the individual pulses overlap. For this purpose the maximal pulse energy $1\mu J$ and the pulse repetition rate 1 KHz were used and the speed of the sample was varied from 0.01 mm/s to 0.5 mm/s. Booth the waveguides of $1\mu m$ in diameter and the index change of the order 10^{-3} which are single mode at the optical frequencies and larger multimode waveguides were produced.

An important task of microfarbication is towards near-surface structures. When the surface is coated by the metallic film the light can be coupled from the bulk to the surface of the material by the surface plasmons. Figure 2a) shows the structures inscribed at $4\mu m$-$5\mu m$ bellow the surface of the fused silica.

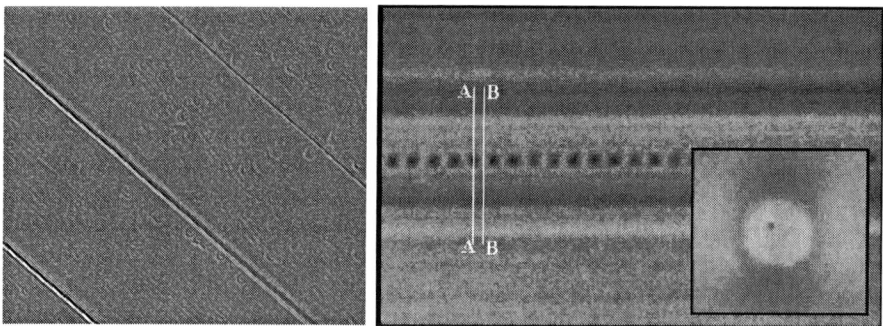

FIGURE 2. a) Structures near the surface of the fused silica. b) Fibre Bragg grating fabricated point-by-point by fs laser.

Fibre Gratings

Figure 2b) shows the fibre Bragg grating (FBG) directly inscribed 'point-by-point' by the femtosecond laser [10]. Another set-up for fs-inscription uses the phase mask [9]. Recently the fs-inscribed FBGs in polymer fibres [12] and PCFs [23] have also been reported. Major advantages of this fabrication method are the possibility of inscription through the fibre coating [24], thermal stability of the gratings [25, 26] and the possibility of fabrication in any kind of fibre, e.g. a fibre laser was made by inscribing two FBGs in the erbium doped fibre [16].

Fs-written long period gratings (LPGs) feature the same advantages as FBGs with the additional functionality obtained by the inscription in cladding. Apart from the fabrication of the standard LPGs the asymmetrisation of the UV-fabricated LPG and its use as the vectorial band sensor were demonstrated [27].

Due to the ellipsoidal index change and the arbitrary displacement of the grating from the fibre axis both FBGs and LPGs can be highly birefringent which can be used for the fabrication of polarisation maintaining devices and the investigation of the index change.

PROFILE OF THE REFRACTIVE INDEX CHANGE

The refractive index change was studied by the systematic irradiation of the glass by single laser pulses focused through the microscope objectives with the magnifications 20x (NA 0.45) or 40x (NA 0.65). A phase microscope Axioskop (Zeiss Inc.) equipped by the software for the quantitative phase microscopy QPm (IATIA) was used for the phase and index profiling.

Two characteristic index profiles were observed. Pulses with the power just above the inscription threshold which is below the critical power of self-focusing produced smooth and positive index change which was barely visible under the microscope. A direct application is the inscription of the waveguides. Pulses with the intensity several times above the threshold produced characteristic structures composed of the central region with the negative index change and the surrounding region with the positive index change, Fig. 3b). The origin of the void is suspected to be a shock wave excited by the heated plasma. It pushes material out of the focus, thus forming a region of densified material around the depleted central part. If the pulse is focused by a high NA objective all the material can be pushed out forming the void in the centre [28], [22]. This regime is useful for the 'point-by-point' fabrication of fibre gratings, optical memories and photonic crystals.

We have studied the dependence of the index change in the fused silica on the pulse energy. A typical profile of the index change produced by a pulse in the high power regime is shown in Fig. 3a) and the dependence of the amplitude of index change on the pulse energy in Fig. 3b). The amplitude appears to saturate for the pulse energies over $1\mu J$ and the oscillations at higher energies were used to approximate the error of the measurement as $7 \cdot 10^{-5}$. It is several times bigger than the precision of QPm, $\Delta n = 10^{-5}$, because the complex shape of the inscribed structures did not allow for the precise focusing required by the QPm.

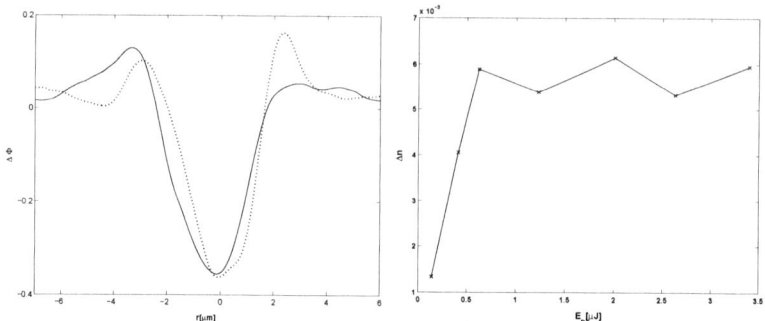

FIGURE 3. a) Phase profile of the structure obtained by the pulse with energy 2µJ. b)Amplitude of the index change versus pulse energy.

In the low energy limit the factor of importance is the inscription threshold which we ad hoc define as the minimal pulse energy required to produce the index change detectable by the described method. The precision of the measurement was increased by the adaptive refinement of the resolution of the pulse power around the threshold detected in the previous step with the final result 80 ± 10 nJ for the pulse energy threshold. The energy transferred to electrons by such a pulse ($\tau_p = 120$fs, NA=0.45) is estimated to be 6.8 eV, which is close to the energy gap of the fused silica 7.6 eV [29], therefore confirming the assumption that the multi-photon ionisation is triggering mechanism of the index change.

We have tested the robustness of the refractive index change induced by femtosecond laser pulses over the time scale of two years and found out that the index profile remained unchanged with the exception of a small increase in the structure diameter. The last can be attributed to the relaxation of the stresses in the region surrounding the void.

NUMERICAL MODELLING

As the multiphoton absorption finishes in a few femtoseconds and the electron recombination has a time scale of several picoseconds, we suggest a two-step model of the induced index change. First, the laser–material interaction is described by the model suggested by Feit and Fleck [30] and adapted by Feng et al. [31]. The result is the distribution of the ionised electrons which is used as an input to the model of the energy transfer from electrons to ions and the subsequent structural modifications of material leading to the refractive index change, e.g. the model by Zhang et al. [20]. In this paper, the first step and its results are explained. The pulse propagation is described by the non-linear Schrödinger equation (NLSE) and the generation of plasma by a plasma balance rate equation. The two equations are coupled and after normalisation form the system

$$iu_z + \kappa \triangle_\perp u - \delta \frac{\partial^2 u}{\partial t^2} + \sigma |u|^2 u = -i\gamma(1 - i\omega\tau)\rho u - i\mu |u|^{2(K-1)} u \qquad (1)$$

220

$$\frac{\partial \rho}{\partial t} = \nu |u|^2 \rho + |u|^{2K} \tag{2}$$

The electric field $\mathscr{E}(\vec{r},t) = \sqrt{I_{MPA}}u(\vec{r},t)$ is normalised to the intensity at which plasma ionization rate becomes very steep $I_{MPA} = \sqrt[K]{\frac{\rho_{BD}K\hbar\omega}{\beta^{(K)}\tau_p}}$ and the plasma density ρ is in units of the breakdown plasma density ρ_{BD}. The first term on the right-hand side of (1) accounts for the pulse defocusing and absorption by plasma and the second for the multiphoton absorption. Since the time necessary for the electron–ion recombination is several nanoseconds, i.e. much longer than the time the pulse is present in the focal region, this term has been omitted from the balance rate equation. The initial condition is a gaussian pulse focused by the lens or the microscope objective. The parameter values are as in [32].

We investigated the propagation of the pulses with the powers below and above the critical power of self-focusing. The weaker pulses were absorbed and defocused by plasma leaving the compact electron cloud with the maximal density below the breakdown density of the material. In the case of the intense laser pulses, strong self-focusing and the plasma effects were balanced out causing the pulse splitting in space and time. As a result, the complex distribution of the plasma density that remains after the pulse has left the focal region was obtained, e.g. Fig. 4b). The calculated plasma distribution can be used as a starting point in the simulations of the permanent refractive index change caused by the energy transfer from the heated electrons to ions, such as is the one based on the thermo-plastic stresses described in [20]. A comprehensive model is an object of the current work.

The qualitative agreement between the plasma density distribution and the index change obtained in experiments is illustrated by the example of a tightly focused pulse with the power above the critical power of self–focusing shown in Fig. 4.

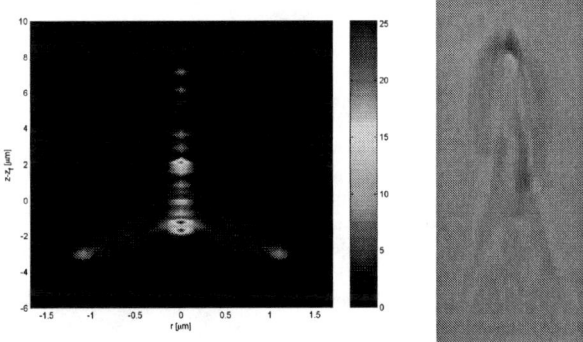

FIGURE 4. b) Simulated spatial distribution of the plasma generated by the pulse with energy 4.6μJ focused by the microscopic objective x40 NA=0.65. b) Correspondent refractive index change in silica.

CONCLUSIONS

Feasibility of the fabrication of versatile and robust optoelectronic components by the focused femtosecond laser pulses has been demonstrated. Since the controllability of the technique depends on knowing the physical processes that lead to the index change, the emphasis of the current research is on experimental and theoretical investigation of these processes. While the model of the pulse propagation and the initial ionisation of electrons has been well established, the comprehensive model of the transition from plasma to the permanent refractive index change does not exist yet. In this paper we suggest that due to the different time scales of the plasma generation and the electron–ion recombination, the problem can be split into the calculation of the distribution of ionised electrons left after the pulse and the calculation of the subsequent index change. Here we have demonstrated the results of the first part that qualitatively agreed with the experimentally induced index change. Efforts have been made to establish the numerical model that will be able to map the plasma density into the profile of the refractive index change.

ACKNOWLEDGMENTS

J. Petrovic would like to acknowledge the help of Yicheng Lai and Thomas Allsop.

REFERENCES

1. D. Du, X. Liu, J. A. Squier, and G. A. Mourou, *Proceedings of SPIE - The International Society for Optical Engineering* **2428**, 422 – 434 (1995).
2. C. B. Schaffer, A. O. Jamison, and E. Mazur, *Applied Physics Letters* **84**, 1441 – 1443 (2004).
3. B. C. Stuart, H. A. M. Feit, N, D, A. N. Rubenchik, B. W. Shore, and M. D. Perry, *Physical Review B* **53**, 1749–1761 (1996).
4. A. M. Streltsov, and N. F. Borrelli, *Journal of the Optical Society of America B: Optical Physics* **19**, 2496 – 2504 (2002).
5. R. Osellame, S. Taccheo, M. Marangoni, R. Ramponi, P. Laporta, D. Polli, S. De Silvestri, and G. Cerullo, *Journal of the Optical Society of America B (Optical Physics)* **20**, 1559 – 67 (2003).
6. A. M. Streltsov, and N. F. Borrelli, *Optics Letters* **26**, 42–43 (2002).
7. W. Watanabe, T. Asano, K. Yamada, K. Itoh, and J. Nishii, *Optics Letters* **28**, 2491 – 2493 (2003).
8. E. N. Glezer, M. Milosavljevic, L. Huang, R. J. Finlay, T. Her, J. P. Callan, and E. Mazur, *Opt. Lett.* **21**, 2023–2025 (1996).
9. S. J. Mihailov, C. W. Smelser, D. Grobnic, R. B. Walker, P. Lu, H. Ding, and J. Unruh, *Journal of Lightwave Technology* **22**, 94 – 100 (2004).
10. A. Martinez, M. Dubov, I. Khrushchev, and I. Bennion, *Electronics Letters* **40**, 1170 – 1172 (2004).
11. T. Allsop, M. Dubov, H. Dobb, A. Main, A. Martinez, K. Kalli, D. Webb, and I. Bennion, *Photonics Europe 2006 (SPIE)* pp. 6193–14 (2006).
12. P. Scully, D. Jones, and D. Jaroszynski, *Journal of Optics A: Pure and Applied Optics* **5**, 92–96 (2003).
13. I. Maxwell, S. Chung, and E. Mazur, *Med. Laser Appl.* **20**, 193–200 (2005).
14. A. Martinez, Y. Lai, M. Dubov, I. Khrushchev, and I. Bennion, *Electronics Letters* **41**, 472 – 474 (2005).
15. T. Allsop, M. Dubov, A. Martinez, F. Floreani, I. Khrushchev, D. Webb, and I. Bennion, *Electronics Letters* **41**, 59 – 60 (2005).

16. Y. Lai, A. Martinez, I. Khrushchev, and I. Bennion, *Optics Letters* **31**, 1672–1674 (2006).
17. S. A. Ahmanov, V. A. Vyslouh, and A. S. Chirkhin, *Optika Femtosecondnih Lasernih Impulsov*, "Nauka", Moskva, 1988.
18. A. Gaeta, *Physical Review Letters* **84**, 3582 – 5 (2000).
19. A. L. Gaeta, G. Fibich, and K. Moll, *Physical Review Letters* **90**, 203902 – 1 (2003).
20. X. Zhang, X. Xu, and A. Rubenchik, *Applied Physics A (Materials Science Processing)* **A79**, 945 – 8 (Sept.-Oct. 2004).
21. A. Zoubir, M. Richardson, L. Canioni, A. Brocas, and L. Sarger, *Journal of the Optical Society of America B: Optical Physics* **22**, 2138 – 2143 (2005).
22. S. Juodkazis, K. Nishimura, S. Tanaka, H. Misawa, E. Gamaly, B. Luther-Davies, L. Hallo, P. Nicolai, and V. Tikhonchuk, *Physical Review Letters* **96**, 166101 (2006).
23. L. Fu, G. Marshall, J. Bolger, P. Steinvurzel, E. Magi, M. Withford, and B. Eggleton, *Electronics Letters* **41**, 638–640 (2005).
24. A. Martinez, I. Khrushchev, and I. Bennion, *Optics Letters* **31**, 1603 – 5 (2006).
25. C. W. Smelser, S. J. Mihailov, and D. Grobnic, *Optics Express* **13**, 5377 – 5386 (2005).
26. A. Martinez, I. Khrushchev, and I. Bennion, *Electronics Letters* **41**, 176 – 8 (2005).
27. T. Allsop, M. Dubov, A. Martinez, F. Floreani, I. Khrushchev, D. Webb, and I. Bennion, *2005 Conference on Lasers and Electro-Optics, CLEO* **3**, 2179 – 2181 (2005).
28. A. Martinez, M. Dubov, I. Khrushchev, and I. Bennion, *2006 Conference on Lasers and Electro-Optics, CLEO* p. JTuD13 (2006).
29. S. Tzortzakis, L. Sudrie, M. Franco, B. Prade, A. Mysyrowicz, A. Couairon, and L. Berge, *Physical Review Letters* **87**, 213902 – 1 (2001).
30. M. Feit, and J. Fleck, J.A., *Applied Physics Letters* **24**, 169 – 72 (1974).
31. Q. Feng, J. Moloney, A. Newell, E. Wright, K. Cook, P. Kennedy, D. Hammer, B. Rockwell, and C. Thompson, *IEEE Journal of Quantum Electronics* **33** (1997).
32. V. Mezentsev, J. Petrovic, J. Dreher, J. Jurgen, and R. Grauer, *Proceedings of SPIE - The International Society for Optical Engineering* .

Modelling of a post-discharge reactor used for plasma sterilization

K. Kutasi*, C. D. Pintassilgo*,†, P. J. Coelho** and J. Loureiro*

*Centro de Física dos Plasmas, Instituto Superior Técnico, 1049-001 Lisboa, Portugal
†Departamento de Física, Faculdade de Engenharia da Universidade do Porto, 4200-465 Porto, Portugal
**Departamento de Engenharia Mecânica, Instituto Superior Técnico, 1049-001 Lisboa, Portugal

Abstract. A three dimensional hydrodynamic model is developed to simulate a post-discharge reactor placed downstream from a flowing microwave discharge in N_2-O_2 used for plasma sterilization. The temperature distribution and the density distributions of $NO(B^2\Pi)$ molecules and $O(^3P)$ atoms, which are known to play a central role in the sterilization process, are obtained in the reactor in the case of discharges at p=8 Torr, f=915 MHz and p=2 Torr, f=2450 MHz, and $N_2-2\%O_2$ mixture composition. Excluding the flow direction, sufficiently low temperatures ideal for sterilization have been found in most part of the reactor. The highest $NO(B)$ and $O(^3P)$ concentrations at the reactor entrance are achieved at the highest pressure values here investigated. However, these larger densities rapidly decrease within a few centimeters below the values obtained at lower pressure. On the contrary, at low pressure the density distributions of $NO(B)$ and $O(^3P)$ are quasi homogenous in most of the horizontal planes.

Keywords: post-discharge, hydrodynamic model, plasma sterilization
PACS: 52.77.-j,52.80.Tn,82.33.Xj

INTRODUCTION

In the last few years the appearing of the polymer based heat sensitive tools in the hospitals brought the need of new sterilization methods. Among the new alternatives, low temperature plasmas became one of the most promising possibilities since the afterglow of a gas discharge provides at relatively low temperatures high concentration of chemically active radicals, such as excited species and UV photons, capable to inactivate microorganisms. Moreover, it has been shown that in plasma sterilization the major role is played by the neutral species [1, 2, 3], which has led the attention to be focused on the post-discharge region, where active radicals and excited species rather than charged particles are present [4]. The post-discharge, compared with the discharge zone, has the great advantage of fulfilling the two conditions of much lower gas temperature and absence of charged species since these could damage the material to be sterilized. Moreau *et al.* [1] have shown that total inactivation of an initial 10^6 spore population can be obtained using the flowing afterglow of a microwave discharge sustained in an $N_2-2\%O_2$ mixture at reduced gas pressure (typically $1-10$ Torr).

Although many conclusions can be derived from experiments concerning the determination of UV light intensities and densities of different active species (*e.g.* N, O) [2, 5], modelling studies can help us to improve the understanding of the complex interplay molecular kinetics that takes place in such medium and to obtain the concentrations of

CP876, *The Physics of Ionized Gases: 23rd Summer School and International Symposium*,
edited by L. Hadžievski, B. P. Marinković, and N. S. Simonović

those species whose densities are difficult or even not possible to measure, and thus to determine the optimal operating conditions for the sterilization reactors.

The modelling of a post-discharge involves the study of three different regions: (i) the plasma source, where the active species are produced, which in our case is a flowing microwave discharge; (ii) the short-lived afterglow connecting the discharge zone with the large post-discharge chamber; and (iii) the long-lived afterglow in the sterilization vessel, where the objects to be sterilized should be placed. In this work we present only the results obtained from a three-dimensional hydrodynamic model for the post-discharge chamber, which allows the determination of the various species concentrations in the whole 3 dimensional reactor. The results obtained for the discharge and early-afterglow region can be found in [4], where the kinetic model for the discharge and the analysis of the time evolution of the species concentrations in the early-afterglow have been considered.

HYDRODYNAMIC MODEL

The self-consistent model for the flowing microwave discharge in N_2-O_2 is based on the solutions for the electron Boltzmann equation coupled to a system of rate-balance equations for the neutral and charged heavy species [4]. The near afterglow that occurs downstream from the discharge, in the same discharge tube, has been considered by cutting-off the excitation by electron impact in the system of master equations. The time-dependent solutions obtained with this model constitute the initial conditions for the present 3D simulation.

The three dimensional hydrodynamic model of the post-discharge reactor is composed of (i) the continuity equations for the different species Eq. 1, (ii) a sole momentum conservation equation Eq. 2, and (iii) the energy conservation equation Eq. 3:

$$\nabla \cdot (\rho y_i \, \mathbf{v} - D_i \, \rho \nabla y_i) = m_i \, S_i \; ; \tag{1}$$

$$\rho (\mathbf{v} \cdot \nabla) \mathbf{v} + \nabla \cdot \mathbb{T} = \rho \, \mathbf{g} \; ; \tag{2}$$

$$\nabla \cdot (\rho \, C_p T \, \mathbf{v} - \lambda \, \nabla T) = 0 \; . \tag{3}$$

Here, ρ denotes the mass density of the gas, $\mathbf{v}$ its velocity, and for each species i, y_i is the relative mass density ($y_i = \rho_i / \rho$), D_i the diffusion coefficient, m_i the mass and S_i the source term. Further, $\mathbb{T}$ is the stress tensor, $\mathbf{g}$ the acceleration due to gravity, T the gas temperature, C_p the specific heat at constant pressure and λ the thermal conductivity. The total mass conservation equation $\nabla \cdot (\rho \mathbf{v}) = 0$ is naturally embodied in Eq. 1, when a sum is performed over all species. The equations are solved in the steady state regime assuming a laminar flow regime. The gas is assumed to be a Newtonian fluid, which means that the $\mathbb{T}$ stress tensor can be written under the following form

$$\mathbb{T} = p \mathbb{I} - \mu [\nabla \mathbf{v} + (\nabla \mathbf{v})^t] + \frac{2}{3} \mu (\nabla \mathbf{v}) \mathbb{I} \; , \tag{4}$$

where p is the static pressure, μ is the dynamic viscosity, $\mathbb{I}$ is the unit tensor and the superscript t means transpose. During our calculations we have neglected the Soret and

TABLE 1. Reactions taken into account in the hydrodynamic model. The rate coefficients can be found in Ref.[4, 7].

Processes	Processes
$N_2(A)+O_2(X)\rightarrow N_2(X)+O_2(X)$	$N_2(A)+O_2(X)\rightarrow N_2(X)+O(^3P)$
$N_2(A)+O(^3P)\rightarrow N_2(X)+O(^3P)$	$N_2(A)+NO(X)\rightarrow N_2(X)+NO(A)$
$N(^4S)+O(^3P)+N_2\rightarrow NO(X)+N_2$	$N(^4S)+O_2(X)\rightarrow NO(X)+O(^3P)$
$N(^4S)+O_2(a)\rightarrow NO(X)+O(^3P)$	$N(^4S)+O(^3P)+O_2\rightarrow NO(X)+O_2$
$N(^4S)+O(^3P)+N_2\rightarrow NO(B)+N_2$	$N(^4S)+O(^3P)+O_2\rightarrow NO(B)+O_2$
$N(^4S)+NO_2(X)\rightarrow N_2(X)+O_2(X)$	$N(^4S)+NO_2(X)\rightarrow NO(X)+NO(X)$
$N(^4S)+NO_2(X)\rightarrow N_2(X)+O(^3P)+O(^3P)$	$O_2(a)+O_2\rightarrow O_2(X)+O_2$
$O_2(a)+O(^3P)\rightarrow O_2(X)+O(^3P)$	$O_2(a)+NO\rightarrow O_2(X)+NO$
$O_2+O_2(X)+O(^3P)\rightarrow O_3+O_2$	$O_2(X)+O(^3P)+O(^3P)\rightarrow O_3+O(^3P)$
$O_2(a)+O_3\rightarrow O_2(X)+O_2(X)+O(^3P)$	$O_3+O(^3P)\rightarrow O_2(X)+O_2(X)$
$O_3+O(^3P)\rightarrow O_2(a)+O_2(X)$	$O_2(b)+O(^3P)\rightarrow O_2(X)+O(^3P)$
$O_2(b)+O(^3P)\rightarrow O_2(a)+O(^3P)$	$O_2(b)+NO\rightarrow O_2(a)+NO$
$O_2(b)+N_2\rightarrow O_2(a)+N_2$	$O_2(b)+O_3\rightarrow O_2(X)+O_2(X)+O(^3P)$
$O(^3P)+NO(X)+N_2\rightarrow NO_2(X)+N_2$	$O(^3P)+NO(X)+O_2\rightarrow NO_2(X)+O_2$
$O(^3P)+O_2(X)+N_2\rightarrow O_3(X)+N_2$	$NO(A)\rightarrow NO(X)+h\nu$
$NO(A)+N_2\rightarrow NO(X)+N_2$	$NO(A)+O_2\rightarrow NO(X)+O_2$
$NO(A)+NO\rightarrow NO(X)+NO$	$NO(B)\rightarrow NO(X)+h\nu$
$NO(B)+N_2\rightarrow NO(X)+N_2$	$NO(B)+O_2\rightarrow NO(X)+O_2$
$NO(B)+NO\rightarrow NO(X)+NO$	$NO(X)+O_3\rightarrow NO_2(X)+O_2(X)$
$NO_2(X)+O(^3P)\rightarrow NO(X)+O_2(X)$	

pressure diffusions, as well as the Dufour effect. The transport data values used in the model are presented in Ref. [7].

In the hydrodynamic model 12 different species are taken into account, namely: the electronic ground-states $N_2(X^1\Sigma_g^+)$ and $O_2(X^3\Sigma_g^-)$; the ground-state atoms $N(^4S)$ and $O(^3P)$; the most populated excited states of both gases, $N_2(A^3\Sigma_u^+)$, $O_2(a^1\Delta_g)$ and $O_2(b^1\Sigma_g^+)$; and the formed species $NO(X^2\Pi)$, $NO(A^2\Sigma^+)$, $NO(B^2\Pi)$, $NO_2(X)$ and O_3. In the hydrodynamic model of the post-discharge reactor the charged species are not taken into account since they recombine in the early-afterglow and at the entrance of the reactor vessel their concentrations are vanishingly small. In the model for the discharge and short-lived afterglow other species have also been considered [4, 6], but here they are not relevant for the 3D simulation, such as: the vibrational molecular distributions $N_2(X^1\Sigma_g^+, v)$ and $O_2(X^3\Sigma_g^-, v')$; other electronic states of molecular and atomic nitrogen $N_2(B^3\Pi_g, B'^3\Sigma_u^-, C^3\Pi_u, a'^1\Sigma_u^-, a^1\Pi_g, w^1\Delta_u)$ and $N(^2D, ^2P)$; the $NO_2(A)$ state; the main positive ions $N_2^+(X^2\Sigma_g^+, B^2\Sigma_u^+)$, N_4^+, O_2^+, O^+ and NO^+; and negative ions O^-. The set of gas phase reactions describing the kinetics of species as considered in the post-discharge chamber is listed in Table. 1. The surface loss/reactions of atoms taken into account in the discharge and early-afterglow regions are neglected in the post-discharge model, since in the late-afterglow the loss probability of atoms on the walls is very low; and since the reactor is large the surface reactions cannot significantly influence the gas phase densities. As already mentioned in Section 1, the active species are created in an N_2-O_2 microwave flowing discharge and after passing through a short afterglow region they are carried into the post-discharge chamber presented in Figure 1. The reactor is similar to the one used by Philip *et al.* [2] in their experiments, however here we consider the exit on the top plane instead of the bottom. From the fluid dynamics point of view there is no difference if the exit is on the top or on the bottom, since the gravity force (the static pressure) at these densities does not play any role. The size of the reactor is $65 \times 25 \times 25$ cm^3. The entrance is situated on the west plane at a distance of 8 cm from the top. The 2.6×2.6 cm^2 square inlet and outlet are symmetrically positioned on the west and top walls, respectively, so that only one half of the chamber needs to be considered in the simulation.

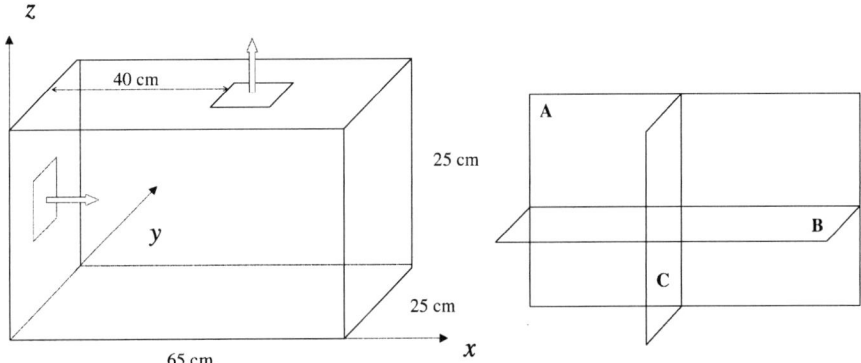

FIGURE 1. Post-discharge chamber (left) and various planes to be considered in the discussion of the results (right).

The hydrodynamical model has been solved by adapting the numerical code developed by Ferziger and Perić [8] to our situation. That code basically solves the steady three-dimensional Navier-Stokes equations, so that the equations for mass density conservation of each species together with the equation for energy conservation have been incorporated in it. The equations are discretized using the finite volume method. The linear algebraic equation system so obtained is then solved with Stone's method iteratively using the multigrid method [8]. In our solution three grid levels are used, the finest grid has $80 \times 40 \times 40$ control volumes.

RESULTS AND DISCUSSION

In this paper the results are presented for two discharge conditions: *i)* $p=8$ Torr, $f=915$ MHz; and *ii)* $p=2$ Torr, $f=2450$ MHz. The discharge is sustained in a tube with $R=1.3$ cm inner radius in both cases, whereas the critical electron densities are set equal to 5.19×10^{10} cm^{-3} and 3.74×10^{11} cm^{-3} for 915 MHz and 2450 MHz, respectively. The temperature at the inlet and at the wall are assumed to be 500 K and 300 K, respectively, the gas flow rate is taken 2×10^3 sccm. The total gas density is considered constant in the post-discharge reactor, at 2 Torr and 8 Torr is taken 6.42×10^{16} cm^{-3} and 2.57×10^{17} cm^{-3}, respectively. The gas flow rate and the length of the afterglow tube determine the flight-time of active species in this region. In order to ensure that the active species may reach the sterilization chamber still reasonably populated, both the gas flow and the length of the afterglow tube should be chosen carefully. Here we consider the length of the afterglow through the flight-time of the species between the instant at which they leave the active discharge zone and the instant they enter in the reactor. All over the calculations we assume a flight-time of 1 ms.

Due to the symmetry of the chamber in y direction we simulate only one half of the chamber, so that the results are presented for half chamber in the $x-y$ and $y-z$ planes. The results are presented for three different views, see Figure 1: (**A**) $x-z$ vertical plane parallel to the inlet flow at constant y; (**B**) $x-y$ horizontal plane at constant z; (**C**) $y-z$ vertical plane perpendicular to the inlet flow at constant x.

Gas temperature

In plasma sterilization an important parameter to control is the value of the gas temperature in the processing zone, which should be lower than $60-70$ °C (typically < 340 K). The temperature distribution depends only on the gas flow rate and on the assumed inlet and wall temperatures, therefore the same distributions are obtained for both pressure cases.

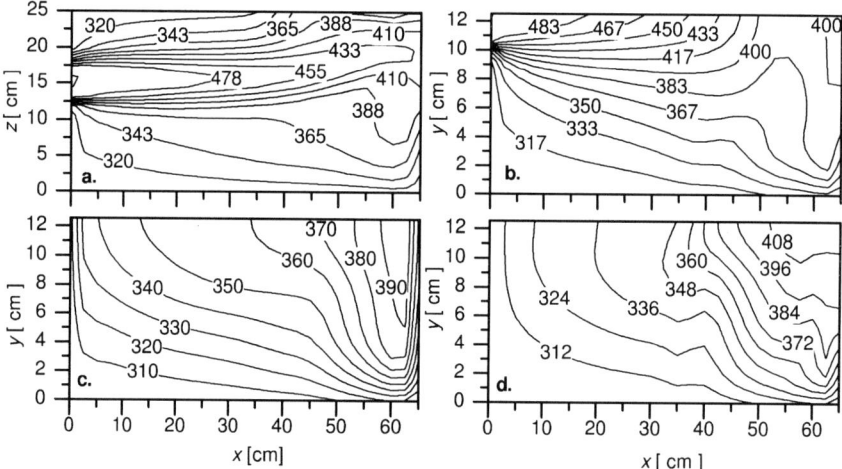

FIGURE 2. Temperature distribution (a) in the x-z vertical plane parallel to the inlet flow at y=12.5 cm, in one-half of the $x-y$ horizontal plane at: (b) z=15 cm, (c) z=10 cm, (d) z=22.5 cm.

Figure 2(a) shows the temperature distribution in the reactor in the $x-z$ plane with y=12.5 cm, that is a vertical plane parallel to the inlet flow (view **A**), when the gas flow is set to 2×10^3 sccm. The gas temperature decreases gradually from 500 K to 300 K towards the reactor's wall and it stays always over 400 K along the flow direction (the entrance is situated between z=14.4 cm and z=17 cm at x=0). Figure 2 (c)-(d) present the temperature in the half $x-y$ horizontal plane for three different values of z coordinate, see view **B**, from z=10 cm up to 22.5 cm height. In the inlet plane z = 15 cm, Figure 2(b), acceptably low temperatures can be found only close to the front wall (y=0 plane), since as shown above in the flow direction the temperature is above 400 K. In the case of the planes, which are positioned with a few centemeters from the inlet, see Figure 2(c,d), in the most part of the plane the temperature is below 360 K. It is worth noting at this point that at the z=22.5 cm plane, Figure 2(d), which is positioned between the inlet and outlet planes, the temperature is roughly the same as in z=10 cm plane, but in the positions where the flow crosses the plane, that is for $x \geq 40$ cm, the temperature becomes larger by about 20 K than in the z=10 cm plane. We note that in planes positioned even lower than the z = 10 cm, the temperature is below 370 K in the whole $x-y$ plane and in the most part of the plane it is even below 340 K.

Concentrations of NO(B) and O(^{3}P)

Moisan *et al.* [3] have shown that in an N_2-O_2 post-discharge, with small O_2 addition, the sterilization occurs as a result of the combined effect of O(^{3}P) atoms and UV photons emitted by NO($B^2\Pi$) molecules in the 250−320 nm spectral range (the so-called NOβ bands). Whereas the O(^{3}P) atoms contribute to the erosion of the protecting layer of the microorganisms, the UV photons associated with the emission of NOβ bands damage irreversibly the genetic material (DNA) of the microorganisms, in an extension that they cannot repair and reproduce themselves. Therefore, here we focuse on the presentation of the density distributions of these two species.

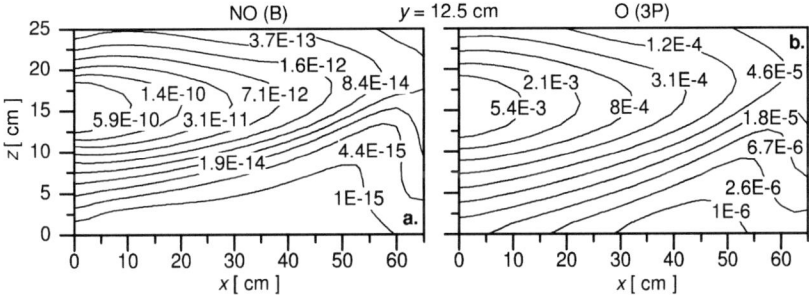

FIGURE 3. Relative mass density distributions of (a) NO(B) and (b) O(^{3}P) in the $x-z$ vertical plane parallel to the inlet flow at y=12.5 cm in the case of p=8 Torr, f=915 MHz.

Figure 3(a) and (b) show in the $x-z$ vertical plane parallel to the inlet flow at y=12.5 cm, view **A**, the relative mass density distributions ($y_i=\rho_i/\rho$) of NO(B) and O(^{3}P), respectively, obtained in the post-discharge reactor at p=8 Torr and f=915 MHz. The densities show a very fast exponential decrease towards the reactor's wall. In the flow direction at $(y, z) = (12.5$ cm, 15 cm) we find a decrease of $\exp(-x/7)$ for NO(B) and of $\exp(-x/11)$ for O(^{3}P) with x in cm. The density of NO(B) decreases by five orders of magnitude from the entrance to the bottom plane, Figure 3(a), whereas for O(^{3}P) this decrease reaches about three orders of magnitude, Figure 3(b). An identical fast decrease can be observed for both species also in the y direction.

Figure 4 shows the relative mass density of NO(B) and O(^{3}P) at p=2 Torr, f=2450 MHz. Figure 4(a) and (d) show these concentrations in the $x-z$ vertical plane with y=12.5 cm. In this plane the decrease of O(^{3}P) concentration, shown in Figure 4(d), is much slower than in the case of 8 Torr. In the flow direction we can identify a second order polynomial decrease contrary to the exponential decrease at 8 Torr. According to this slower decrease at 2 Torr, the density of $O(^3P)$ diminishes from the entrance to the bottom plane just by a factor of 4. On the other hand, the NO(B) density at 2 Torr, shown in Figure 4(a), decreases by one order of magnitude. The same

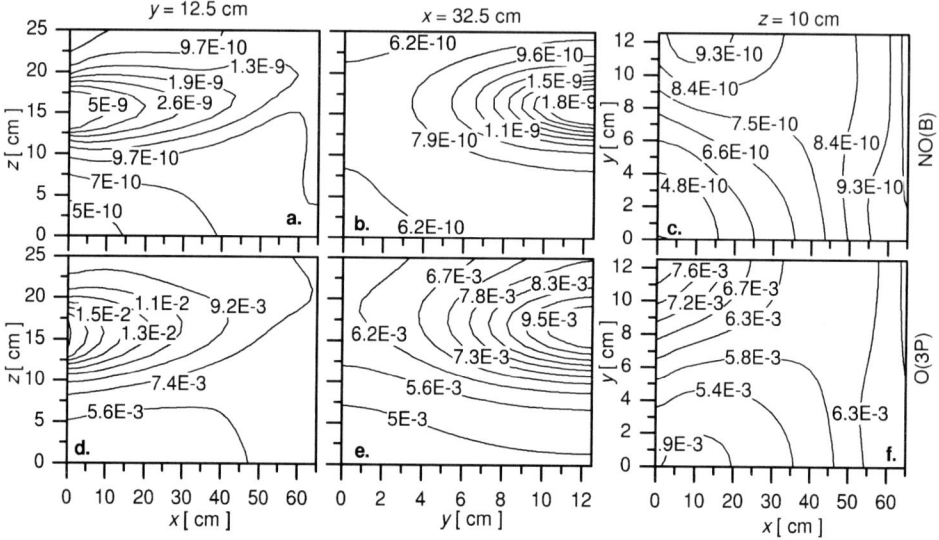

FIGURE 4. Relative mass density distributions of NO(B) and $O(^3P)$, respectively, in the case of p=2 Torr, f=2450 MHz, for the following views: (a, d) $x-z$ vertical plane parallel to the inlet flow at y=12.5 cm (view **A**); (b, e) $y-z$ vertical plane perpendicular to the inlet flow at x=32.5 cm (view **C**); (c, f) the $x-y$ horizontal plane at z=10 cm (view **B**).

behaviour can be also observed in the y direction, see Figure 4(b) and (e), where the NO(B) and $O(^3P)$ relative mass density distributions are presented in the $y-z$ vertical plane perpendicular to the inlet flow with x=32.5 cm (view **C**). On the contrary, these distributions are close to homogenous in the $x-y$ horizontal plane with z=10 cm height (view **B**), see Figure 4(c) and (f). Quasi homogeneous distributions can be found also in most of the $x-y$ horizontal planes, *e.g.* if we consider the z=5 cm plane at $y = 12.5$ cm in Figure 4(a) we can see that in the x direction the NO(B) relative mass density only changes slightly from 5×10^{-10} to 9.7×10^{-10}, and that in the y direction at $x = 32.5$ cm, Figure 4(b), the variation is from 6.2×10^{-10} to $\approx 7\times10^{-10}$.

The faster decrease of the $O(^3P)$ density at 8 Torr can be explained by larger losses than at 2 Torr. The losses are governed by the reaction of $O(^3P)$ with $NO_2(X)$; by the three body reactions of $O(^3P)$ with $N(^4S)$ and NO(X) in the presence of N_2 and O_2, leading to the formation of NO(X, B) and $NO_2(X)$ states; and by the three body reactions for O_3 formation. Since at 8 Torr the gas density is four times larger than at 2 Torr, the losses due to the three body reactions are considerably higher, which can result in a faster decrease of $O(^3P)$ density. In the case of p=8 Torr we also find at the entrance of the post-discharge chamber an $NO_2(X)$ relative concentration of about two orders of magnitude larger than at 2 Torr. This difference in the $NO_2(X)$ concentration at the entrance and in the flow direction results in larger losses for $O(^3P)$ and contribute to a faster decrease of the atomic oxygen concentration at 8 Torr. On the other hand, the losses of NO(B) are governed, besides the radiative decay, by quenching by NO, O_2 and N_2. At 8 Torr the relative densities of NO(X) and O_2(X) at the entrance of the chamber are one order of magnitude larger than at 2 Torr, resulting in higher NO(B) losses at 8 Torr, which explain the much faster decrease of NO(B).

CONCLUSIONS

In this work we have been investigated the distribution of the temperature and of the density of the active species in a post-discharge reactor for sterilization purposes by means of a three dimensional hydrodynamic model. The active species entering the reactor have been created in $N_2-2\%O_2$ flowing microwave discharges generated in a tube of 1.3 cm inner radius. The calculations have been carried out for two discharge conditions: i) p=8 Torr, f=915 MHz; and ii) p=2 Torr, f=2450 MHz, when the gas flow rate has been set to 2×10^3 sccm.

The calculated temperature distributions have shown that, in the case of 2×10^3 sccm gas flow, the temperature decreases gradually from 500 K to 300 K towards the walls and remains always over 400 K in the flow direction. In the case of the horizontal planes the temperature is lower than 370 K or 340 K depending on the proximity to the inlet.

The calculated relative mass densities of the most relevant species for sterilization, namely $O(^3P)$ and NO(B), present an exponential decrease in the chamber at 8 Torr, while at 2 Torr the density decrease is much slower. At this lower pressure, a second order polynomial decrease has been identified in the flow direction. We have also found that at 2 Torr the density distributions of $O(^3P)$ and NO(B) are close to homogeneous in most of the horizontal planes.

For more results the reader can refer to Ref. [7], where the effect of the pressure, gas flow rate and gas mixture composition on the density distributions is also presented.

ACKNOWLEDGMENTS

This work has been supported by the Portuguese Science Foundation FCT through project POCTI/FAT/44221/2002 and through the post-doc fellowship of KK.

REFERENCES

1. Moreau S, Moisan M, Tabrizian M, Barbeau J, Pelletier J, Ricard A, Yahia L'H 2000 *J. Appl. Phys.* **88** 1166
2. Philip N, Saoudi B, Crevier M-Ch, Moisan M, Barbeau J, Pelletier J 2002 *IEEE Trans. Plasma Sci.* **30** 1429
3. Moisan M, Barbeau J, Crevier M-Ch, Pelletier J, Philip N, Saoudi B 2002 *Pure Appl. Chem.* **74** 349
4. Pintassilgo C D, Loureiro J, Guerra V 2005 *J. Phys. D: Appl. Phys.* **38** 417
5. Ricard A, Moisan M, Moreau S 2001 *J. Phys. D: Appl. Phys.* **34** 1203
6. Guerra V, Loureiro J 1999 *Plasma Sources Sci. Technol.* **8** 110
7. Kutasi K, Pintassilgo C D, Coelho P J, Loureiro J 2006 *J. Phys. D: Appl. Phys.* to be published
8. Ferziger J H and Peric M *Computational Methods for Fluid Dynamics* 2002, 3rd rev. ed., Springer

SECTION 3

LOW TEMPERATURE PLASMAS

Invited Lectures
Topical Invited Lectures
Progress Reports

Spectroscopic Studies of Atomic and Molecular Processes in the Edge Region of Magnetically Confined Fusion Plasmas

J. D. Hey[1], S. Brezinsek[2], Ph. Mertens[2] and B. Unterberg[2]

[1]*School of Physics, University of KwaZulu-Natal, Westville Campus, Private Bag X54001,Durban 4000, South Africa (hey@ukzn.ac.za, j.d.hey@sympatico.ca)*
[2]*Institut für Plasmaphysik, Forschungszentrum Jülich, Association EURATOM-FZJ, Trilateral Euregio Cluster, Germany, www.fz-juelich.de/ipp*
(s.brezinsek@fz-juelich.de, ph.mertens@fz-juelich.de, b.unterberg@fz-juelich.de)

Abstract. Edge plasma studies are of vital importance for understanding plasma-wall interactions in magnetically confined fusion devices. These interactions determine the transport of neutrals into the plasma, and the properties of the plasma discharge. This presentation deals with optical spectroscopic studies of the plasma boundary, and their rôle in elucidating the prevailing physical conditions. Recorded spectra are of four types: emission spectra of ions and atoms, produced by electron impact excitation and by charge-exchange recombination, atomic spectra arising from electron impact-induced molecular dissociation and ionisation, visible spectra of molecular hydrogen and its isotopic combinations, and laser-induced fluorescence (LIF) spectra. The atomic spectra are strongly influenced by the confining magnetic field (Zeeman and Paschen-Back effects), which produces characteristic features useful for species identification, temperature determination by Doppler broadening, and studies of chemical and physical sputtering. Detailed analysis of the Zeeman components in both optical and LIF spectra shows that atomic hydrogen is produced in various velocity classes, some related to the relevant molecular Franck-Condon energies. The latter reflect the dominant electron collision processes responsible for production of atoms from molecules. This assignment has been verified by gas-puffing experiments through special test limiters. The higher-energy flanks of hydrogen line profiles probably also show the influence of charge-exchange reactions with molecular ions accelerated in the plasma sheath ('scrape-off layer') separating limiter surfaces from the edge plasma, in analogy to acceleration in the cathode-fall region of gas discharges. While electron collisions play a vital rôle in generating the spectra, ion collisions with excited atomic radiators act through re-distribution of population among the atomic fine-structure sublevels, and momentum transfer to the atomic nuclei via ion-induced dipole collisions with the bound electrons. The ions are thus important in randomising and equilibrating the velocity distribution of atomic products of molecular dissociation.

Keywords: Fusion edge plasmas, tokamak boundary, atomic & molecular processes, optical and LIF spectroscopy, Zeeman (Paschen-Back) effect, electron-atom collisions, ion-atom collisions.
PACS: 32.10.Dk, 32.30.-r, 32.50.+d, 32.60.+i, 33.20.-t, 34.50.Dy, 34.50.-s, 34.80.Dp, 34.80.Ht, 52.40.Hf, 52.55.Fa, 52.70.-m, 68.49.-h, 79.20.Rf

1. INTRODUCTION

Spectroscopic studies [1-3] of the plasma boundary in magnetically confined fusion plasmas provide information of vital importance for understanding the plasma-wall interactions which govern fuelling and recycling processes [4-8]. This improved

CP876, *The Physics of Ionized Gases: 23rd Summer School and International Symposium,*
edited by L. Hadžievski, B. P. Marinković, and N. S. Simonović
© 2006 American Institute of Physics 978-0-7354-0377-2/06/$23.00

understanding also leads to optimisation in the design of plasma-facing wall components through use of the concept of local edge cooling [5-7,9]. While originally spectroscopic investigations were largely limited to the spectra of ions and atoms [1,2], molecular spectroscopy [3,9-11] has played an increasingly important rôle in recent years in shedding light on additional details of the atomic and molecular processes. Both passive and active spectroscopic methods have been applied successfully [1-3]. In analysing the results from both methods, it is essential to take account of the presence of a strong toroidal magnetic field (magnetic induction of the order of a few T), and a weaker poloidal field (of the order of a few tenths of 1T). This magnetic field is responsible for an appreciable Zeeman (Paschen-Back) effect on the observed spectra, which can often be used to advantage in the spectral analysis, owing to the uniqueness of particular Zeeman patterns [12]. Thus, well-resolved Zeeman patterns facilitate the identification of impurity species in the plasma [12-15]. For the typical edge plasma parameters of interest, the Zeeman effect is, apart from Doppler broadening, the primary spectral line broadening mechanism in atomic spectra [13-17], the Stark effect [18-20] playing a rôle mainly in divertor plasmas (see Table 1 for typical plasma parameters applicable to the TEXTOR tokamak). Spectra from highly excited levels of hydrogen and its isotopes are examples of the combination of both broadening mechanisms Zeeman and Stark effects), and in the edge regions of certain devices the electron density may be sufficiently high for Stark broadening to predominate [21]. In that case, charged particle densities derived from the Inglis-Teller limit [18] may be checked against values derived from other diagnostic methods.

TABLE 1. Some typical parameters for the TEXTOR tokamak [2,14,25-27]. This device is versatile, and can provide a range of magnetic field strengths and plasma conditions.

Toroidal magnetic field	Major radius	Minor radius	Last closed flux surface (LCFS)
$\left|\vec{B}_{tor}\right| = 2.25\,\text{T}$	1.75 m	0.50 m	0.46 m
Poloidal magnetic field	$\bar{n}_e$ (central)	n_e (LCFS)	kT_e (LCFS)
$\left|\vec{B}_{pol}\right| = 0.16\,\text{T}$	$5 \times 10^{19}\,\text{m}^{-3}$	$4 \times 10^{18}\,\text{m}^{-3}$	50 eV
Toroidal current	Heating Power	kT_i (LCFS)	Integrated particle flux (LCFS)
360 kA	1.5 MW	100 eV	10^{22} / s

From the Doppler broadening of individual Zeeman components, an ion or atom temperature can be derived from fits to measured spectra [2,13-16]. In certain special cases, such as the O I triplet transition at 844.6 nm, or the C I triplet transition at 909.5 nm (see Table 2), a small number of π components can be isolated by means of a linear polariser, when observations are made orthogonally to the magnetic field [2,22,23]. For longitudinal observations (tangential to the magnetic flux surfaces), a quarter-wave plate and linear polariser can be used to eliminate either σ_+ or $\sigma-$ components, e.g. in the case of the weaker C IV line from the doublet listed in Table 2. [14]. Spectroscopic studies of Zeeman components emitted near special limiter surfaces have provided valuable information on the plasma conditions applicable to chemical and physical sputtering in the tokamak edge plasma [15,22,23].

A related application of well-understood Zeeman patterns has been the derivation of atom and ion temperatures for successive ionisation stages of the same element. The temperatures of such 'test' particles are, however, often very different from the local 'ion temperature' [24,25,27], i.e. that of the local protons and deuterons (the 'field' particles), being produced through competition between heating by collisions with these field particles and ionisation by electron impact [14,15,24].

The most complicated atomic spectra are emitted by hydrogen (deuterium) and by other hydrogenic species excited by charge-exchange recombination [25], consisting of many, partially overlapping Zeeman components, which can no longer be isolated by optical means, and whose Doppler broadening corresponds to several distinct kinetic energy ranges. Special techniques have been developed to determine these atomic radiator temperatures, which have been found to correspond to particular velocity classes of atoms [26,28-30]: 'fast' atoms arising from charge-exchange recombination with hot 'field' particles transported outwards from the plasma interior, 'warm' atoms arising from 'usual' molecular dissociation processes and from ion-atom collisional heating, and 'cold' atoms produced by special molecular dissociation processes, viz. electron impact-induced molecular dissociation and ionisation [31-38]. Already in the first systematic study of this kind on TEXTOR [26], the conclusion was drawn that a substantial population of hydrogen (deuterium) exists in molecular form in the scrape-off layer and within the plasma edge, as required to sustain a significant fractional population of 'cold' atoms. The neglect of this molecular population would lead to incorrect interpretation of atomic processes in the boundary layer, as well as incorrect estimation of certain rate coefficients [11], a conclusion reached many years earlier by Engelhardt [39] in connection with a high-voltage theta pinch plasma.

These conclusions regarding the molecular hydrogen population and substantial population of associated 'cold' atoms in TEXTOR were confirmed by two different experimental studies. The first of these (laser-induced fluorescence [40,41]) provided quantitative support for the spectroscopic results, but without yielding direct evidence for the proposed production mechanism of the 'cold' atoms (since only atoms already in the ground state are 'interrogated' directly by the laser beam) [42,43]. However, the new data helped to resolve a 'puzzle' already noted by Bogen [2]: whereas a Doppler width corresponding to 0.7 eV had been recorded by fluorescence measurements on L_α, emission of H_α appeared to yield about 1 eV. Now, significantly lower temperatures had been recorded both for the atomic velocity distribution [42,43] and for the Balmer lines [26,28-30]. The second independent approach has been through Fulcher- α band (600-640 nm, $3p\ ^3\Pi_u \rightarrow 2s\ ^3\Sigma_g^+$) spectroscopy [9-11,44,45], which, combined with measurements of Balmer-α, provides both molecular rotational temperatures and total (molecular plus atomic) particle fluxes into the plasma.

In summary, therefore, spectral line profile studies on the plasma boundary have made enormous strides since the earlier use of only the far line wings for the determination of ion temperatures from charge-exchange neutrals [46-49]. We proceed to discuss several important topics outlined in the Introduction, in more detail below.

2. PRINCIPLES

Tokamak spectra are radiated from regions of plasma where the value of the magnetic induction is a significant fraction of the atomic unit (in SI):

$$B_0 = \frac{\alpha^3 m_e c}{e\, a_0} = 12.5168 \text{ T}.\tag{1}$$

The symbol α denotes the fine-structure constant, m_e the electron mass, and a_0 the Bohr radius [50,51]. Atomic spectra are more strongly affected by the magnetic field than are molecular spectra. An appreciable distortion of Doppler-broadened atomic line profiles by the Zeeman effect occurs for radiator temperatures below

$$kT(\text{eV}) \leq M(\text{u}) \left(\frac{n^4 B}{40\, Z^2} \right)^2,\tag{2}$$

where the radiator mass M is expressed in amu, Z is the effective nuclear charge (ionisation stage: 1 for neutrals, 2 for singly-ionised atoms, ...) and n the (effective) principal quantum number pertaining to the upper level of an $n-\alpha$ line [17,25]. Note that, while the importance of the Doppler effect does not depend *per se* upon the spectral region, the Zeeman effect is enhanced towards longer wavelengths. This is immediately seen on comparing the relative Doppler width (FWHM)

$$\frac{\Delta\lambda^D_{1/2}}{\lambda} = 2\sqrt{2\ln 2\, \frac{kT}{M c^2}}\tag{3}$$

with the relative splitting from the zero-field position of the $\sigma_\pm$ components in a simple Lorentz triplet [52]:

$$\frac{\Delta\lambda_\pm}{\lambda} = \mp \frac{\mu_B}{hc}\lambda B = \mp 46.68645 \left[\text{m}^{-1}\,\text{T}^{-1} \right] \lambda B,\tag{4}$$

where μ_B denotes the Bohr magneton. Broadening of hydrogenic spectra by the first-order Stark effect (microscopic electric field) may be neglected in relation to the Zeeman effect, provided that the electron concentration lies below the limit [26,51]

$$n_e \leq 3 \times 10^{19} \left(\frac{Z B}{n} \right)^{3/2} \text{m}^{-3}.\tag{5}$$

We see that Stark broadening first becomes relevant to higher series members. On the other hand, both the (fully developed) Paschen-Back effect and the 'diamagnetic', quadratic Zeeman effect [53] may need to be accounted for, the latter for magnetic field-strengths in excess of [17]

$$B \geq 5 \times 10^4 \frac{Z^2}{n^4} [\text{T}].$$

(6)

Generating precise theoretical spectra which include all three effects, for strong magnetic fields and high n values relevant to tokamak studies, presents a challenge.

Another common source of line broadening is opacity [18]. For the purposes of tokamak spectroscopy, however, this is often negligible, even in the line cores, except for the Lyman lines from the divertor. Relevant criteria for checking this are found in [17,54]. An 'extreme' example discussed in [26], is the opacity in the line cores of the α-γ lines of deuterium for an atomic temperature of $kT_a = 0.4 \, \text{eV}$ and an effective path length of 10 cm in a quasi-homogeneous boundary layer. The population density in level $n = 2$ required to produce an opacity $\tau = 1$ in the line cores is above the values typical for the range of conditions in TEXTOR.

In summary, for spectra from non-hydrogenic atoms, line broadening often occurs mainly through splitting by the Zeeman effect, and a convolution of Doppler and instrumental broadening. In the case of the spectra of the hydrogen isotopes, this is also true of the lower-lying series members over a wide range of conditions in the boundary plasma, with the added complication that several velocity classes of atoms may contribute to the line radiation. While the Paschen-Back effect may be weak for non-hydrogenic spectra in relation to the linear Zeeman effect, it is very important in hydrogen spectra, and cannot be ignored, since the individual line strengths are now very different from their unperturbed (zero-field) values. As a guide, the following simple criterion may be written down for the magnetic field at which the magnetic perturbation becomes comparable with the fine-structure splitting within a multiplet, i.e. where a strong Paschen-Back effect may be expected:

$$B \geq B_{\text{P-B}} = \frac{1}{46.68645 \left[\text{m}^{-1} \right]} \frac{\Delta \lambda_{FS}}{\lambda^2} [\text{T}].$$

(7)

In applications, a typical value for the unperturbed fine-structure splitting ($\Delta \lambda_{\text{FS}}$) is selected. As an illustration, we compare in Table 2 the case of Balmer H_α with that of some typical impurity lines of carbon and oxygen atoms and ions. This comparison shows that, whereas the Paschen-Back effect is strongly developed for the hydrogen isotopes and some other atomic transitions (e.g. in O I [23]) occurring in the magnetic fields of many tokamak devices (see Figures 1a and 1b), it is comparatively weak in many non-hydrogenic ions, particularly those in higher charge states. However, it is prudent to include the Paschen-Back effect in all calculations yielding data for diagnostic purposes on tokamaks, especially for spectra in the visible and infra-red.

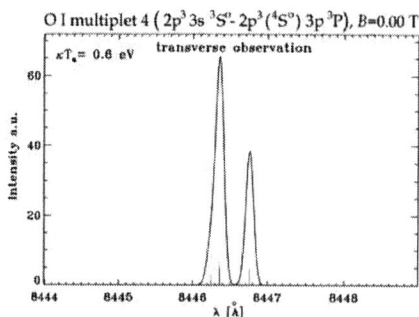
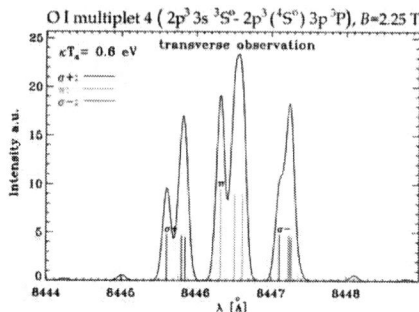

FIGURES 1A AND 1B. An oxygen triplet transition (left) showing a strong influence of the perturbing magnetic field of 2.25 T (right), yielding 19 Zeeman components in the 3 possible polarization states.

It should be emphasised that most of the formulae (viz. the inequalities) given in this section are intended as rough criteria only for judging the importance of particular effects. These are in no sense substitutes for precise calculations.

TABLE 2. Some values for the characteristic magnetic field-strength B_{P-B} **for which a significant Paschen-Back effect may be expected in the spectra listed. The fine-structure splitting ($\Delta\lambda_{FS}$) is representative of the multiplet in question (wavelength λ_{mult}).**

Radiator	Transition	λ_{mult} [nm]	$\Delta\lambda_{FS}$	B_{P-B} [T]
H (D)	$H_\alpha(n = 3 \rightarrow n' = 2)$	656.3	14 pm	0.7
C I	$2p\,3s\;^3P_2^o - 2p\,3p\;^3P_2$	909.5	0.63 nm	16
C II	$2s^2\,3s\;^2S - 2s^2\,3p\;^2P^o$	658.0	0.48 nm	24
C III	$2s\,3s\;^3S - 2s\,3p\;^3P^o$	464.9	0.12 nm	12
C IV	$1s^2\,3s\;^2S - 1s^2\,3p\;^2P^o$	580.5	1.07 nm	68
O I	$2p^3\,3s\;^3S^o - 2p^3\,3p\;^3P$	844.6	0.04 nm	1.2
O II	$2p^2\,3s\;^4P - 2p^2\,3p\;^4D^o$	465.2	0.30 nm	30

3. THE ZEEMAN (PASCHEN-BACK) EFFECT

For present purposes, this is considered as the main interaction causing additional line broadening. In the following, we assume the validity of the LS-coupling scheme. In the presence of a magnetic field, the perturbed atomic Hamiltonian is written [53]

$$\hat{H} = \hat{H}_0 + \hat{H}_{nucl} + \hat{H}_{mag}^I + \hat{H}_{mag}^{II}, \tag{8}$$

where contributions have been included from the isolated atom, the nuclear spin interactions (hyperfine structure splitting), and two different atom-magnetic field interactions, respectively. The first (I, 'paramagnetic') arises from the permanent moments of the bound electron(s), while the second (II, 'diamagnetic') accounts for magnetic moments induced by the magnetic field. It is usually permissible to ignore the hyperfine structure contributions as of minor importance. The term (I) is responsible for the Paschen-Back effect, and hence the departure of the line strengths from their zero-field values. Moreover, it results in the appearance of ΔJ-forbidden lines in the spectrum [55] at intermediate field strengths.

The atomic matrix representing the magnetic interaction operator $\hat{H}_{mag}^{I}$ has both diagonal and off-diagonal elements [52,56,57]:

$$\langle SLJM | \hat{H}_{mag}^{I} | SLJM \rangle = \mu_B \, g_J \, BM \,, \tag{9}$$

$$\langle SLJM | \hat{H}_{mag}^{I} | SL, J \pm 1, M \rangle = \mu_B \, B \left(g_s - g_\ell \right)$$
$$\times \sqrt{\frac{(L+S+1+J_>)(L+S+1-J_>)(L+J_>-S)(S+J_>-L)}{4 J_>^2 (2J_>+1)(2J_>-1)} \left(J_>^2 - M^2 \right)} \,, \tag{10}$$

where the Landé g-factor is given by

$$g_J = \frac{g_\ell \left[J(J+1) + L(L+1) - S(S+1) \right] + g_s \left[J(J+1) + S(S+1) - L(L+1) \right]}{2 J(J+1)} \,. \tag{11}$$

For non-hydrogenic atoms and ions (with the possible exception of He), it is usually (owing to the limited precision of data on the unperturbed fine-structure levels) quite sufficient to use for the orbital and spin g-values, respectively, of the optical electron: $g_\ell = 1.0$ and $g_s = 2.0$. On the other hand, since the eigen-energies for hydrogenic atoms and ions may be computed to arbitrary precision, we employ more precise values for these quantities in such cases, e.g. [50, 58]

$$g_\ell = 1 - \frac{m_e}{M_{nucl}} \qquad\qquad g_s = 2 \left[1 + \frac{\alpha}{2\pi} - 0.328 \left(\frac{\alpha}{\pi} \right)^2 \right] \tag{12}$$

For the multiplet transition $S\,L \rightarrow S'\,L'$ (with $S = S'$ in LS-coupling), the permitted ranges of the respective magnetic quantum numbers are therefore

$$|L-S| \leq M \leq L+S \qquad\qquad |L'-S'| \leq M' \leq L'+S' \tag{13}$$

The emitted photon appears to be either linearly polarised in a plane containing both the observation direction and the imposed magnetic field direction (π polarisation, $\Delta M = M - M' = 0$), or polarised perpendicularly to this plane ($\sigma_\pm$ polarisation, $\Delta M = M - M' = \pm 1$). A detailed analysis of intensities of these components, and their measurement by polarisation separation optics for arbitrary directions is given in [59].

A perturbation matrix is set up in the quantum number J for each M-value. This is a tridiagonal, square matrix of dimension $n^2 = (L+S+1-M) \times (L+S+1-M)$, containing the matrix elements calculated from equations (9)-(11) above. The n roots λ_i $(i=1,2,...,n)$ of the secular determinant now yield the perturbed eigen-energies for each particular M-value corresponding to the original fine-structure levels, assigned to the zero-field values in accordance with the non-crossing theorem of von Neumann and Wigner [60]. For each root λ_i, the corresponding eigenvector of the perturbation matrix is derived, subject to the normalisation condition [56]. This yields the combination of zero-field states $|SLJM\rangle$ into which the perturbed eigenstate may be decomposed. This procedure is repeated for the perturbation matrix of the lower fine-structure levels. In this way, both the Paschen-Back wavelengths and individual line strengths (S_{PB}) may be derived for the relevant field-strength. $|\vec{B}|$. From these, an overall 'theoretical' line profile for the particular multiplet is constructed for fitting to experimental data. The electric dipole case takes the following form [18,20,26,51,56]:

$$I(SL \to S'L') = \frac{4}{3} \frac{m_e}{c} a_0^3 \alpha^2 \sum_{JM,J'M'} \omega^4 F(\omega) \frac{n_{JM}}{g_{JM}} \frac{S_{PB}(JM - J'M')}{a_0^2 e^2} P_{\Delta M}(\theta_p) \frac{\Delta\Omega}{4\pi}, (14)$$

which includes the normalised polarisation function (a function only of the angle θ_p subtended by the magnetic field and observation directions)

$$P_{\Delta M}(\theta_p) = \frac{3}{8\pi} \sin^2 \theta_p \quad (\pi : \Delta M = 0); \qquad \frac{3}{16\pi} \left(1 + \cos^2 \theta_p\right) \quad (\sigma_\pm : \Delta M = \pm 1). (15)$$

Expression (14) accounts for radiation emitted per unit angular frequency interval into a solid angle $\Delta\Omega$ from unit volume of the plasma. The angular frequency normalised line shape (including an assumed instrumental response function) is denoted by $F(\omega)$, the population density of an individual fine-structure sublevel (JM) is denoted by n_{JM}, with $g_{JM} = 1$. This formulation is simplified in practice in at least two ways:
(i) The angular frequency factor is put equal to its value for the multiplet in the absence of fine structure, ω_0^4 (ignoring a 'trivial' source of asymmetry [18]).
(ii) It is usually assumed that ion-radiator collisions are sufficiently frequent to impose a statistical distribution among the fine-structure sublevels [50], also in the presence of the external magnetic field, and even if the original population mechanism actually favours certain of the sub-levels. We then have the simplified expression:

$$I(SL \to S'L') = \text{constant} \times \sum_{JM,J'M'} F(\omega) \frac{S_{PB}(JM - J'M')}{a_0^2 e^2} P_{\Delta M}(\theta_p) \frac{\Delta\Omega}{4\pi}. \qquad (16)$$

An interesting feature of this régime is the appearance of ΔJ-forbidden components in the spectrum, which disappear again for field-strengths sufficiently high for the Paschen-Back effect to complete the decoupling of spin and orbital angular momenta [55]. Formulae for evaluating the numbers of allowed and forbidden $\sigma_\pm$ and π components in transitions between principal quantum numbers n and n' in hydrogenic atoms, and for multiplets of various types in non-hydrogenic atoms, may be found in [14,25,26,51]. We conclude by mentioning a few practical aspects of such calculations. Firstly, a sufficiently good knowledge of the unperturbed atomic fine structure is assumed. For this purpose, the best available atomic energy levels should be used, supplemented, if necessary by fine-structure estimates [52] of adequate precision. Note from equation (7), that knowledge of the local magnetic field-strength to within 0.001 T is consistent with that of the fine structure to within 0.05 cm^{-1}. When fitting, careful attention should be paid to the transmission characteristics of any interference filters used in the spectral recordings, or otherwise the possible existence of spectra of different diffraction orders within the observation 'wavelength window' [28]. Then, there is the question whether particular states of polarisation have been favoured (without polarisation selection), or altered in relative intensity, e.g. by the reflection grating, or by reflection at windows along the optical path [14,61].

4. ROLE OF MAGNETIC FIELD IN PARTICLE COLLISIONS

While the magnetic field can be of major importance in determining the appearance of particular spectra of interest, its rôle in various collision processes can often be neglected. This includes the evaluation of elastic scattering rates, population transfer between fine-structure sublevels by ion collisions, and atomic excitation by electron collisions. Two criteria are relevant for checking whether the magnetic field may be neglected for these purposes. Firstly, the perturber gyro-radius [62] should exceed the local plasma Debye length [18,63]. For a charged ('field') particle (charge $+Q_f e$, mass M_f), this implies that the electron concentration should exceed [15,29,30,51]

$$n_e > \frac{m_e}{8\pi a_0^3 M_f} \alpha^4 \left(\frac{Q_f B}{B_0} \right)^2 . \tag{17}$$

in terms of the field-strength B_0 from equation (1). For example, for deuterons in an edge plasma of electron concentration $n_e = 1.0 \times 10^{18}$ m^{-3}, B should not exceed 27.5 T. This criterion is therefore generally well fulfilled for ion but not necessarily for electron perturbers. For electron perturbers, the relevant spatial scale for atomic excitation processes is the impact parameter for strong collisions [18]

$$\rho = \frac{\hbar}{m_e u} \left| \langle i | \frac{\vec{r}}{a_0} | j \rangle \right| \tag{18}$$

(electron-radiator collisions, with $u=|\vec{u}|$ denoting the magnitude of the relative velocity). Since the thermal average of this length is considerably smaller than the electron gyro-radius for relevant field-strengths, the role of the magnetic field may be ignored for both ion and electron collisions, for most practical purposes.

5. ATOMIC PRODUCTS OF MOLECULAR DISSOCIATION

Atomic hydrogen and its isotopes, and atomic oxygen, arise through molecular dissociation processes, of which a large variety are relevant, each with its own characteristic Franck-Condon energy E_0 [37,64]. The kinetic energy of the parent molecules is usually too small to influence the atomic kinetic energy in the laboratory frame [65]. Ion-atom collisions rapidly randomise and equilibrate the atomic kinetic energies into a Maxwellian with kinetic temperature $kT_a \approx E_0$, the main collision mechanism being the ion-induced dipole interaction [28,29,65-67]. Expressions for estimating characteristic time-scales in such processes, by analogy with studies by Spitzer [62] for Coulomb collisions, are now available [65].

For these atomic species, various Franck-Condon energies exist, each with its own threshold energy for production. The availability of the corresponding dissociation channel therefore depends upon the edge plasma parameters; thus, spectroscopic atomic 'temperatures' can vary appreciably [30]. Striking examples of controlled variations in the local values of n_e and T_e are provided by gas puffing experiments through test limiters [68]. While the case of O I [69] is discussed in [15,51], we consider H and D production here. Letting X denote either isotope, we find in table 3 data for the main dissociation mechanisms of the parent molecules by electron impact: the threshold kinetic energies for the electron perturbers, and Franck-Condon energies per atom produced. The data suggest the apparent paradox, that hotter edge plasmas result in the release of 'colder' atoms, as has been verified spectroscopically [30]. Reactions (i) - (iv) are of interest as important sources of Balmer line radiation (n = 3, 4, 5), processes (i) and (iv) requiring in addition electron impact excitation, for which characteristic time-scales may be found in [28,65]. An (impurity) ion can take the place of the electron, the inverse process being a three-body collision with the energy of molecular formation taken up by the ion and released as an emission line. This is the basis of chemiluminescence in long-period variable stars [71,72]. Here we require a close resonance in an ionic level with the molecular dissociation energy. One candidate for the inverse of (i) is Mg^+, with a 4.4 eV resonance excitation energy [72].

TABLE 3. Production mechanisms for atomic hydrogen (deuterium) by electron impact-induced molecular dissociation, with the corresponding threshold energies and Franck-Condon energies per atom released [33,34,36,70].

	Reaction	ε_{th} (eV)	E_0 (eV)
(i)	$X_2 + e^- \rightarrow X(1s) + X(1s) + e^-$	8.9 ± 0.1	2.24
(ii)	$X_2 + e^- \rightarrow X(n\,\ell) + X(1s) + e^-$	17.0 ± 0.2	0.2
(iii)	$X_2 + e^- \rightarrow X(2s) + X(1s) + e^-$	14.6 ± 0.3	0.3
(iv)	$X_2 + e^- \rightarrow X(1s) + X^+ + e^- + e^-$	18.0 ± 0.2	0.2

244

In addition to the so-called 'narrow component'(NC) in the atomic energy spectrum arising from (ii)-(iv), there is also a 'broad component' (BC) with E_0 in the range of 4-8 eV, and threshold energy of some 25 eV and above [31-34,38,73,74] which would contribute to the flanks of the Balmer line profiles. This component also arises from electron collisions. However, while the range of production mechanisms for the NC is limited (see table 3), the BC can arise in other ways, including charge-exchange recombination and collisions with hot protons transported outwards from the plasma interior [26]. Other interesting mechanisms are familiar from gas discharge studies of the phenomenon known as 'excessive Balmer line broadening' [75-79].

The suggestion having been made that gas discharge studies could provide important clues to the Balmer line profiles in tokamaks [28,51,61], certain strong similarities in the two cases are immediately apparent from [75-79]. These include strong and often asymmetrical line wing production, and up to three or more radiator 'velocity classes' [28,51,75-79]. Apart from electron impact dissociative excitation and ionization of molecules, higher atomic kinetic energies are obtained from molecular collisions with protons and, especially with the molecular ion H_3^+ (see e.g. [75,80] for production mechanisms of this ion) accelerated in the cathode fall region through asymmetrical charge-exchange. Other molecular ions, such as X_5^+, could also play a role besides H_3^+, whose spectrum in absorption is well known [80,81]. The corresponding neutral molecule (H_3 and all its isotopomers) has been studied in detail in discharge tube spectra [82-88]. However, whether such molecules can survive in the relatively harsh environment of the tokamak edge plasma is not yet known.

It remains to show the relevance of discharge tube physics [28,61,75-79] to the tokamak, i.e. to point out the analogous role to the cathode fall region. The plasma sheath ('scrape-off layer') near limiter surfaces provides a potential drop [89]:

$$eU_s = -\frac{1}{2}kT_e \ln\left[\left(\frac{2\pi m_e}{M_i}\right)\left(1+\frac{T_i}{T_e}\right)\right]. \qquad (19)$$

(This formula allows for inequality of ion and electron temperatures near the plasma edge, as observed in plasmas of lower density [25,27].) The kinetic energy of an ion would therefore be increased by some $3kT_e$ on its way to the limiter, which would be adequate to provide the population of energetic H_2^+, and hence H_3^+, ions to facilitate the production of 'excessive Balmer line broadening' of the type discussed here.

In conclusion, we mention some fascinating spectroscopic observations on tubes of the Plücker (Geissler) type, in the presence of magnetic fields of order 1 T: partial depolarization and rotation of Zeeman components, and the production of huge, asymmetrical wings to the Balmer lines, including evanescent asymmetries which are apparently dependent upon the polarity of the electromagnet providing the field [28,61,90]. These observations are practically only qualitative in nature, and lack quantitative theoretical verification, apart from a clearly shown connection to the process of atomic excitation transfer with the noble gases or mercury vapour.

6. CHEMICAL AND PHYSICAL SPUTTERING

Chemical sputtering is associated with dissociation of molecules such as D_2O, CO, C_2, CH_4 at limiter surfaces. The corresponding impurity atoms therefore appear with their characteristic Franck-Condon energies (rapidly equilibrated to $kT_a \approx E_0$), as well as a directed velocity provided by the release mechanism. In dense, cold (detached) plasmas ($\bar{n}_e > 3.5 \times 10^{19}$ m^{-3}), chemical sputtering is the predominant release mechanism for both C and O, and appears to be the main mechanism for O over a wide range of plasma conditions [2,23]. For example, since physical sputtering in the case of C has a threshold energy of about 30 eV, equation (19) predicts chemical sputtering of C alone for $kT_e \leq 10$ eV, as found in [22,23].

The formula of Thompson [23,91] for physical sputtering contains as parameter the surface energy $E_s = \frac{1}{2} M v_s^2$, of the order of several eV, but often not well known, because of the complexity of the ion bombardment process. This quantity may be derived from fits to the corresponding flux distribution formula ($\Gamma =$ total flux)

$$\Gamma(v)\,dv = 2\Gamma \left\{ \left(\frac{v}{v_s} \right) \left[1 + \left(\frac{v}{v_s} \right)^2 \right]^{-2} \right\} \frac{dv}{v_s}, \tag{20}$$

yielding experimental values of $E_s = 9.3$ eV for C I and $E_s = 5.5$ eV for O I from measurements on TEXTOR of line profiles (π components only) corresponding to the transitions listed in Table 1 [22,23]. There is enormous scope for further work here.

7. COLLISIONAL HEATING OF ATOMS AND IONS

Once released into the plasma, atoms are subjected to sequential ionization processes (mainly by electron collisions), and simultaneously to heating by ion collisions, the two processes acting 'in competition' during the lifetime of a particular ionization stage [14,24]. Thus, the observed temperatures for the lower charge states may lie well below that of the background deuterons (protons). If the initial atomic velocity distribution is assumed to be quasi-Maxwellian, we may use the expression for the temporal evolution of the 'test' atomic temperature through elastic binary collisions with field ions (subscript f), as derived in [67,92,93]:

$$\frac{dT}{dt} = \frac{16}{3\sqrt{\pi}} n_f \frac{M_r}{\Sigma M} (T_f - T) \beta^{5/2} \int_0^\infty u^5 q_s(u) \exp\left(-\beta u^2\right) du, \tag{21}$$

where

$$\beta = \frac{M M_f}{2k(M T_f + M_f T)}, \qquad (22)$$

with M_r denoting the reduced mass of the colliding partners, and $\Sigma M = M + M_f$. The momentum transfer (diffusion) cross-section $q_s(u)$ is a function of the relative speed $u = |\vec{v} - \vec{v}_f|$ (a conserved quantity in the collision [94]), and the interaction potential, which we express for the dominant ion-induced dipole type as

$$V(r) = -\frac{Q_f^2}{2} \hbar c \, \alpha \frac{\alpha_{pol}}{r^4}, \qquad (23)$$

in terms of the (scalar) atomic polarisability α_{pol} [28,29,66]. The calculation of $q_s(u)$ proceeds by either classical or quantum mechanical (partial wave) scattering methods, the two results being sufficiently close for present purposes [30]. The classical method, developed by Maxwell [95] for the repulsive and by Langevin (see [96]) for the attractive r^{-4} potential, reveals a variety of interesting trajectories in the centre of mass frame [65,96]. However, since the atomic state is easily disrupted by short-range collisions, the heating process for neutrals is less efficient [30] than originally estimated [28], despite the rapid increase in atomic polarisability with principal quantum number. After the atom has lost its first electron through electron impact ionisation, the long-range Coulomb interaction comes into play, with a significant increase in heating rate [14,62]. The reduction of the original heating rate [62] by a factor of 2 in [14] is in effect consistent with a later treatment of Coulomb collisions, showing that the Coulomb logarithm in [62] is too large by just this factor [97]. It is often important to use an effective collisional ionization rate coefficient ξ_Q^{eff} including collisional excitation to and ionization from higher levels [15,17,20].

8. CONCLUSIONS

For successful functioning of a fusion reactor based upon the magnetically confined fusion plasma, control of the plasma boundary region is essential. This in turn requires a detailed understanding of the important atomic and molecular processes. While much remains to be learned, optical spectroscopic methods have played a major rôle in improving our present knowledge of microscopic interactions at the plasma edge [98].

9. ACKNOWLEDGMENTS

This text was written within the framework of a German (WTZ)-South African (NRF) co-operation programme 'Radiative processes for fusion edge plasmas.' Financial support from the University of KwaZulu-Natal is gratefully acknowledged. One of us (JDH) wishes to thank Dr. C. C. Chu for important contributions to this research project, and to acknowledge informative discussions with the late Dr. P. Bogen, Prof. E. Hintz, and the late Prof. A. D. Thackeray.

REFERENCES

1. P. Bogen and E. Hintz, "Plasma Edge Diagnostics using Optical Methods," in *Physics of Plasma-Wall Interactions in Controlled Fusion*, edited by D. E. Post and R. Behrisch, New York: Plenum 1986, pp. 211-280.
2. P. Bogen, *Physica Scripta* **T47**, 102-109 (1993).
3. A. Pospieszczyk, "Diagnostics of Edge Plasmas by Optical Methods," in *Atomic and Plasma-Material Interaction Processes in Controlled Thermonuclear Fusion*, edited by R. K. Janev and H. W. Drawin, Amsterdam: Elsevier 1993, pp. 213-242.
4. U. Samm, P. Bogen, H. Hartwig et al., *J. Nucl. Mater.* **162-164**, 24-37 (1989).
5. U. Samm, P. Bogen, H. A. Claassen et al., *J. Nucl. Mater.* **176-177**, 273-277 (1990).
6. U. Samm, P. Bogen, G. Esser et al., *J. Nucl. Mater.* **220-222**, 25-35 (1995).
7. B. Unterberg, S. Brezinsek, G. Sergienko et al., *J. Nucl. Mater.* **337-339**, 515-519 (2005).
8. S. Brezinsek, G. Sergienko, A. Pospieszczyk et al., *Plasma Phys. Control. Fusion* **47**, 615-634 (2005).
9. Ph. Mertens, S. Brezinsek, P. T. Greenland et al., *Plasma Phys. Control. Fusion* **43**, A349-A373 (2001).
10. A. Pospieszczyk, Ph. Mertens, G. Sergienko et al., *J. Nucl. Mater.* **266-269**, 138-145 (1999).
11. S. Brezinsek, Ph. Mertens, A. Pospieszczyk, G. Sergienko et al., *Contrib. Plasma Phys.* **42**, 668-674 (2002).
12. J. D. Hey, C. C. Chu and Ph. Mertens, *Contrib. Plasma Phys.* **42**, 635-644 (2002).
13. G. M. McCracken, U. Samm, S. J. Fielding at al., *J. Nucl. Mater.* **176-177**, 191-196 (1990).
14. J. D. Hey, Y. T. Lie, D. Rusbüldt and E. Hintz, *Contrib. Plasma Phys.* **34**, 725-747 (1994).
15. J. D. Hey, C. C. Chu, S. Brezinsek et al., *J. Phys. B: At. Mol. Opt. Phys* **35**, 1525-1553 (2002).
16. R. C. Isler, R. W. Wood, C. C. Klepper et al., *Phys. Plasmas* **4**, 355-368 (1997).
17. J. D. Hey, *Trans. Fusion Technol.* **25** (2T, part 2), 315-325 (1994).
18. H. R. Griem, *Plasma Spectroscopy*, New York: McGraw-Hill 1964.
19. H. R. Griem, *Spectral Line Broadening by Plasmas*, New York: Academic 1974.
20. H. R. Griem, *Principles of Plasma Spectroscopy*, Cambridge 1997.
21. B. L. Welch, H. R. Griem, J. Terry at al., *Phys. Plasmas* **2**, 4246-4251 (1995).
22. P. Bogen and D. Rusbüldt, *Nucl. Fusion* **32**, 1057-1061 (1992).
23. P. Bogen and D. Rusbüldt, J. Nucl. Mater. **196-198**, 179-183 (1992).
24. G M McCracken and U. Samm, *Proc. 8th Topical Conf. on Atomic Processes in Plasmas* (Portland, Maine), edited by E. S. Marmar and J. L. Terry, New York: AIP 1992, pp. 144-153.
25. J. D. Hey, Y. T. Lie, D. Rusbüldt and E. Hintz, 20th *Eur. Phys. Soc. Conf. on Controlled Fusion and Plasma Physics*, edited by J.A.C. Cabral, M.E. Manso, F. M. Serra & F. C. Schüller, **17C** (part III), 1111-1114 (1993).
26. J. D. Hey, M. Korten, Y. T. Lie, A. Pospieszczyk et al., *Contrib. Plasma Physics* **36**, 583-604 (1996).
27. A. Huber, A. Pospieszczyk, B. Unterberg et al., *Plasma Phys. Control. Fusion* **42**, 569-578 (2000).
28. J. D. Hey, C. C. Chu and E. Hintz, *J. Phys. B: At. Mol. Opt. Phys.* **32**, 3555-3573 (1999).
29. J. D. Hey, C. C. Chu and E. Hintz, *Contrib. Plasma Phys.* **40**, 9-22 (2000).
30. J. D. Hey, C. C. Chu, Ph. Mertens et al., *J. Phys. B: At. Mol. Opt. Phys.* **37**, 2543-2567 (2004).
31. R. S. Freund, J. A. Schiavone and D. F. Brader, *J. Chem. Phys.* **64**, 1122-1127 (1976).
32. K. Ito, N. Oda, Y. Hatano and T. Tsuboi, *Chem. Phys.* **17**, 35-43 (1976).
33. K. Ito, N. Oda, Y. Hatano and T. Tsuboi, *Chem. Phys.* **21**, 203-210 (1977).
34. H. Tawara, Y. Itikawa, H. Nishimura and M. Yoshino, *J. Phys. Chem. Ref. Data* **19**, 617 - 636 (1990).
35. G. H. Dunn and L. J. Kieffer, *Phys. Rev.* **132**, 2109-2117 (1963).
36. A. Crowe and J. W. McConkey, *J. Phys. B: At. Mol. Phys.* **6** 2088-2107 (1973).
37. D. Reiter, P. Bogen and U. Samm, *J. Nucl. Mater.* **196-198**, 1059-1064 (1992).
38. H. Kubo, H. Takenaga, T. Sugie et al., *Plasma Phys. Control. Fusion* **40**, 1115-1126 (1998).
39. W. Engelhardt, *Phys. Fluids* **15**, 2074-2075 (1972).
40. Ph. Mertens and P. Bogen, *Appl. Phys. A* **43**, 197-204 (1987).
41. Ph. Mertens and P. Bogen, *16th Eur. Conf. on Controlled Fusion and Plasma Physics* (Venice), edited by S. Segre, H. Knoepfel and E. Sindoni,. **13B** (part III), 983-986 (1989).
42. Ph. Mertens and M. Silz, *J. Nucl. Mater.* **241-243**, 842-847 (1997).
43. Ph. Mertens and A. Pospieszczyk, *J. Nucl. Mater.* **266-269**, 884-889 (1999).
44. A. Pospieszczyk, G. Sergienko and D. Rusbüldt, *Contrib. Plasma Phys.* **40**, 162-166 (2000).
45. S. Brezinsek, P. T. Greenland, Ph. Mertens at al., *J. Nucl. Mater.* **313-316**, 967-971 (2003).
46. K. Höthker, *Nucl. Fus.* **16**, 253-261 (1976).
47. P. Bogen, K. J. Dietz and A. Pospieszczyk, in *High Beta Plasmas*, edited by D. E. Evans, Proceedings of the Third Topical Conference, Culham, Oxford: Pergamon, 1976, pp. 197-201.

48. R. D. Bengtson, J. Boedo and W. L. Rowan, *Rev. Sci. Instr.* **57**, 2026-2028 (1986).
49. J. A. Boedo, R. D. Bengtson, A. Ouroua and P. M. Valanju, *Rev. Sci. Instr.* **59**, 1494-1496 (1988).
50. H. A. Bethe and E. E. Salpeter, *Quantum Mechanics of One- and Two-Electron Atoms*, Berlin: Springer 1957.
51. J. D. Hey, C. C. Chu and Ph.. Mertens, *Spectral Line Shapes: Proc. 16th International Conf. on Spectral Line Shapes* (Berkeley, California), edited by C. A. Back, New York: AIP Conf. Proc. 645, 2002, pp. 26-39.
52. E. U Condon and G. H. Shortley, *The Theory of Atomic Spectra*, Cambridge University Press 1970.
53. R. H. Garstang, *Rep. Prog. Phys.* **40**, 105-154 (1977).
54. W. A. Cilliers, J. D. Hey and J. P. S. Rash, *J. Quant. Spectrosc. Rad. Transfer* **15**, 963-978 (1975).
55. F. Paschen and E. Back, *Physica* **1**, 261-273 (1921).
56. C. C. Kiess and G. Shortley, *J. Res. Nat. Bur. Standards (USA)* **42**, 183-207 (1949).
57. R. I. Semenov and V. I. Tuchkin, *Opt. Spectrosc.* **88**, 147-150 (2000); **89**, 493-497 (2000).
58. W. E. Lamb, Jr., *Phys. Rev.* **85**, 259-276 (1952).
59. A. Iwamae, M. Hayakawa, M. Atake et al., *Phys. Plasmas* **12**, 042501 (6 pp) (2005).
60. J. von Neumann and E. Wigner, *Phys. Zeit.* **30**, 467-470 (1929).
61. C. C. Chu and J. D. Hey, *Contrib. Plasma Phys.* **40**, 597-606 (2000).
62. L. Spitzer, Jr., *Physics of Fully Ionized Gases* 2nd edn, New York: Interscience 1962.
63. G. Mathys, *Astron. Astrophys.* **139**, 196-210 (1984).
64. P. M. S. Blackett and J. Franck, *Z. Physik* **34**, 389-401 (1925).
65. J. D. Hey, C. C. Chu and Ph. Mertens, *J. Phys. B: At. Mol. Opt. Phys.* **38**, 3517-3534 (2005).
66. E. A. Mason and E. W. McDaniel, *Transport Properties of Ions in Gases*, New York: Wiley 1988.
67. E. W. McDaniel, J. B. A. Mitchell and M. E. Rudd, *Atomic Collisions: Heavy Particle Projectiles*, New York: Wiley 1993.
68. B. Unterberg, S. Brezinsek, G. Sergienko et al., 30th *Eur. Phys. Soc. Conf. on Controlled Fusion and Plasma Physics* (St. Petersburg), **27A**, P-3.166 (2003) [CD-ROM version only].
69. R. S. Freund, *J. Chem. Phys* **54**, 3125-3141 (1971).
70. M. Misakian and J. C. Zorn, *Phys. Rev. A* **6**, 2180-2196 (1972).
71. K. Wurm, *Z. Astrophysik* **10**, 133-153 (1935).
72. A. D. Thackeray, *The Observatory* No. 763, 327-329 (1937).
73. T. Ogawa and M. Higo, *Chem. Phys.* **52**, 55-64 (1980).
74. N. Kouchi, M. Ukai and Y. Hatano, *J. Phys. B: At. Mol. Opt. Phys.* **30**, 2319-2344 (1997).
75. M. Kuraica and N. Konjević, *Physica Scripta* **50**, 487-492 (1994).
76. M. R. Gemišić Adamov, B. M. Obradović et al., *IEEE Trans. Plasma Sci.* **31**, 444-454 (2003).
77. N. Konjević and M. M. Kuraica, *The Physics of Ionized Gases: Proc. 22nd Summer School and International Symposium* edited by L. Hadžievski, T. Grozdanov and N. Bibić, New York: AIP, pp. 268-281 (2004).
78. N. Cvetanović, M. M. Kuraica and N. Konjević, *J. Appl. Phys.* **97**, 033302 (8 pp) (2005).
79. N. M. Šišović, G. Lj. Majstorović and N. Konjević, Eur. Phys. J. D. **32**, 347-354 (2005).
80. T. Oka, *Phys. Rev. Lett.* **45**, 531-534 (1980).
81. A. R. W. McKellar and J. K. G. Watson, *J. Mol. Spectrosc.* **191**, 215-217 (1998).
82. G. Herzberg, *J. Chem. Phys.* **70**, 4806-4807 (1979).
83. I. Dabrowski and G. Herzberg, *Can. J. Phys.* **58**, 1238-1249 (1980).
84. G. Herzberg and J. K. G. Watson, *Can. J. Phys.* **58**, 1250-1258 (1980).
85. G. Herzberg, H. Lew, J. J. Sloan and J. K. G. Watson, *Can. J. Phys.* **59**, 428-440 (1981).
86. G. Herzberg, J. T. Hougen and J. K. G. Watson, *Can. J. Phys.* **60**, 1261-1284 (1982).
87. I. Dabrowski and G. Herzberg, "The Electronic Emission Spectra of Triatomic Hydrogen: The 6025 Å Bands of H_2D and HD_2," in *Amazing Light: A Volume Dedicated to Charles Hard Townes on his 80th Birthday*, edited by R. Y. Chiao, New York: Springer 1996, Ch. 16, pp. 173-190.
88. M. Vervloet and J. K. G. Watson, *J. Molec. Spectrosc.* **217**, 255-277 (2003).
89. P. C. Stangeby, *The Plasma Boundary of Magnetic Fusion Devices*, Bristol: IOP 2000.
90. C. C. Chu, J. D. Hey and Ph. Mertens, *Spectral Line Shapes: Proc. 17th Int. Conf. on Spectral Line Shapes (Paris)* edited by E. Dalimier, Paris: Editions Frontier Group, pp. 412-414 (2004).
91. M. W. Thompson, *Phil. Mag.* **18**, 377-414 (1968).
92. E. A. Desloge, *Phys.Fluids* **5**, 1223-1225 (1962).
93. A. M. Cravath, *Phys. Rev.* **36**, 248-250 (1930).
94. L. G. H. Huxley and R. W. Crompton, "The Motions of Slow Electrons in Gases," in *Atomic and Molecular Processes* edited by D. R. Bates, New York: Academic 1962, pp. 335-373.
95. J. C. Maxwell, *The Scientific Papers of James Clerk Maxwell* vol II edited by W. D. Niven, Cambridge University Press, pp. 26-78 (1890).
96. P. Langevin, in translation in E. W. McDaniel, *Collision Phenomena in Ionized Gases*, New York: Wiley, appendix II (1964)
97. D. Li, *Nucl. Fusion* **41**, 631-635 (2001).
98. Ph. Mertens and S. Brezinsek, *Fusion Science and Technology* **47**, 161-171 (2005).

Atmospheric Pressure Glow Discharges

W G Graham and G Nersisyan

International Research Centre for Experimental Physics, Queens University Belfast, BT7 1NN, Northern Ireland

Abstract. Their relative engineering simplicity, plasma uniformity and chemistry make Atmospheric Pressure Glow Discharges (APGD) very attractive for plasma processing applications. Here some of the basic characteristics of glow discharges are introduced. The basic dielectric barrier discharge and how it can be operated in a uniform glow rather filamentary mode is described. Electrical and laser-based measurements that throw light on the underlying physics of APGDs are presented, along with a model which seeks to explore the plasma chemistry of these discharges.

Keywords: Atmospheric pressure glow discharge, sheath, dielectric barrier discharge, laser induced fluorescence, plasma chemistry, helium/air.
PACS: 52.80.s, 52.80. Hc, 52.70.-m

INTRODUCTION

Atmospheric pressure glow discharges (APGD) are pulsed, radially uniform plasmas that can be created in gas at around atmospheric pressure. They can be created under certain operating conditions in dielectric barrier discharge (DBD) systems which generally produce non-uniform filamentary discharges. APGDs can produce reactive radials from relatively inert feed gases and can be operated at near-room temperature and so coupled with their uniformity they are finding increasing application in surface treatment technologies particularly for textiles, polymers and biological material. Here a brief overview of the APGD is given, describing in general terms how they are produced and describing some recent research activity in the area by our group. While there are reports of APGD and APGD-like behaviour in a number of gases and gas mixtures, here the focus will be on helium with some air-related impurities present. It is worth emphasizing that there is a plethora of different high pressure plasma systems and many of the generalizations that are made here may not apply in specific plasma devices.

CP876, *The Physics of Ionized Gases: 23rd Summer School and International Symposium,*
edited by L. Hadžievski, B. P. Marinković, and N. S. Simonović

PLASMAS: AN OVERVIEW

Processes and Constituents

The main collisions processes of applications interest in plasmas are: Ionization e.g. $X + e \rightarrow X^+ + 2e$, leading to plasma creation, excitation e.g. , $X + e \rightarrow X^* + e$, leading to light emission and hence the use of the term glow discharge and dissociation e.g. $XY + e \rightarrow X + Y + e$, leading to plasma chemistry. The interactions of the electrically charged particles with each other, the neutral gas and contact surfaces produces the unique physical and chemical environment of the plasma.

The plasma environment used in processing is complex and in most circumstances, unless special precautions are taken it must be anticipated that the following will be present in the plasma.

- Electrons
- Positive ions (atomic and molecular)
- Negative ions (atomic and molecular)
- Excited atoms, molecules and ions.
- Dissociation products (atomic, molecular, neutral, positive, negative)
- Process products (atomic, molecular, neutral, positive, negative)
- Em radiation
- Clusters
- Dust
- Contact surfaces

While this environment creates the huge potential for plasma applications it also makes getting an understanding of how they are ignited, sustained and extinguished very challenging. There is a huge amount of literature on the topic. Recent textbooks on low pressure [1] and high pressure discharges [2] provide the best route into the topic.

Discharge Classification

There are many ways of classifying plasmas. This review focuses on electrically produced relatively low plasma density ($< 10^{18}$ m^{-3}) and electron temperature (< 10 eV) plasmas, often called low temperature or technological plasmas. So this precludes discussion of laser-produced, nuclear fusion-related and some astrophysical plasmas. The low temperature plasmas can be further categorised, in terms of the temperature of the constituents as, either thermal or non-thermal plasmas.

In thermal plasmas e.g. arcs, jets and torches, the plasma constituents are in thermal equilibrium with one another i.e. $kT_e \cong kT_i \cong kT_{gas} \cong kT_{surf}$. The currents driving the plasma range from about 10 to 1000 A and the gas temperatures from about 5,000 to 50,000K. The plasmas can be from 1 to 100% ionised.

Non-thermal plasmas were, until recently, predominantly identified with low pressure dc, rf or microwave driven plasmas but now they can also be found in higher pressure   dielectric barrier discharges (DBD) such as   atmospheric pressure glow discharges (APGD) also in some plasma jets. In these plasmas $kTe \gg kTi \cong kTgas = kTsurf$. The currents driving the plasma range from about 1 mA to 10's A and the gas temperatures from about 300 to 2,000K.   The plasmas can be from 10^{-7} to 100% ionised and over 50% dissociated. These non-thermal plasmas are of current interest since the combination of hot ionising and dissociating electrons with cold ions and gas allows the delivery of hot chemistry to cool surfaces.

The Sheath

Low pressures discharges often show clear evidence of the formation of a sheath at contact surfaces, particularly at electrodes. Sheath creation is associated with the collective behaviour of plasmas due to the coulomb interaction between the charged particles. Initially both electrons and ions move to the contact surfaces. In the steady state the plasma acts to maintain quasi-neutrality ($n_i = n_e$) so the ion and electron fluxes to the surfaces must be equal but since the electron velocities are greater than the ion velocities and since their densities are equal, the electron flux at the plasma edge is greater than that of the ions. The plasma therefore self adjusts to balance the loss rate of the ions and electrons. The positive charge left behind after the initial loss of electrons concentrates near the solid in a thin layer called the plasma sheath. This region is usually quite thin, is not quasi-neutral and so has potential gradients. These act to reflect electrons back into the plasma and accelerate the ions to the wall, creating a flux balance so that in the plasma $n_i = n_e$ and only very weak electric fields exist but in the sheath $n_i > n_e$ and here electric fields do exist.

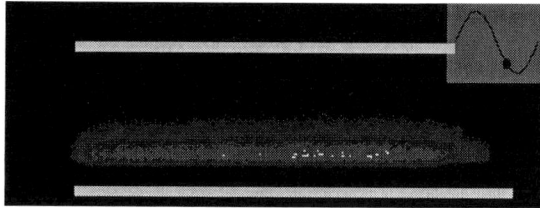

FIGURE 1. The light emission from a low pressure, parallel plate, capacitively coupled, rf driven argon discharge captured at the phase of the applied voltage cycle indicated in the inset but representative of the time averaged emission.

In low pressures glow discharges the sheath can often be seen as a thin dark layer between a luminous region in the plasma and the contact surface. This is illustrated in fig. 1 where the light emission from a low pressure parallel plate, capacitively coupled discharge is shown. Details of the plasma device and this measurement are given in reference 3. The bright emission occurs at the sheath edge since electrons gain sufficient energy, through acceleration in the sheath fields, to excite (and ionize) the gas. Only discharges that replicate this electric field structure, which can be observed as radially uniform and axially non-uniform emission patterns, can be described as glow discharges.

DIELECTRIC BARRIER DISCHARGES

Generally in plasma production the voltage required to breakdown a gas gap and form a plasma depends with the product p x d, where p is the gas pressure and d is the gap distance. The dielectric barrier discharge, shown schematically in fig. 2, resembles a low pressure glow discharge assembly. It consists of two planar metal electrodes but these are coated on the plasma-facing side with a dielectric material to prevent arcing at the high voltages used. The electrode gap is about 1 to 10 mm to minimize the voltages required. The electrodes are usually powered by a sinusoidal or pulsed high voltage with an amplitudes ranging from about of 1 to 20 kV and frequencies from 3 to 50 kHz.

The discharge in a DBD has generally been reported as consisting of many filaments randomly distributed across the electrode gap. They persist for 1 to 10 ns and have a diameter of ~100μm. The mechanisms by which they are initiated and sustained are thought to be quite well understood [4]. In fig. 3 a fast (few ns) ICCD image of a DBD in an air-helium mixture captures the creation of two filaments. However in the same device, but under specific conditions, instead of the non-homogenous filamentary DBD, a radially uniform, pulsed discharge can be generated as also shown in fig. 3. This mode of the DBD is known as an atmospheric pressure glow discharge (APGD).

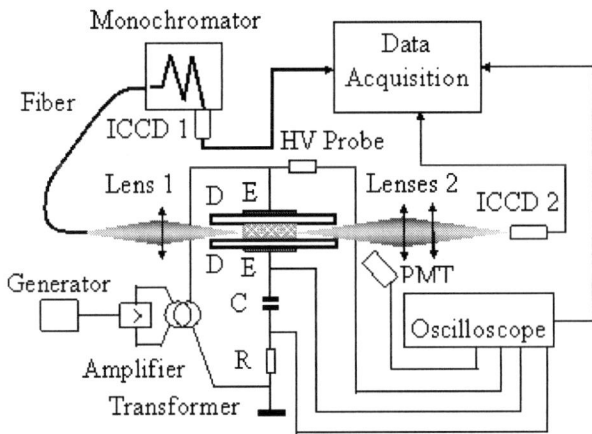

FIGURE 2. A schematic of a typical DBD discharge along with the electrical and optical diagnostics used in the studies reported here.

It is now understood that the establishment of the uniform APGD is a complex process, which strongly depends on the elementary processes in the inter-electrode gap and interactions with the charged surfaces of the dielectric barriers. The creation of the uniform APGD depends on the gas in the gap, the gap size and the dielectric material. The driving voltage and frequency required to sustain an APGD depends on the particular combination of these three. Within this broad outline there are a number of important issues. The gas which appears to give the widest operating window is helium but it requires some concentration of impurity to be present [5-8] and similarly with the other noble gases although the operating windows are much more constrained. While a uniform discharge can be created in nitrogen gas too this appears to be a Townsend rather than glow discharge [9]. The weaker Townsend glow has insufficient plasma density to create a sheath and so the potential difference is maintained across the electrode gap rather than localised at the sheaths.

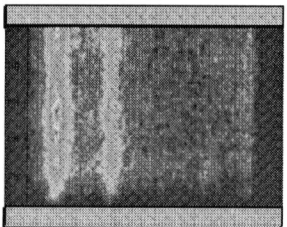 

FIGURE 3. Fast gated (~ 2 ns) images of a DBD system at the discharge current maximum operating under, on the left, conditions which produce a filamentary discharge and, on the right, conditions that produce a radially uniform discharge.

There are models for He-based APGDs. These use a number of one and two-dimensional simulations [5, 8, 10-14]. The models suppose that the filament-like breakdown of the gap is avoided when there are "seed" electrons in the gap allowing discharge creation before an increasing gap voltage reaches the non-seeded breakdown value. It is estimated that a density of about 10^6 cm^{-3} such "seed" electrons would be sufficient [5]. The models usually determine that there are insufficient electrons trapped from the previous discharge and the density must be supplemented in the late afterglow by Penning ionization of impurities (usually assumed to be nitrogen) in the gap by He metastables. They also need to assume some contribution from He metastable or ion bombardment of the dielectric materials. Some models assume trapping of ions in the gap [15].

CHARACTERISATION OF AN APGD

There have been many reports of experimental measurements in APGDs, they have been mainly of electrical waveforms and fast imaging of the gas gap to try to distinguish between filamentary and glow modes. However fully distinguishing between an APGD and a many filamented DBD and what appear to be a number of micro-APGDs in the gap is not straightforward. Time-resolved emission spectroscopy is used to provide insight into the plasma kinetics.

Here the focus will be on measurements from our group where interest is in providing experimental validation, benchmarking of the physics models and providing input to inform the development of a model of plasma chemistry in He-based APGDs.

Two APGD systems feature in the work reported, one (QUB1) [16] was developed for sample processing but with adequate access for diagnostics and sits in air with flowing helium. The system could have an electrode area of up to 200 cm^{-3}.

The second system (QUB 2) [17] was developed specifically for plasma light emission and laser-based diagnostic studies and is contained in a vacuum chamber which could be evacuated to a base pressure of less than 10^{-6} Torr before helium gas was introduced to pressures of up to 1000 Torr. Here the electrode area was ~ 50 mm diameter. Typically the in both glass dielectric material was used but optical-quality quartz was also used. The dielectrics were usually separated by 5 to 10 mm. The measurements reported here made in QUB2 unless otherwise stated.

The power supply was a signal generator, audio power amplifier and step-up transformer combination. The electrical characterisation of each system was based on time resolved measurements of the applied voltage, the discharge current, and the charge, transferred across the electrode gap and the voltage across the dielectric barrier gap. It was then possible to obtain the value for the instantaneous voltage across the dielectric gap which differs from the applied voltage because of charge on the dielectric surface A fast PMT tube was used to monitor the total emission while space and time resolved emission measurements were obtained by imaging the gap or a portion of it onto an ICCD, triggered with a time delayed pulse generated initially by the power supply and with gate durations as low as 20 ns.

A typical electrical signal from an APGD is shown fig. 4. This is characterized by a single discharge current peak each half cycle of the applied voltage and occurring when the gap voltage reaches a value of 1.65 kV, for this particular p x d condition.

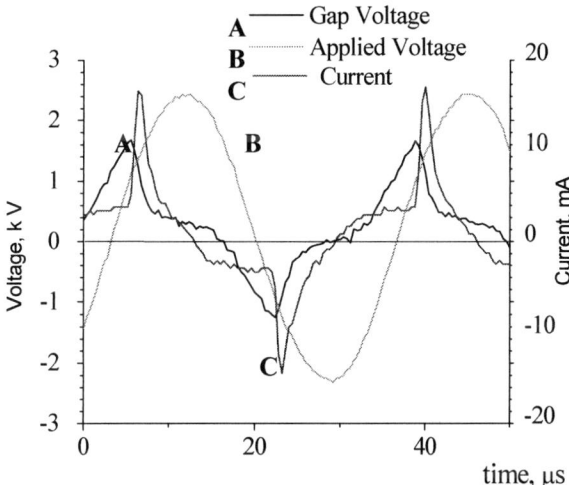

FIGURE 4. Typical applied voltage, discharge current and gap voltages of a He-based APGD.

As mentioned above the use of helium is critical to the production of APGDs. The helium acts to reduce the breakdown voltage. This is because of the electronic structure of helium which allows electrons to be accelerated more effectively in electric fields up to energies required to ionise it. It is also assisted by its long-lived metastable states (He_m). It is also believed that the He contributes to continued ionisation of the gas in the gap which allows the gas to breakdown uniformly at a voltage below that required to produce filaments and so producing an APGD.

This requires the presence of some impurity gas in the gap so that the long lived metastable states of He can create ion-electron pairs through Penning Ionisation. For example the process,

$$He_m + N_2 = N_2^+ (B^2\Sigma_u^+) + He$$

(1)

is particularly effective, since this is an almost resonant. This is also true for the interaction of N_2 with the helium ion He_2^+ which is created in high pressure He discharges. The $N_2^+ (B^2\Sigma_u^+)$ state rapidly decays with the emission of a 391.4 nm photon which acts as an indicator for process 1.

We have set out to measure the time dependence of the densities of the He_m and N_2^+ in the QUB 2 system. A laser induced fluorescence technique (LIF) was used. A laser beam was propagated through the central region of the discharge. For He_m detection a 388.86nm beam excites the $He(2^3S)$ state to the $He(3^3P)$ state, collisions transfer the excitation to the $He(3^3S)$ and $He(3^3D)$ states and emission from the $3^3D \rightarrow 3^3P$ transition at 587.56 nm was observed. For N_2^+ the exciting beam was at 391.4 nm and the detected emission a 427.8 nm. The emitted light was observed at 90° to the laser beam direction by imagining the discharge volume onto the slit of imaging spectrograph/ICCD combination. A 15 ns ICCD gate was trigger from a pulse generated at a specific phase of the applied voltage. The trigger pulse could be delayed so that the time dependence of the species density through the discharge could be measured. At the same time time-resolved emission spectra were measured. Typical results for He_m are shown in fig. 5.

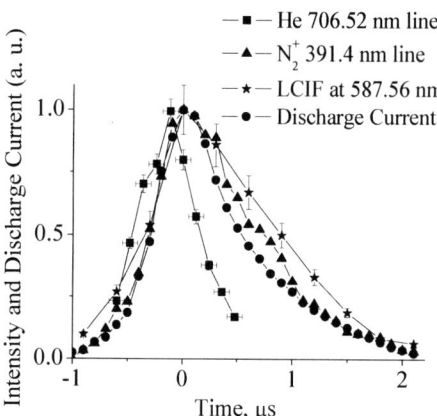

Figure 5. Time dependence of normalized He_m density (LCIF), the line intensities of N_2^+ 391.4 nm and He 706.5 nm emission and the discharge current, at 300 Torr with a 3.8kV (peak to peak) applied voltage at a frequency of 30 kHz and the gas gap of 5 mm.

The relative time dependences of the LCIF signal, the HeI 706.52 nm line and N_2^+ 394.1 nm line intensities and the discharge current are shown in fig. 5. A pressure of 300 Torr was used here since at that pressure the DBD has a uniform radial distribution over a large range of applied voltages and frequency. All the signals exhibit the same rise time to within the accuracy of measurement. The time dependence of the 706.52 nm HeI line intensity indicates the presence of energetic electrons. It is more rapid than the decay of the discharge current, indicating other sources of ionisation at this stage of the discharge. The LCIF signal indicates that the He(2^3S) density grows with the discharge current but decays with a longer time constant, ~2.4μs, similar to the decay of N_2^+ emission. This shows that, as expected, Penning ionisation is the major contributor to the excited state N_2^+ production. Using known collisional transfer rates the absolute He metastable state densities can be determined. In our system at atmospheric pressure and a power density of 150 mWcm^{-3} this is found to be 1.5 x 10^{10} cm^{-3}. By extrapolating the time dependence the metastable density at the beginning of the next pulse is estimated to be ~ 10^4 cm^{-3}.

The N_2^+ 394.1 nm line behaviour is indicative of the production of the excited B$^2\Sigma_u^+$ state of the ion. LIF measurements allowed us to study the time dependence of the N_2^+ ground state. The measurements were made in the QUB2 system but with optically flat quartz dielectrics rather than glass. An APGD plasma similar to that produced with glass dielectric was created with applied voltages around 30 kHz but at lower gap voltages and with higher light intensity. As can be seen in Figure 6 the N_2^+ ground state LIF signal peaks about 5 μs after the maximum in the discharge current. The reason for this behaviour has not yet been fully understood but could be associated the decay of the excited B$^2\Sigma_u^+$ state of the ion to the ground state and the production of N$_4^+$ during the discharge current pulse,

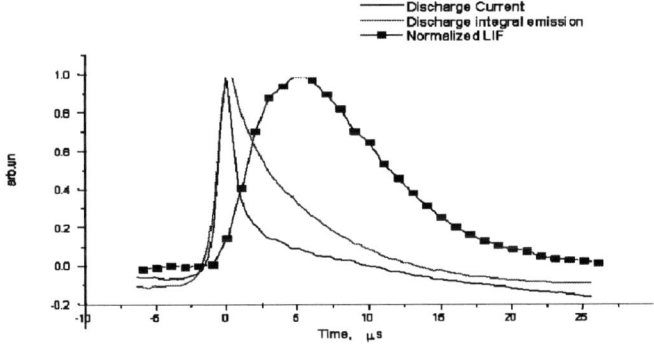

FIGURE 6. Time dependence of the relative discharge current, total light integrated light emission and N_2^+ LIF signal.

SIMULATION CODE

In applications a critical feature is the chemical composition of the gas in the discharge. In the case of He-air mixtures this is very complex. A computational model to study the evolution of species concentrations under prescribed excitation conditions and with prescribed gas mixtures specifically to enable a variety of different scenarios to be explored rapidly is being developed [18]. An important element is close interplay with experimental results.

The model is based on the electron-beam generated plasmas code developed by Vidmar [19] which has been refined and augmented for use in helium-air mixtures [20]. It is zero dimensional and therefore assumes an isotropic gas mixture and does not consider diffusion and convective transport. The inputs include the pressure, the air impurity levels, the ionization rate, as well as the pulse on- and pulse off-times, and the value of the reduced electric field imposed on the system. The code employs a lookup table to determine T_e from the specified value of the reduced electric field. The code currently includes 461 reactions and tracks 58 species including neutral atoms and molecules, metastable species, vibrationally-excited N_2 and O_2, electrons, positive and negative ions and various water cluster ions.

The ionization rate, pulse duration and repetition rates were chosen to be close to those used in the experiment. However the impurity species and their concentration are not known. Therefore the air fraction in the simulations was adjusted to find agreement between the simulation and experimental time dependence and magnitude of the metastable helium atoms. We believe the level of impurities required is entirely reasonable, given the experimental configuration and mode in which the experiments were conducted.

As shown in fig. 7 the model indicates that the dominant radical species are N, O, H and OH. It also indicates that after the first 5 or 6 cycles there is little or no modulation of the radical species with the ionisation pulse. The model also indicates that the positive ion species are dominated by NO^+, which shows only slight modulation with the ionisation pulse, and O_2^+, $O_2^+.H_2O$ and N_4^+, which are more heavily modulated.

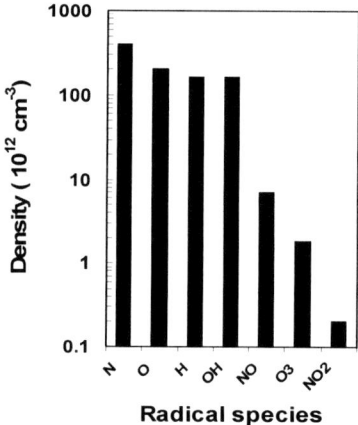

FIGURE 7. The predicted density of radical species after 50 ionisation pulses in a helium-air glow discharge plasma at 300 Torr helium with 850 ppm air, pulse duration = 800 ns, pulse repetition rate = 60 kHz.

CONCLUSION

It has been demonstrated that discharges that display many of the characteristics of low pressure glow discharges can be produced at atmospheric pressure. They are intrinsically pulsed plasmas. The transition from the normal atmospheric pressure filamentary mode to glow mode appears to need some residual ionisation in the gas allowing uniform breakdown before the otherwise anticipated gap breakdown voltage, in a filamentary form, is reached. This probably explains the need for inert gases to achieve the ASPGD mode. There is a basic understanding of how they work but still much of intriguing science to be done. Their relative engineering simplicity, plasma uniformity and chemistry make APGDs very attractive for plasma processing applications, not only for large dimension and high throughput applications, but also in applications where vacuum cannot be tolerated e.g. biomedical technology.

ACKNOWLEDGMENTS

The authors wish to acknowledge the ongoing contributions of T Morrow (QUB) and .K R Stalder (Stalder Technologies and Research) to the work reported here.

REFERENCES

1. M.B.Lieberman and A.Lichtenberg, *Principles of Plasma Discharges and Materials Processing second edition*,John Wiley 2005
2. *Non-Equilibrium Air Plasmas at Atmospheric Pressure,* Edited by K. Becker *et al.* IOP Publishing, 2005
3. C. M.O. Mahony, R. Al Wazzan and W. G.Graham, *Appl. Phys. Lett.* **71**, 608-610 (1997)
4. B. Eliasson and U.Kogelschatz, *IEEE Trans. on Plasma Science* **19**, 309 (1991)
5. F. Massines and G. Gouda, *J. Phys. D: Appl. Phys* **31**, 3411-20 (1998)
6. Yu. Golubovskii, V. Maiorov, J. Behnke and J. F. Behnke *J. Phys. D: Appl. Phys.* **36**, 39-49 (2003)
7. M.G .Kong and X.T. Deng, *IEEE Trans. on Plasma Science* **31**, 7-18 (2003)
8. L..Mangolini, C.Anderson, J.Heberlein and U.Kortshagen, *J. Phys. D: Appl. Phys.* **37**, 1021-1030 (2004)
9. F. Massines, P.Sègur, N.Gherardi, C.Khamphan and A.Ricard, *Surf. and Coat. Tech.* **174-175**, 8-14 (2003)
10. X.T. Deng and M.G.Kong, *IEEE Trans. on Plasma Science* **32**, 1709-1715 (2003)
11. F.Tochikubo, T. Chiba and T. Watanabe, *Jpn J. Appl. Phys.***38**, 5244-5250 (1999)
12. A. Rabehi, P. Ségur, F.Massines, R. Ben Gadri and M.C.Bordage, *XXIII ICPIG*, Toulouse, 1997, pp 44-45
13. H-E. Wagner, R. Brandenburg, K.V. Kozlov and A.M. Morozov *Contributed papers, XV Symposium on Applications of Plasma Processes* Podbanské, Slovakia, 2005, pp 67-70
14. P. Zhang , C.Anderson, J. Heberlein and U. Kortshagen *Abstracts and Papers, IX Int. Symposium on High Pressure, Low Temperature Plasma Chemistry* Padova, Italy 2004, paper 2O-03.
15. J.R. Roth, *Industrial Plasma Engineering, Applications to Nonthermal Plasma Processing*, IOP Publishing, 2001
16. G. Nersisyan and W.G. Graham, *Plasma Sources Sci. Technol.* **13**, 582-587 (2004)
17. G. Nersisyan, T. Morrow and W.G. Graham, *Appl. Phys. Lett.* **85**, 1487-1489 (2004)
18. K R Stalder, R J Vidmar, G Nersisyan and W G Graham, *J.Appl. Phys.***99**, 093301 (2006).
19. R. J. Vidmar, IEEE Trans. Plasma Sci. **18**, 733 (1990)
20. R. J. Vidmar, K. R. Stalder, "Electron-Beam Generated Plasma in Air: Pulsed and Continuous Generation", *42nd AIAA Aerospace Sciences Meeting and Exhibit*, Reno, NV, January 2004, paper 2004-359 (unpublished).

Diagnostics for the Dynamics of Power Dissipation in Technologically used Plasmas

T. Gans, D. O'Connell, J. Schulze, V.A. Kadetov, U. Czarnetzki

*Institute for Plasma and Atomic Physics, Center for Plasma Science and Technology (CPST),
Ruhr-University Bochum, 44780 Bochum, Germany*

email: timo.gans@web.de

Abstract. Radio frequency (rf) discharges are widely used for technological applications. Despite this, power dissipation mechanisms in these discharges are not yet fully understood. The limited understanding is mainly caused by the complexity of underlying phenomena and very restricted experimental access. Recent advances in phase resolved optical emission spectroscopy (PROES) in combination with adequate modeling of the population dynamics of excited states allow deeper insight into underlying fundamental processes. This paper discusses the application of PROES in a variety of rf-discharges, such as: capacitively coupled plasmas (CCP), dual-frequency CCP (2f-CCP), inductively coupled plasmas (ICP), and magnetic neutral loop discharges (NLD).

Keywords: Low-temperature plasmas, radio-frequency discharges, power dissipation, plasma ionization, electron dynamics, electron heating, plasma diagnostics, phase resolved optical emission spectroscopy (PROES)

PACS: 52.70.-m, 52.70.Gw, 52.80.Pi, 52.50.Qt

INTRODUCTION

Non-equilibrium low temperature plasmas, in particular radio-frequency (rf) discharges, are widely used for technological applications. Increased demands on plasma technology, particularly from the micro-electronics industry, have resulted in the development of various types of discharges based on different power coupling mechanisms. Despite this, the complexity of these mechanisms is not yet fully understood. Insight into power dissipation requires temporal resolution on various time scales; in particular the dynamics within the rf cycle is of importance. Detailed investigations are, therefore, a challenge for diagnostics. Recent advances in phase resolved optical emission spectroscopy (PROES) provide a non-invasive access with excellent spatial and temporal resolution on a nano-second time scale [1-6].

The optical emission from rf discharges exhibits temporal variations within the rf cycle. These variations are particularly strong in capacitively coupled plasmas (CCPs), but also easily observable in inductively coupled plasmas (ICPs). Neglecting these variations in classical time averaged optical emission spectroscopy (OES), based on balance equations, can result in serious misinterpretation [1]. The effect of neglecting temporal changes is not as pronounced in ICPs as in CCPs [5]. However, even these relatively small modulations can be exploited for insight into power dissipation. Using modeling of the dynamics of various excited states compared with phase resolved measurements of the optical emission yields detailed information on the electron dynamics.

ANALYTICAL MODEL FOR THE POPULATION DYNAMICS OF EXCITED STATES

In contrast to the standard corona model commonly used for OES of stationary low density plasmas, PROES requires a time dependent model based on rate equations to take into account the transient character of the exciting electrons.

The temporal modulations of the optical emission are caused by temporal changes in the electron energy distribution function (EEDF). Highly excited states in atoms or molecules are excited by electrons in the tail of the EEDF where the modulation is particularly strong. Temporal variations are, therefore, observable in the optical emission from these high electronically excited atoms or molecules.

Electron impact excitation out of the ground state is described by the excitation function $E(t)$. For an excited state i, not populated through cascade processes or step-wise excitation, the excitation function $E_i(t)$ can be determined directly from the measured number of photons per unit volume and unit time $\dot{n}_{Ph,i}(t)$ [4]:

$$E_i(t) = \frac{1}{n_0 A_{ik}} \left(\frac{d\dot{n}_{Ph,i}(t)}{dt} + A_i \dot{n}_{Ph,i}(t) \right). \tag{1}$$

Here, $\dot{n}_{Ph,i}(t) = A_{ik} n_i(t)$ is given by the transition probability A_{ik} of the observed emission and the population density of the investigated state $n_i(t)$; n_0 is the ground state density. The effective decay rate A_i takes into account spontaneous emission, radiation trapping and quenching [2]:

$$A_i = \sum_k A_{ik} g_{ik} + \sum_q k_q n_q , \tag{2}$$

where g_{ik} is the so-called escape factor [7, 8] and k_q the quenching coefficient with the species q of density n_q.

For quantitative investigations cascade processes can be substantial [9-12]. The population density $n_i(t)$ of the investigated state i can be described by the following rate equation including cascades from state c:

$$\frac{dn_i(t)}{dt} = n_0 E_i(t) - A_i n_i(t) + A_{ci} n_c(t) . \tag{3}$$

261

The population density $n_c(t)$ obeys a rate equation analogous to eq. 3, without cascade processes:

$$\frac{dn_c(t)}{dt} = n_0 E_c(t) - A_c n_c(t).$$ (4)

These coupled differential equations for the investigated state i and the cascade state c can be solved in a general manner for the periodic boundary conditions of the rf discharge $\left(n_{i,c}(t) = n_{i,c}(t + T_{RF})\right)$:

$$
\begin{aligned}
n_i(t) &= n_0 \left(\frac{\widetilde{E}_i(T_{RF}, A_i)e^{-A_i T_{RF}}}{1 - e^{-A_i T_{RF}}} + \widetilde{E}_i(t, A_i) \right) e^{-A_i t} \\
&+ \frac{n_0 A_{ci}}{A_i - A_c} \left[\left(\frac{\widetilde{E}_c(T_{RF}, A_c)e^{-A_c T_{RF}}}{1 - e^{-A_c T_{RF}}} + \widetilde{E}_c(t, A_c) \right) e^{-A_c t} - \left(\frac{\widetilde{E}_c(T_{RF}, A_i)e^{-A_i T_{RF}}}{1 - e^{-A_i T_{RF}}} + \widetilde{E}_c(t, A_i) \right) e^{-A_i t} \right]
\end{aligned}
$$ (5)

Here, the substitution $\widetilde{E}_x(t, A_y) = \int_0^t E_x(t')e^{A_y t'} dt'$ has been used.

The time dependence of the excitation function, reflecting the electron dynamics, is strongly dependent on the power coupling mechanisms. These time dependencies are discussed for various types of rf discharges in the following.

SET-UP FOR PHASE RESOLVED OPTICAL EMISSION SPECTROSCOPY

The set-up for a typical PROES experiment using a fast, gate-able, intensified CCD camera (ICCD) is shown in fig. 1. A spectral filter or spectrometer allows for spectral discrimination of emission lines. The CCD camera is synchronized with a signal from the rf generator powering the discharge. Measurements with a defined gate width can be made, typically 2 ns for the used radio-frequencies. A variable delay between the camera gate and the rf voltage allows for phase resolved measurements within the rf cycle (fig. 1). The intensities measured in the camera gate time are integrated over many rf cycles, for a certain phase setting, to obtain good signal to noise ratio. Fainter emission lines can be integrated for longer. The ICCD camera allows spatially resolved measurements in the various discharge systems.

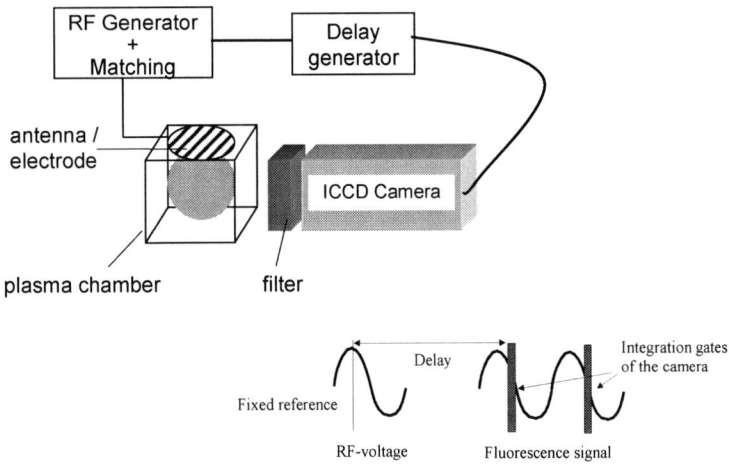

FIGURE 1. Typical experimental set-up for PROES measurements.

CAPACITIVELY COUPLED PLASMAS (CCP)

Single Frequency CCP

Single frequency CCPs are often considered as the simplest configuration for an rf discharge. The optical emission, however, exhibits a very pronounced and complex dynamics. The basic excitation mechanisms of a plan parallel CCP are discussed in the following. The electrode gap is 25 mm and the electrode diameter is 100 mm.

Fig. 2(a) displays the phase and space resolved optical emission from a hydrogen CCP (100 W, 142 Pa) with a small admixture (1%) of Neon as tracer gas. The observe Ne $2p_1$ state is practically free of population through cascades [12]. The abscissa comprises one rf cycle and the transverse axis indicates the distance from the powered electrode.

Eq. 1 is used to determine the excitation dynamics (fig. 2(b)) from the measured phase resolved optical emission (fig. 2(a)). Different electron impact excitation processes can be clearly identified. The first process is caused by a field reversal across the space charge sheath, typical for hydrogen rf discharges. During this phase electrons are accelerated towards the powered electrode and induce a strong impact excitation. The second excitation process is due to sheath expansion heating of electrons moving to the plasma bulk, when the sheath potential becomes negative again. The third process results from secondary electrons created by ion impact. Due to the small mass of hydrogen ions they are able to follow the applied electric field. Thus, time dependent ion bombardment determines the creation of secondary electrons at the electrode surface. During the phase of maximum bombardment the high sheath potential of several hundred Volts results in an acceleration of electrons and electron multiplication through ionization. The high energetic directed electrons created in the sheath region also induce excitation when they enter the plasma bulk.

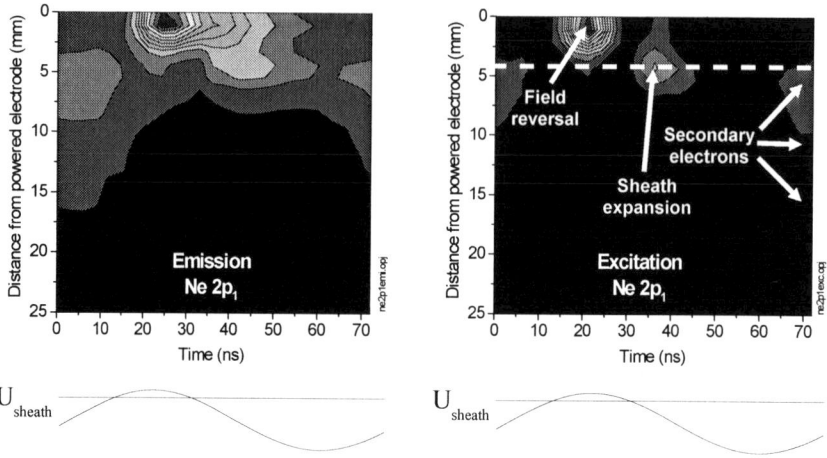

FIGURE 2. Phase and space dependent emission (a) and electron impact excitation (b) in a capacitively coupled hydrogen rf discharge. A small admixture of Neon is used as tracer gas.

Dual Frequency CCP

CCPs operated simultaneously with two radio-frequencies are used to achieve separate control of plasma density and ion impact energy onto the substrate [13-20]. The plasma density is expected to be mainly controlled by the higher frequency while the ion energy is determined by both the lower and higher frequency. Here, PROES with temporal resolution within the low frequency rf cycle is discussed.

The investigated plasma is a modified industrial dual frequency CCP (2f-CCP) etch-reactor (Exelan®, Lam Research). The gap between the two, plane, parallel plate, electrodes is 13 mm and the radius is 110 mm. The bottom electrode is powered with both frequencies (2 MHz and 27.12 MHz) simultaneously and the top electrode is grounded. Two quartz rings surrounding the electrode gap produce a symmetric discharge.

Fig. 3 shows the time and space resolved optical emission from a discharge operated at $P_{27} = 800$ W and $P_2 = 200$ W. The emission recorded is from the He 3^3S-state at $\lambda = 706.5$ nm ($\tau = 36.1$ ns) in a He–O$_2$ discharge (1500 sccm He, 1000 sccm O$_2$ at 490 mTorr). The low frequency cycle is scanned with a resolution of 36.88 ns, averaging over the dynamics within the high frequency cycle. In fig. 3 the abscissa shows the phase of the 2 MHz cycle, whereas the ordinate corresponds to the distance between bottom (y = 0) and top (y = 1) electrode.

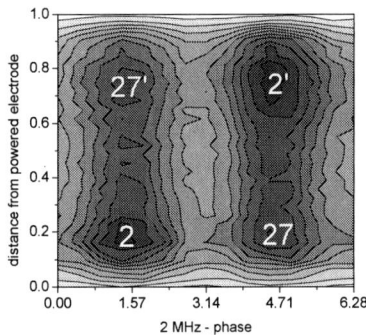

FIGURE 3. Space and phase resolved optical emission illustrating the electron dynamics within the low-frequency rf-cycle. The pronounced dynamics exhibits a strong coupling of both frequencies. The emission maxima indicated as 2 and 2'scale with the 2 MHz power relative to the 27 MHz power, while the maxima indicated as 27 and 27' scale vise versa.

The emission and corresponding excitation exhibit a pronounced dynamics within the low frequency cycle. The confined discharge is symmetric with similar excitation mechanisms occurring in front of both electrodes. Two double peak structures can be easily identified: two peaks (indicated as 2 and 27) at the bottom electrode at different phases in the rf cycle, and two peaks (indicated as 27' and 2') close to the top electrode at the same phases. The two peaks at each electrode are separated by half a low frequency cycle. Separate power variations of both frequency components show a 'diagonal correlation' of the peaks. With increasing 27 MHz power the peaks indicated as 27 and 27' increase in relation to the other two peaks; and for increasing 2 MHz power the peaks indicated as 2 and 2' increase correspondingly. These dependencies illustrate that the excitation mechanisms in front of both electrodes are of the same nature and hence 180° out of phase. The pronounced maxima around the sheath edge are typical for CCPs in α-mode, where energetic electrons are created in the rapidly moving sheath and penetrate into the plasma bulk [3].

The dynamics in the dual frequency discharge can be understood in the following picture. The rapid oscillations of the sheath-edge, determined by the high-frequency component, drive the energy gain of electrons. The velocity of these oscillations depends on the spatial movement of the sheath-edge and, therefore, on the local ion density. The spatial structure of the sheath is predominantly governed by the large low-frequency voltage. Thus, the position of the sheath-edge oscillation is strongly determined by the low-frequency. Ion flux conservation in the sheath results in a decrease of ion density towards the electrode. Depending on the phase of the low-frequency voltage, and the corresponding ion density in the vicinity of the sheath-edge, the same high-frequency voltage change results in different spatial movements of the sheath-edge. The lower the ion density the larger the spatial movement, resulting in higher sheath velocities and, hence, increased energy gain for electrons.

This explains the strong coupling of both frequencies. Maximum energy gain for electrons can be expected around the minimum voltage of the low-frequency component, when the sheath-edge is close to the electrode, corresponding to low ion densities and extremely fast instantaneous sheath-edge velocities. Thus, the peaks 2 and 2', corresponding to minimum low-frequency sheath voltages, are strongly dependent on the 2 MHz power, which governs the spatial structure of the sheath. Excitation through energetic electrons, created during these phases, decreases with penetration through the plasma bulk. Additional energy gain and increased excitation can be observed close to the opposite electrode at maximum sheath extension (peaks 27' and 27). This energy gain, through high-frequency oscillations, is less dependent on the 2 MHz power since the spatial structure of the sheath is not as relevant.

INDUCTIVELY COUPLED PLASMAS (ICP)

The rf inductively coupled plasma (ICP) is an electrodless discharge type. The planar coil configuration is the most commonly used for material processing. The rf power is coupled into the plasma by means of an induction coil through a dielectric window so that the plasma is not in contact with the antenna. The power coupling mechanism is analogous to a transformer, whereby the antenna acts as the primary coil of the transformer and the plasma as the secondary coil. The oscillating currents in the antenna generate a time dependent magnetic field. This time varying magnetic field induces an electric field in the plasma, according to Faraday's law.

However, the antenna can also act as an electrode. Therefore, both inductive and capacitive power coupling mechanisms can coexist in ICPs. Capacitive coupling drives the discharge at low plasma densities and is referred to as E-mode. During plasma ignition and at low rf powers the discharge is capacitive. As the plasma density increases, and the discharge can support the induced currents, a transition to inductive mode occurs, referred to as H-mode. The transition from capacitive mode to inductive mode, known as the E-H transition, can often be identified by an obvious increase in luminosity of the plasma.

Investigations of an ICP have been carried out in a modified GEC reference cell. Fig. 4 shows PROES measurements under a power variation in a hydrogen ICP operated at 10 Pa. Two rf cycles are resolved, with a 2 ns gate width in steps of 1 ns. The absolute intensity of the measured H_α-emission varies over orders of magnitude while the discharge transits from capacitive to inductive power coupling. The plotted modulation, however, is normalized to the time averaged intensity at each power. The normalized modulation clearly illustrates the capacitive coupling at low powers and inductive coupling at higher powers. The capacitive coupling is characterized by a pronounced modulation, with one emission maximum per rf cycle, while the inductive mode exhibits smaller modulation amplitude with twice the rf frequency.

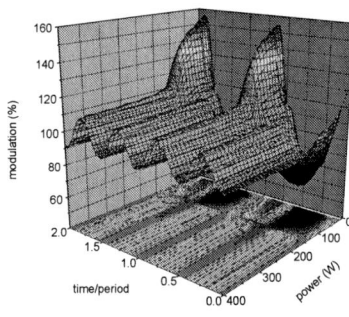

FIGURE 4. Normalized time dependent optical emission in an inductively coupled plasma as a function of rf power.

Fourier-components in H-mode and E-mode

Electrons in rf discharges acquire energy from the oscillating electric field and lose energy through elastic and inelastic collisions with the background gas. The electron energy distribution function can be described by the Boltzmann equation. In rf discharges the EEDF shows time dependent modulations. In capacitive discharges the electrons gain energy mainly due to the sheath dynamics. The electrons are accelerated into the plasma bulk region, and thus gain energy, once in every rf cycle. Thus, the EEDF is modulated with a frequency corresponding to the rf frequency. While, in ICPs the electrons gain energy due to the induced electric current. This current oscillates forward and backward in one rf cycle. Since the direction does not matter, electrons gain energy twice in each rf cycle. Therefore, in ICPs, the EEDF time dependence is with twice the rf frequency. The modulation in inductively coupled plasmas is not as pronounced as in capacitively coupled plasmas. ·

The time dependence of the EEDF in rf discharges can be observed in the optical emission. The different power coupling mechanisms exhibit different signatures in the temporal modulation of the emission, as could be already identified in fig. 4. A Fourier-analysis of the optical emission allows one to distinguish between these mechanisms. The correlation of the various Fourier-components gives insight into electron motion and heating in the discharge.

As discussed above, the EEDF and optical emission in ICPs are modulated by the induced electric field, and thus with twice the rf frequency. Fig. 5 shows a 2d-spatially resolved measurement of the second harmonic component of H_α-emission ($\lambda = 656$ nm). It shows maxima close to the quartz at around two-thirds the antenna radius, where the induced electric field has its maximum.

As discussed above, capacitive coupling can be identified through the 1ω-component. However, in capacitive mode, the mechanisms are more complex than in inductive mode. A variety of power coupling mechanisms can be present in capacitive plasmas; these different mechanisms are then overlapped in the 1ω-Fourier component of the emission. The field reversal present in hydrogen plasmas and the sheath expansion are determined by the non-linear sheath dynamics and thus exhibit, in addition to a 1ω-component, also higher harmonics. Secondary electrons are determined by ion bombardment and also the subsequent acceleration of electrons produced by secondary emission. These processes follow a $(\cos \omega t + 1)^2$ function [3]. Thus secondary electrons can be expected in both the first and second harmonic component. Fig. 6(a) and 6(b) show the first and fourth harmonic of the space resolved emission, respectively. The first harmonic is a combination of a variety of capacitive power coupling mechanisms, thus making it difficult to distinguish individual processes. However, the fourth harmonic reveals detailed insight into power dissipation in capacitive-mode. Two distinct excitation mechanisms can be identified. The field reversal, at the quartz can be clearly observed and distinguished from the sheath expansion, at a position further into the plasma at maximum sheath extension.

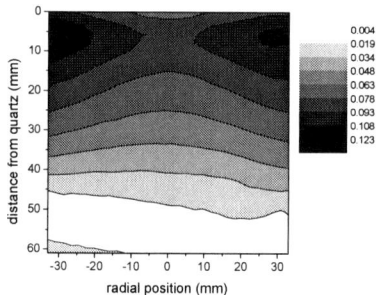

FIGURE 5. 2d-spatially resolved measurement of the 2ω-component of the normalized modulation in H-mode.

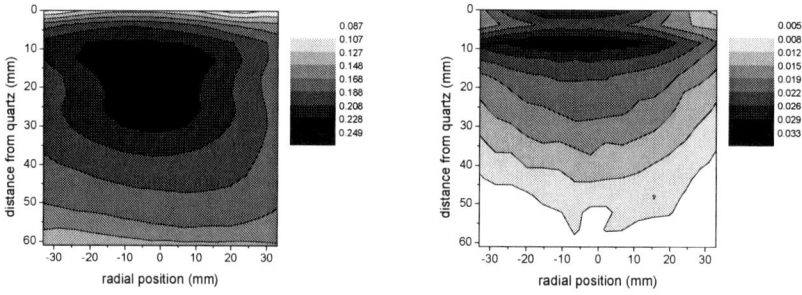

FIGURE 6. 2d-spatially resolved measurement of the 1ω-component (a) and 4ω-component (b) of the normalized modulation in E-mode.

268

MAGNETIC NEUTRAL LOOP DISCHARGES (NLD)

The NLD concept utilizes an inhomogeneous static magnetic field configuration, with a neutral loop (NL) region where the magnetic field vanishes [21]. An inductive radio-frequency electric field is superimposed on this magnetic NL. Three coaxial coils, of different diameters, surrounding the chamber produce the desired magnetic field configuration. The current in the top and bottom coil flow in the same direction and the current in the middle coil is opposite in direction to the other two. This arrangement produces a quadrupole magnetic field configuration bent into a torus structure with a magnetic null along a ring in the torus - the so called NL. The magnetic null ring, with relatively strong magnetic field gradients, is located just below the planar ICP antenna for efficient plasma production. An oscillating rf electric field is induced along the NL. The antenna is operated at 13.56 MHz and separated from the plasma by a quartz dome. The discharge can also be operated as a conventional ICP, without magnetic fields.

Such a discharge allows operation at significantly lower pressures than conventional ICPs due to electron confinement and more efficient collisionless electron heating. Low process pressure decreases ion scattering through collisions in the sheath, in front of the substrate, resulting in better etch anisotropy. In addition uniform plasma surface treatment over large areas can be achieved by varying the diameter of the NL radius.

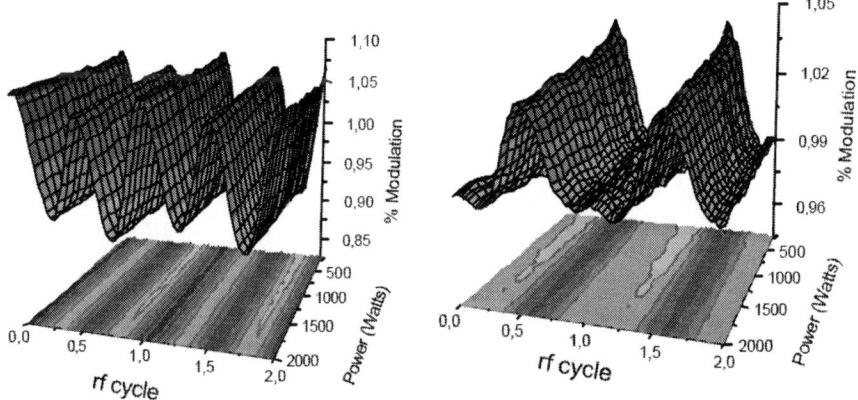

FIGURE 7. Normalized time dependent optical emission in an ICP (a) and NLD (b) as a function of rf power.

The phase resolved Ne $2p_1$ emission ($\lambda = 585.2$ nm) from a pure Neon plasma at 1 Pa for varying powers is observed. Fig. 7(a) shows the modulation of the emission measured in pure ICP mode of the discharge without magnetic fields. The optical modulation is normalized to the time averaged intensity. Modulation with twice the rf frequency is clearly visible. Fig. 7(b) shows the modulation measured with magnetic field in the center of the NL under the same conditions as in fig. 7(a). Compared to pure ICP, two differences are apparent: The modulation in the NLD is significantly

269

lower and the two peaks within one rf cycle become asymmetric with increasing power, producing a first harmonic component. Fig. 8(a) and 8(b) show the spatially resolved measurement of the first and second harmonic components, respectively, in NLD-mode at 2000 W.

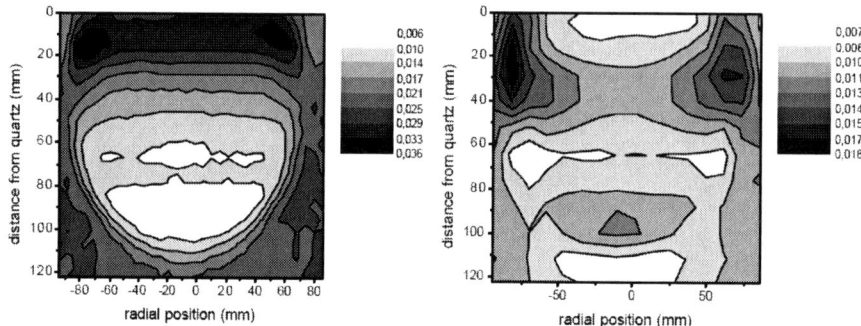

FIGURE 8. 2D-image of the 2ω-amplitude (a) and 1ω-amplitude (b) using a Fourier analysis of the normalized time dependent emission in a NLD.

In NLD-mode the power dissipation in the discharge is more complex than in pure ICP-mode, with no magnetic fields. In contrast to ICP-mode a first harmonic component dominates the modulation of the emission. The comparatively small second harmonic component shows similar structures as in ICP-mode (fig. 5). The main difference is that the observed structures representing the induced electric field are closer and more confined to the quartz surface. This can be attributed to the higher plasma densities in NLD-mode.

The structure of the first harmonic component in fig. 8(b) is complex. It shows features around the NL region as well as along the separatrices. A structure along the separatrix can be observed close to the bottom of the plot where electrons following the separatrix move towards the observation window. The first harmonic component in the modulation of the optical emission can be correlated to a time independent drift component of electrons coupling with the induced electric field.

CONCLUSIONS

Recent developments in PROES have made detailed experimental investigations of electron dynamics and power coupling mechanisms in rf plasmas possible. The application of PROES in various rf plasmas based on different power coupling mechanisms has been discussed. The basis is a time dependent model for the population dynamics of excited states within the rf cycle. Characteristic signatures of different power coupling mechanisms in Fourier-components of the measured temporal modulation have been shown and understood. Despite the already achieved success, PROES is still a relatively new diagnostics technique which gives a lot of promise for future developments.

ACKNOWLEDGMENTS

Funding by the DFG, in the frame of the SFB 591 and the GRK 1051, and the EU within FP 5 is gratefully acknowledged.

REFERENCES

1. T. Gans, V. Schulz-von der Gathen, H.F. Döbele, Plasma Sources Sci. Technol. **10**, 17 (2001)
2. T. Gans, Chun C. Lin, V. Schulz-von der Gathen, H.F. Döbele, Phys. Rev. A **67**, 012707 (2003)
3. T. Gans, V. Schulz-von der Gathen, H.F. Döbele, Europhysics Letters **66**, 232 (2004)
4. T. Gans, V. Schulz-von der Gathen, H.F. Döbele, Contr. Plasma Phys. **44**, 535 (2004)
5. M. Abdel-Rahman, T. Gans, V. Schulz-von der Gathen, H.F. Döbele, Plasma Sources Sci. Technol. **14** (2005) 51
6. T. Gans, J. Schulze, D. O'Connell, U. Czarnetzki, R. Faulkner, A.R. Ellingboe, M.M. Turner, Appl. Phys. Lett., accepted
7. R.H. Huddlestone and S.L. Leonard, Plasma Diagnostic Techniques, Academic Press (1965)
8. A. Anders, A Formulary for Plasma Physics, Akademie-Verlag, Berlin (1990)
9. I.P. Bogdanova and S.V. Yurgenson, Opt. Spektrosk. **61**, 241 (1986)
10. J.E. Chilton, J.B. Boffard, R.S. Schappe, C.C. Lin, Phys. Rev. A **57**, 267 (1998)
11. J.E. Chilton, M.D. Stewart, Jr., C.C. Lin, Phys. Rev. A **62**, 32714 (2000)
12. J.E. Chilton, M.D. Stewart, Jr., and C.C. Lin, Phys. Rev. A **61**, 52708 (2000)
13. E. Kawamura, M.A. Lieberman, A.J. Lichtenberg, Phys. of Plasmas **13**, 53506 (2006)
14. T. Kitajima, Y. Takeo, Z. Lj. Petrovic, T. Makabe, Appl. Phys. Lett. **77**, 489 (2000)
15. T. Denda, Y. Miyoshi, Y. Komukai, T. Goto, Z. Lj. Petrovic, T. Makabe, Jour. Appl. Phys. **95**, 870 (2004)
16. P.C. Boyle, A.R. Ellingboe, M.M. Turner, J. Phys. D: Appl. Phys. **37**, 697 (2004)
17. P.C. Boyle, A.R. Ellingboe, M.M. Turner, Plasma Sources Sci. Technol. **13**, 493 (2004)
18. H.C. Kim and J.K. Lee, Phys. Rev. Lett. **93**, 085003 (2004)
19. J. Robiche, P.C. Boyle, M.M. Turner, A.R. Ellingboe, J. Phys. D: Appl. Phys. **36**, 1810 (2003)
20. H.C. Kim, J.K. Lee, J.W. Shon, Phys. Plasmas **10**, 4545 (2003)
21. Z. Yoshida, T. Uchida, Jap. J. Appl. Phys. **34**, 4213 (1995)

Aerodynamic Effects in Weakly Ionized Gas: Phenomenology and Applications

S. Popović and L. Vušković

Department of Physics, Old Dominion University
Norfolk, Virginia, USA

Abstract. Aerodynamic effects in ionized gases, often neglected phenomena, have been subject of a renewed interest in recent years. After a brief historical account, we discuss a selected number of effects and unresolved problems that appear to be relevant in both aeronautic and propulsion applications in subsonic, supersonic, and hypersonic flow. Interaction between acoustic shock waves and weakly ionized gas is manifested either as plasma-induced shock wave dispersion and acceleration or as shock-wave induced double electric layer in the plasma, followed by the localized increase of the average electron energy and density, as well as enhancement of optical emission. We describe the phenomenology of these effects and discuss several experiments that still do not have an adequate interpretation. Critical for application of aerodynamic effects is the energy deposition into the flow. We classify and discuss some proposed wall-free generation schemes with respect to the efficiency of energy deposition and overall generation of the aerodynamic body force.

Keywords: electric wind, weakly ionized gas, shock wave, dispersion.
PACS: 52.30.-q, 52.35.Tc, 52.77.-j, 52.80.Hc

INTRODUCTION

Interaction of plasma with gas flow is a rather broad subject and misinterpretations are rather frequent, especially when a particular problem is set into an interdisciplinary context. Therefore, we will start with defining the terminology and for that purpose we will divide the flow-speed domain into several groups: (a) low-speed gas flow, with flow speed of the order of 0.1-10 m/s, (b) subsonic flow, with flow speed above 10 m/s, but lower than one third of speed of sound, approximately, (c) compressible subsonic flow (flow speed below speed of sound), (d) supersonic flow, with flow speed 1-5 times speed of sound, and (d) hypersonic flow, with flow speeds exceeding 5 times the speed of sound. Each region of flow-speed domain has its own phenomenology, mechanisms, and applications. For instance, the low-speed gas flow can be generated by purely electro-aerodynamic mechanisms, whereby drifting charged particles set the neutral particles in motion by elastic collisions, while the high-speed flow is best modified by non-uniform (local) gas heating.

CP876, *The Physics of Ionized Gases: 23rd Summer School and International Symposium*,
edited by L. Hadžievski, B. P. Marinković, and N. S. Simonović

In this presentation we describe electro-aerodynamic mechanisms that lead to formation of electric or corona wind and its modern cumulative versions in repetitive pulsed discharges that have been promised to generate more substantial flow velocities. These elegant, but delicate low-power techniques are compared to more robust thermal (global or local) mechanisms that are driving dispersion of shock waves, by generating density gradients and body forces.

Discussions and controversies around the interpretations of modern experiments in plasma aerodynamics look remarkably similar to the old arguments, only modern debates last much shorter. Therefore, we first give a brief historical account on the aerodynamic effects found in gas discharges. Further, we establish the links between the old and new observations of purely electrical or purely thermal effects by placing them into perspective of future plasma aerodynamic applications. We accomplish this by describing the modern experiments and dilemmas on propagation and dispersion of aerodynamic and acoustic waves through ionized gases. Finally, we discuss the ways to use the observed laboratory effects in order to bridge the gaps between desktop experiments, wind tunnel tests, and in-flight validations. Here the crucial role is played by energy deposition into wall-free discharges, confined by electromagnetic beams, diffusion, and gas flow. In each step of the outlined presentation we state fundamental questions that have to be addressed and resolved.

HISTORICAL PERSPECTIVE

Electric Wind

William Gilbert wrote in Chapter II of the Second Book of *De Magnete* [1] about the attraction exerted by amber "or, more properly, the attachment of bodies to amber" and proceeded to define "material effluvium emitted by electrics" as the mechanism of attraction [1]. In the first recorded attempt to explain electrostatic repulsion, Niccolo Cabeó [2] described the phenomenon by supposition that a charged body "produces an effluvium, which drives off the surrounding air..." [3]. Even though inaccurate interpretation of the electrostatic attraction and repulsion, these were the first suggestions that an electric fluid (effluvium) in motion could be associated with electrified objects. These suggestions of fluid motion associated with electrified bodies were so much alike and so close to physical reality that one finds difficult to take these as purely intuitive accounts. It is almost like they were based on someone's perception, which was not yet clear enough to allow precise written description. The two accompanying phenomena, electrostatic attraction and repulsion, and motion of air around a charged object were *de facto* separated by Robert Boyle [4] who refuted the "effluvium mechanism" of repulsion by demonstrating the electrostatic repulsion in vacuum. Probably the first to actually observe and describe the electric wind was Francis Hauksbee [5] in 1709 who reported ".. a weak blowing sensation by holding the charged tube close to his face...". He was an ingenious experimentalist, who left behind many electrostatic machines and accurate descriptions of many electrostatic effects. Although he was able to describe accurately the phenomenon of "barometric light" (probably the first experimental account of an electrical gas discharge),

Hauksbee was not able to see the extremely weak glow produced by the tube's tip. However, he described quite accurately the phenomenon we now know as *corona wind*. One can only speculate that the observation of the electric wind has led him to conclude that the electrical effluvium really existed, especially after he determined that barometric light persisted even at, for that time, rather substantial vacuum of about 30 Torr.

Since Cavallo [6] established the correspondence between charging of air around a charged object and subsequent observation of air flow, many founders of the modern concept of electricity, including Faraday and Maxwell, have occasionally touched the subject of electric wind [3]. They referred to the experiment depicted schematically in Fig. 1. Some of their accounts illustrate how the concept of electric wind evolved along the concept of electricity. For instance, in 1838 Faraday wrote: "... the part [of the air] which is charged may be but a small portion of that which is ultimately set in motion..." [7]. Indeed, part of the problem with corona wind is that the electrostatic energy transferred to charged particles is ultimately redistributed over orders of magnitude larger concentrations of neutral particles and the average speed is thereby reduced. Up to about the middle of 20th century the experiment shown in Fig. 1 was included in the textbooks on electromagnetism in various alternative forms [8], but since then it has somehow omitted of modern educators.

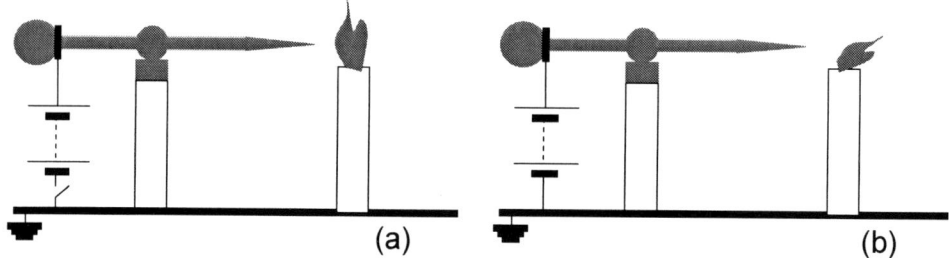

FIGURE 1. Corona wind causes candle flame to bend away.

In 1873 J. C. Maxwell gave a remarkably accurate description of corona breakdown and subsequent appearance of corona wind. He wrote "... as soon as the resultant [electric intensity] in the neighborhood of the discharge point has reached certain limit the insulating power of the air gives way, so that the air close to the point becomes a conductor ... but the charged particles of air being free to move under the action of electrical force, tend to move away from the electrified body ... and thus produce a certain current of air from the point, consisting of charged particles, and probably of others carried along by them..." [9].

In 1899 Chattock [10] derived the first quantitative relationship between the differential pressure p driving the corona wind and the current I to be

$$p = \frac{I\,x}{A\,\mu},$$ (1)

where x is the distance from the corona tip, A is the area of discharge cross section, and μ is the ion mobility. It is easily shown that this result is equivalent to the better known formula of electrostatic pressure [11]:

$$p = \frac{\varepsilon\, E^2}{2} \tag{2}$$

where E is the electrical field strength and ε is the dielectric constant. Assuming that only dynamic pressure is important one obtains the wind speed as "paraelectric" [12] effect:

$$p = \frac{\rho\, v^2}{2} = \frac{\varepsilon\, E^2}{2} \Rightarrow v = E\sqrt{\frac{\varepsilon}{\rho}} \tag{3}$$

where v is the flow speed and ρ is the gas density.

Flow speed in the continuous corona discharge linearly increases with electric field strength. It is relatively small, generating wind velocity typically not larger than several meters per second. Still, its applications are numerous in electrostatic precipitation, drying technology, and other technologies associated with low-speed flow.

A possibility of the cumulative effect generated by pulsed, repetitive discharges has been studied by several groups [12-14]. The general idea is to exploit effect of residual space charge in the afterglow between repetitive pulses to increase momentum transfer from ions to neutral particles with the aim to approach ionic velocities, that are about 100 m/s or higher. Once the ion velocity is efficiently transferred to neutral velocities this could be an elegant and efficient method of flow modification.

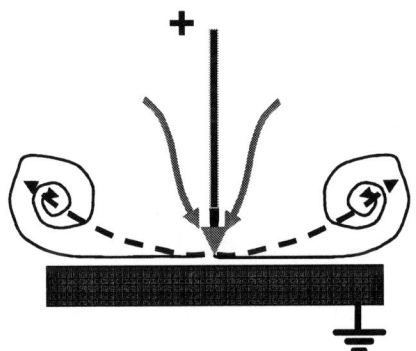

FIGURE 2. Electric wind generated in pulsed, repetitive positive corona.

Typical flow field around the pulsed, repetitive positive point-to-plane corona is shown schematically in Fig. 2 from the results of Refs. [13, 14]. Corona wind was accelerated in the gap between positive and grounded electrode and forms a doughnut-shaped vortex that expands in the direction of dashed line and dissipates when reaches the edge of the grounded electrode. By adjusting the repetition rate of the corona, it was possible to reach local speeds of 40 m/s [13, 14].

This approach was received with substantial curiosity, but it is still not clear how to bridge the gap from low-speed to high-speed regime. One has to note that a number of claims have been risen in the recent times, but so far have not resulted in a significant application. The results with flying "asymmetric capacitors" from 1930s until now, which evolved from the Brown-Biefeld effect, seem to suggest that flow velocities closer to ion drift can be achieved [15]. Similar promising reports have stirred interest for corona loudspeakers, plasma sound amplifications, and sonic boom mitigation [16, 17]. In all these cases the promised performance required much more power than it was possible to deliver, and the interest for them gradually faded out.

High-Speed Plasma Aerodynamics

Generation of Shock Waves

In comparison with the low-speed effects such as corona wind, high-speed plasma aerodynamics is a relatively new discipline. A group of phenomena that are relevant to the flow in ionized gas are the shock waves generated by electrical discharges. First detailed description of a phenomenon of this kind, which could be explained in the terms of an ideal shock tube, was given by Lord Rayleigh [18]. He described a strong flow of ionized gas from an induced electrodeless discharge in hydrogen into an adjacent supply tube. The flow of partially ionized gas consisted of highly energized atomic hydrogen, which was subsequently used for study of Balmer lines. Dominance of atomic hydrogen in the shock front testified of abundant number of high energy electrons, usually not present in thermal plasmas. The original Rayleigh experiment (See Fig. 3) has led to development of the electrically driven shock tubes capable of producing supersonic and hypersonic flow.

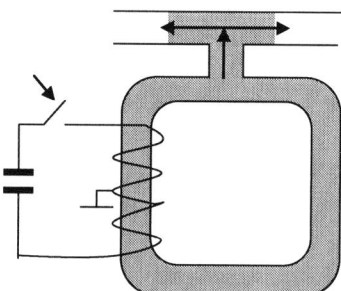

FIGURE 3. Scheme of Rayleigh experiment.

Dispersion and Propagation of Shock Waves in Weakly Ionized Gas

Since late 1970s several research groups in Russia have conducted plasma-aerodynamic experiments to modify shock wave behavior. Results of these experiments can be formulated in the claims that plasma-aerodynamic effects can cause "anomalous relaxation" of a bow shock wave and that plasma effects can cause the speed of a shock wave to increase, the shock structure to disperse, and the shock wave amplitude to dissipate. Although verified experimentally, the phenomena were not completely understood.

There were two generic types of experiments. In the first type [19, 20] shown schematically in Fig. 4a, an electrically-driven shock wave was generated at one end of a long tube filled with gas. Typically, a d.c. glow discharge was generated in the middle section of the tube where the interaction between the shock wave and ionized gas took place. Repeated observations have shown that the time of flight was shorter, the shock signature was longer, and the amplitude was lower in the presence of the discharge. In the second type of experiments [21], conducted in ballistic ranges (see Fig. 4b), it was observed that the standoff distance of the bow shock was longer in the discharge compared to the experimental results in a neutral gas.

In the middle of 1990s the subject got attention of the aerodynamic community outside Russia, and immediately the debate broke out on the underlying mechanisms of the observed "anomalies", which has already simmered in Russian plasma aerodynamic community. The debate question was set as "thermal versus non-thermal effects", which turned to be barely relevant to applications, and in some cases, an altogether inadequate proposition. Arguments in favor of thermal nature of the phenomenon have been based on the fact that the paradigm results of Ref. [20] were due to thermal gradients that were present ordinarily in glow discharges, and created well-documented "pump effects" and gas convection effects. Arguments in favor of non-thermal effects were based on measured super-thermal propagation velocities [22-24], observed polarity or electromagnetic mode dependence of the effect [25, 26], measured effects of lateral magnetic field [27], and ultraviolet radiation [28]. Neither of these measurements have been disproved so far.

(a)

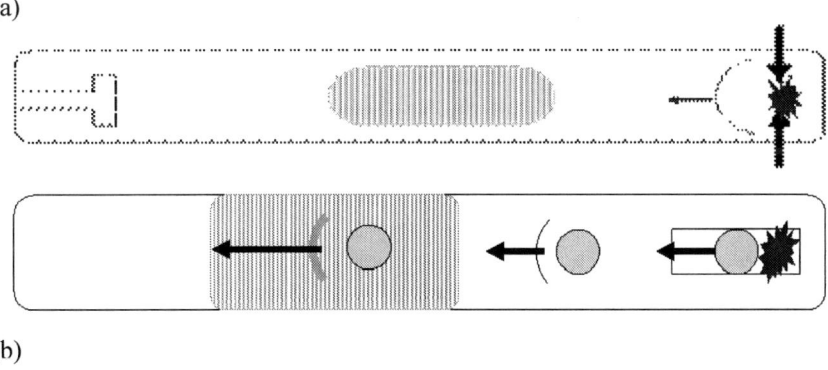

(b)

FIGURE 4. Two generic shock-tube plasma-aerodynamic experiments: (a) electrically-driven shock propagating through a plasma discharge tube; and (b) ballistic range experiment with a discharge plasma section.

The debate subsequently obtained a more appropriate formulation: global versus local thermal effects [29]. First group of effects are related to the specific morphology of the electrical discharges, where thermal gradients do develop independently of the interaction with a shock wave. However, local thermal effects are the product of interaction between the shock and ionized gas. They follow from the formation of double layer at the shock due to separation of charges, which is a well-known effect [30, 31], though still poorly understood. Some experimental evidence [25, 29] suggests that the double layer may be an order of magnitude stronger than expected, the reason of which is still not quite clear.

EXPERIMENTS

Traveling Waves

Dispersion and Propagation of Shock Wave in d.c. Glow Discharge

Ganguly, Bletzinger, and Garscadden [20] designed an experiment that combined the d.c. glow discharge and a simplified electrically-driven shock tube. In the original form the apparatus consisted of a long Pyrex tube, which was evacuated and filled with argon at 30 Torr. Pressure in the tube was kept constant with a weak gas flow. Two 5 cm long hollow cylindrical electrodes were placed 30 cm apart in the middle of the tube. A dc electrical discharge was produced between the electrodes. The discharge was placed between two extensive regions of cold gas. At the end of one region a spark gap source generated an acoustic shock wave, which then propagated through cold gas at an average shock velocity of 1.7 times the upstream local speed of sound. A scheme of the set up is given in Fig. 5a.

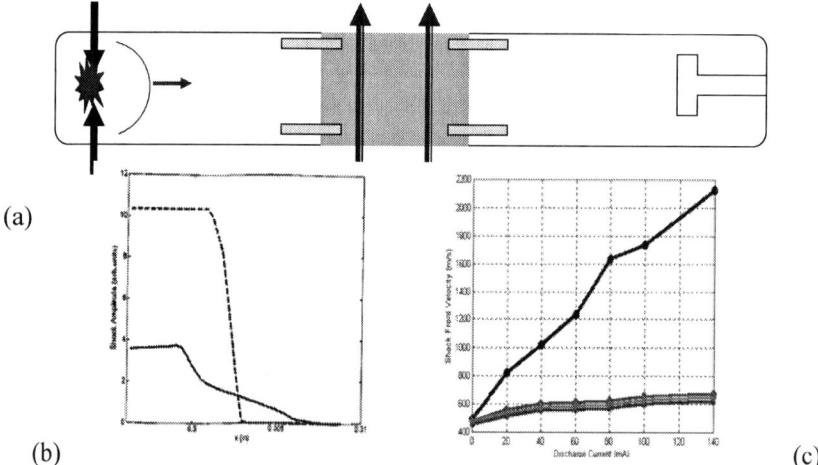

(a)

(b) (c)

FIGURE 5. Experiment in the d.c. glow discharge: (a) Scheme of the shock dispersion experiment in the d.c. glow discharge [20]; (b) Reconstructed shock profile at discharge axis; (c) Propagation velocity in cathode region and in positive column [22].

This experiment was intended to exemplify the generic problem of a planar shock wave propagating through a spatially uniform weakly ionized nonequilibrium gas. In reality, the planar description is correct only in a small region around the axis of glow discharge. In the major portion of the discharge strong radial gradients distort the idealized one-dimensional structure of the shock wave. However, the propagation of shock wave along the axis can still be treated as the one-dimensional problem.

In its propagation through the tube, the shock wave crossed two pairs of laser beams of a photo-acoustic deflection system placed at two different positions inside the discharge region as shown in Fig. 5a. Shock structures corresponding to the integrated photo acoustic signals are shown in Fig. 5b, presented in the coordinate frame of reference frozen at the time when the middle of shock wave front was crossing the observation plane. In the discharge, the shock front was split into two or more weaker shocks distributed over at least one order of magnitude larger spatial interval. Also, the total amplitude of the shock was reduced to a much smaller value in comparison to the shock amplitude in the absence of the discharge. In order to obtain maximum information from this experiment a multi-step procedure was developed. The procedure is described in detail in Refs. [22, 26]. After recording the raw data, time of flight and laser deflection signal waveforms, the analysis proceeds in several steps. In the first step, the propagation velocity at each position is calculated as a ratio of the distance between two laser beams and the time of flight between them. Also, the amplitude of shock is obtained by integration of laser deflection waveforms. Then the Rankine-Hugoniot relations are applied to evaluate simultaneously Mach number, temperature, and density at the shock. Analysis of three parameters determines thermal or super-thermal character of propagation velocity increase or shock amplitude decrease. As shown in Fig. 5c, the propagation velocity in the cathode region was found to be super-thermal.

In a follow up experiment with pulsed glow discharge [32] the dispersion of shock wave was not found. This led a large part of research community to conclude that the dispersion is purely a thermal effect. In the microwave cavity experiment [26], which has a plasma column that is much more homogeneous in comparison to the d.c. glow discharge, we also did not obtain deflection signal splitting. In our opinion, however, a more correct question was posed in Ref. [29]: are the dispersion effects global (in other words – related to the gradients developed in the discharge per se) or local (related to the transient gradients developed in the interaction between shock wave and plasma)? In the experiment of Ref. [32] there was no time for radial gradients to develop, hence a part of dispersion effect was eliminated. In the local thermal effects the key role is played by the double layer formed across the shock structure. In one measurement [25, 29] the strength of the double layer was found to be order of magnitude higher than expected. Our preliminary results with standing shock wave indicate excessive population of excited states at the shock, which suggests a local presence of energetic electrons that is a signature for a strong double layer.

Microwave Discharge in a Resonant Cavity

We report recent results of our work on experiments with a combination of a microwave cavity discharges and an electrically driven shock tube [26]. Compared to the d. c. glow discharge experiments [20, 24, 25] the microwave cavity discharge is electrodeless, hence without any solid obstacle to the incident shock wave. Also, metal electrodes are always suspected to contaminate the plasma. Impurities in the weakly ionized gas may modify in an uncontrollable manner the characteristics of the gas and deteriorate the quality of information obtained from the shock modification. In addition, this discharge is more homogeneous than the d.c. glow discharge and therefore more suitable for simulating the dispersion of a planar shock wave. Scheme of the experimental set-up is shown in Fig. 6a. The tube made of quartz has inner diameter 3 cm and length 1.2 m. Shock waves are generated in a spark gap and spark generator energy was between 32 and 100 J, stored in the 1 μF capacitor. Spark was triggered with an auxiliary spark gap switch, typically with pneumatic control. The anode-to-cathode voltage of the magnetron was half-wave rectified, the rectifying components being taken from the hardware of microwave oven. The cavity was cooled with a fan. Ten pairs of circular windows and three pairs of slits were cut through the cavity wall to provide optical diagnostic paths. The windows were located at equal distance along the cylinder. Three pairs of slits were used for measurements of radial distribution.

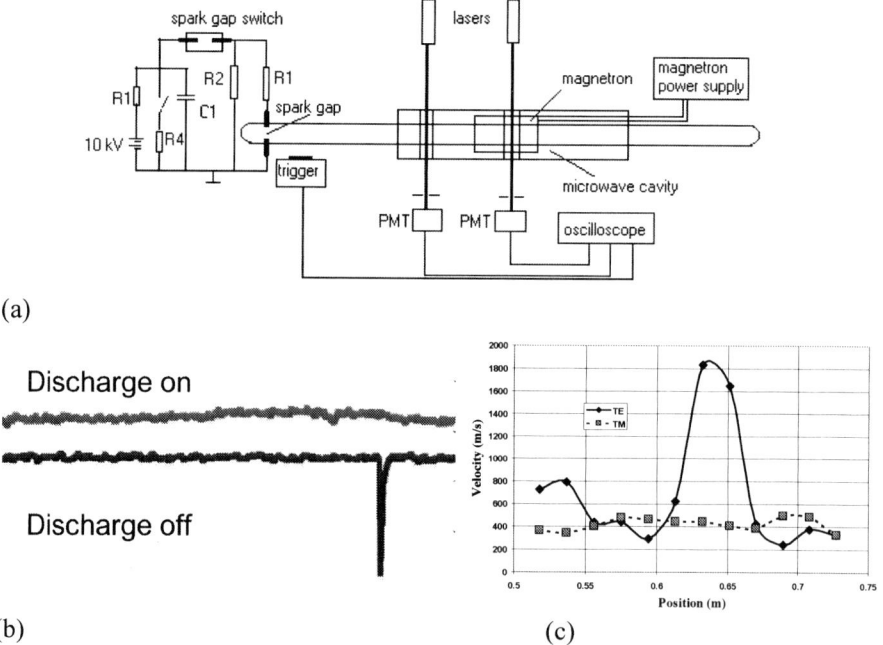

(a)

(b)

(c)

FIGURE 6. Microwave cavity experiment: (a) Scheme of the microwave cavity experiment; (b) Waveforms of deflection signals; (c) Velocity propagation when dominant mode was TE or TM.

Figures 6b and 6c show representative results of the experiment, (b) strong dispersion of the shock wave and (c) mode-sensitive propagation velocity, where the super-thermal velocities were achieved when dominant mode was transverse electric, TE where electric field was perpendicular to the direction of propagation.

Standing Waves

Spectral analysis of double layer associated with a traveling wave is difficult to perform due to transient character of the optical signal. Standing waves in front of stationary blunt models are more convenient for optical spectroscopy. Experimental set up to study standing waves is a combination of supersonic flow tube and a microwave cavity discharge. Supersonic flow was generated with cylindrical convergent-divergent (De Laval) nozzle. The nozzle was made of non-conductive microwave-transparent ceramics. Nozzle was designed nominally for M = 3.5 but actual measurements of stagnation and static pressure yielded M =3.28. Test section is located downstream the discharge section of the tube. Flowing afterglow from the microwave cavity discharge provides the necessary ionization and conductivity. The specific feature of the microwave discharge used in our experiments is that it sustains high level of ionization relatively far away from the resonant cavity. In this configuration, quartz tube acts as microwave waveguide that drives microwave field far into the test chamber, contrary to the d.c. flowing afterglow, where the external d.c. field is nonexistent outside the inter-electrode volume. Although we still call it flowing afterglow, in reality most of the ionization is sustained by the microwave field extending along the tube.

Shock wave was formed in front of a blunt model that was positioned in the center of the quartz tube using three symmetrical glass spacers. Distribution of line intensity along z direction was measured by moving the model along z-axis using the external holding magnets. Each position of the model was measured with a ruler. In the region of standing shock wave a double-peak line intensity increase by a factor of three to five was observed, which indirectly confirms the findings described in Ref. 25.

APPLICATIONS

Aside of the dilemmas related to the nature of observed effects, implementation of the findings have already shown the potential for developing the high-speed flow control technology and defining it an enabling technology for high-speed aero transportation. In order to become an applicable technology, at least on the level that the corona wind has achieved at low-speed flows, quite a few challenging tasks remain to be solved. In spite successful experimental results on electrohydrodynamic body forces and promising results on magneto-fluid effects, there are no mechanisms other than thermal to generate substantial plasma-aerodynamic effects at supersonic speed. In high-speed flow, the most promising are mechanisms leading to local thermal effects, because the energy required to excite global thermal effects is prohibitively high. On the other hand, successful use of local thermal effects will have to rely on fast, precise, and efficient control of double layers and ionization waves, which are the plasma engines for local heating in the free flow.

Historically, the list of most important potential applications of the physics of shocks and plasmas to the supersonic and hypersonic flow included reduction of overall vehicle drag, asymmetric modification of the flow field in order to provide flow control with no or minimum addition to the fuselage, generation of electrical power, plasma breaking or plasma propulsion, and variable control of shock locations to assist inlet design and Mach number transitions [29]. In every application the key element is efficient energy deposition into the free flow, shock or boundary layer. Many discharge schemes have been proposed for these and other applications. They can be classified in four groups:

(1) Miniature, point-like sources (0-D) typically generated in the focus of a laser beam [33];

(2) Arrays of point-like sources or filamentary plasmas (1-D) formed by constriction due to non-uniform gas heating [34];

(3) Plane-parallel filamentary discharges and planar surface discharges (2-D) [35-37];

(4) Volumetric arrays of plasmoids (3-D) [38, 39].

There are three primary effects that favor the microwave filamentary and certain types of point-like discharges to all other types of high E/N plasmas: (*i*) high power efficiency, (*ii*) possibility to reduce breakdown threshold, and (*iii*) feasibility of implementation without addition of mechanical elements to the fuselage.

Although most of the flow-plasma interaction physics has been studied in diffuse discharges, this type was discarded as too reflective to be efficient for energy deposition. Filamentary discharges are driven by thermal ionization instability to obtain precisely what is needed for the efficient energy deposition into the flow field. Contrary to common approach in ballpark estimations, it is not the high degree of ionization with large penalty of 100 eV per ion-electron pair, but rather a limited localized ionization combined with non-uniform gas heating that is efficiently absorbing microwave energy. The excessive absorption of microwave radiation leads to generation of filaments of hot ionized gas that are closer to the thermal equilibrium than the weakly ionized gas of diffuse discharges. Thus, combined ionization and heating mechanisms lead to practical absorption levels up to 70% of input microwave power.

CONCLUSION

Phenomenology of ionized gas flow is so complex that a separate name "plasma aerodynamics" is justified. Phenomenon of electric wind has a long history in parallel to the development of static electricity, it is fairly well understood, and has been experiencing a revival in numerous applications of low–speed flows. At the other side of the speed domain, the dispersion of shock waves in interaction with ionized gas has a potential to provide an important enabling technology for high-speed transportation. Although the fundamental aspects of the mechanisms of dispersion are not completely understood, research effort on applying the phenomenon to supersonic and hypersonic flow is underway. Key element of this work is the development of wall-free discharges as the devices that provide energy deposition with suitable geometry to generate an asymmetric modification of the free flow. It has been proven that wall-free discharges can be shaped into forms that mimic the conventional actuators by using various forms of electromagnetic fields. In this way the wall-free discharges justify the meaning of "plasma" by acting as an ultimate molding material.

REFERENCES

1. W. Gilbert, *De Magnete,* (translated by Mottelay, P.F.), p. 85, Ch. 2, Book II, (Dover Publications Inc 1958).
2. N. Cabeo, *Philosophia magnetica* (Cologne 1629) p. 35.
3. M. Robinson, *Am. Jour. Physics* **30**, 366 (1962).
4. R. Boyle, *Experiments and notes about the mechanical origine or production of electricity* (London, 1675) p. 29.
5. F. Hauksbee, *Physico-Mechanical Experiments on Various Subjects* (London 1709, reprint New York: Johnson Reprint Corp, 1970) p. 46.
6. T. Cavallo, *A Complete Treatise of Electricity in Theory and Practice,* Dilly, London, (1777).
7. M. Faraday, *Experimental Researches in Electricity,* v. 1, Paragraphs 665, 1442-1444, 1535, 1592, 1595, (London, 1839, reprinted by Courier Dover Publ. 2004).
8. R. P Feynman, R. B. Leighton, and M. Sands, *The Feynman Lectures on Physics,* v. II, p. 9-8.
9. J. C. Maxwell, *Treatise on Electricity and Magnetism,* v.1, p. 52, (Oxford 1873, reprinted by Oxford University Press 1998).
10. A. P. Chattock, *Phil. Mag.* **48**, 401 (1899).
11. O. M. Stuetzer, *J. Appl. Phys.* **30**, 984 (1959).
12. J. R. Roth, *Physics of Plasmas* **10**, 2117 (2003).
13. J. Batina, F. Noel, S. Lachaud, R. Peyrous, J. F. Loiseau, *J. Phys. D:Appl. Phys.* **34**, 1510, (2001).
14. J. F. Loiseau, J. Batina, F. Noel, R. Peyrous, *J. Phys. D:Appl. Phys.* **35**, 1020 (2002).
15. T. B. Bahder and C. Fazi, ARL Tech Rep. (2003).
16. M. S. Cahn and G. M. Andrew, *AIAA Paper 68-24* (1968).
17. D. S. Miller and H. W. Carlson, NASA TN-D-5582 (1969).
18. R. J. S. Lord Rayleigh, *Proc. Roy. Soc.* **183**, 26 (1944).
19. A. I. Klimov, A. N. Koblov, G. I. Mishin, Yu. L. Serov, and I. P. Yavor, *Sov. Tech. Phys.Lett.* **8**, 192 (1982).
20. B. N. Ganguly, P. Bletzinger, and A. Garscadden, *Physics Letters A* **230**, 218 (1997).
21. G. I. Mishin, Yu. L. Serov, and I. P. Yavor, *Sov. Tech Phys. Lett.* **17**, 413 (1991).
22. S. Popović and L. Vušković, *Physics of Plasmas* **6**, 1448 (1999).
23. V. Basargin and G. I. Mishin, *Sov. Tech. Phys.Lett.* **11**, 85 (1985).
24. P. Bletzinger and B. N. Ganguly, *Physics Letters A* **258**, 342 (1999).
25. P. Bletzinger, B. N. Ganguly, and A. Garscadden, *Physics of Plasmas* **7**, 4341 (2000).
26. P. Kessaratikoon, S. Popović, and L. Vušković, *AIAA Paper 2004-1021.*
27. V. A. Gorshkov, A. I. Klimov, A. N. Koblov, G. I. Mishin, and K. V. Kodathaev, *Sov. Phys. Tech. Phys.* **29**, 595 (1984).
28. A. Yu. Gridin, A. I. Klimov, G. I. Mishin, *Sov. Tech. Phys. Lett.* **16**, 295 (1990).
29. P. Bletzinger, B. N. Ganguly, D. Van Wie, and A. Garscadden, *J. Phys. D: Appl. Phys.* **38**, R33 (2005).
30. M. Y. Jaffrin, *Physics of Fluids* **8**, 606 (1965).
31. J. Kwan and B. Ahlborn, *Physics of Fluids* **27**, 499 (1984).
32. S. Macheret, Y. Z. Ionikh, N.V. Chernysheva, A. P. Yalin, L. Mertinelli, and R. B. Miles, *Physics of Fluids* **13**, 2693 (2001).
33. H. Yan, R. Adelgren, M. Boguszko, G. Eliot, and D. Knight, *AIAA Paper 2003-1051.*
34. Y. F. Kolesnichenko, V. G. Brokvin, D. V. Khmara, V. A. Lashkov, I. Ch. Mashek, and M. I. Ravkin, *AIAA Paper 2003-0361.*
35. S. Popović, L. Vušković, I. I. Esakov, L. P Gratchev, and K. V. Khodataev, *Appl. Phys. Lett.* **81**, 1964 (2002).
36. S. Popović, R. J. Exton, and G. C. Herring, *Appl. Phys. Lett.* **87**, 061502 (2005).
37. R. J. Exton, S. Popović, G. C. Herring, and M. Cooper, *Appl. Phys. Lett.* **86**, 121403 (2005).
38. R. J. Exton, R. J. Balla, B. Shirinzadeh, G. J. Brauckmann, G. C. Herring, W. C. Kelliher, J. Fugitt, C. J. Lazard, and K. V. Khodataev, *Physics of Plasmas* **8**, 5013 (2001).
39. R. J. Exton, R. J. Balla, G. C. Herring, S. Popović, and L. Vušković, *AIAA Paper 2003-4181.*

Spectra of Ions Produced by Corona Discharges

J. Skalný[1], G. Hortváth[1] and N. J. Mason[2]

[1]Department of Experimental Physics, Comenius University, Mlynska dolina F-2, 84248 Bratislava, Slovakia
[2]Open University, Department of Physics and Astronomy, Walton Hall, Milton Keynes MK7 6AA, United Kingdom

Abstract. A mass spectrometric study of ions extracted from both positive and negative DC corona discharges, initiated in point-to plane electrode system, has been carried out in ambient air at low air pressure (5 – 30) kPa. The average relative humidity of air was typically 40-50 %. Ions were extracted through a small orifice in the plane electrode into an intermediate gap where the low pressure prevented further ion-molecule reactions. Mass analysis of negative ions formed in the negative corona discharge using ambient air has shown that the yield of individual ions is strongly affected by trace concentrations of ozone, nitrogen oxides, carbon dioxide and water vapour. In dry air the CO_3^- ion was found to be dominant. In presence of water this is converted very efficiently to cluster ions $CO_3^-.(H_2O)_n$ containing one and more water molecules. The yield of $O_3^-.(H_2O)_n$ clusters or core ions was found to be considerably lower than in some other studies at atmospheric pressure. The mass spectrum of ions extracted from drift region of a positive corona discharge was simpler being dominantly cluster ions $H_3O^+.(H_2O)_n$ most probably formed from O_2^+ ions, a two step process being active if water molecules are present in the discharge gap even at relatively low concentration

Keywords: Corona discharge, Ion-molecule reactions, Mass-spectrometry
PACS: 52.80.Hc, 82.33.Xj, 82.30 Fi, 32.10 Bi

INTRODUCTION

Corona discharges, with both positive and negative polarities, are known to be efficient sources of unipolar ions. However there is little information on the mass spectra of product ions. A more detailed knowledge of the chemical composition of ions produced by corona discharges will significantly contribute to towards our understanding of the role of negative and positive ions in many devices using corona discharges for example; electrostatic precipitators, air cleaners, ionisers, xerocopy machines and the devices used for surface treatment of polymers. Recently there has been considerable interest in the use of corona discharges to replace radioactive sources in IMS. Most of these applications employ the corona discharges in ambient air. Therefore there is an urgent need to measure ion spectra from such discharges and understand how these may affect the use of corona discharges in the aforementioned technologies.

CP876, *The Physics of Ionized Gases: 23rd Summer School and International Symposium,*
edited by L. Hadžievski, B. P. Marinković, and N. S. Simonović

Mass spectrometric analysis of the ions formed in a corona discharge is a complicated diagnostic technique due to the inherent difficulties arsing from the large pressure difference between the corona source and the pressure needed to operate the mass spectrometer as well as problems arising from mass discrimination due to the small orifice used for extraction of ions. Accordingly This there have only been a few studies reporting mass spectrometric analysis of the ions extracted from corona discharges is low [1-14]. It should also be noted that such previous results contradict one another.

The aim of presented paper is to discuss some of the parameters that are likely to be the source of the differences between the results of previous authors. Both positive and negative corona discharges in ambient air will be analysed.

EXPERIMENTAL APPARATUS

The apparatus used in these experiments is that of the ELION group at NTH University of Trondheim, Norway and is shown schematically in Figure 1.

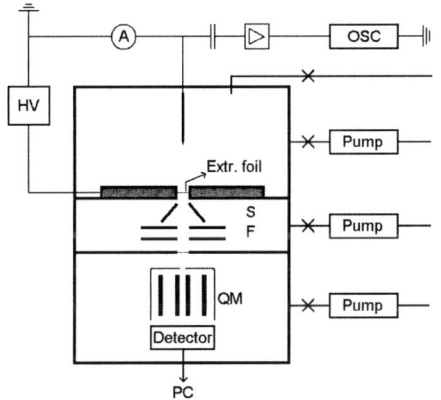

FIGURE 1. Schematic diagram of the experimental apparatus used in the present experiments.

The apparatus consists of three main parts: (i) a discharge volume, (ii) an intermediate volume containing an electrode system (S, F) to focus the extracted ions into (iii) the mass spectrometer (QM). The discharge chamber contained point-to-plane electrodes for generating the corona discharge. A Pt wire ending in a tip of radius of 0.1 mm was positioned on axis separated from a gold plated brass plane electrode at a distance 10 mm. An extraction foil was placed in the central part of the plane electrode. The extracted ions expand through an orifice into an intermediate region pumped by diffusion pump while the ions are focused onto a skimmer with a 2 mm opening leading to a differentially pumped Balzers OMG 101 quadrupole spectrometer (QM). Ambient air of relative humidity of around 50 % was used in experiments. A flow rate 50 cm^3/s was maintained by an electronically regulated valve. A constant air pressure (between 5-30) kPa was maintained in the discharge gap during experiments.

EXPERIMENTAL RESULTS AND DISCUSION

In positive corona discharges operated with dry air the dominant ionic species extracted from the discharge were NO^+ and O_2^+. However the addition of even a small amount of water vapors to the feed gas dramatically changed the spectrum with cluster ions $H^+(H_2O)_n$ appearing and these quickly dominated the spectrum. The yield of each individual ion was found to be strongly dependent on the discharge current (and hence the voltage applied to electrodes), this is evident from comparison of the two selected shown in Figure 2.

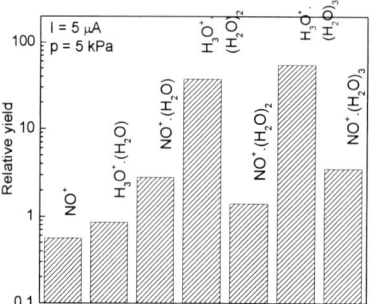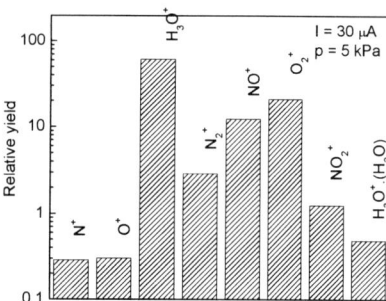

FIGURE 2. The relative yield of positive ions extracted from positive corona discharge fed by humid air at a pressure 5 kPa with two different discharge currents (5 and 30 μA).

We observed two distinct modes in the positive corona discharge, (i) observed at low currents where only molecular and cluster ion species were observed and (ii) a high pressure mode in which atomic ions N^+ and O^+ appeared in the extracted ion spectra. The critical current required for transfer from a low to high current mode increases with pressure. The existence of these two modes is likely to be a consequence of the broadening of the glow region towards the plane electrode as the voltage applied to the electrodes increases. This is also the reason for the dramatic reduction in the abundance of $H^+(H_2O)_n$ clusters containing more than two water molecules. The increase of electric field is nearly linear with increasing voltage and this is intensified by the effect of the space charge in drift region, which increases with the discharge current. The increase in the electric field also reduces the time necessary for the transport of parent positive ions, formed in glow region, across the drift region. This transport effect and the broadening the glow region towards the plane electrode ensures that the reaction time for ion-molecule processes is dramatically reduced in high electric fields.

In dry air the N_2^+ and O_2^+ ions are most probably generated by direct electron impact ionisation accompanied by formation of N^+ and O^+ via dissociative electron impact ionisation. N_2^+ ions are efficiently converted either to O_2^+

$$N_2^+ + O_2 \rightarrow O_2^+ + N_2 \tag{1}$$

$k_1 = (3.8 - 6.2) \times 10^{-11}$ cm^3/sec [15]

or undergo a reaction with atomic oxygen in glow region of the discharge gap

$$N_2^+ + O \rightarrow NO^+ + N \qquad (2)$$

$k_2 = (1.4 - 6.2) \times 10^{-10}$ cm^3/sec [15].

The presence of N_2^+ ions recorded in the high current mode is further evidence that process (1) is active predominately in drift region and any contributions from process (2) will be marginal in this region.

In dry air O_2^+ ions can be removed from the discharge through the formation of $O_2^+.O_2$ clusters (mass 64)

$$O_2^+ + O_2 + M \rightarrow O_2^+.O_2 + M \qquad (3)$$

$k_3 = 2.5 \times 10^{-28}$ cm^6/sec [16].

A second potential sink for O_2^+ ions is their reaction with NO_2, which may be formed in the glow region at high discharge currents by

$$O_2^+ + NO_2 \rightarrow NO_2^+ + O_2 \qquad (4)$$

$k_4 = 6.6 \times 10^{-10}$ cm^3/sec [15].

At low values of the discharge current NO_2^+ ions are absent from the spectra recorded in moist air due to the high efficiency of the three body process

$$O_2^+ + H_2O + M \rightarrow O_2^+.H_2O + M \qquad (5)$$

$k_5 = (2.3\text{-}2.8) \times 10^{-28}$ cm^6/sec [16].

This reaction is predominately active in the low electric field of the drift region. Therefore in high current mode, when the drift region is reduced the efficiency of process (4) prevails and ion NO_2^+ is present in spectra but no $NO_2^+ (H_2O)_m$ clusters are observed. Process (5) dominates over (4) if

$$[H_2O] \geq \frac{[NO_2]}{10\delta} \qquad (6)$$

where δ is the relative density of air. In a flowing air discharge the concentration of nitrogen oxides in the low current mode is small so condition (6) can be easily fulfilled even if there are only traces of water vapors in the discharge gap. If, however, the discharge is operated in the high discharge current mode, NO_2^+ ions can be formed at the extraction orifice and transported to the mass spectrometer without formation of any clusters with water.

Reaction (5) is also the origin of the water clusters $H^+(H_2O)_n$ since (5) is followed by a very fast two-body process

$$O_2^+.H_2O + H_2O \rightarrow H_3O^+ + OH + O_2 \tag{7}$$

$k_7 = 2.2 \times 10^{-9}$ cm^3/sec [16].
The H_3O^+ ion is the first member of group of clusters $H^+(H_2O)_n$ $n = (1, 2,...)$, which are formed by sequential reactions

$$H^+.(H_2O)_n + H_2O + M \rightarrow H^+.(H_2O)_{n+1} + M \tag{8}$$

$k_8 = 1 \times 10^{-28}$ cm^6/sec [16].
$O_2^+.O_2$ ions observed in dry air discharges disappeared in moist air due to their reaction with water molecules

$$O_2^+.O_2 + H_2O \rightarrow O_2^+.(H_2O) + O_2 \tag{9}$$

$k_9 = 2.2 \times 10^{-28}$ cm^6/sec [16].

Our results (Figure 2) confirm these predictions of the substantial effect of water molecules on the spectra of ions generated in positive corona discharges in air. Moreover the transport time across the drift region, which is considerably affected by applied voltage, determines the abundance of individual ions extracted from discharge at plane electrode position.

Since three body processes dominate in the drift region of the positive corona discharge fed by wet air and the rate of such processes is proportional to the air pressure we would expect considerable differences in the mass spectrum of extracted ions as a function of pressure. This is evident in Figure 3 where the relative yield of sum of clusters $H^+(H_2O)_n$ $n = (1, 2,...)$ in the spectra is shown as a function of the discharge current at three different pressures of wet air. Hence we can conclude that the pressure is the third factor influencing the mass spectrum. Moreover it can be easily surmised from the present measurements that $H^+(H_2O)_n$ $n = (1, 2,...)$ the cluster ions will dominate the spectrum of ions produced by positive corona at atmospheric pressure.

Figure 3 also illustrates the effect of the discharge current on the abundance of the cluster ions $H^+(H_2O)_n$ $n = (1, 2,...)$. Due to increasing current, initiated by an increase in the applied voltage, the electric field in the drift region is increased and the cluster ions obtain a higher energy while moving across the drift region and consequently in collisions with neutral molecules weakly bounded water molecules are released from larger cluster complexes. Therefore the abundance of clusters $H^+(H_2O)_n$ is observed to reduce with increasing current.

.

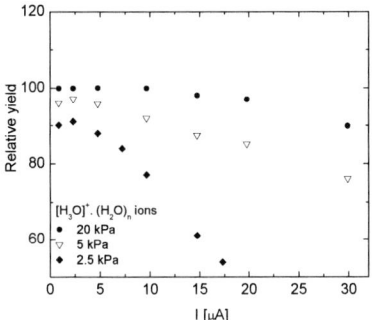

FIGURE 3. The relative yield of $H^+(H_2O)_n$ clusters n = (1, 2,...) measured in the extraction spectrum as a function of the discharge current at three different values of air pressure.

The relative yield of negative ions produced in negative corona discharge fed by moist air is shown in Figure 4 at a constant pressure of 20kPa but for two different discharge currents.

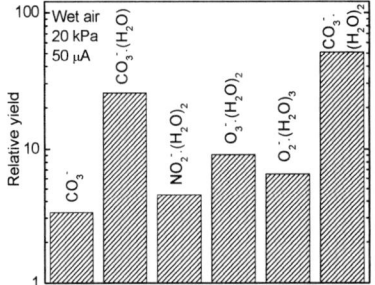 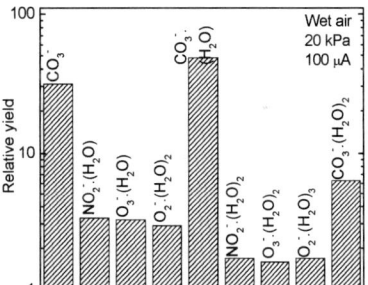

FIGURE 4. The relative yield of negative ions extracted from a negative corona discharge fed by moist air at a pressure 20 kPa for (a) 50 µA and (b) 100 µA.

The dominant anion O^- is generated at high electric fields, typical for glow region, (reduced electric field E/N≈100Td) by the dissociative electron attachment reaction

$$e + O_2 \rightarrow O^- + O \tag{10}$$

The rate constant k_{10} is an increasing function of reduced electric field E/N, where N is the air density [17]. The contribution to the total O^- flux produced by dissociative electron attachment to carbon dioxide will be marginal due to low concentrations of carbon dioxide in ambient air (300 ppm). Similarly few O^- anions are likely to be formed through the process of dissociative attachment to water molecules.

Although the concentration of carbon dioxide is so low, O^- may be converted via the fast three body ion-molecule reaction with CO_2 to form CO_3^- ions

$$O^- + CO_2 + M \rightarrow CO_3^- + M \tag{11}$$

$k_{11} = 1 \times 10^{-29}$ cm^6/sec [18].

CO_3^- anions were found to be most abundant anion in mass spectra reported by several previous authors [1, 2, 3, 4, and 7] for dry air. In the most detailed experiment reported to date, Ross and Bell [9] observed CO_3^- anions to be dominant when using a high flow rate of air in the same direction as the drift velocity. When the gas flow was in the opposite direction to the drift velocity CO_3^- anions were only dominant at low flow rates while at higher flow rate the O_2^- and CO_4^- anions were dominant. The results of this experiment [9] provide strong evidence for the influence of chemical species produced in corona discharges, even at low concentration, an phenomena we have recently also observed in corona discharges using CO_2 [7].
The anion mass spectrum is also strongly affected by content of CO_2 in the nascent air. Earlier experiments of Gravendeel and de Hoog [5] demonstrated that as the concentration of carbon dioxide was reduced (value undefined), the abundance of O^- and O_3^- ions in ambient air increases while the abundance of CO_3^- anion is suppressed in experiments performed at pressure 25 Torr [1]. However in the experiment of Gravendeel and de Hoog [9] conducted using synthetic air containing only minor traces of CO_2 O_3^- anions were dominant over a wide range of pressures.

These experiments [1,5] and those of Ross and Bell [9] also provide evidence that there is an efficient conversion of O_3^- to CO_3^-

$$O_3^- + CO_2 \rightarrow CO_3^- + O_3 \tag{12}$$

$k_{12} = 6 \times 10^{-10}$ cm^3/sec [18].
O_3^- anions are formed via a three body process, which is in competition with (11)

$$O^- + O_2 + M \rightarrow O_3^- + M \tag{13}$$

$k_{13} = 3.3 \times 10^{-31}$ cm^6/sec [17].
The characteristic reaction time for processes (11) and (13) at the pressure of 20 kPa is $1/k_{11}[CO_2][M] \approx 12$ µs and $1/k_{13}[O_2][M] = 5.2$ µs respectively. It is therefore clear that O^- ions generated in the glow region will be preferentially converted to O_3^- ions in this region.
A second source of O_3^- ions is the charge exchange process occurring during the collision of molecular oxygen ions with ozone molecules. The formation of ozone in a negative corona discharge fed by air is a well known process and its role has been discussed extensively recently [7]. The charge exchange is performed as a two-body process

$$O_2^- + O_3 \rightarrow O_3^- + O_2 \qquad (14)$$

$k_{14} = 3 \times 10^{-10}$ cm^3/sec [17].
This is active almost in the drift region where ions O_2^- are formed vi.

$$e + O_2 + M \rightarrow O_2^- + M \qquad (15)$$

The rate constant k_{15} is increasing when the reduced electric field E/N is decreasing [17]. Therefore reaction is efficient predominately in the drift region. The process (12) fed by both processes (13) and (14) is the primary source of the CO_3^- ions. The direct conversion of O^- ions to CO_3^- (11) is dominating if the concentration of carbon dioxide exceeds the value of around 1000 ppm.
In dry air a part of molecular ions O_2^- is converted to CO_4^-

$$O_2^- + CO_2 + M \rightarrow CO_4^- + M \qquad (16)$$

$k_{16} = 1 \times 10^{-29}$ cm^6/sec [5].
This is the source of ions CO_4^- observed only in spectra of negative corona discharge fed by dry air [7].
The last sink of O_2^- ions, which is active at higher discharge current, is fast charge exchange with nitric oxide NO_2.

$$O_2^- + NO_2 \rightarrow NO_2^- + O_2 \qquad (17)$$

$k_{17} = 2 \times 10^{-9}$ cm^3/sec [5].
The neutral NO_2 molecules are formed only in the glow discharge at concentration of 10 % of that of the ozone concentration. However, due to one range higher rate constant of process (17) the rate of processes (14) and (17) is comparable. Taking into account the fact that in dry air the abundances of O_3^- and NO_2^- were found to be comparable we can surmise that the role of conversion of ions O_2^- produced via (15) is most likely only marginal. Our hypothesis is supported also by the nearly same concentrations of water clusters of mentioned two molecular ions found in wet air (Figure 4.). As we have reported earlier the role of ozone and nitrogen oxides starts to be important at concentration of ozone higher than around 25 ppm [7].
As it is evident for Figure 4., the spectra of ions is changed dramatically if water molecules are present in the discharge gap at sufficient amount. Most of primary molecular ions form the clusters

$$O_2^- + H_2O + M \rightarrow O_2^-.H_2O + M \qquad (18)$$

$k_{18} = 2 \times 10^{-28}$ cm^6/sec [5]

$$CO_3^- + H_2O + M \rightarrow CO_3^-.H_2O + M \qquad (19)$$

$k_1 = 2 \times 10^{-28}$ cm^6/sec [5]

$$O_3^- + H_2O + M \rightarrow O_3^-.H_2O + M \qquad (20)$$

$k_{20} = 2.7 \times 10^{-28}$ cm^6/sec [19].

$$NO_2^- + H_2O + M \rightarrow NO_2^-.H_2O + M \qquad (21)$$

$k_{21} = 2 \times 10^{-28}$ cm^6/sec [5].

All processes (18) – (21) are followed by further bounding of water molecules in clusters. The number of molecules in cluster is considerably affected by discharge current, hence by electric field in the drift region where all of them are active. The higher voltage (current discharge) the smaller number of water molecules the clusters contain. The effect is evident from two plots in Figure 4.

Only the water clusters of primary ion CO_4^- formed in dry air absent in the spectra measured in wet air because of the efficient process active in presence of water

$$CO_4^- + H_2O \rightarrow O_2^-.H_2O + CO_2 \qquad (22)$$

$k_{22} = 2.5 \times 10^{-10}$ cm^3/sec [19].

The process (22) considerably contributes to increase in abundance of $O_2^-.(H_2O)_n$ clusters. This is higher than could be expected from low abundance of primary ions O_2^- found in dry air.

Processes producing the water clusters are of three body character; hence one can easily surmise that the pressure of air must considerably affect rate of those and therefore the abundance of individual clusters. This is evident from Figure 5.

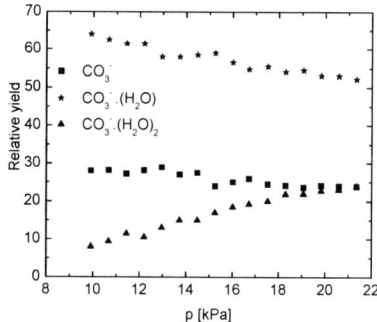

FIGURE 5. The relative yield of negative ion CO_3^- and its clusters with water molecules in wet air at the discharge current I = 100 μA.

Therefore the pressure is an important factor influencing the spectra of ions, as well as the content of water and concentration of electronegative chemical species produced in discharge gap is. The typical products are ozone and nitrogen oxides. The concentration of those is influenced by the flow rate of air through the reactor. The higher the flow rate is the lower concentration of such products is in discharge gap and consequently also the effect on mass spectra [7].

CONCLUSIONS

The clusters $H^+(H_2O)_n$ n = (1, 2, 3), were found absolutely dominant in the spectra of positive ions extracted from positive corona discharge fed by wet air in pressure range of (5 – 20) kPa. The composition of ions is considerably effected by the discharge current. Two well distinguished modes of positive corona discharge, (i) low current when only molecular and cluster ion species were observed and (ii) high pressure mode when atomic ions N^+ and O^+ were observed.

The single ion CO_3^- and their water clusters $CO_3^-.(H_2O)_n$ n = (1, 2) dominated in spectra of negative ions. The abundance of the individual group of registered ions was evidently influenced by the air pressure. Besides of this as well as the discharge current, the mass spectra is affected by the flow rate determining the concentration of electronegative chemical species in the discharge, as ozone and nitrogen oxides are.

ACKNOWLEDGMENTS

This research project was partially supported by Slovak Grant Agency VEGA 1/1267/04, ESF projects COST P9 and EIPAM. This work was supported by Science and Technology Assistance Agency under the contract No. APVT-20-007504. One of authors, (J.S.), is grateful to Prof. S. Sigmond for enabling him to perform experiments at ELION group of NTH.

REFERENCES

1. M. M. Shahin, *Appl. Opt., Supplement on Electrophotography* **3**, 106 (1969).
2. P. S. Gardiner and J. D. Cragss, *J. Phys. D: Appl. Phys.* **10**, 1003 (1977).
3. N. L. Allen, P. Coxon, R. Peyrous and Y. Teisseyre, *J. Phys. D: Appl. Phys.* **14**, L207 (1981). p.554.
4. J. Skalny, *Acta Physica Univ. Comen.* **27**, 161 (1987).
5. B. Gravendeel and F. J. de Hoog, *J.Phys.B: At. Mol. Phys.* **20**, 6337 (1987).
6. B. Held, R. Peyrous, *Czech. J. Phys.* **49**, 301 (1999).
7. J. D. Skalny, T. Mikoviny, S. Matejcik and N. J. Mason, *Int. J. Mass Spectrom.* **233**, 317 (2004).
8. K. Nagato, Y. Matsui, T. Myiata, and T. Yamauchi, *Int. J. Mass Spectrom.* **248**, 142 (2006).
9. S. K. Ross and A. J. Bell, *Int. J. Mass Spectrom.* **218**, L1 (2002).
10. M. M. Shanin, *J. Chem. Phys.* **45**, 2600 (1966).
11. F. W. Karasek, D. W. Denney and E. H. de Decker, *Anal. Chem.* **46**, 970 (1974).
12. G. E. Sprangler, K. N. Vora and J. P. Carrico, *J. Phys. E: Sci. Instrum.* **18**, 191 (1986).
13. Y. H. Chen, H. H. Hill Jr., D. P. Wittmer, *Int. J. Mass Spectrom. Ion Processes* **154**, 1 (1996).
14. M. Pavlik and J. D. Skalny, *Rapid Commun. in Mass Spectrom.* **11**, 1757 (1997).
15. V. G. Anicich, *J. Phys. Chem. Data* **22**, 1489 (1993).
16. D. L. Albritton, *Atom. Data and Nucl. Tab.* **22**, 36 (1978).
17. B. Elliason, "Electrical Discharge in Oxygen Part 1: Basic Data, Rate Coefficients and Cross Sections", Report KLR 83/40 C, Brown Boveri Forschungszentrum Baden-Dättwil, Switzerland, 1985.
18. H. Hokazono, M. Obara, K. Midorikawa and H. Tashiro, *J. Appl. Phys.* **69**, 6850 (1991).
19. M. L. Huertas, J. Fontan and J. Gonzales, *Atm. Environment.* **12**, 2351 (1978).

Study on the asymmetry of the Balmer lines

Marco Antonio Gigosos* and Manuel Ángel González†

*Departamento de Óptica, Universidad de Valladolid. 47071 Valladolid (Spain)
†Departamento de Física Aplicada, Universidad de Valladolid. 47071 Valladolid (Spain)

Abstract. A comparison between computer simulated Balmer lines and recent experimental profiles is shown in this work. The influence of different effets on the calculated profiles is discussed. Though several points must still be clarified a little deeper, the agreement between experimental and calculated profiles guarantees a good understanding of the relevant physical effects.

Keywords: Plasma diagnosis, Stark broadening, computer simulation.
PACS: 32.70.Jz; 52.70.-m; 52.65.-y; 32.60.+i.

INTRODUCTION

It is well known that Stark broadened profiles are a very useful tool to do plasma diagnosis. Besides, hydrogen lines have proven to be very useful for many plasma conditions due to their large linear Stark broadening. Though the behaviours of line widths upon density and temperature have been extensively studied and excellent agrements between experiment and calculations can be found, some noticeable discrepancies remain for the line shifts and asymmetries.

One of the more interesting and noticeable asymmetries that can be observed in the hydrogen lines is that of the H_β line. From this point of view, the comparison between H_β experimental profiles and calculated profiles may give some insight on the relative importance of different effects on the asymmetries and shifts of spectral lines.

In this sense, a recent work [1] supplies new experimental H_β data and discusses some comparisons between the asymetries obtained experimentally and calculated theoretically [2, 3]. Besides, that work considers three different asymmetry parameters that permit to charaterize the asymmetry behaviour of the experimental profiles as well as to compare easily the trends along the full profile of the experimentally obtained parameters with the ones obtained through a theoretical calculation.

Computer simulated Balmer-alpha, -beta, -gamma and -delta lines have been calculated using the technique described in this work. Due to the recent interesting available data supplied in [1] for the H_β line, this paper will be focussed on the comparison between the mentioned experimental results and calculated profiles in order to check the influence of different effects considered in the simulations. It must be remembered that computer simulations may be considered as an ideal laboratory that permit to study the influence of different independent effects in the line profile.

There is, besides, another very interesting point in the comparison between experimental and calculated H_β profiles, and not only the full widths at half maximum (FWHM), as is usualy done in plasma diagnosis. In many cases the noise under the line wings as well as the influence of neighbouring lines are not easily substracted, what

CP876, *The Physics of Ionized Gases: 23rd Summer School and International Symposium*,
edited by L. Hadžievski, B. P. Marinković, and N. S. Simonović

makes difficult to accurately stablish the FWHM of the line. Then a good description of the line center, accounting for asymmetries and shifts could provide an alternative and accurate diagnosis method for the cases in which the FWHM could not be measured with enough accuracy.

CALCULATION TECHNIQUE

As it is well known since Anderson works [4], the dipolar emission profile is obtained as the Fourier transform of the emitter dipole moment autocorrelation function :

$$I(\omega) = \frac{1}{\pi} \int_0^\infty dt \, \cos(\omega t) \, \{C(t)\} \,, \tag{1}$$

$$C(t) = \text{tr} \left[\mathbf{D}(t) \cdot \mathbf{D}(0) \right] \,, \tag{2}$$

$$\mathbf{D}(t) = U^+(t) \mathbf{D}(0) U(t) \,, \tag{3}$$

where $\mathbf{D}$ is the dipole moment of the transition under study —normalized so that $C(0) = 1$— and $U(t)$ is the time evolution operator of the system, that obeys the Schrödinger equation :

$$i\hbar \frac{d}{dt} U(t) = [H_0 + q\mathbf{E}(t) \cdot \mathbf{R}] U(t) \,. \tag{4}$$

H_0 is the unperturbed emitter hamiltonian, $\mathbf{E}(t)$ is the electric field sequence undergone by the emitter, and $q\mathbf{R}$ is its dipole moment.

In a computer simulation the perturbers —ions and electrons— movement is reproduced numerically, and the electric microfield $\mathbf{E}$ at the emitter position is calculated. Once this perturber field is calculated, the equations (4) are solved numerically and, by using equation (3), the evolution of dipole moment is obtained. This calculation is repeated a large number of times with a representative set of microfield temporal sequences. In expression (1) symbols { } mean an average of emitters in the plasma, what in our case means an average of the emitter dipole autocorrelation functions, each of them obtained from a sequence of the perturber microfield $\mathbf{E}(t)$.

In our simulations we have considered a weakly coupled, homogeneous and isotropic plasma. Then in the simulation we assume that the particles are independent and move along straight line paths with constant velocity. Those velocities satisfy a Maxwell-Boltzmann distribution. In the simulation, a finite spherical volume is assumed with the same number of ions and electrons. The emitter is placed at rest at the centre of the sphere. The relative movement between the heavy perturbers and the emitter is described using the so-called μ−ion model [5]. The reinjection method of the particles that due to their movement reach the edge of the sphere, which is the most delicate aspect of the simulation technique, is detailed in the reference [6]. This method guarantees that the statistical distributions used, homogeneity and isotropy of the particles positions, isotropy of the paths as well as Maxwellian distribution for the velocity, are steady along the simulation [7] and that there is no correlation between the outgoing and the incoming particles.

As we have considered that the perturbers are independent particles, the electric field of the ensemble of simulated ions and electrons is evaluated at the emitter position according to the expression of the Debye shielded field, in order to take into account, at least in an approximate way, the correlation effects between charged particles of different sign [8].

In the computer simulation, the perturbers movement as well as the emitter evolution are carried on with discrete *time steps* Δt. These steps are chosen so that the electric field $\mathbf{E}(t)$ may be considered static for the time duration of the step. In this case, the solution of the differential equation (4) is :

$$
\begin{aligned}
U(t+\Delta t) &= M(t+\Delta t, t)U(t) \\
&\simeq \exp\left[-\frac{i}{\hbar}(H_0 + q\mathbf{E}(t)\cdot\mathbf{R})\Delta t\right] U(t).
\end{aligned}
\tag{5}
$$

In our case, the operator $\mathbf{R}$ includes all the transitions between all the states with principal quantum number $n = 1$ to $n = 5$. Then, quadratic Stark effect is obtained in a natural way, accounting then of the main cause of the lines asymmetry and shift. Quadrupolar effects, that have not been taken into account here, would be noticeable at densities higher than those considered here.

To calculate exponential (5), it is necessary to obtain the eigenvalues and the eigenvectors of the hamiltonian. In order to easy this calculation, we have used the cartesian basis in which all the matrix elements of the three components of $\mathbf{R}$ operator are real numbers. For the diagonalization process we have used the Jacobi method [9].

For each time step in the simulation one gets the numerical expression of matrix $M(t+\Delta t, t)$ —see expression (5)— and it is then multiplied by $U(t)$ to go to next time step. Once the evolution operator $U(t)$ is obtained, the dipole autocorrelation function must be calculated —expressions (2) and (3)—.

RESULTS

In order to check the simulation results, the calculations have been focussed on the conditions of the experiments described in [1]. Here a comparison between our results and those shown in [1] will be shown.

In figure 1 a comparison between an experimental profile and a simulated one is shown. The most interesting point is the agreement of both profiles in points close to the center of the line.

The authors of [1] use three asymmetry parameters in order to characterize the shape of the experimental profiles. These parameters permit a quantitative comparison of the shapes and asymmetries of the measured and calculated profiles beyond the graphical comparison shown in figure 1. The next expressions give the asymmetry parameters used in [1], and that will also be analysed here :

$$
A_1(\Delta\lambda) = \frac{I_R(\Delta\lambda) - I_B(\Delta\lambda)}{I_R(\Delta\lambda) + I_B(\Delta\lambda)}
\tag{6}
$$

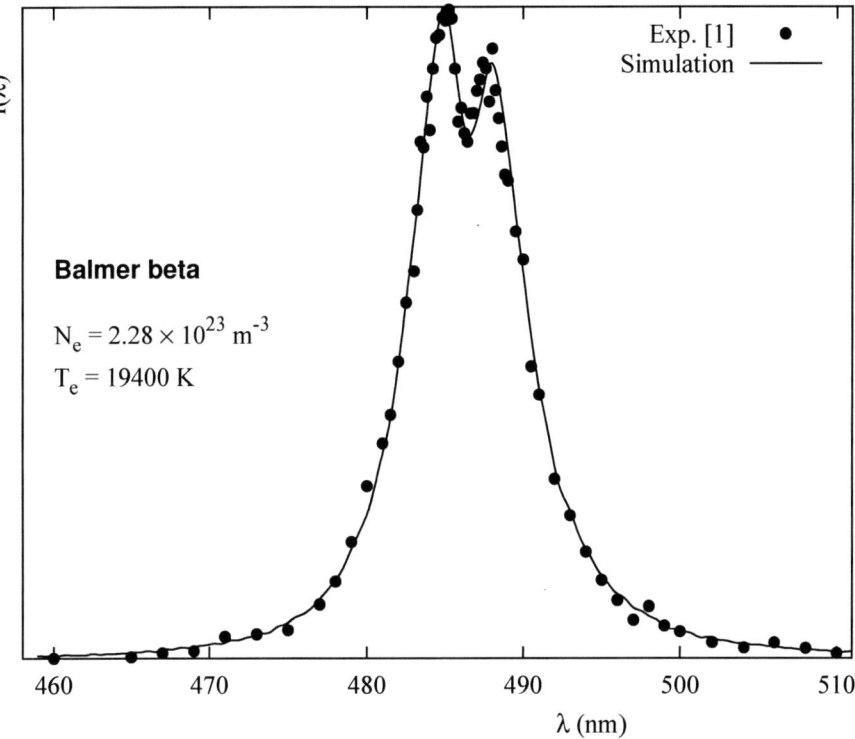

FIGURE 1. Comparison of an experimental H_β profile with the corresponding profile obtained in the simulation for the conditions of electron density, N_e, temperature, T, and plasma composition pointed out by the authors of [1].

$$A_2(\varphi) = \frac{\Delta\lambda_R(\varphi) - \Delta\lambda_B(\varphi)}{\Delta\lambda_R(\varphi) + \Delta\lambda_B(\varphi)} \tag{7}$$

$$A_3(\varphi) = \frac{\Delta\lambda_R(\varphi) - \Delta\lambda_B(\varphi)}{2} \tag{8}$$

$$\varphi(\Delta\lambda) = I(\Delta\lambda)/I_0 \tag{9}$$

In these expresions, subindexes R and B denote the red and blue sides of the line, respectively; $\Delta\lambda_R(\varphi)$ is the detuning (in this case in the red side) from the center of gravity of the line for which the normalized intensity φ takes a given value. I_0 is the average value of the line peaks. For a discussion of these parameters see section II.F and figure 4 of [1].

Figures 2 to 4 show the comparisons between the asymmetry parameters corresponding to the profiles shown in figure 1. The asymmetry parameters obtained in the simulations show good agreement with the experimental ones both in values and in their trends. Similar results are obtained for the other experimental conditions shown in [1].

Computer simulated H_β profiles have been calculated taken into account the coupling

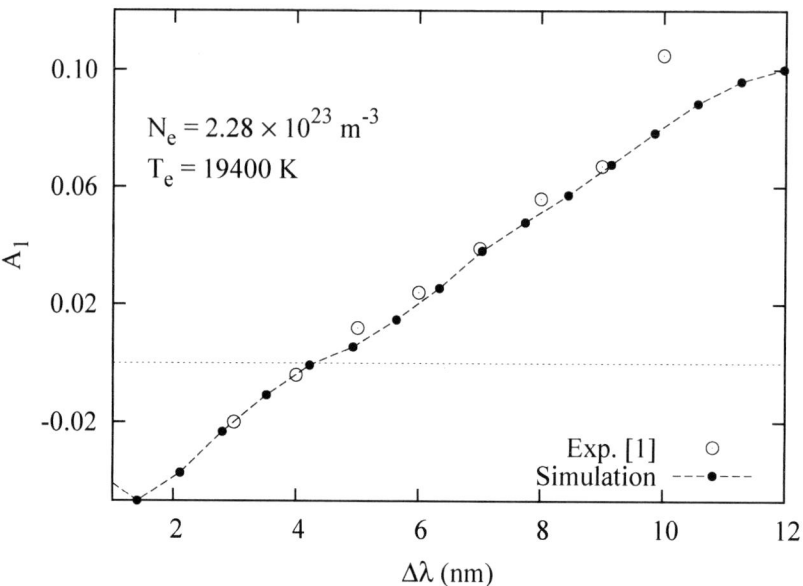

FIGURE 2. Comparison of the result corresponding to asymmetry parameter A_1 for the profile shown in figure 1.

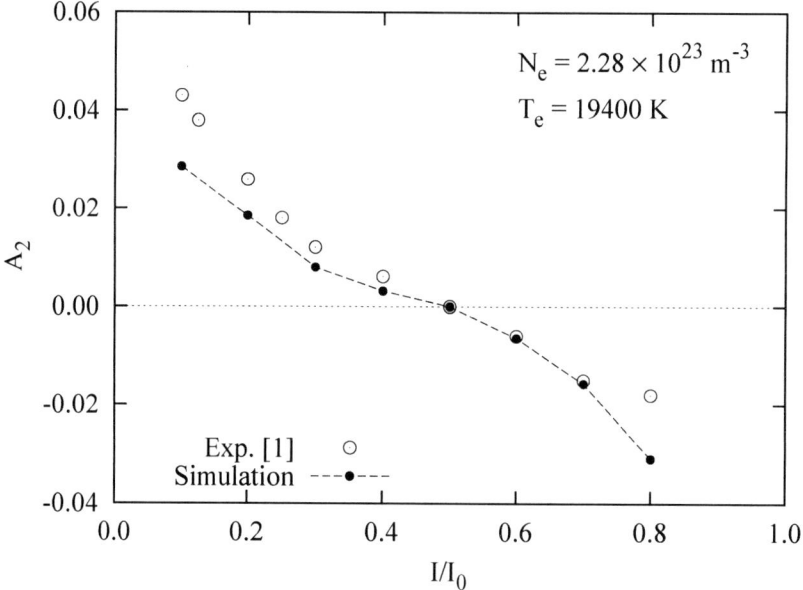

FIGURE 3. Comparison of the result corresponding to asymmetry parameter A_2 for the profile shown in figure 1.

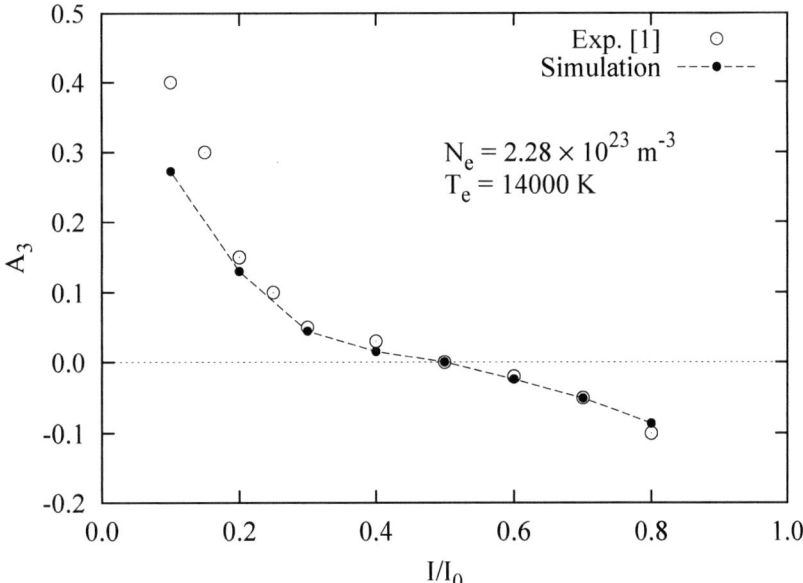

FIGURE 4. Comparison of the result corresponding to asymmetry parameter A_3 for the profile shown in figure 1.

between all the states corresponding to the levels $n = 1$ to $n = 5$. A overall good agreement between the calculated profiles and the measured ones is observed. It is interesting to point out that the calculation has been done without taking into account the effects of field gradients. These effects would only be noticeable for higher values of electron density.

Following the data given in [1], in this calculation we have considered that the ions and the electrons in the plasma have the same kinetic temperature. Our calculation technique also allows us to consider two temperature plasmas [10], and the effects due to this unbalance can also be easyly analysed, though it has not been necessary in this work.

As in other works, our simulation technique has the limitation of considering independent particles. However, in this case, as we are dealing with a neutral emitter, this limitation is not very important [8]. Take into account that the central part of the line is dominated by the dynamical shape of electric fields due to the close perturbers, and that dynamics can be well described using straight line trajectories.

Finally, the good agreement obtained in this calculation opens the possibility of improving the diagnosis tables [10] and the informatic tools [11] that permit to do plasma diagnosis comparing the full profiles, as the central part is, in general, what is easiest of being measured.

ACKNOWLEDGMENTS

The authors thank Prof. S. Djurović for supplying some unpublished experimental data and for his fruitful discussions. Extensive discussions on line asymmetries with Dr. A. Demura are also aknowledged as one of the origins of this work. This work has been partially supported by the Spanish Ministerio de Educación y Ciencia under grant ENE2004-05038/FTN and by the Consejería de Educación, Junta de Castilla y León (project num. VA032A06).

REFERENCES

1. S. Djurović, D. Nikolić, L. Savić, S. Sörge and A. V. Demura 2005 Phys. Rev. E **71** 036407-1.
2. A. V. Demura and G. V. Sholin 1975 J. Quant. Spectrosc. Radiat. Transf. **15**, 881.
3. A. Demura, V. V. Pleshakov, and G. V. Sholin, Preprint No. IAE-5349/6, I. V. Kurchatov Institute of Atomic Energy, Mos- cow, 1991, 98 p. in Russian, translation to French by Dr. M. Busquet, 1993.
4. P. W. Anderson 1949 Phys. Rev. **76** 647.
5. J. Seidel and R. Stamm 1982 J. Quant. Spectrosc. Radiat. Transfer **27**, 499.
6. M. A. Gigosos, V. Cardeñoso 1996 J. Phys. B: At. Mol. Opt. Phys. **29** 4795.
7. M. A. Gigosos and V. Cardeñoso 1987, J. Phys. B **20**, 6006.
8. E. Dufour, A. Calisti, B. Talin, M. A. Gigosos, M. Á. González, T. del Río Gaztelurrutia and J. W. Dufty 2005 Phys. Rev. E **71** 066409-1–9.
9. W. W. Press, S. A. Teukolsky, W. T. Vetterling and B. P. Flannery 1997 *Numerical Recipes un C* (New York : Cambridge University Press)
10. M. A. Gigosos, M. Á. González and V. Cardeñoso 2003 Spectrochimica Acta Part B **58**, 1489.
11. R. Žikić, M. A. Gigosos, M. Ivković, M. Á. González and N. Konjević 2002, Spectrochimica Acta Part B **57**, 987.

Optical Emission Spectroscopic Techniques for Low Electron Density Diagnostics

M.Ivković

Institute of Physics, 11081 Belgrade, P.O.Box 68, Serbia

Abstract. This paper comprises an analysis of optical emission spectroscopy (OES) techniques and results of their application for diagnostics of middle and low electron densities in low temperature plasmas. The following OES diagnostic techniques based on: 1) line merging along spectral line series, 2) use of line shapes and Stark halfwidths of hydrogen Balmer lines, 3) line shape of helium lines with forbidden components and 4) use of molecular nitrogen bandhead intensities are studied, discussed, tested and applied and in some cases ugraded for electron density measurements. The overall comparative analysis is performed also.

Keywords: Optical emission spectroscopy; electron density diagnostics

PACS: 52.25Ya, 32.70.Jz,52.70Kz

INTRODUCTION

Low and medium electron density plasmas are extensively used in analytical atomic spectroscopy as a light sources for optical emission spectroscopy (OES), for plasma processing and in various technologies, such as laser ablation, thin film deposition, creation of different nanostructures and nanocomposite etc. Therefore, the interest for plasma diagnostics is growing, and the need for improvement of old and development of new techniques is a constant task. Due to their non-perturbative nature, high spatial resolution and variety of different methods, the OES techniques are of particular interest.

In this study, the discussion will be limited primarily to the diagnostics of electron density, N_e, in low temperature plasmas by techniques based on: 1) line merging along spectral line series, 2) use of line shapes and Stark halfwidths of hydrogen Balmer lines, 3) line shape of helium lines with forbidden components and 4) use of molecular nitrogen band head intensities. For other plasma parameters measurements, more details can be found in several recent review articles and textbooks [1-4] and references cited therein. Within this work all these techniques are applied and tested in different plasma sources and theirs advantages and drawback discussed.

2. EXPERIMENT

A schematic diagram of the experimental setup is presented in Fig.1. Instead of low-current, 8A, U shaped argon stabilized arc plasma at atmospheric pressure, shown

CP876, *The Physics of Ionized Gases: 23rd Summer School and International Symposium*,
edited by L. Hadžievski, B. P. Marinković, and N. S. Simonović

in this Figure and described in details elsewhere [5, 6], a so-called "open capillary" configuration MIP (single 30 mm long Al oxide tube with 6 mm external and 2 mm inner diameter, inserted in the center of cavity) and Mini MIP torch were also used. More details about MIP sources and experimental procedure one can find in [7].

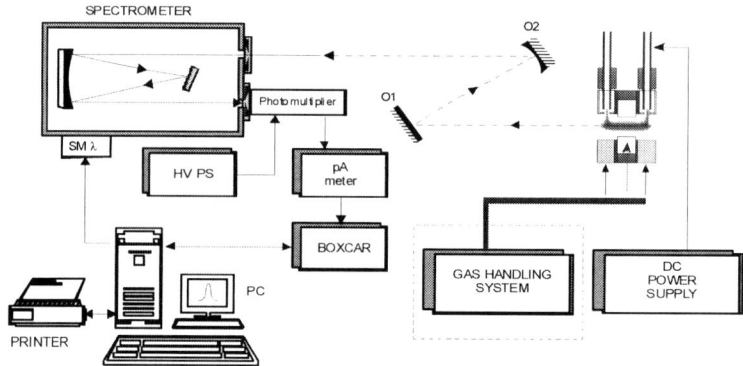

FIGURE 1. Schematic diagram of the experimental setup
SM - step motor; HVPS - high voltage power supply, O_1 - plane mirrors; O_2 - mirror focal length 50 cm

3. RESULTS

Within this section short review of OES techniques and results of theirs application for N_e diagnostics of low electron density plasmas were presented.

3.1. Series Limit

The oldest one is the so-called Inglis-Teller method for N_e determination from the line merging along spectral series [8], see also [9]. This formula relates the upper level principal quantum number, n_{max}, of the last detected line along spectral series with N_e:

$$\log (N_e + N_i) = 23.26 - 7.5 \log n_{max} + 4.5 \log z \qquad (1a)$$

where N_i is plasma ion density and z-effective nuclear charge ($z = 1$ for singly charged ions), when electron velocity is sufficiently small, i.e. when condition [9]

$$T < \frac{4.6 \times 10^5 \; z}{n_{max}} [K] \qquad (1b)$$

is fulfilled. By assuming $N_e = N_i$, the electron density may be determined by identifying the last spectral line in the series and introducing n_{max} in Eq.(1). For example, for H_ε line of the Balmer series $n_{max} = 7$. In [9] the envelope of merging lines is used to increase the accuracy of Eq. (1) for N_e determination. The Inglis-Teller method offers remarkable possibilities, especially at low N_e, where the accuracy of the method increases (larger number of lines is detected), and the error may be as low as a few percents [9]. The possibilities and difficulties in the determination of N_e are illustrated in Fig.2, see also [10]. Although neutral argon lines interfere with hydrogen lines belonging to the Balmer series, the latter are clearly discernible and the series limit may be easily determined. For $n_{max} = 10$, see Fig.2, and assuming $N_e = N_i$, the electron density in the range of 1.4 to 2.9 x 10^{21} m^{-3} (11 <n_{max}< 10) is estimated.

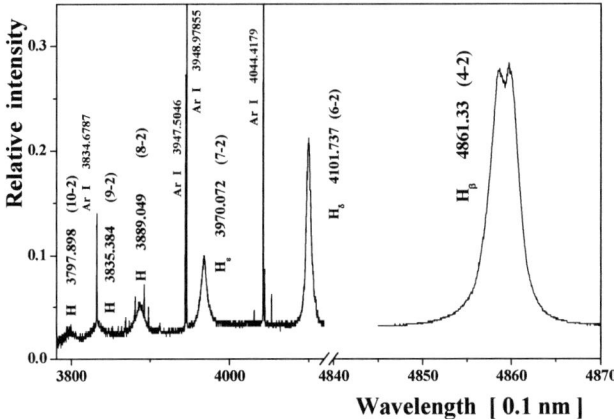

FIGURE 2. Spectra recording of some hydrogen Balmer lines at a axis of the U shaped DC argon arc at atmospheric pressure with current of 8 A

3.2. Hydrogen Balmer Beta Line

Since the beginning of the sixties, the most frequently used techniques for N_e determination is based on the half-width and shape of the hydrogen Balmer beta (H_β = 486.13 nm) spectral line. On the basis of a numerous tests we proposed [5] use of approximate formula between N_e and line Stark width W_S determined by Wiese [11],

$$N_e \ [m^{-3}] \ = \ 10^{22} \cdot \left(\frac{W_S}{4.7333} \right)^{1.49} \tag{2a}$$

with W_s previously determined from Keleher approximate deconvolution formula [12]

$$W_S \ = \ \left(W_m^{1.4} \ - \ W_{DI}^{1.4} \right)^{\frac{1}{1.4}}$$

$$W_{DI} \ = \ \left(W_D^2 \ + \ W_I^2 \right)^{0.5}$$

$$W_D \ = \ 3.58 \cdot 10^{-7} \ \lambda \left(\frac{T_g}{M} \right)^{0.5} \tag{2b}$$

where, W_m is the measured H_β HWHM, while W_I and W_D are the instrumental and Doppler broadening HWHM (all in 0.1 nm units), T_g – gas temperature and M is the ion mass.

After analysis of a numerous programs for N_e determination from the fitting of whole profiles, a new one is made [13]. This program enables use of: area or intensity

303

normalization and excluding a different parts of experimental profiles during comparison with one of three theories. Other manipulation of input data such as smoothing, shift, export etc. and graphic presentation were enabled in Windows version, which main screen was shown in Figure 3a. Program was tested with most accurate experiments with independatly determined N_e in range of two order of magnitude and some of the obtained data are shown in Figure 3 and summarized in Table 1.

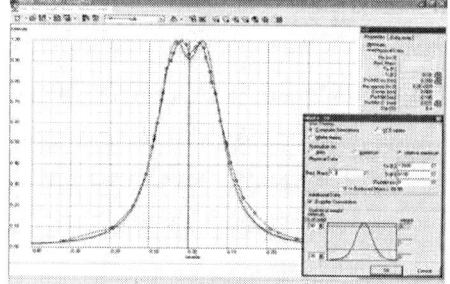 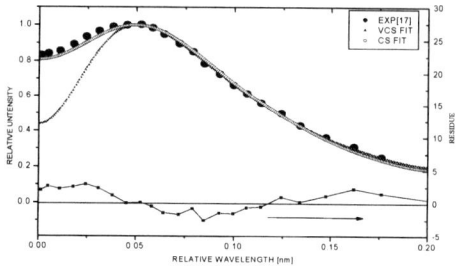

FIGURE 3. a) Illustration of windows main screen of "Hbeta" program
b) Example of comparison of experimental and theoretical data, se Table 1.

TABLE 1. Experimental conditions and results of comparison.

Figure	3a	3b
Reference	[14]	[11]
Reduced mass μ	1.3	0.5
Electron temperature [K]	13500	13400
Ion temperature[K]	8100	13400
Instrumental FWHM[nm]	0 unknown	0.03
Experimental N_e[m^{-3}]	8.3 E+20	8.3 E+22
Approximate N_e Eqs.(2) [m^{-3}]	8.01 E+20	7.85 E+22
VCS FIT N_e [m^{-3}]	8.77 E+20	8.17 E+22
MMM FIT N_e [m^{-3}]		7.66 E+22
CS FIT N_e [m^{-3}]	7.98 E+20	7.41 E+22

3.3. Higher Members of Hydrogen Balmer Series

The electron density can be detrmined from the line profiles of higher members of spectral series. It is already known that these lines are for the same N_e broader than H_β and influence of the other broadening mechanism smaller. The use of line wings of these lines [15] is troublesome due to the existance of neithboroug atom or molecular lines. Therefore, another method for N_e determination that requires, for example, comparisons of theoretical and experimental half width only would be more appropriate. To achieve this goal, Bengston et all. [16] used impact approximation for electrons and quasistatic approximation for ions to calculate hydrogen $H_6 - H_{12}$ Balmer lines Stark widths. With theoretical data of line full halfwidth, summarized in Table 2, N_e from the lines presented in Figure 2, is evaluated using formula (3)[16]:

TABLE 2. Data [16] and results [10] for several Balmer lines measured in U shaped Ar arc (see Fig.1): $\alpha_{1/2}^{n}$ – normalized line width, W_m – measured FWHM and W_g –contribution of Gausian part.

Transition	$\alpha_{1/2}^{n}$ [13]	W_m [nm]	W_g [%]	N_e [m^{-3}]
6 – 2	0.150	0.73	5.6	2.71 x 10^{21}
7 – 2	0.184	0.86	4.7	2.56 x 10^{21}
8 – 2	0.283	1.30	3.1	2.49 x 10^{21}
9 – 2	0.345	1.56	2.5	2.43 x 10^{21}
10 – 2	0.458	2.30	1.7	2.84 x10^{21}

$$N_e \; [m^{-3}] = 8.0 \; x \; 10^{18} \left(\frac{w \, [\text{Å}]}{\alpha_{1/2}^{n}} \right)^{3/2} \quad (3)$$

With the instrumental FWHM of 0.03 nm, and assuming gas temperature $T_g \approx T_{exc}$, (where electron excitation temperature T_{exc} = 9000 K is determined from the Boltzmann plot of Ar I spectral lines), the Gaussian w_g part of line profiles is estimated, see Table 1. The comparison of the Gaussian part with the total line width shows that it may be neglected. The comparison of the N_e = 2.54 x 10^{21} m^{-3} determined from the H_β profile in Fig.1. and values 1.4 to 2.9 x 10^{21} m^{-3} (11 <n_{max}< 10) estimated by Eq.1a, with those in Table 2 shows that widths of the H_6 - H_{10} lines can be used with a reasonable accuracy $\pm$ 12 - 15% for N_e diagnostics.

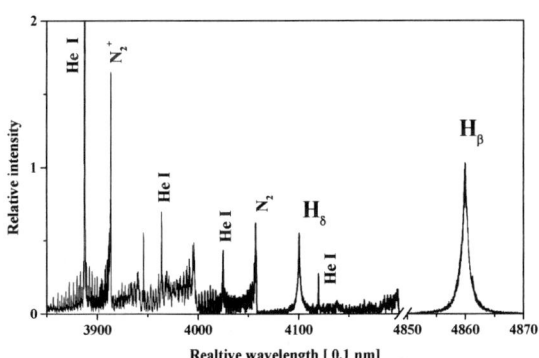

The interference with molecular bands causes the main difficulty for application of these lines, see example in Figure 4 . Besides facts that the end - on recorded spectral line shapes of helium lines and Balmer lines are distorted due to superposition of line profiles emitted from layers with different plasma parameters, the determined electron densities from H_β: N_e = 3.1; H_δ: N_e = 3.9; He I 447.1: N_e = 3.1; and HeI 492.2 nm (see Fig 5b): N_e = 3.4 x 10^{20} m^{-3} are in a good agreement.

FIGURE 4. Spectra recordings from open capillary MIP at atmospheric pressure; Power input 100 W and flow rate of He 0.7 l/min

3.4. Helium lines with Forbidden Component

In the case of helium plasmas, electron density can be determined by using the shape of some visible He I lines with forbidden components. The complex structures of these lines are extensively studied both theoretically and experimentally.The inclusion of ion dynamic effects in theoretical descriptions of helium lines with forbidden components greatly improves agreement with experimental results. However, the discrepancy of predicted forbidden line intensity with the experiment [17] still remains for all three theoretical approaches. This is a main reason why experimentally determined formulas relating N_e with the parameters of helium lines with forbidden component such as F/A - forbidden (F) to allowed (A) line maximum intensity, D/A - deep (D) i.e. minimum intensity between forbidden and allowed line and A line intensity) and s - wavelength separation between F and A determined by Czernichowski and Chapelle [18] are mainly used. Due to the fact that parameter s is not sensitive to distortion of the strong allowed line due to the possible presence of a self-absorption effect, the following relation was used in this work

$$\log N_e \, [m^{-3}] = 23.056 + 1.586 \log (s[nm] - 0.156) + 0.225 \, [\log (s - 0.156)]^2 \qquad (5)$$

where 0.156 nm in Eq.(6) is the separation between unperturbed F and A line.

This procedure can be applied for low N_e determination, but for more precise F and s determination, with a different amplification when recording forbidden or allow component must be used, see Fig.5a. It should be pointed out that great care must be also taken when using He I lines with forbidden components for the determination of N_e lower than few times 10^{14} [cm^{-3}]. At these densities, the low intensity of the forbidden component (less than few percents of the allowed one), may be masked by noise or in presence of nitrogen by molecular lines from 6-8 and 8-10 bands of the first negative system of N_2^+.

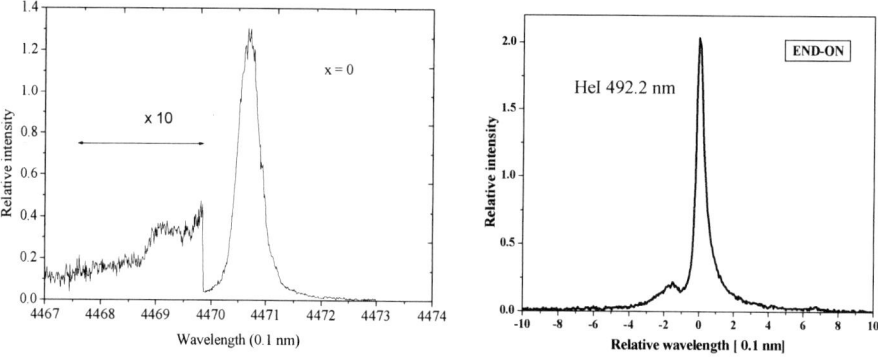

FIGURE 5. a) Illustration of line shape recording with different amplification of the signal. Line emitted from the center of the Mini MIP at height 1mm from the torch orriface. The flow rate of He was 0.6 l/min, and He+ 3% H_2 0.2 l/min through the outer and inner capillary. b) Same as in Fig.4.

3.5. Intensity of the N_2 and N_2^+ Molecular Band Heads

According to [19-21], the electron density in nitrogen and nitrogen/He plasmas can be determined from the intensity of N_2 second positive system (SPS) band head (0-0) at 337.1 nm (C $^3\Pi_u \rightarrow$ B $^3\Pi_g$) and N_2^+ first negative system (FNS) band head of (0-0) at 391.4 nm (B$^2\Sigma_u^+ \rightarrow$ X$^2\Sigma_g^+$).

Namely, by using the simplified kinetic model of the N_2 (C$^3\Pi_u$) state, see e.g. [20] and assuming that the steady-state population of the upper energy state is equal zero, a linear relation between the 337.1 nm band intensity and the electron density at constant pressure and constant electric field may be obtained. This relation is confirmed with 5% accuracy in a volumetric near field microwave plasma [21].

A similar discussion can be applied to the band head of FNSystem of nitrogen ion [20]. Due to an additional excitation process for the upper level population, the intensity of the 391.4 nm line has quadratic dependence upon electron density, i.e. $I_{(391.4\ nm)} \sim N_u \sim A\ N_e^2 + B\ N_e$, where N_u is the population of the $B^2\Sigma_u^+$ state, while A and B are constants, which must be independently determined. According to the authors [19], this method can be applied even in plasmas with non-Maxwellian electron energy distribution.

The band head intensity method is tested in MIP at atmospheric pressure with power input of 100 W and at constant He flow rate of 0.7 l/min. The radial distributions of the band head intensities and hydrogen Balmer beta line shapes – H_β are obtained by Abel inversion procedure from lateral side-on recordings. From the determined H_β line shapes electron densities are calculated using approximate experimental formula, see Eq.2. Finally, dependence of the molecular nitrogen band head intensities versus log N_e for different values of radius are presented in Figure 6.

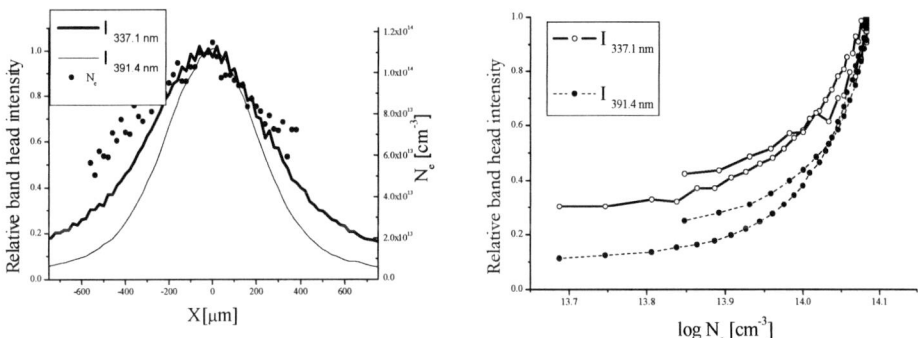

FIGURE 6. a) Radial dependence of the molecular nitrogen band head intensities and N_e and
b) dependence of the band head intensities versus log N_e for different values of radius,
in MIP at atmospheric pressure; Power input 100 W and flow rate of He 0.7 l/min

It is evident that the application of this method for N_e determination requires further elaboration and experimental verifications in different plasma sources and gas mixtures. It should be noticed that the calibration of log N_2 band intensity vs. log N_e determined using another independent diagnostic technique enables slope parameter determination. The extrapolation of intensity vs. N_e plot may be used for lower N_e plasma diagnostics.

4. CONCLUSIONS

The use of the Balmer beta line is still a best method till the low electron densities, when the influence of other broadening mechanism and fine structure splitting became important. In that case other analyzed methods can be favorable.

The techniques based on line series merging is still attractive especially at very low densities, but in cases when interfering lines are weak or absent.

The use of Stark FWHM of higher members of Balmer series is simple and successfully applied even in diagnostics of tokamak plasmas [22], but is always accompanied with relatively large uncertainty of $\alpha_{1/2}^n$ parameter; see Table 1in [16]. In order to increase accuracy of the use of higher members of Balmer series numerical procedure for electron density determination by fitting whole experimental profiles of hydrogen Balmer series with theoretical ones is under development.

The helium lines with forbidden components can be used in a very broad range of electron densities and even at lower than 10^{15} cm^{-3}. However, for lower electron density diagnostics the more complicated procedure must be used and further theoretical studies will be welcomed.

It should be stressed out that the molecular nitrogen band heads intensities offer a greatest possibility for diagnostics of very lower densities, but both theoretical and experimental studies in different plasma conditions are needed.

ACKNOWLEDGMENTS

This work is partially supported by the Ministry of Science and Environmental Protection of Serbia under projects 141032 and by Spanish Ministry for Education and Science under project ENE2004 – 05038.

REFERENCES

1. Q.Jin, Y.Duan, and J.A.Olivares, *Spectrochim.Acta* **B 52**, 131-161 (1997).
2. W.Locchte-Holtgreven, *Plasma Diagnostics*, American Institute of Physics, New York 1995.
3. J.Ropke, P.B.Davies, M.Kaning and B.P.Lavrov, Diagnostics of non-equilibrium molecular plasmas using emission and absorption spectroscopy, in *"Low temperature plasma physics - Fundamental Aspects and Applications"* Eds. R.Hippler, S.Pfau, M.Schmidt, K.H. Schonbach. Wiley-VCH, Berlin, N.Y. Toronto, 2001.
4. H.R.Griem, *Plasma Spectroscopy*, Academic Press, New York 1964.
5. M.Marinković, T.Vickerrs, *J.Appl.Spectr.* **25**, 319-324 (1971).
6. M.M. Kuzmanović, M.S. Pavlović, J.J. Savović, M.Marinković, *Spectrochim. Acta* **B 58**, 239-248 (2003).
7. M.Ivković, S.Jovićević and N.Konjević, *Spectrochimica Acta* **B 59**, 591-605 (2004).
8. D.Inglis and E. Teller, *Astrophys.J.* **90**, 439 (1939).
9. C.R.Vidal, *J. Quant. Spectrosc.Radiat. Transfer* **6** 461- 477 (1966).
10. M.Ivković, S.Jovićević and N.Konjević, *BPU: Fifth General Conference of the Balkan Physical Union*, August 25-29, Vrnjačka Banja, Serbia and Montenegro, Book of abstracts, p.220, (2003).
11. W.L. Wiese, D.E. Kelleher and D. R Paquette, *Phys. Rev.* **A, 6** 1132-1153(1972).
12. D.E. Kelleher, *J. Quant. Spectrosc.Radiat. Transfer* **25**, 191 (1981).
13. R.Zikić, M.A.Gigosos, M.Ivković, M.A.Gonzalez and N.Konjević, *Spectrochim. Acta* **B 57**, 987-998 (2002)
14. C. Thomsen and V. Helbig, *Spectochim. Acta,*. **B 46**, 1215-1225 (1991).
15. C.R.Vidal, Stark broadening of the Pashen lines, Proc. 7th Int. Conf. Phenomena in Ionized Gases, Gradjevinska knjiga Publishing, Belgrade (1965) 168 –173.
16. R.D.Bengtson, J.D.Tannich and P.Kepple, *Physical Review* **A 1**, 532-533 (1970).
17. H. Richter and A.Piel, *Quant. Spectrosc.Radiat. Transfer* **33**, 615 (1985).
18. A.Czernichowski and J.Chapelle, *J. Quant. Spectrosc.Radiat. Transfer* **33**, 427–435 (1985).
19. K.Behringer and U.Fantz, *J.Phys.D: Appl. Phys.* **27**, 2128– 2135 (1994)..
20. S.D.Popa, *J.Phys.D: Appl.Phys.* **29**, 416-418 (1996).
21..Exton R.J., Jeffrey Balla R., Hering G.C, Popovic S and Vuskovic L. *Proc. 34th AIAA Plasmadynamics and Lasers Conference*, 23–26 Jun 2003, Orlando, Florida, USA, paper 4181.
22. L. Welch, H. R. Griem, J. Terry, C. Kurz, B. LaBombard, B. Lipschultz, E. Marmar and G.McCracken, Phys. Plasmas 2 (1995) 4246 – 4251 .

Laser-induced plasma spectroscopy: principles, methods and applications

Violeta Lazic, Francesco Colao, Roberta Fantoni, Valeria Spizzichino, Sonja Jovicevic*

ENEA, FIS-LAS, V. E. Fermi 45, Frascati (RM), Italy
Tel. 06 94005885, Fax 06 94005400, lazic@frascati.enea.it
*Institute of Physics, 11080 Belgrade, Pregrevica 118, Serbia

Abstract. Principles of the Laser Induced Plasma Spectroscopy and its advances are reported. Methods for obtaining quantitative analyses are described, together with discussion of some applications and the specific problems.

Keywords: LIBS, LIBS, laser, plasma, quantitative, matrix, double-pulse
PACS: 52.70.-m

INTRODUCTION

Laser Induced Plasma (or Breakdown) Spectroscopy - LIPS (or LIBS) is a powerful tool for rapid in-situ analyses of solid, liquid or gaseous samples in different surroundings transparent for the laser radiation used [1]. Spectrally resolved plasma emission generated by laser ablation of solid targets or by breakdown in medium (gaseous or liquid) is exploited to determine qualitatively the elemental sample composition. Development of wide range, high-resolution spectrometers such as Echelle or compact spectrometer arrays, allows for multi-elemental measurements also by applying a single laser shot. Concentration of the elements in the sample could be also retrieved by applying an appropriate calibration on matrix-similar reference materials, and in some cases, also by Calibration-Free approach [2]. Methods for LIBS quantitative analyses for fluctuating plasma parameters and for variable sample ablation rates have been also developed [3]. A wide range of LIBS applications has been demonstrated in past two decades [4]. The examples include: on-line monitoring in steal industry, control of materials and leakages in power plants, environmental analyses (soils, waters), analyses of the objects related to Cultural Heritage, protection against terrorism (explosive identification), planetary exploration (surface characterization) etc. Significant improvement of LIBS sensitivity, which is generally in order of ppm, has been obtained by applying double-pulse laser excitation, feasible also by using a single laser source [5]. In the case of underwater analyses, the Limits Of Detection (LOD's) are so reduced down to ppb level [6].

CP876, *The Physics of Ionized Gases: 23rd Summer School and International Symposium*,
edited by L. Hadžievski, B. P. Marinković, and N. S. Simonović

QUANTITATIVE ANALYSES

Common approach for obtaining quantitative material analyses by LIBS is based on use of the calibration curves generated on reference samples. The main limitation of this approach is related to the so-called matrix effect i.e. strong influence of the material composition on the produced plasma emission and on the final analytical results. Due to this problem, the initial calibration should be performed with standards having matrices similar to those of the samples to be analyzed, and the accurate concentration values usually can not be retrieved on a priori unknown samples. Even for well matched standard matrices, different sources of analytical errors could be present, such as slight variation of the experimental conditions and differences in the coupling of the laser radiation with the samples. In order to improve LIBS analytical accuracy, different kinds of the LIBS signal normalization have been proposed, mainly to compensate variability in the ablation rate in the case of solid samples. The examples include lines intensities normalization on the acoustic signal produced by the laser-induced shock waves [7], on the continuum plasma emission [3], on the overall plasma emission [8] and on a line intensity of some major matrix element [9]. The latter type of normalization is often used in the relative measurements of the element concentrations, which absolute values can be retrieved only for well characterized matrices with a fixed content of the element used for the normalization. Typically, LIBS analyses of solid samples after calibration on well-matched standards have uncertainty 5-10% and limits of detection are in ppm range.

Another approach for retrieving the element concentrations by LIBS is the so-called Calibration-Free (CF) procedure [10], which is based on a simultaneous detection of all major elements in the sample and the assumption of plasma in Local Thermal Equilibrium (LTE). The plasma temperature T can be determined from Boltzmann plot applied on unsaturated lines of one plasma species, and usually atomic lines from one of the major sample constituent are used.

For the plasma in LTE, neglecting re-absorption of the plasma emission, the spectrally integrated line intensity corresponding to the atomic or ionic transition between energy levels E_k and E_i of the element α is given by:

$$I^{ki}{}_\alpha = a'_\alpha N_\alpha \frac{g_k A_{ki} e^{-E_k/kT}}{U_\alpha(T)} \tag{1}$$

where a'_α is a constant depending on the experimental conditions, N_α is species number density in plasma, g_k is level degeneracy, A_{ki} is the transition probability, k is Boltzmann constant and $U_\alpha(T)$ is partition function.

In typical LIBS plasmas, during the observation window, the presence of the second and higher ionization stages could be neglected. The number density ratio of neutral (N_α^I) and first ionized species (N_α^{II}) of each element α in the plasma depends on plasma temperature and electron density N_e through Saha equation:

$$\frac{N_\alpha^{II}}{N_\alpha^I} = \frac{1}{N_e} \cdot \frac{U_\alpha^{II}(T)}{U_\alpha^I(T)} B(kT)^{3/2} e^{-\frac{E_\alpha}{kT}} \equiv f_{2\alpha}(N_e, T) \tag{2}$$

where: $U_\alpha^I(T)$ and $U_\alpha^{II}(T)$ are partition functions of atomic and the first ionization stage respectively; E_∞ is the effective ionization energy in the plasma surrounding; B is a constant with a value of 6.05E+21 cm^{-3}.

Knowing the plasma temperature, its electron density could be measured from Stark broadening of the emission lines or determined from the population ratio of neutrals and ions of one element, through eqns. *(1-2)*.

Other elements in the plasma could be determined from the emission of one neutral (or ionic) line *(eqn. 1)* and the number density of the corresponding ions (i.e. atoms) is then calculated from eqn. *(2)*. If the sample stochiometry is conserved in the LIBS plasma, the concentration of the element α in the sample may be written as:

$$C_\alpha = b_\alpha{}'(N_\alpha^I + N_\alpha^{II}) \tag{3}$$

Where $b_\alpha{}'$ is an experimental constant for one element, which is assumed to be equal for all the elements in the CF normalization:

$$\sum_\alpha C_\alpha = 1 \tag{4}$$

The CF procedure requires that all the major sample elements are simultaneously detected by LIBS and was successfully applied for analyses of aluminum and precious alloys [10]. However, there are elements that are difficult to be measured by LIBS, such as sulphur and chlorine in presence of oxygen, and elements that suffer from interference from the surrounding atmosphere, such as oxygen for the measurements in air. If such elements are present in major quantities inside the sample, the normalization *(eqn. 4)* could lead to large concentration errors. Whenever the samples contain oxides which form is known for the major matrix elements, oxygen contribution could be calculated [11] and inserted in *(eqn. 4)*. Further limitations of CF-LIBS accuracy are due to uncertainty of available databases for less studied elements, which in particular could contribute to analytical errors through the calculation of partition functions. Finally, the CF procedure assume the preservation of the sample stochiometry in the gas phase, which is often missing for some classes of materials, as for example copper alloys containing the zinc.

Beside the two above methods for obtaining quantitative LIBS results, a mixed approach including both the initial calibration and CF procedure have been also reported [11]. Here, the initial calibration was used to retrieve the coefficients b'_α for each analyzed element, which were supposed to be different from one element to another. Then the CF normalization was applied following the revised formula:

$$\sum_\alpha C_\alpha / b'_\alpha = 1 \tag{5}$$

In this way, the uncertainties in the atomic data bases, the matrix effect, an incomplete plasma stochiometry, as well as influence of different time/space species distribution in the plasma, are partially compensated.

However, in many cases a simplified calibration procedure is used and reported in the literature, where the element concentration in the sample is simply considered proportional to the intensity of the chosen analytical line (eqn. 1). Such calibration coefficients could be considered valid only for a certain set of the plasma parameters even at the fixed experimental conditions. In some cases, the plasma parameters on the examined sample differs significantly from those obtained on the standards, and the element concentrations retrieved from the calibration graphs should be corrected for a factor that was derived in [3] both for the atomic and ionic analytical lines. Otherwise, the calibration should include both the atomic and ionic species concentration in the plasma [8] (see also eqn. 3).

APPLICATIONS

Analyzes Of Soils and Sediments

Characterization of sediments and soils is particularly important for geological studies, for environmental protection (example - determination of pollution by heavy elements) and in agriculture. The use of LIBS for soil analyses was recently proposed also for planetary exploration, in particular for the next missions to Mars [12].

Spectra from the soils/sediments are particularly rich as containing the emission lines from numerous elements (Fig. 1). Due to spectral interferences, a number of the useful analytical lines for some elements are limited. While for the major sample constituents the resonant transitions should be possibly avoided, the strong, resonant transitions are useful for detecting the trace elements, such as copper line at 328 nm. Often, the soil/sediment samples are first pressed into pellets and then analyzed by LIBS. For obtaining the quantitative results a laborious calibration for each element, by using the certified samples must be first performed. Due to a great variability of existing soil/sediment matrices, of their granulometry and other physical properties, very different signal intensities could be observed under the same experimental conditions [13]. One approach for LIBS analyses on a wide range of the soil samples is to include into calibration very different types of the soil standards and to apply some kind of normalization [3,7-9,14] which partially compensates the variable ablation rates. Also, if the calibration curves include both ionic and atomic species, one of them could be calculated from Saha equation after determining the plasma temperature and the electron density, better overall accuracy could be achieved on a wider span of soil matrices. In this way, the analytical accuracy better than 20% was obtained for most of the elements in various soil samples [8].

LIBS was also applied for scanning the vertical element distribution on sediments slices, directly analyzed after drying [15]. LIBS measured stratigraphical distribution of the elements related to the past bio-activity in waters (Mn and Ba) result well correlated with the available data for the global temperature in the past and it was also possible to determine an average sedimentation rates for the periods after the last glaciations.

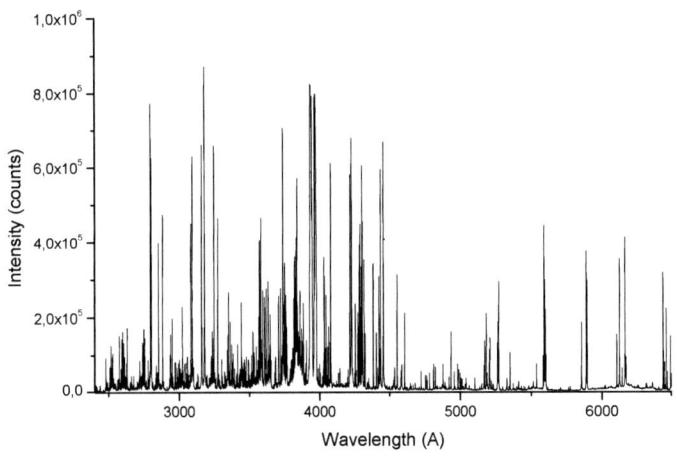

FIGURE 1. Example of the LIBS spectra from a soil sample.

Applications for Cultural Heritage

Determination of material composition has a great importance in the field of Cultural Heritage, both for cataloguing and conservation purposes. The LIBS applicability in-situ without sample preparation, its multi-elemental analytical capability and a low invasiveness, makes it one of the most interesting techniques for characterization of the historical and artistic objects [3,11,16]. This technique also allows for a micro-stratigraphical measurements [3,16], which are important for characterization of multi-layered structures (example paints, binding materials and protective layers), including outer, degraded layers formed due to aging of the artifacts. In the latter case, the LIBS stratigraphical analyses not only helps to establish the optimal restoration procedure, but could be also implemented to monitor a cleaning process [3], thus to avoid an excessive material removal from the surface.

Accurate LIBS analyses of unknown archeological material are not always feasible because of matrix effects and lack of certified standards with similar composition. However, only qualitative or semi/quantitative LIBS analyses, which give information also about minor and trace elements, are often sufficient to recognize a material, to determine its age and provenience, and to establish the manufacturing process [11,16].

As for example, we analyzed by LIBS different parts of a fragment (finger) belonging to one roman sculpture from Frascati Museum (Italy) [11]. The measured element concentrations are reported in Tab. 1. From these analyses, the sculpture results clearly made of leaded tin bronze, as expected for the Roman Republic period. The internal part of the sculpture, which shows a gray color and porous aspect probably due to an incompletely baked core, results to be rich of Sn and Pb, where Pb was added to increase fluidity of molten material. The sculpture surface, which is compact and very dark, has a higher amount of Cu while Sn and Pb are significantly reduced with respect to the internal part. Lead also could be reduced on the surface due to oxidation and corrosion processes, which cause the disappearance of the lead grains characteristic for polyphase alloys. On the external parts of sculpture, also carbon was detected (at 247.9 nm), attributed to the copper carbonate patina. A higher presence of Si, Ca and Al on the sculpture surface can be related to the manufacturing process where moulds of fired clay were used. Vanadium, detected only on the external parts of the sculpture, surely origins from the material used for the mould. Such relatively high V concentration indicates probable sculpture provenience from Middle East, where soil (used for moulds) is generally rich of this element. Zinc was detected in traces on the sculpture surface, but its higher presence on the nail area may also indicate residuals of a treatment with gold like plating, that was common on some parts of sculptures (lips, eyes etc) from this period.

TABLE 1. Element concentration measured by LIBS on different parts of the Roman sculpture fragment

Element	LIBS measured concentration (%wt)		
	Internal	External	Nail
Cu	47.5	67.9	60.7
Fe	0.32	0.18	0.26
Si	0.14	2.2	3.2
Pb	26.9	13.4	11.8
Sn	18.1	12.6	14.5
Ni	0.027	0.020	0.018
Zn	0	0.0042	0.018
Ca	1.7	3.25	4.6
Al	0.062	2.3	1.1
V	0	0.43	0.22
Total	94.7	99.5	95.0

Underwater analyzes

One of the growing requests for LIBS technique regards in-situ analyses of water solutions, which is important for environmental control, for geological and marine researches, for study of chemical reactions in the liquid phase etc. In presence of water-air interface, for most of the measured elements, LOD's obtained by ablation of steady-state water surface [17], are in order of 1-100 ppm. Improvement of the detection limit has been obtained by ablating a surface of liquid jets [18], where the splashing effects were reduced. However, in some cases direct measurements on bulk water are required, so in absence of liquid-gas interface. Examples include detection of leakages in industrial plants, measurements of biological activity, determination of nutrients and pollution in deep waters, characterization of sub-glacial waters, etc.

Focusing a short laser pulse with sufficient energy into a liquid, a dielectric breakdown takes place generating plasma. Rapid heating of the liquid is followed by its explosive expansion and formation of a gas bubble. Intensity of the plasma emission produced in bulk water is generally lower than at water-air interface due to several factors that include: water absorption of the laser and plasma emission and their scattering on suspended particles and micro-bubbles, radiation shielding by the high density plasma and fast quenching in the dense medium. Furthermore, the emission lines are strongly enlarged by a high electron plasma density and all the mentioned effects lead to a relatively poor signal in single pulse LIBS measurements.

Much better analytical performances of underwater LIBS could be obtained by applying a double-pulse laser excitation technique [6,19-21]. In such case, the first laser pulse produces a cavitations bubble in water, while the second, probing pulse excites the plasma inside the bubble. After the second laser pulse, a relatively intense and narrow spectral emission can be observed due to gaseous state inside the bubble and consequently reduced plasma quenching. As a result, the LOD's bellow 1 ppm were obtained for some elements directly analyzed from bulk waters [6,19]. An additional LOD improvement (up to one order of magnitude) could be obtained by a proper signal post-processing, as it was demonstrated recently [6]. The maximum LIBS signal could be achieved when the second pulse hits the sample when the gas bubble produced by the first pulse reaches its expansion maximum [20].

Underwater LIBS was also applied for recognition of the archeological materials [21] with aim to catalogue and recuperate only valuable finding from the sea bottom. Recently, feasibility for LIBS analyses of submerged sediments has been also demonstrated [22], which is important for fast mapping of the accumulation of heavy metal pollutants, as well as for geological and biological screening measurements.

CONCLUSIONS

Although the LIBS technique is not extremely sensible and not very accurate, its recent strong development might be attributed to a high versatility, to the applicability in-situ with low invasiveness and in real time, as well as to a possibility for the instrument miniaturization. Presently, only a few commercial LIBS instruments exist in a market, but they still require highly skilled operators and a scientific support. On the other side, a great number of the LIBS systems have been already developed and implemented for the specific user applications. Further LIBS developments also regard an establishment of a standard procedure for accurate quantitative analyses, valid for a wide range of the materials, as well as a development of user-friendly instrument software with reduced false responses relative to the material identification.

REFERENCES

1. L. J. Radziemski, "From LASER to LIBS, the path of technology development", Spectrochim. Acta B, 57, 2002, pp. 1109-1113.
2. V. Lazic, L. Caneve, F. Colao, R. Fantoni, L. Fornarini, V. Spizzichino, "Quantitative elemental analyses of archaeological materials by laser induced breakdown spectroscopy (LIBS) – an overview", SPIE Vol. 5857, 2005, (Eds. R. Salimbeni and L. Pezzati), pp. 58570H 1-12.
3. V. Lazic, R. Fantoni, F. Colao, A. Santagata, A. Morone, V. Spizzichino, "Quantitative Laser Induced Breakdown Spectroscopy analysis of ancient marbles and corrections for the variability of plasma parameters and of ablation rate", J. Anal. Atom. Spectrom. 19, 2004, pp. 429-436.
4. J. D. Winefordner, I. B. Gornushkin, T. Correll, E. Gibb, B. W. Smith, N. Omenetto, "Comparing several atomic spectrometric methods to the super stars: special emphasis on laser induced breakdown spectrometry, LIBS, a future super star", J. Anal. At. Spectrom., 19, 2004, pp. 1061-1083.

5. F. Colao, V. Lazic , Fantoni, S. Pershin, "A comparison of single and double pulse laser induced breakdown spectroscopy of aluminium samples", Spetrochim. Acta B 57, 2002, pp. 1167-1179.
6. V. Lazic, F. Colao, R. Fantoni, V. Spizzichino "Laser Induced Breakdown Spectroscopy in water: improvement of the detection threshold by signal processing", Spectrochim. Acta B, Vol 60, 2005, pp. 1002-1013.
7. C. Chaleard, P. Mauchien, N. Andre, J. Uebbing, J. L. Lacour, C. Geertsen, "Correction of Matrix effects in Quantitative Elemental Analyses With Laser Ablation Optical Emission Spectrometry", J. Anal. At. Spectrom. 12, 1997, pp. 183-188.
8. V. Lazic , R. Barbini, F. Colao, R. Fantoni, A. Palucci, "Self absorption model in quantitative Laser Induced Breakdown Spectroscopy measurements on soils and sediments", Spectrochim. Acta B, 56, 2001, pp. 808-820.
9. L. St-Onge, M. Sabsabi, P. Cielo, Quantitative Analyses of Additives in Solid Zinc Alloys by Laser-Induced Plasma Spectrometry, J. Anal. At. Spectrom. 12, 1997, pp. 997-1004.
10. M. Corsi, G. Cristoforetti, V. Palleschi, A. Salvetti, E. Tognoni, "A fast and accurate method for the determination of precious alloy carratage by Laser induced Plasma Spectroscopy", Eur. Phys. J. D13, 2001, pp. 373-377.
11. F. Colao, R. Fantoni, V. Lazic, V. Spizzichino, "Laser Induced Breakdown Spectroscopy for semi-quantitative analyses of artworks - application on multi-layered ceramics and copper based alloys", Spectrochim Acta B57, 2002, pp. 1219-1234.
12. F. Colao, R. Fantoni, V. Lazic , A. Paolini, "LIBS application for analyses of martian crust analogues: search for the optimal experimental parameters in air and CO2 atmosphere", Appl. Phys. A 79 (2004) 143-152.
13. B.C. Castle, K. Talbardon, B. W. Smith,. J. D. Winefordner, Variables Influencing the Precision of Laser-Induced Breakdown Spectroscopy Measurements, Appl.Spectrosc. 52 Vol. 5 (1998) 649-657.
14. D. Body, B. L. Chadwick, Optimization of the spectral data processing in a LIBS simultaneous elemental analysis system, Spectrochimica Acta B, 56 (2001) 725-736.
15. R. Barbini, F. Colao, R. Fantoni, V. Lazic, A. Palucci, M. Angelone, "On board LIBS analysis of marine sediments Collected during the XVI Italian campaign in Antartica", Spetrochim. Acta B 57 (2002) 1203-1218.
16. D. Anglos, Laser-Induced Breakdown Spectroscopy in Art and Archeology, Appl. Spectrosc. 55 (2001) 186A.-205A.
17. P. Fichet, P. Mauchien, J. F. Wagner, C. Moulin, Quantitative elemental determination in water and oil by laser induced breakdown spectroscopy, Anal. Chim. Acta, 429 (2001) 269-278.
18. P. Yaroschyk, R. J. S. Morrison, D. Body, B. L. Chadwick, Quantitative determination of wear metal in engine oils using laser-induced breakdown spectroscopy: a comparison between liquid jets and static liquids, Spectrochim. Acta B 60 (2005) 986-992
19. D. A. Cremers, L. J. Radziemski, T. R. Loree, Spectrochemical analyses of liquids using the laser spark, Appl. Spectrosc. 38 (1984) 721-729.
20. Casavola, A. De Giacomo, M. Dell'Aglio, F. Taccagna, G. Colonna, O. De Pascale, S. Longo, Experimental investigation an modeling of double pulse laser induced plasma spectroscopy under water, Spectrochim. Acta B 60 (2005) 975-985.
21. V. Lazic, F. Colao, R. Fantoni, V. Spizzichino "Recognition of archeological materials underwater by laser induced breakdown spectroscopy", Spectrochim. Acta B, Vol 60 (2005) 1014-1024
22. V. Lazic, F. Colao, R. Fantoni, V. Spizzichino, S. Jovicevic, "Underwater sediment analyses by Laser Induced Breakdown Spectroscopy and calibration procedure for fluctuating plasma parameters", Spectrochim. Acta B, submitted.

Investigation of the statistical nature and structure of the electrical breakdown time delay in gas diodes filled with neon

Čedomir A. Maluckov[1], Jugoslav P. Karamarković[2] and Miodrag K. Radović[3]

[1] Technical Faculty in Bor, University of Belgrade, Vojske Jugoslavije 24, 19210 Bor, Serbia
[2] Faculty of Civil Eng. and Architecture, University of Niš, Beogradska 14, 18000 Niš, Serbia
[3] Faculty of Sciences and Mathematics, University of Niš, P.O.B.224, 18001 Niš, Serbia

Abstract. The electrical breakdown time delay in gas diodes filled by neon at the low pressures is investigated experimentally and theoretically. Experimental results are obtained measuring the characteristics of gas diodes filled by spectroscopically pure neon. In order to discard any systematic trend during the measurement procedure, checking of the measured values randomness preceded the statistical analysis of the experimental results. Novel theoretical model is established for interpretation of obtained experimental results on the breakdown time delay. The model is based on the assumptions of the exponential distribution of the statistical time delay and Gaussian distribution of the formative discharge time. Therefore, the density distribution of the breakdown time delay is assumed to be convolution of the statistical and formative time delay distributions. Parameters of the statistical and formative time delay, as stochastic variables, are modeled by the numerical Monte Carlo method. Numerical distributions are tested to the corresponding experimental distributions of the breakdown time delay by varying the distribution parameters. In addition, the asymmetry coefficient and skewness coefficient of the breakdown time delay distribution, and coefficients of the statistical and formative time delay distributions are analyzed. Numerically calculated time delay distributions fit well to the corresponding experimental distributions in gas diodes filled with neon at low pressures.

Keywords: Breakdown voltage distributions, Convolution based statistical model.
PACS: 52.80.Hc, 02.50.

INTRODUCTION

The electrical breakdown in gases represents the transition of the discharge in the gas from non-self-sustaining to self-sustaining form [1]. The electrical breakdown in gases can be interpreted as a macroscopic event with a stochastic nature. Investigation of the stochastic nature of the electrical breakdown dates from von Laue [2], and develops in papers of many researches [3–5]. The statistical theory of electrical breakdown is established on the presumption of the Townsend breakdown mechanism [1]. This presumption is equivalent to the small pressure and small overvoltage regimes in gases, when the influence of the space charge is neglected. The breakdown criterion in Townsend theory is:

CP876, *The Physics of Ionized Gases: 23rd Summer School and International Symposium*,
edited by L. Hadžievski, B. P. Marinković, and N. S. Simonović
© 2006 American Institute of Physics 978-0-7354-0377-2/06/$23.00

$$\gamma\left[\int_0^d \exp(\alpha dx) - 1\right] = 1 \qquad (1)$$

where α is the primary ionization coefficient, and γ is the effective secondary ionization coefficient which includes secondary processes induced by positive ions γ_i, metastable atoms (molecules) γ_m, and photons γ_{ph}, i.e., $\gamma = \gamma_i + \gamma_m + g_{ph}$.

The electrical breakdown mechanism in gases can be considered as a combination of two distinct processes. The first process corresponds to the occurrence of one or several physical events leading to the creation of an initial free electron. This process is the Poisson random process, and time for the electron appearance is the statistical time delay t_S. Thus, the breakdown statistical time delay t_S is characterized with the exponential distribution [1, 2]. The second process is the process of ionization and carrier multiplication in gas, which results in the development of a low impedance conducting plasma. Corresponding characteristic time is the formative time delay t_F. In most cases, the ionization events, which lead to breakdown, are the predictable sequences of action. This implies that the multiplication process has a high probability and the formative time delay is well described by a normal distribution [6]. Therefore, the electrical breakdown time delay t_D consists of the breakdown statistical time delay t_S and the formative time delay t_F ($t_D = t_S + t_F$) [1].

The shape of breakdown time delay distribution depends on experimental conditions. In many experiments $t_F \ll t_S$ [1], the breakdown time delay distributions are determined only with the breakdown statistical time delay, and have exponential shape. However, in some cases formative time delay cannot be neglected. More exact analysis assumes t_F as a single value, and the distribution of the breakdown time delay is a shifted exponential distribution [1]. This is a good approximation for $t_F < t_S$, even for comparable t_S and t_F if t_F is characterized with a sufficiently small variance. In the opposite cases, when the formative time delay is a dominant part of the total time delay, the breakdown time delay distributions can be approximated by Gaussian distributions. This corresponds to the conditions with well defined discharge initialization process (the high pressure [6] or the large electron yield conditions [7]).

In present paper, we try to establish a new model which takes into account the Gaussian distribution of formative time delay and exponential distribution of statistical time delay. Therefore we modeled the breakdown time delay as a sum of two random variables, one of them with exponential distribution and the other with the Gaussian distribution. We analyzed that model in details using statistical methods and tested it on the breakdown time delay distributions recorded in the neon filled tubes, for various experimental conditions, relaxation times, overvoltages, the intensity of UV radiation and auxiliary glow current.

318

CONVOLUTION BASED STATISTICAL MODEL

The convolution approach for description of the electrical breakdown time delay distribution is based on the following presumptions. The total time delay is a random variable t_D, with values t_D (i.e., measured values of total time delay), and the random variable t_D is the sum of two independent random variables t_S and t_F ($t_D = t_S + t_F$). The first random variable t_S (with values t_S) is the breakdown statistical time delay characterized with the exponential distribution f_S:

$$f_S = \frac{1}{E(t_S)} \exp\left(-\frac{t}{E(t_S)}\right) \tag{2}$$

The distribution parameter is $E(t_S) = 1/(YP)$, where Y is the electron yield, and P is the probability that breakdown is initialized by an initial electron [1]. The product YP can be defined as the rate of the triggering events which initiate the discharge.

The second variable t_F (with values t_F) is formative time delay characterized with Gaussian distribution f_F:

$$f_F = \frac{1}{\sqrt{2\pi\sigma(t_F)}} \exp\left(-\frac{(t - E(t_F))^2}{2\sigma(t_F)}\right) \tag{3}$$

The distribution parameters are mathematical expectation $E(t_F)$ and standard deviation $\sigma(t_F)$. The density distribution of the random variable t_D can be obtained as the convolution of the density distributions of the random variables t_S and t_F:

$$f(t_D) = \int_0^{t_D} f_S(t) f_F(t_D - t) dt \tag{4}$$

In the present paper the density distributions of t_D are obtained by the numerical integration of equation (4) adopting the "mechanical quadrature" method [8]. In order to obtain the appropriate values of exponential and Gaussian distribution parameters, the Monte Carlo simulation algorithm is developed for generation of the distribution of t_D. The numerical subroutines for pseudo random numbers, Gaussian distribution (the Box-Muller method), and exponential distribution (the method of the inverse function) are taken from [9]. The parameters of the statistical time delay and formative time delay distribution $(E(t_S), E(t_F)$ and $\sigma(t_F))$ were considered as fitting parameters associated with the experimental time delay mean value $\overline{t_D}$ and experimental standard deviation σ_E,

$$\overline{t_D} = E(t_S) + E(t_F)$$
$$\sigma_E^2 = E(t_S)^2 + E(t_F)^2 \tag{5}$$

The consistency of the numerically generated and experimental distributions is checked by the χ^2 test [10]. The fitting procedure is reduced on variation of only one parameter until the χ^2 sum is minimized. For more details about the numerical procedure, and calculation of the distributions parameters $E(t_S)$, $E(t_F)$ and $\sigma(t_F)$, it is worth consulting [10–20].

EXPERIMENT

The measurements of the time delay are performed on a neon filled diodes at the pressure from 1 mbar to 13.3 mbar and room temperature. The tube bulb is made of molybdenum glass, with molybdenum electrode carrier. Before the neon has been admitted, the tubes were baked out at 350 °C and pumped down to the pressure of 10^{-7} mbar in a process similar to that used for the production of x-ray and other electron tubes. The tube was filled with Matheson research grade neon.

The measurements of the breakdown time delay have been performed using the electronic systems schematically illustrated in Fig. 1 (a) and Fig. 1 (b). Details about these systems are given in [13] (Fig. 1 (a)) and [16] (Fig. 1 (b)).

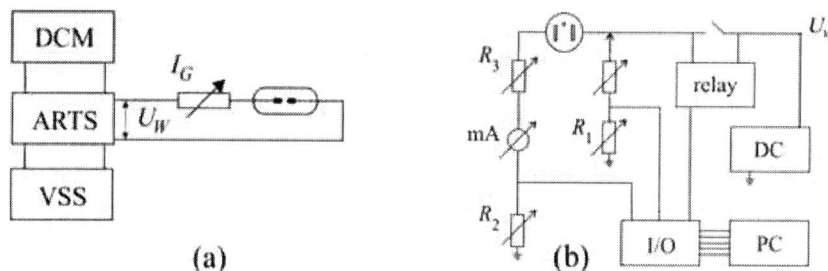

(a)   **(b)**

Figure 1. A schematic diagram of the experimental layouts.

APPLICATION OF STATISTICAL MODEL ON BREAKDOWN TIME DELAY DISTRIBUTIONS

The convolution model is tested by analyzing experimentally obtained the time delay distributions for various values of experimental parameters as the relaxation time, overvoltage, the intensity of γ radiation, distance between electrodes (minimum and right part of Paschen curve) and auxiliary glow current.

Time delay distributions for various relaxation times

The convolution model of the time delay distributions is tested to experimentally obtained distributions for various the relaxation time from 3 ms to 600 ms in neon filled diode at pressure of 6.5 mbar (listing of measuring parameters is given in reference [16]). The overvoltage of 11% is kept constant. Theoretical delay density distributions are shown in Fig. 2 for noted values of the relaxation time. The

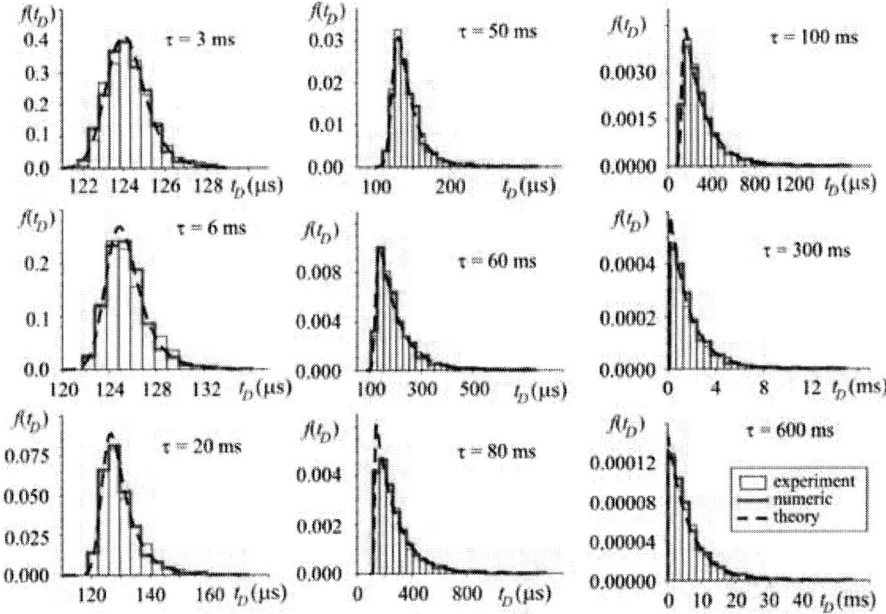

Figure 2. The breakdown time delay density distributions for indicated relaxation times.

corresponding experimental density distributions are given by histograms from 1000 successive and independent measurements. In addition, in Fig. 2 the numerically generated the time delay distributions are shown by histograms without classes due to simplicity. The dashed curves correspond to the theoretical density distributions calculated by relation (4).

In Fig. 3 the distribution coefficients skewness γ_3 and kurtosis γ_4, with respect to the Gaussian distributions, are plotted as functions of the relaxation time. The corresponding distribution parameters $E(t_S)$, $E(t_F)$ and $\sigma(t_F)$ are shown in Fig. 4. The agreement of theoretically modeled and experimentally obtained the time delay distributions is obvious in figures from Fig. 2 to Fig. 4. In addition, the dependence of the shape of the time delay distributions on the distance between electrodes and the

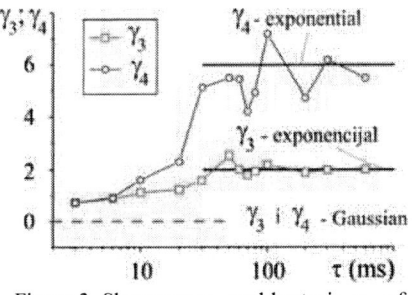

Figure 3. Skewness γ_3 and kurtosis γ_4 of experimental distributions vs τ.

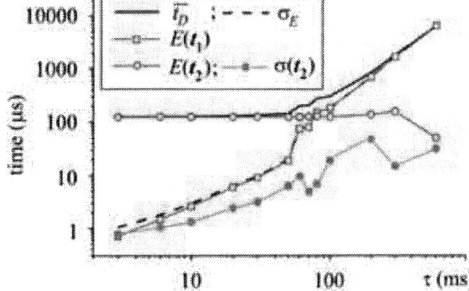

Figure 4. Distribution parameters and mean values and experimental standard deviations vs τ.

formative time delay is observed. Generally, the time delay distributions tend to Gaussian for small values of the relaxation time, where total time delay is determined by the formative time. For relaxation time from 50 to 200 ms, the statistical and formative time delay are comparable, and for higher values of the relaxation time the statistical time prevails in total time delay. In the last case the time delay distribution is the exponential one.

Investigations of the convolution method applicability for various values of the relaxation time are presented in references [10, 12, 14, 17, 18, 20]. The model validity is confirmed. In addition, the dominant influence to the shaping of the distribution is attached to the neutral active states of neon atoms independent of their origin. In other words, the active states which lasted from the previous discharges and these created after the voltage was applied to the initialized discharge equally influence the distribution shape.

Time delay distribution for various overvoltages and distances between electrodes

Test of the convolution model for various values of the overvoltage is published in references [10, 11, 13, 15-18, 20]. Analysis of the distribution parameters, skewness and kurtosis, shown that the smaller values of overvoltage corresponds to more symmetric distributions. It is related to the formative time delay dominance at low overvoltages. In addition, it is shown that distribution shape depends on the relation between the statistical and formative time delay [10, 11, 13, 15-18, 20].

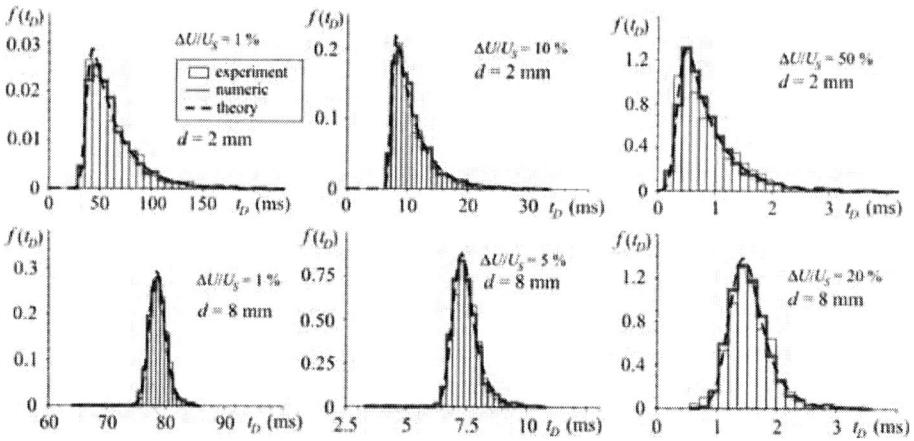

Figure 5. The breakdown time delay density distributions for indicated ovrvoltages and gaps.

The validity of the convolution method is considered with respect to various distances between electrodes and overvolatages from 1% to 50%, in neon gas diode at 13.3 mbar [10, 11, 13]. The agreement between theoretical and experimental results is confirmed. In the area of the minimum of the Paschen curve the time delay distribution is exponential, and in the right from the minimum the time delay distribution is Gaussian. This is associated with the ratio of the statistical and formative time in the total time delay.

Time delay distribution for various values of auxiliary glow current

Test of the convolution model for various values of auxiliary glow current in diode filled with neon at 13.3 mbar is done in reference [17]. The corresponding density distributions are illustrated in Fig. 6. It is confirmed that with increasing values of the auxiliary glow current the more symmetric distributions are obtained. This is due to the increasing influence of the active neutral states from the auxiliary discharge. In addition, the time delay distributions from the convolution method nicely agree with the experimentally obtained distributions.

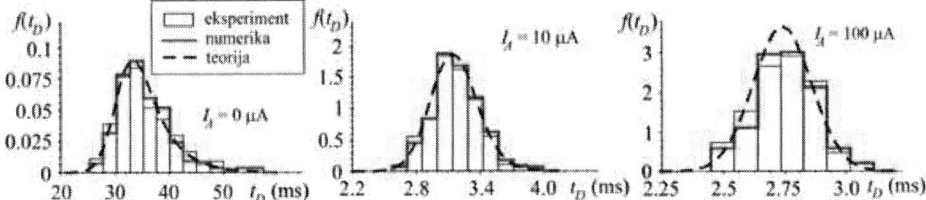

Figure 6. The breakdown time delay density distributions for indicated auxiliary current.

Time delay distribution in the presence of the UV and γ radiation

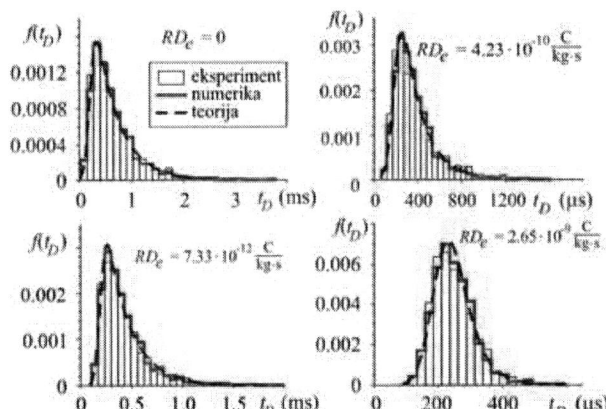

Figure. 7 The breakdown time delay density distributions for indicated exposition dose rate RD_e.

Testing of the convolution model for various rates, RD_e, of *UV* and γ radiation is given in [10, 16, 19]. The correspondding total breakdown time delay density distributions are shown in Fig. 7. The $^{60}_{27}$Co radioactive source is used for radiating the gas diode filled with neon at 6.5 mbar [19]. The analysis of distribution parameters [19] shows the decreases of the statistical time delay with increasing RD_e, while the formative time is not significantly changed. In addition, the electron yield grow is observed exposing the diode to *UV* radiation what is associated with the decrease of the statistical time delay and the symmetry of distribution is more established with the increasing RD_e.

CONCLUSION

In this paper the convolution-based statistical model of the time delay distribution time is established for interpretation of obtained experimental results on the breakdown time delay in gas diode filled by neon at the low pressure. This model is compared with the experimental distributions of the breakdown time delay for various values of the relaxation time, the overvoltage, the distance between electrodes, the auxiliary current and exposition doses of γ and UV radiation. The numerical Monte Carlo analysis is based on calculation of the coefficients of the statistical and formative time delay distributions. It is shown that the numerically calculated time delay distributions fit well to the corresponding experimental distributions. This confirms the assumption of the exponential distribution of the statistical time delay and Gaussian distribution of the formative discharge time, and quality of the applied convolution based statistical model for the breakdown time delay distribution.

ACKNOWLEDGMENTS

This work was supported by the Ministry of Science and Environmental Protection of the Republic of Serbia under Contract 141008.

REFERENCES

1. J. M. Meek and J. D. Craggs, *Electrical Breakdown of Gases*. New York: Wiley, 1978.
2. M. von Laue, Annu. Phys. (Leipzig) **76**, 261 (1925).
3. L. B. Loeb, Rev. Mod. Phys. **20**, 151 (1948).
4. R. A. Wijsman, Phys. Rev. **75**, 833 (1949).
5. R. L. Farquhar, B. Ray, and J. D. Swift, J. Phys. D **13**, 2067 (1980).
6. J. Moreno, M. Zambra, and M. Favre, *IEEE Trans. Plasma Sci.* **30**, 417–422 (2002).
7. I. V. Spasić, M. K. Radović, M. M. Pejović, and Č. A. Maluckov, J. Phys. D **36**, 2515 (2003).
8. D. E. Knuth, Seminumerical Algorithms, 2nd ed, The Art of Computer Programming (Addison-Wesley, Reading, 1981), Vol. 2.
9. W. H. Press et all, Numerical Recipes in Fortran 77: The Art of Scientific Computing, 2nd ed., Fortran Numerical Recipes (Cambridge University Press, Cambridge, 1997), Vol. 1.
10. Č. A. Maluckov, Investigation of the statistical nature and structure of the electrical breakdown time delay in gas diodes filled with neon, PhD Thesis, Faculty of Electrical energineering, University of Nis (2004).
11. Č. A. Maluckov, J. P. Karamarković and M. K. Radović, "Statistical analysis of electrical breakdown time delay in neon at 13.3 mbar pressure", Contributed papers of 21st International Symposium on the Physics of Ionized Gases, (Soko Banja 2002), Editors M. K. Radović and M. S. Jovanović, pp. 414-17
12. Č. A. Maluckov, J. P. Karamarković and M. K. Radović, "Statistical analysis of electrical breakdown time delay in neon filled diode at 13.3 mbar", Proceedings of 5th General Conference of the Balkan Physical Union, (Vrnjačka Banja 2003), Edited by: S. Jokić, I. Milošević, A. Balaž and Z. Nikolić, pp. 1093-1096.
13. Č. A. Maluckov, J. P. Karamarković, and M. K. Radović, IEEE Trans. Plasma Sci. **31**, 1344–1348 (2003).
14. Č. A. Maluckov et all, "Application of Convolution Model on the Electrical Breakdown Time Delay Distribution in Neon Filled-Diode at 6.5 mbar", Contributed papers of 22nd International Symposium on the Physics of Ionized Gases, (National Park Tara, 2004), Editor Lj. Hadžijevski, pp. 381-84.
15. Č. Maluckov, M. Radović i J. Karamarković, "Konvolucioni model raspodele vremena kašnjenja u neonu na 13.3 mbar", Zbornik radova 11. Kongresa fizičara Srbije i Crne Gore (Petrovac na Moru 2004), Urednici N. Konjević, B. Vujičić i P. Miranović, (in Serbian), sekcija 3, str. 79-83.
16. Č. A. Maluckov et all, Phys. Plasmas **11**, 5328–5334 (2004).
17. Č. A. Maluckov, J. P. Karamarković and M. K. Radović, Contrib. Plasma Physics, **45**, 118–129 (2005).
18. Č. A. Maluckov et all, "The application of convolution-based statistical model on the breakdown time delay distributions in krypton", Conference Record-Abstracts of 2005 IEEE International Conference on Plasma Science, (Monterey, California, 2005), p. 201.
19. Č. A. Maluckov et all, IEEE Trans. Plasma Sci., **34**, 2–6 (2006).
20. Č. A. Maluckov et all, Physics of Plasmas, in press (August 2006).

Spatial Structure and Basic Kinetic Processes in Low-Pressure Gas Discharges

Dragana Marić

Institute of Physics Belgrade, Pregrevica 118, 11080 Zemun, Serbia

Abstract. In this paper we present an analysis of time and space resolved development of all typical regimes of low-pressure DC discharge in argon – low current Townsend discharge, oscillations and constrictions of discharge and high current glow discharges. Our work is based on ICCD recordings of discharge structure, synchronized with current and voltage measurements. Special care is given to radial effects and influence of dielectric walls during development of glow discharge structure.

Keywords: Townsend discharge, oscillations, cathode fall, constriction, normal glow, abnormal glow

PACS: 51.51.+v, 52.80.

INTRODUCTION

This work represents a continuation of extensive studies of electrical properties of low-pressure DC discharges and their spatial structure, started by Phelps and coworkers (e. g. [1-3]). Our idea was to follow time resolved structure of the discharge in order to gain a better understanding of processes leading to the development of constriction, cathode fall and other features of glow discharges. This work is based on systematic temporally resolved imaging of light emission from discharge by fast ICCD camera, supported by voltage-current measurements.

Recently, similar studies of spatiotemporal development of cathode-fall dominated DC discharges in argon [3] and of ignition of high current glow [4], have been reported. Our aim was to extend the studies to all typical modes of low-pressure discharges and to extend the knowledge of kinetics of formation and maintenance of these discharges in a simple geometry.

Due to scaling laws, typical for space charge dominated discharges, these studies can be very useful in investigations of micro discharges. They can also be extended to more complex systems used in specific applications and even to transients in cathode dominated high frequency discharges. Ultimately, our goal is to combine these data with 2D models, which will provide further information on basic kinetic processes in the discharge.

CP876, *The Physics of Ionized Gases: 23rd Summer School and International Symposium,*
edited by L. Hadžievski, B. P. Marinković, and N. S. Simonović

EXPERIMENTAL SETUP

In our experiment, the discharge is established in a simple plane-parallel geometry. The cathode is made of copper, while the anode is made of quartz with transparent yet conductive thin film of platinum deposited on its surface. This way it is possible to record both radial and axial profiles of emission. The diameter of the electrodes is 5.4 cm while the electrode separation can be set at three different values – 1.1 cm, 2.1 cm and 3.1 cm.

We run a discharge at very low current (1-2 μA) by applying a DC voltage to resistors connected in series with electrodes. Triggering part of the circuit produces short voltage pulses superimposed on dc voltage. This technique enables us to reduce heating and conditioning of the cathode during the measurements.

The delay generator built into the ICCD camera (Andor, iStar DH720-18U-03) enables us to synchronize recording of the light emission with pulse development and voltage-current measurements.

The details of experimental setup are presented in our previous papers (e.g. [5,6]).

RESULTS AND DISCUSSION

We performed measurements at pd = 250 Pacm, 150 Pacm and 45 Pacm for three different electrode gaps, that covered formation and maintenance of different modes of discharge – low current diffuse (Townsend) discharge, constricted normal glow and abnormal glow discharge.

Pressure times electrode gap values that are close to Paschen curve minimum (150 Pacm) were used as a test case. Higher pd-s were interesting to observe significantly constricted modes of a discharge, and lower pd-s to study contribution of heavy particles to discharge operation.

Townsend discharge

To describe temporal development of Townsend discharge, we have selected operating conditions – pd = 150 Pa cm, d = 1.1 cm. Voltage and current waveforms are shown in Fig. 1. Labels 1-5 indicate the times at which the images were taken. Corresponding axial emission profiles are shown in Fig. 2. After application of a low voltage pulse, the discharge first oscillates for a while, until a stationary state is established. Discharge voltage initially follows the shape of the pulse that would be expected in vacuum (dashed line). Increase in voltage is followed by the discharge current increase. As the circuit capacitance starts loosing the charge through the discharge, voltage decreases, while the current still increases. This kind of behavior is typical for the decreasing part of the voltage-current characteristics at low currents. Throughout the development of the discharge, axial emission profiles exhibit exponential increase from the cathode towards the anode (Fig. 2), which is characteristic of low current Townsend discharges in a homogeneous field. The emission intensity follows the intensity of the discharge current, while the slope of the profile in semi-logarithmic scale remains the same. Only at the highest current, small

326

decrease of intensity near the anode can be observed. This indicates the onset of space charge induced deviation from homogeneous electric field.

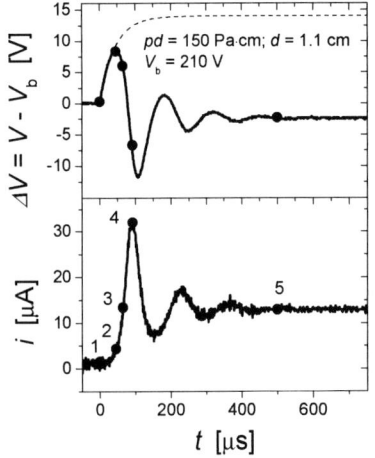

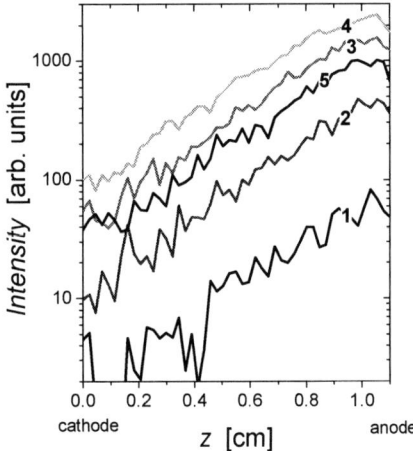

FIGURE 1 Discharge voltage and current waveforms for development of Townsend discharge. (pd = 150 Pa·cm, d = 1.1 cm).

FIGURE 2 Axial profiles of emission that correspond to labels 1-5 in Fig. 1.

Oscillations

In the region of transition from Townsend regime to normal glow, discharge oscillations can occur [1,2]. Fig. 3 shows voltage and current waveforms of free running oscillations. Labels 1-11 indicate the times at which the images were taken. Corresponding axial emission profiles are shown in Fig. 4 (for points labeled 1-6). 2D images of the discharge are presented in Fig. 5.

At early times, after the application of the voltage pulse, emission intensity exponentially increases from the cathode towards the anode (curve 1), which is typical for low current Townsend discharges. Further on, formation of the peak of emission, induced by the space charge, can be observed (curve 2). This is consistent with the formation of the cathode fall, where position of the peak indicates the edge of the cathode fall [5, 6]. The peak of emission rapidly moves towards the cathode (curve 3). At this point, the discharge is centered and curved towards the cathode (Fig. 5-3). This kind of behavior is characteristic of rapid cathode fall development. The discharge profile then broadens radially and the peak intensity moves further towards the cathode (curve 4). The peak intensity of emission is now somewhat decreased, indicating that the discharge has switched to a more economic regime due to the change in electric field profile. Following the decrease of current, the intensity of emission decreases and the peak of emission moves away from the cathode (curve 5), until Townsend-like profile develops (curve 6). In subsequent oscillation periods, axial behavior of the discharge is quite similar. However, radial behavior gradually changes. The discharge gets even more constricted and the peak of emission moves closer to the

327

wall. In Fig. 5 (lower set of pictures), discharge development in the third period of oscillation is shown. In subsequent periods, the development remains the same.

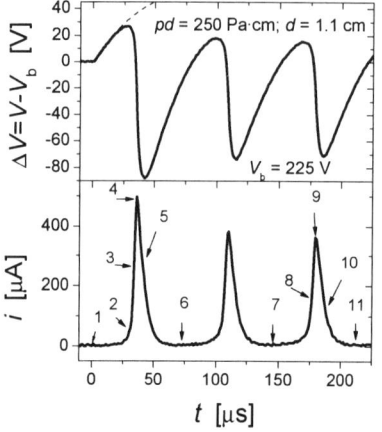

FIGURE 3 Voltage and current signals after the application of voltage pulse.

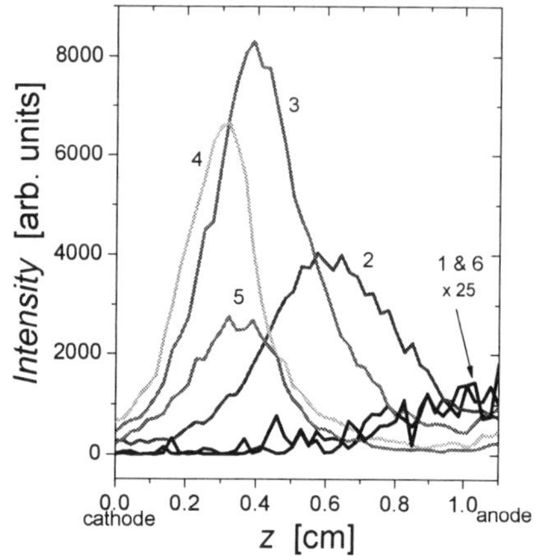

FIGURE 4 Axial emission profiles that correspond to labels 1-6 in Fig. 3.

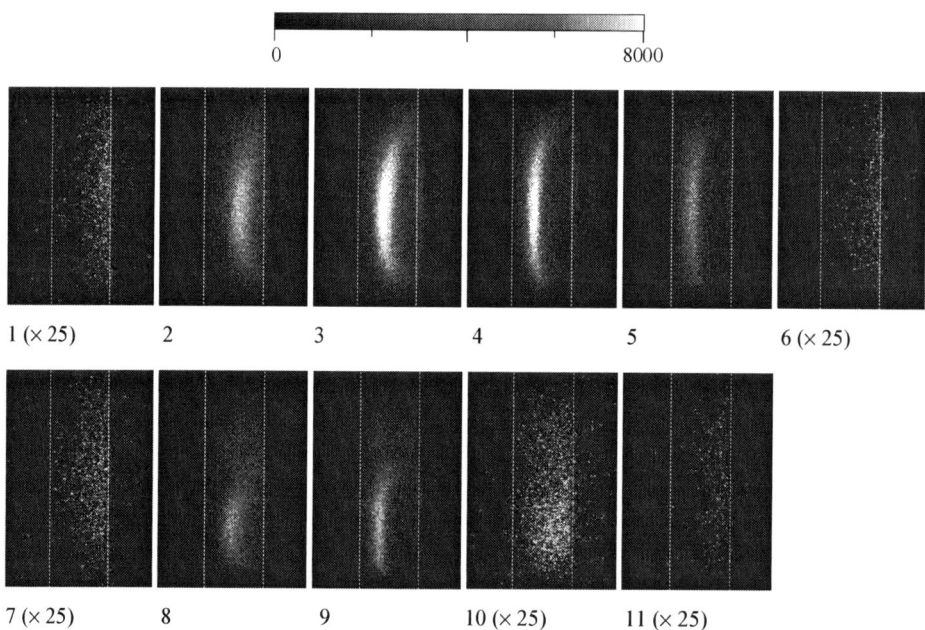

1 (× 25) 2 3 4 5 6 (× 25)

7 (× 25) 8 9 10 (× 25) 11 (× 25)

FIGURE 5 2D scans of discharge development. Labels 1-11 correspond to labels at Fig. 3. Dotted lines indicate positions of the cathode (left) and the anode (right). Intensity of emission in pictures labeled by 1, 6, 7, 10, and 11 is multiplied by factor 25, for a better visibility.

As we have shown, throughout the voltage and current oscillations, ionized gas oscillates both radially and axially. Furthermore, it takes several periods for the discharge to reach the final form which is sustained in subsequent oscillations.

Constrictions

Fig. 6 shows current and voltage waveforms that correspond to development of constricted regime of discharge. 2D images of the discharge are presented in Fig. 7.

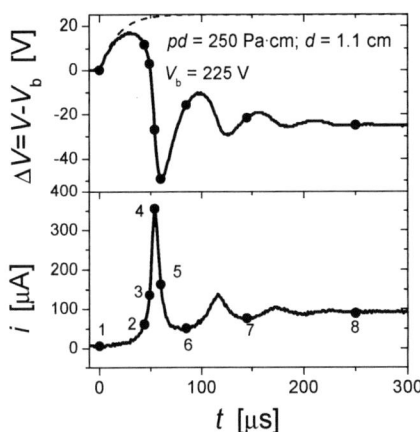

Starting from Townsend's diffuse regime, the discharge gradually exhibits space charge effect. Formation of the peak of emission can be observed, which is consistent with the formation of the cathode fall. The peak of emission rapidly moves towards the cathode. At this point, the discharge is centered and curved towards the cathode (label 3 in Fig. 7), which is typical for rapid cathode fall development. Detailed analysis of this kind of discharge development was presented in [7]. We have shown that during the cathode fall development the discharge can be observed as a group of parallel channels that operate independently.

FIGURE 6. Current and voltage waveforms during formation of constriction of discharge.

As the discharge current reaches maximum, the discharge profile broadens radially and the peak intensity moves further towards the cathode (label 4). Following the decrease of the current, the intensity of emission decreases and the peak of emission moves away from the cathode (label 5), until Townsend-like profile develops (label 6). As the discharge approaches the steady state, formation of constriction can be observed. During this phase of discharge development, the peak of emission gradually moves axially towards the cathode and radially towards dielectric wall of the discharge chamber.

Further analysis, based on the assumption that the discharge can be represented by parallel channels [7], has shown that during establishment of the stationary state, in channel that corresponds to the highest emission intensity (solid squares) the discharge gradually develops from Townsend-like regime to normal glow regime of discharge operation. On the other hand, the remaining discharge channels retain Townsend-like behavior with a gradual decrease of emission intensity. Axial emission profiles of the three selected channels in the stationary state are shown in Fig. 8. Under the given operating conditions, electric field is too low for the discharge to operate in Townsend regime. This mode of discharge clearly operates in the non-self-sustained mode thanks to diffusion of charged particles from the constricted channel. Obviously, throughout formation of the constricted regime of the discharge, different discharge cannels become dependent. The current growth in one of the channels leads to turning off of the remaining channels due to the decreased operating voltage.

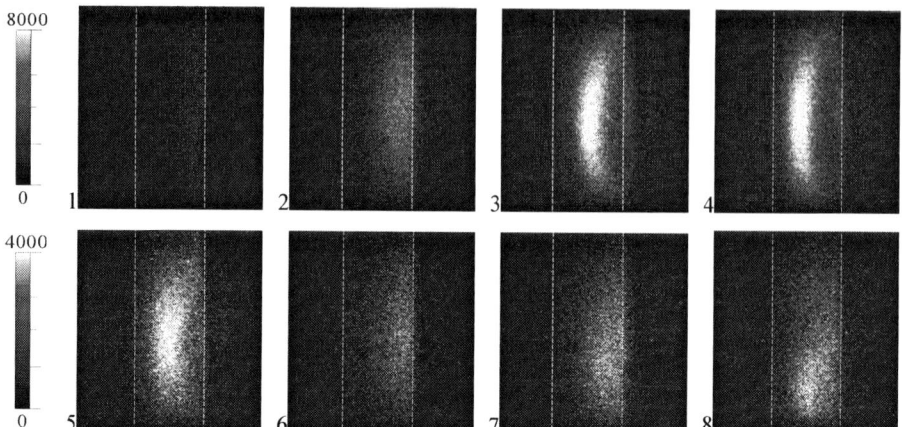

FIGURE 7. 2D scans of temporal development of discharge constriction. Labels 1-8 correspond to labels at Fig. 6. Dotted lines indicate positions of the cathode (left) and the anode (right).

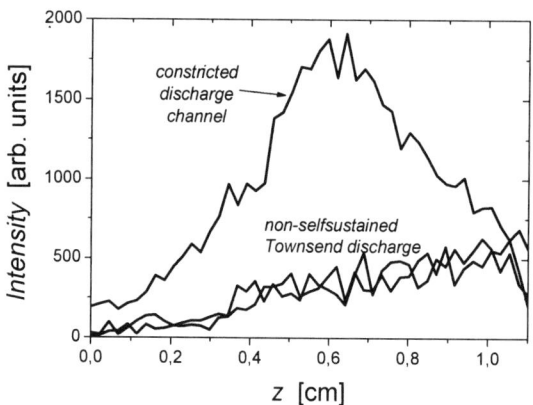

FIGURE 8. Axial emission profiles of three selected discharge channels.

Abnormal glow

Finally, we will present the formation of abnormal glow with fully developed cathode fall. We selected the lowest pressure covered in this study and the widest electrode gap, in order to show influence of heavy particles, which is the most pronounced in our system, under these conditions. Figures 9 and 10 show current-voltage waveforms and selected axial emission profiles, respectively, for $pd = 45$ Pa·cm and $d = 3.1$ cm. Emision profile for initial moment is not shown in Fig. 10, because of too low signal-to-noise ratio. During the first ~40 μs, the discharge is in Townsend regime, the emission profile exhibits typical exponential increase towards the anode due to electron induced excitation. Another peak of emission can be observed near the cathode, which is contributed to excitation by heavy particles – ions (Ar^+ and fast atoms (Ar^f). Further on, both, contribution of excitation by electrons and by heavy particles increase. As the cathode fall develops, peak of emission moves away from the anode. At the current maximum, there is a small drop in peak of emission induced by the electrons, while the contribution of heavy particles rises due to a fast change in axial electric field distribution. This kind of "wave of excitation" can be expected for the rapid change in electric field distribution throughout formation of cathode fall. As the discharge approaches the steady state, intensity of emission decreases, while the profile shape remains the same.

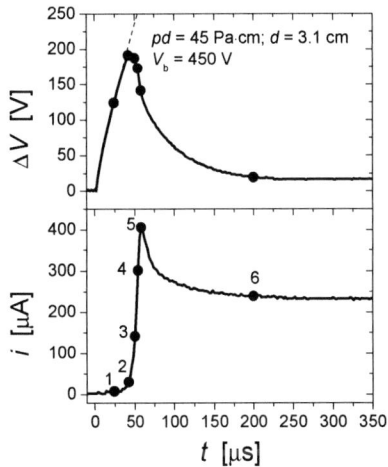

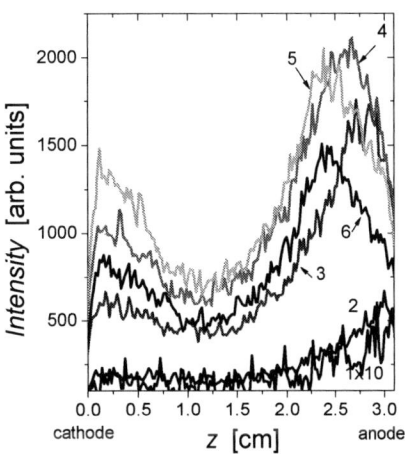

FIGURE 9 Voltage and current signals during formation of abnormal glow discharge.

FIGURE 10 Axial emission profiles that correspond to labels 1-6 in Fig. 9.

SUMMARY

We presented spatiotemporal development of most of the characteristic modes of low-pressure low-current discharges in argon.

– During development of the Townsend regime, the emission intensity follows intensity of the discharge current. Axial profile is typical for low current discharges in homogeneous field, with small deviations at the highest currents
– Throughout the voltage and current oscillations, ionized gas oscillates both radially and axially. After several periods of oscillations, the spatial structure of the discharge adjusts itself to the more stable state.
– In the range of normal glow, analysis of spatial structure development in time enabled us to follow kinetics of fast cathode fall formation. Significantly constricted form of discharge establishes, as the discharge approaches to the steady state. We have

331

shown that in this regime two (even three) different modes of discharge can coexist at the same time.

– Through the abnormal glow discharge formation, we were able to follow the kinetics of electron and heavy particle induced excitation. Time resolved development of discharge structure has shown that rapid redistribution of electric field distribution through the formation of cathode fall initially leads to distinct increase of electron induced excitation. This is then followed by an increase of heavy particle induced excitation. This kind of behavior has been expected, but never experimentally confirmed.

Results shown here represent necessary basis for development of plasma models, especially twodimensional models that include radial effects in discharge and influence of dielectric walls.

ACKNOWLEDGEMENTS

This work was performed as a part of requirement for a PhD thesis which was completed under supervision of Dr. Zoran Lj. Petrović. The work was supported by the project 141025 of the MNZZS of Serbia.

REFERENCES

1. Z. Lj .Petrović and A. V. Phelps, Phys. Rev. E **47** 2806 (1993).
2. Z. Lj. Petrović, A.V. Phelps, *Phys. Rev. E* **56**(5) 5920 (1997).
1. B. Jelenković and A. V. Phelps, *J. Appl. Phys.* **85** 7089 (1999).
4. E. Wagenaars, M. D. Bowden and G. M. W. Kroesen, *IEEE Trans. Plasma Sci.* **33**, 254 (2005).
5. D. Marić, P. Hartmann, G. Malović, Z. Donko and Z. Lj. Petrović, *J. Phys. D: Appl. Phys.* **36** 2639 (2003).
6. D. Marić, K. Kutasi, G. Malović, Z. Donko and Z. Lj. Petrović, *Eur. Phys. J. D* **21** 73 (2002).
7. D. Marić and Z. Lj. Petrović, *Proceedings 23rd SPIG*, (Edited by Nenad S. Simonović, Bratislav P. Marinković and Ljupčo Hadžievski), Kopaonik, Serbia, 2006, pp. 415.

Electric field measurements by Doppler-free Stark spectroscopy of the low-excited levels of atomic hydrogen

Minja Gemišić Adamov, Andreas Steiger and Joachim Seidel

Physikalisch-Technische Bundesanstalt, Abbestraße 2-12, 10587 Berlin, Germany

Abstract. A laser spectroscopic method for electric field measurements that observes the Stark spectra of the low excited levels $n = 2$ and $n = 3$ of atomic hydrogen has been explored. As advantage these levels can be excited Doppler-free from the ground state by a single tuneable pulsed UV laser and the highly resolved Stark spectra are easy to understand and to be calculated. Using hydrogen and deuterium the Stark spectra of the $n = 2$ level are detected as optogalvanic signal. For three different cases of laser polarization the $n = 3$ spectra of hydrogen are measured simultaneously with optogalvanic and laser induced Balmer alpha fluorescence detection. Electric fields down to 200 V/cm can be determined from the Stark spectra of $n = 2$ level, while the spectra of $n = 3$ level enable measurements of electric fields as small as 50 V/cm in each of the three cases of laser polarization.

Keywords: electric field measurements, Stark effect, hydrogen atom
PACS: 32.60+i, 52.70Kz, 42.62Fi

INTRODUCTION

One of the major requirements in gas discharge diagnostics is the determination of electric field strengths, which is often the key parameter for understanding the plasma behavior. Many techniques have been developed for electric field measurements, but non intrusive optical spectroscopy provides the best sensitivity [1]. In particular, laser spectroscopy provides possibility of precise measurements below the Doppler broadening limit, which is usually severe border for sensitivity of the measurements. Absorption of the two photons, at the same wavelength, from two counter-propagating laser beams cancels the first order Doppler effect. Besides laser spectroscopy allows for measurements with a few micrometer spatial and a few nanosecond temporal resolution. As field sensitive probes He and Ar atoms as well as some molecules (BCl, NaK, BH) can be used. But the hydrogen atom is the prominent candidate. Its excited levels are highly sensitive to the electric field, the theory of its Stark splitting is well-known since long ([2] and references therein), and it is present in many plasmas of industrial interest.

In contrary to other methods [3] which use two different lasers to excite Rydberg levels, spectra of the low levels $n = 2$ and $n = 3$ excited with a single laser two-photon excitation are observed. Good sensitivity is achieved by the high performance of the special pulsed UV-lasers.

CP876, *The Physics of Ionized Gases: 23rd Summer School and International Symposium*,
edited by L. Hadžievski, B. P. Marinković, and N. S. Simonović
© 2006 American Institute of Physics 978-0-7354-0377-2/06/$23.00

EXPERIMENTAL SET-UP

The principles of tunable, pulsed, solid state, UV laser systems based on the concept of nonlinear frequency conversion and developed in our group are described in Refs. [4, 5]. The laser system at 243 nm uses improved commercial Mirage 500 laser system with final part for sum frequency generation of UV radiation added in our laboratory and mounted in the same housing.

The exceptional laser system at 205 nm consists of an optical parametric oscillator (OPO) with KTP crystal within plan-parallel cavity pumped by the second harmonic of a Nd:YAG laser and seeded with a cw diode laser, which provides single-longitudinal-mode radiation tuneable at 820 nm. The pulsed OPO radiation is amplified in two Ti:saphire crystals and transformed into its fourth harmonic in three sum frequency generation processes in BBO crystals. This system delivers UV radiation at 205 nm with about 5 mJ pulse energy in less then 4 ns and a spectral bandwidth of about 300 MHz, Fig. 1.

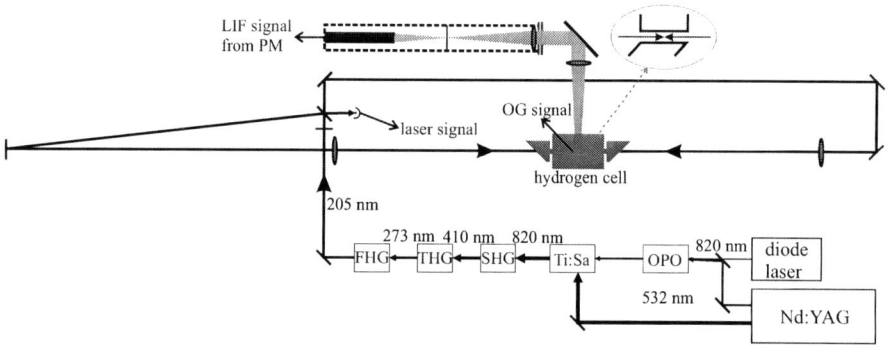

FIGURE 1. Scheme of the laser system at 205 nm and the experimental set-up.

In the experiment hydrogen atoms are produced in a special low pressure cell by thermal dissociation at the hot surface of a tungsten filament. The electric field is controlled by a voltage applied between the filament and another wire set parallel to it. The atoms are excited from the ground state by two photons from counter-propagating laser beams in the center between the two wires. The experimental set up provides Doppler-free spectra and allows to do Stark splitting measurements either by an optogalvanic signal from the cold wire after ionization by a third laser photon or by detecting the fluorescence photons emitted by the excited atoms. For the latter purpose a window is installed in the side-wall of the hydrogen cell and fluorescence photons are detected, after spectral and spatial filtering, by a photo multiplier optimized for pulsed operation.

RESULTS AND DISCUSSION

Electric fields of 200 V/cm or more can be measured by excitation of the $n = 2$ level with two 243 nm laser photons [5]. The measured spectra, Fig. 2 of both hydrogen and deuterium atoms have been shown to provide reliable results for electric field

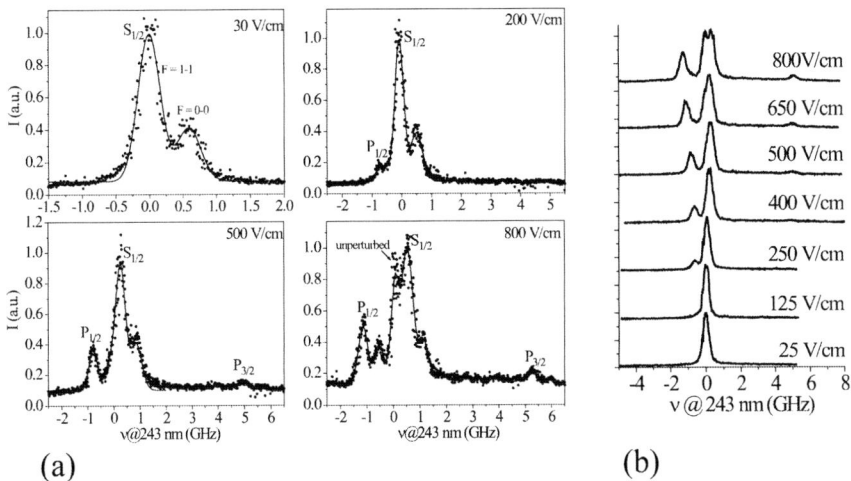

FIGURE 2. Spectra of the two-photon excited $n = 2$ level of (a) hydrogen and (b) deuterium for different values of electric field.

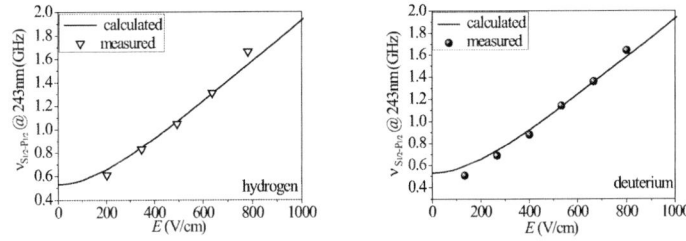

FIGURE 3. Measured and calculated frequency separation between the two selected Stark components of $n = 2$ level as function of the electric field strengths for hydrogen and deuterium.

measurements, Fig. 3. The deuterium spectra exhibit a simpler structure due to the unresolved small hyperfine splitting and allow for more straightforward data processing. With 243 nm excitation, only optogalvanic detection is simply applicable, because fluorescence radiation is emitted only at Lyman alpha in the vacuum ultraviolet spectral region. On the other hand, controlled tunable pulsed laser radiation is easier to generate at 243 nm than at 205 nm. Hence, this measurement method is preferable in cases where optogalvanic detection is possible and electric fields are high enough.

Better electric field sensitivity is obtained by observing the Stark splitting of the $n = 3$ level of atomic hydrogen [6]. Excitation of this level is followed by emission of Balmer alpha fluorescence radiation in the visible at 656 nm, which can be measured simultaneously with the optogalvanic detection. Examples of measured spectra are presented in Fig. 4. Both spectra give good results in comparison with calculated values based on the $n = 3$ level Stark splitting, Fig. 5. Therefore, the detection method can be chosen according to the characteristics of the gas discharge of interest.

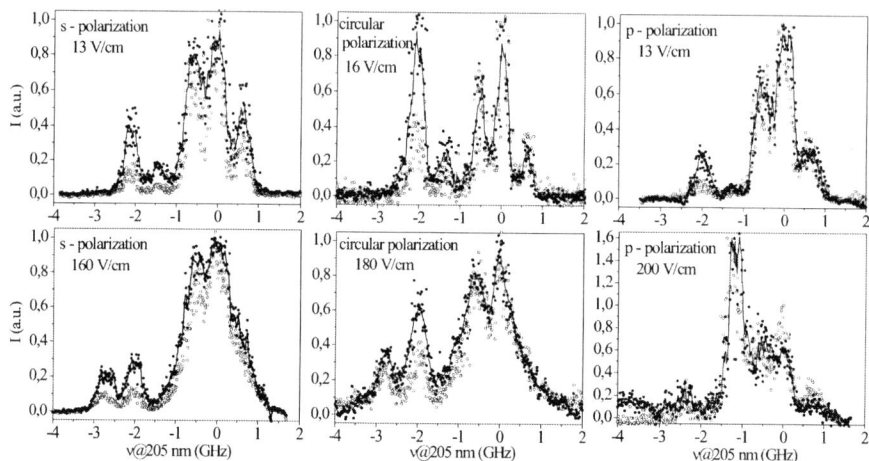

FIGURE 4. Spectra of the two-photon excited $n = 3$ level of hydrogen measured for three different laser polarizations in weak and strong electric field.

Analysis of the measured Stark spectra revealed the frequency separation of two Stark components as the parameter most suitable for electric field determination in all three cases of laser polarization investigated here. For linear polarization of the two laser beams perpendicular to the external field (s-polarization) and also for circularly polarized radiation exciting $\Delta m = 0$ transitions by 1s-3s/d, as well as in the case of 1s-2s excitation in general, this parameter is the frequency separation between the line components corresponding to the two-photon transitions to the upper states evolving from the field-free Lamb shifted $S_{1/2}$ state and from the $P_{1/2}$ state, respectively. The latter transition is forbidden in the field-free case, but due to the level mixing in an electric field the forbidden $P_{1/2}$ component appears with increasing intensity and shifts away from the $S_{1/2}$ component with increasing electric field strength. This frequency shift has been proven as reliable parameter for electric field determination in the cases mentioned above.

In the case of linear polarization of the two laser beams parallel to the electric field (p-polarization), the $3P_{1/2}$ component remains very weak and is practically undetectable in the measured spectra. In this case, however, only one of the $3D_{5/2}$ and one of the $3D_{3/2}/3P_{3/2}$ Stark components have significant excitation probability, in contrast to the other cases of laser polarization. The frequency separation between these two individual Stark components increases with the electric field strength and can be used as a measuring parameter for the electric field in the case of p-polarization. The transitions are allowed also at zero field, but due to the decreasing slope of their frequency shift, the low field detection limit for these two components was also found at field strengths of about 50 V/cm, Fig. 5. This value thus represents the practical lower limit for all three cases of laser polarization investigated here. Comparison of the different cases

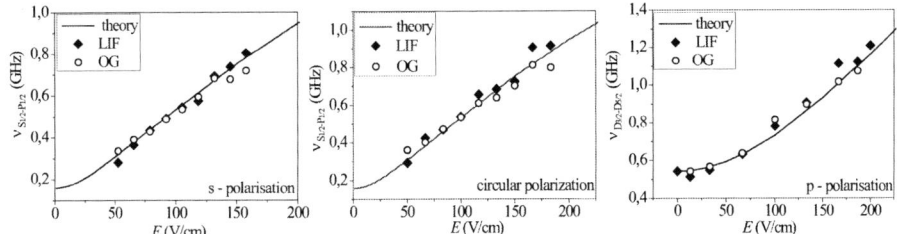

FIGURE 5. Measured and calculated frequency separation between the two selected Stark components of $n = 3$ level as function of the electric field strengths for the three cases of laser polarization.

shows that preference should be given to the circular polarization, because the measured Stark spectrum does not depend on the direction of the electric field. In addition, the selection rules yield more pronounced $3S_{1/2}$ and $3P_{1/2}$ components in the spectrum, which facilitates the electric field determination.

CONCLUSION

As compared to other methods that require comparison of a whole measured spectrum with a set of theoretically calculated spectra, our approach is simpler without a loss of sensitivity. The advantage results from the fact that the determination of the frequency difference does not require critical line intensity information such as varying fluorescence yields or collisional mixing effects for different Stark components.

The results of the performed investigations show that the one-step laser excitation method is a very convenient tool for use in the electric field diagnostics of plasma sources containing hydrogen. The laser pulse energies made available by both solid state laser systems open the possibility for sheet diagnostics as well, and the narrow laser bandwidths allow to perform electric field measurements with about 10 V/cm uncertainty, nanosecond time resolution, and a spatial resolution limited by the fluorescence detection system.

REFERENCES

1. J. Lawler, and D. Doughty, *Adv. At. Mol. Opt. Phys* **34**, 171 (1994).
2. G. Lüders, *Ann. Phys.* **6**, 301 (1951).
3. U. Czarnetcki, D. Luggenhölscher, and H. Döbele, *Phys. Rev. Lett.* **81**, 4592 (1998).
4. A. Steiger, K. Grützmacher, and M. de la Rosa, "Efficient Generation od Pulsed Single-Mode Radiation Tunable in the UV-C Region," in *12th International Congress Laser 95, Laser in Research and Engineering*, edited by W. Waidelich, Springer-Verlag, Berlin Heidelberg, 1996, pp. 308–311.
5. M. Gemišić-Adamov, A. Steiger, K. Grützmacher, and J. Seidel, "Measurements of Local Electric Fields by Doppler-Free Spectroscopy of Atomic Hydrogen," in *22nd Summer School and International Symposium on the Physics of Ionized Gases*, edited by L. Hadžijevski, Vinča Institute of Nuclear Sciences, Belgrade, 2004, pp. 329–332.
6. J. Booth, J. Derouard, M. Fadlallah, L. Cabaret, and J. Pinard, *Opt. Comm.* **132**, 363 (1996).

Radiation Transfer In Arc Plasmas

Vladimir Aubrecht and Milada Bartlova

Brno University of Technology, Faculty of Electrical Engineering and Communication,
Technicka 8, 616 00 Brno, Czech Republic

Abstract. In this paper, attention is given to the theoretical prediction of radiative heat transfer in arc plasmas of air. We present new databank of partial characteristics and net emission coefficients at various plasma pressures. In addition, results of calculations of the fraction of the ultraviolet region are presented as a function of a temperature, pressure and arc radius.

Keywords: Arc plasma, plasma radiation, method of partial characteristics, net emission coefficient.
PACS: 52.25.Os

INTRODUCTION

From the technical point of view, one of the most exploited gas discharge is the electric arc. It differs from the other type of discharges by its high current density and small voltage drops near the electrodes. It is also characterized by intensive radiation. The role of radiation rises rapidly with increasing pressure and arc current.

Difficulties in investigating radiative heat transfer in real systems are related to the nonlinear nature of the equations that describe the phenomenon and to the strong dependence of the characteristics of the radiation field on the frequency and properties of the arc plasma. The non-local nature of radiation does not permit exact integration of the transport equations over frequency and angles in general form.

Several approximate methods of accounting for radiative transfer in the arc plasmas have been developed. A computationally convenient method is the use of net emission coefficients of radiation defined by Lowke [1]. Application of this method for the prediction of temperature profiles gives good results for central arc temperatures, but it very roughly predicts temperature profiles at the low temperatures near the edge of the arc, because of the absorption of ultra violet radiation emitted at the centre of the arc at high temperatures.

The re-absorbing of radiation can be taken into account more precisely by using other approximate method of partial characteristics (MPC), formulated by Sevast'yanenko [2]. The possibilities of MPC are discussed worldwide, especially with intent on using the method in complex CFD problems of switchgear modeling. Beside of our work [3, 4], large scale of contributions has also been made by University of Toulouse under supervision of Dr. Gleizes [5].

In this paper, we present new databank of partial characteristics and net emission coefficients for air plasma at various plasma pressures and temperatures.

CP876, *The Physics of Ionized Gases: 23rd Summer School and International Symposium*,
edited by L. Hadžievski, B. P. Marinković, and N. S. Simonović
© 2006 American Institute of Physics 978-0-7354-0377-2/06/$23.00

ABSORPTION PROPERTIES OF PLASMA

Theoretical calculations of radiation field are based on the knowledge of the plasma composition and subsequently absorption coefficients. The dry air was assumed US standard atmosphere from sea level [6] consisting of N_2, O_2, Ar and CO_2. An equilibrium composition of the air plasma was computed using Tmdgas code [7]. Input data for the composition calculation are specific enthalpy and standard thermodynamic functions of all accounted species; we assume atoms and up to the triple ions of N, O, Ar, C elements, respectively, and diatomic molecules O_2, N_2, N_2^+, NO, NO^+. In Fig. 1, we show the particle densities as a function of the plasma temperature at a pressure of 0.101 MPa.

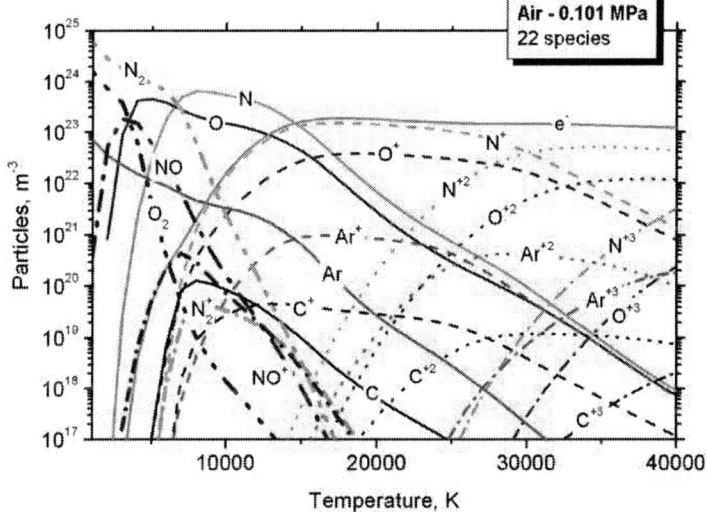

FIGURE 1. Composition of dry air as a function of temperature at a pressure of 0.101 MPa.

Prediction of the absorption coefficients is a very difficult task due to very complex structure of atomic and molecular spectra. One must deal both with continuous radiation made by photo-recombination and "bremsstrahlung" processes and discrete radiation which consists of hundreds of spectral lines. When experimental data is lacking, calculation of absorption coefficient represents a formidable task since the radial wave functions of all free and bound electronic states must be known. However, simplifications can be made by using various semi-empirical methods.

Radiation continuum may be divided into photon emission from free-bound or recombination radiation and free-free (bremsstrahlung) radiation. The spectral absorption coefficient of the process is related to the photon absorption cross section $\sigma_{v,i}^a$ by

$$\kappa_{v,i}^{bf} = \sigma_{v,i}^a . N_i^a \tag{1}$$

where N_i^a is the population density of the i-th electronic state E_i^a of the absorbing species "a". Thus, at a given spectral frequency, plasma temperature and pressure, the total bf spectral absorption coefficient is

$$\kappa_\nu^{bf} = \sum_a \sum_i \sigma_{\nu,i}^a . N_i^a \qquad (2)$$

where the summation is over all atoms and ions having energy states such that $h\nu \geq E_\infty^a - E_i^a$ where E_∞^a is the ionization potential.

The cross sections of photon absorption of ions were treated using Coulomb approximation for hydrogen-like species [8]. The same approximation was used for treatment of free-free transitions. In the case of neutral atoms, the photon absorption cross sections were calculated using the quantum defect method of Burgess and Seaton [9].

In the discrete radiation calculation spectral lines broadening and their complex shapes have to be carefully considered. The lines are broadened due to numerous phenomena. The most important are Doppler broadening, Stark broadening, and resonance broadening. For each line we have calculated the values of half-widths and spectral shifts. Due to the structure of the lines, very fine integration step has to be chosen in all considered computations which lead to the enormous computation times.

Example of the continuum and line absorption coefficient of air (without influence of molecule absorption) is presented in Fig. 2 for the temperature 9 000 K and the pressure of 1 bar.

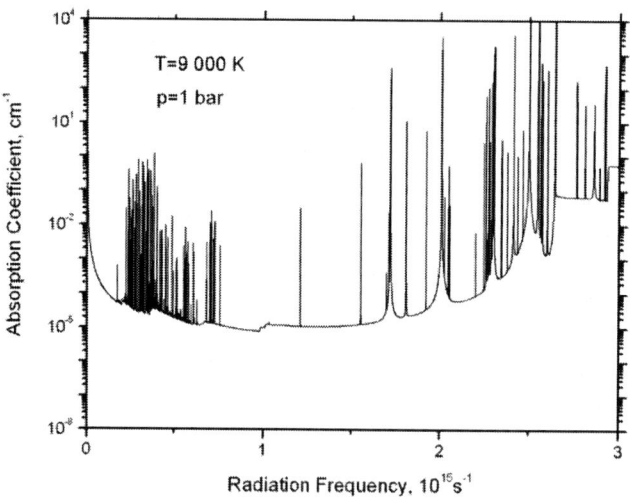

FIGURE 2. Absorption coefficient of continuum and line radiation of air arc plasma

340

Molecular band contributions to the total absorption must be taken into account for plasma when the temperature is below 8 000 K. Energy levels in a molecule are given respectively by the whole molecule rotation, by a vibration of consisting atoms and due to changes in its electron configuration. Energetic gap between consecutive rotational states is quite small (~10^{-3} eV), their radiation spectra lay in microwave region. Vibrational states are separated by energy gaps in the range of 0.1 eV and these states correspond to infrared region of the radiation. Electron states have higher energies with typical energy gaps in the order of an eV, therefore the transitions between electron states give the main contribution to the radiative heat transfer.

In the calculation of absorption coefficient for the molecular band system we have used Franck-Condon principle. For approximate calculation of radiative properties, it is useful to use the absorption coefficient averaged through the rotational spectrum, and also partially smeared through the vibration structure [10]. An example of calculation of photo absorption cross section for NO^+ molecule is shown in Fig. 3 as a function of radiation frequency for various temperatures.

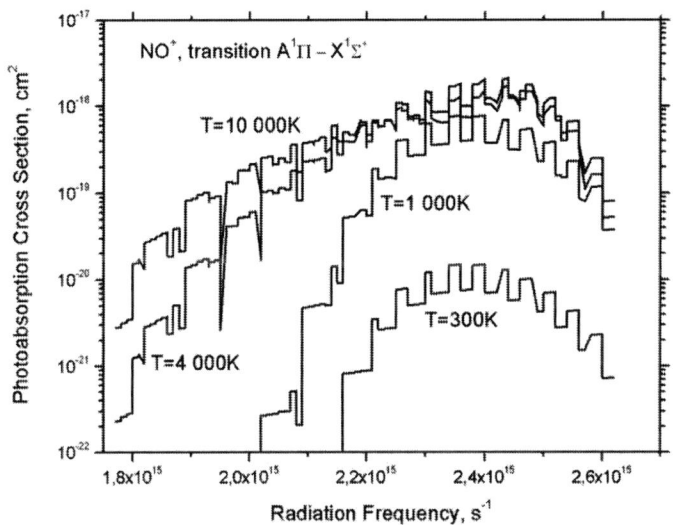

FIGURE 3. Photo-absorption cross section of NO^+ molecule for transition $A^1\Pi \rightarrow X^1\Sigma^+$ as a function of radiation frequency for various temperatures.

Calculated total spectral coefficients of absorption (continuum, lines and molecular electron states) at various plasma temperatures are presented in Fig. 4. From the Fig. 4, it can be seen quite large contribution of molecular species to the absorption of the visible and ultraviolet radiation at low temperatures. For the temperatures above 10 000 K, the effect of molecules to the radiative transfer is negligible.

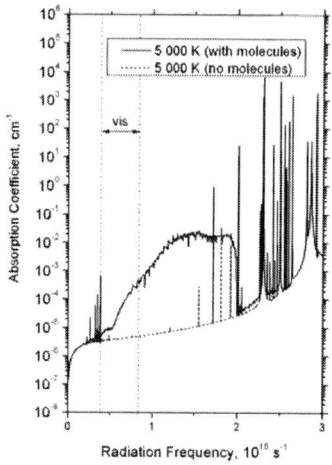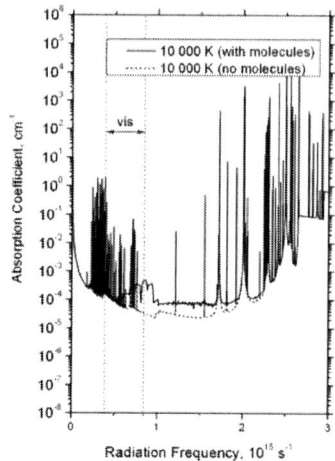

FIGURE 4. Spectral absorption coefficients of radiation in air plasma at various temperatures as a function of radiation frequency at the plasma pressure of 1 bar.

METHOD OF PARTIAL CHARACTERISTICS

Calculated absorption coefficients are used in further computation of the net emission coefficients and partial characteristics of radiation. The basic quantity in radiation transport of energy is intensity of radiation I ($\mathbf{n}$, X) which is the radiation power per solid angle per unit area perpendicular to the direction $\mathbf{n}$ at a point X. Intensity is the function of frequency and temperature. The approximate method of partial characteristics consists of "pre-computing" the most time consuming integration over frequencies in the form of data tables which can be then easily used for the prediction of radiation quantities. Details of the method of partial characteristics have been described in our previous paper [3]. Total radiation intensity at point X can be calculated by summation of all partial intensities between point X and plasma boundary R

$$I(X) = \int_{X}^{R} \Delta I\left(T_X, T_\xi, |X - \xi|\right) d\xi \tag{3}$$

Explanation of the meaning and symbols of partial intensities is in Fig. 5.

Net emission of radiation $\nabla I(X)$ along a line segment $\overline{XR}$ can be evaluated as

$$\nabla I(X) = Som(X, R) - \int_{X}^{R} \Delta Sim\left(T_X, T_\xi, x\right) d\xi \tag{4}$$

342

where the functions *Som* and Δ*Sim* are the partial characteristics defined in [3]. These two functions are pre-calculated in advance with parameters T_X, T_ξ, $\overline{XR} = |X - R|$ and they are used in the form of tables for prediction of radiation intensities, fluxes and their divergences.

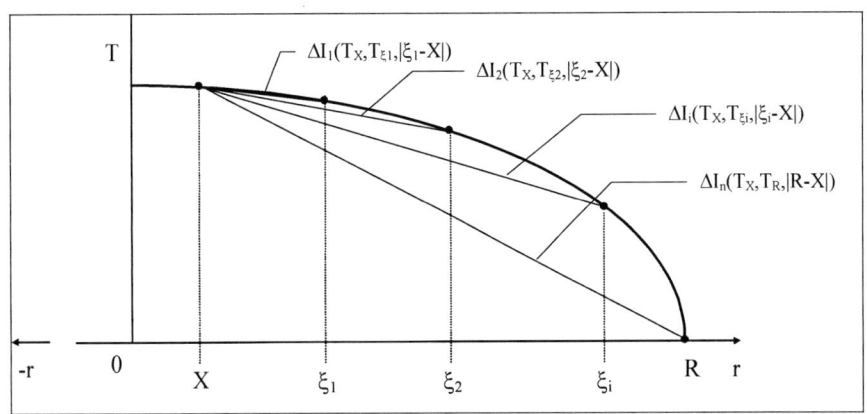

FIGURE 5. Schematic drawing with symbols explaining principle of the method of partial characteristics.

The relationship between the radiation intensity and the radiation flux is given by equation

$$\mathbf{F}_R(X) = \int_{4\pi} I(X,\mathbf{r})\mathbf{r} \, d\Omega \tag{5}$$

where $d\Omega$ is a unit solid angle. Divergence of radiation flux is given by

$$\nabla \cdot \mathbf{F}_R(X) = \int_{4\pi} \nabla I(X,\mathbf{r})\mathbf{r} \, d\Omega \tag{6}$$

If an isothermal temperature distribution is assumed in the plasma, the function *Som* corresponds to the net emission coefficient of radiation ε_N at a given temperature T

$$\varepsilon_N = \int_0^\infty B_\nu(T)\kappa_\nu(T,p)\exp[-\kappa_\nu(T,p)R]d\nu \tag{7}$$

where κ_ν is absorption coefficient, B_ν is Planck function and R represents radius of an isothermal plasma cylinder.

RESULTS

Examples of using the tables of partial characteristics for the air plasma are given in Figs. 6 and 7, respectively. Radial distributions of the radiation flux density and divergence of the radiation flux density (net emission) for the air plasma at a pressure of 1 bar are plotted in the figures. Chosen temperature profile (thin dashed line)has its maximum at 10 000 K with the temperature at the wall of 1 000 K. Resulting curves correspond to the middle of the plasma cylinder with the height of 4 cm and radius of 0.5 cm. Results are presented with and without molecular species taken into account. Influence of molecular species on radiation flux is obvious from the Figure 6. With molecules taken into account radiation flux is higher and more radiation escapes from the arc column.

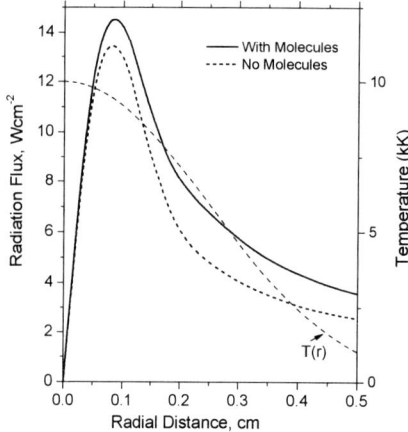
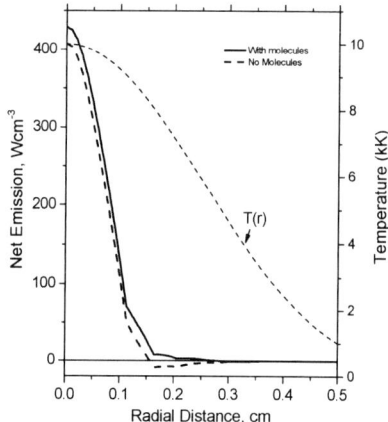

FIGURE 6. Radial distribution of radial flux density in the air plasma cylinder with the height of 4 cm at a pressure of 1 bar.

FIGURE 7. Radial distribution of net emission (divergence of radiation flux density) in the air plasma cylinder with the height of 4 cm at a pressure of 1 bar.

An effect of molecular species on the net emission coefficients is shown in Fig. 8. Correspondingly to Figure 4, molecules contribute significantly to radiative transfer at the temperatures up to 10 000 K.

An important quantity in predictions of arc properties is the fraction of the arc radiation that is in various region of the spectrum. An ultraviolet radiation is of central importance. It is emitted from the central parts of the arc and re-absorbed by the narrow region of the cold gas surrounding the arc plasma. In Fig. 9 we show the fraction of the net emission coefficient of wavelength less then 300 nm for various arc column radii. It can be clearly noted a domination role of the ultraviolet radiation at higher temperatures – for radius of 0.01 cm and temperature 20 000 K this fraction is about 0.97 whereas for 3 000 K it is about 0.05. Thinner lines indicate the fraction of the net emission coefficients calculated without molecular species taken into account. Strong effect of the molecules at temperature below 10 000 K can be seen.

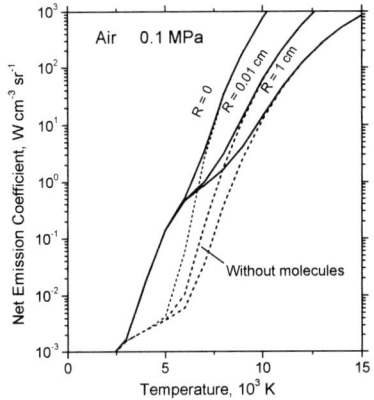
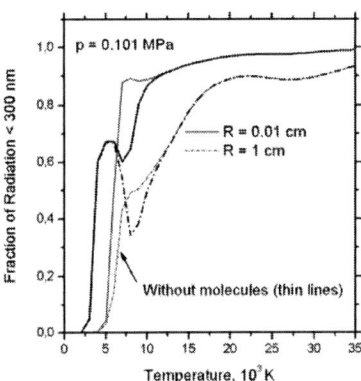

FIGURE 8. Net emission coefficient of air plasma for various radii at a pressure of 0.1 MPa with/without molecular species taken into account

FIGURE 9. Fraction of net emission coefficients of wavelength less than 300 nm for various radii at a pressure of 0.1 MPa.

CONCLUSION

Calculations have been performed for the partial characteristics and the net emission coefficients of radiation in air plasmas as a function of temperature for pressures 1, 5 and 10 bars. Special attention was given to the contribution of selected molecular species through their photo-absorption and photo-dissociation. It has been shown that molecules have significant influence on radiative heat transfer at temperatures below 10 000 K. Examples of using partial characteristics for prediction of radial distributions of radiation flux density and divergence of radiation flux density in a plasma cylinder with an arbitrary temperature distribution are presented.

ACKNOWLEDGMENTS

Authors gratefully acknowledge financial support from Grant Agency of Czech Republic under projects GA 202/06/0898, 102/04/2090 and from Ministry of Education under project No. MSM 0021630503.

REFERENCES

1. J. J. Lowke, *J. Quant. Spectrosc. Radist. Transfer* **14**, 111 (1974).
2. V. G. Sevast'yanenko, *J. Eng. Phys.* **36**, 138 (1979).
3. V. Aubrecht and J. J. Lowke, *J. Phys. D: Appl. Phys.* **27**, 2066 (1994).
4. V. Aubrecht and M. Bartlova, *IEEE Trans. On Plasma Sci.* **25**, 815 (1997).
5. G. Raynal and A. Gleizes, *Plasma Sources, Sci. Technol.* **4**, 152 (1995).
6. D. R. Lide (Editor in Chief), *Handbook of Chemistry and Physics.* 80th Edition 1999 – 2000. Bocca Raton: CRC Press, 2000, Table 14 – 16.
7. O. Coufal, *Acta Technica CSAV* **37**, 209 (1992).
8. I. I. Sobel'man, *Atomic Spectra and Radiative Transitions,* Berlin: Springer Verlag, 1992.
9. A. Burges and M. Seaton, *Rev. Mod. Phys.* **30**, 992 (1958).
10. A. Ch. Mnacakanjan, *TVT* **6**, 236 (1968).

SECTION 4

GENERAL PLASMAS

Invited Lectures
Topical Invited Lectures
Progress Reports

Frontier of Fusion Research: Path to the Steady State Fusion Reactor by Large Helical Device

Osamu Motojima

National Institute for Fusion Science, Toki-shi, Gifu-ken, 509-5292 Japan

Abstract. The ITER, the International Thermonuclear Experimental Reactor, which will be built in Cadarache in France, has finally started this year, 2006. Since the thermal energy produced by fusion reactions divided by the external heating power, i.e., the Q value, will be larger than 10, this is a big step of the fusion research for half a century trying to tame the nuclear fusion for the 6.5 Billion people on the Earth. The source of the Sun's power is lasting steadily and safely for 8 Billion years. As a potentially safe environmentally friendly and economically competitive energy source, fusion should provide a sustainable future energy supply for all mankind for ten thousands of years. At the frontier of fusion research important milestones are recently marked on a long road toward a true prototype fusion reactor. In its own merits, research into harnessing turbulent burning plasmas and thereby controlling fusion reaction, is one of the grand challenges of complex systems science.

After a brief overview of a status of world fusion projects, a focus is given on fusion research at the National Institute for Fusion Science (NIFS) in Japan, which is playing a role of the Inter University Institute, the coordinating Center of Excellence for academic fusion research and by the Large Helical Device (LHD), the world's largest superconducting heliotron device, as a National Users' facility. The current status of LHD project is presented focusing on the experimental program and the recent achievements in basic parameters and in steady state operations. Since, its start in a year 1998, a remarkable progress has presently resulted in the temperature of 140 Million degree, the highest density of 500 Thousand Billion/cc with the internal density barrier (IDB) and the highest steady average beta of 4.5% in helical plasma devices and the largest total input energy of 1.6 GJ, in all magnetic confinement fusion devices.

Finally, a perspective is given of the ITER Broad Approach program as an integrated part of ITER and Development of Fusion Energy project Agreement. Moreover, the relationship with the NIFS' new parent organization the National Institutes of Natural Sciences and with foreign research institutions is briefly explained.

Keywords: LHD, steady state fusion reactor,
PACS: 52.55.Hc, 28.52.-s

1. INTRODUCTION

This paper summarizes the research activities at NIFS and Japan. The objective of the fusion research is to realize to the Sun on the Earth. The mass of the Sun is 2×10^{27} tons, which is 333 thousand times larger than the Earth. Its central temperature is 15 Million °C. Energy Source of the Sun is well known "Hydrogen Nuclear Fusion". It is burning for 8 Billion Years, stably and steadily! In the history of the modern science,

CP876, *The Physics of Ionized Gases: 23rd Summer School and International Symposium*,
edited by L. Hadžievski, B. P. Marinković, and N. S. Simonović
© 2006 American Institute of Physics 978-0-7354-0377-2/06/$23.00

Dr. Hans Bethe was first to propose the burning mechanism of the Sun in 1939 [1]. He was awarded the Nobel Prize for Physics in 1967. The target of our research is to clearly understand the fusion reactions in the magnetized plasmas in order to reach the goal, as soon as possible, within the next 30 years. Helical system at is exploring the

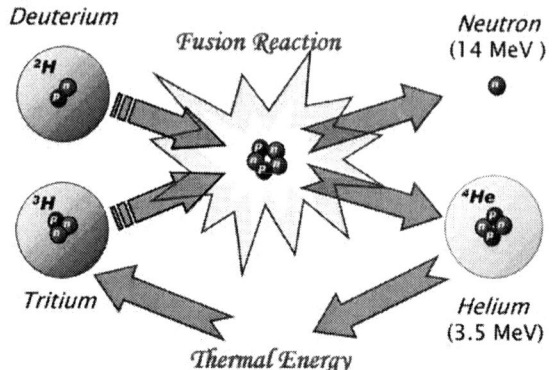

FIGURE 1. D-T fusion reaction.

steady state physics of 100 Million °C plasmas (10 keV). The ITER, the International Thermonuclear Experimental Reactor project [2] has finally started this year, 2006; to be built in Cadarache in France. Since the thermal energy produced by fusion reactions divided by the external heating power, i.e., the Q value, will be larger than 10, this is a big step of the fusion research for half a century trying to tame the nuclear fusion energy for the 6.5 Billion people on the Earth. As a potentially safe, environmentally friendly and economically competitive energy source fusion should provide a sustainable future energy supply for all mankind for ten thousands of years. At the frontier of fusion research important milestones are recently marked on a long road toward a true prototype fusion reactor. In its own merits, research into harnessing turbulent burning plasmas and thereby controlling fusion reaction is one of the grand challenges of complex systems science.

The reaction utilized on the Earth is Deuterium - Tritium (D–T) reaction, which is a thermal and chain reactions as shown in Fig.1, where eventually, three necessary conditions, called the Lawson's Criterion come up, which are summarized in Fig. 2 [3]. At the beginning of fusion research in the decade of 1950, there was an optimistic view that it was possible to realize the Lawson Criterion in a short period. However, looking back to the history of the fusion research we can easily understand that this optimism was largely unfounded. We had to wait for the development of plasma physics and fusion engineering for more than 30 years. Also, we have learned that the peaceful usage of the fusion energy would require a large scientific effort.

At the temperature of 10 keV, the hydrogen isotopes are ionized to form plasmas, which are the typical example of the nonlinear and complex matter. It is the target of our fusion research to understand the physical properties of high temperature plasma

Lawson Condition (by Dr. J. Lawson in 1957)
1, Temperature: $\mathbf{T} > 100$*Million* $°C$ $\Rightarrow$ *Plasma*
2, Density: $\mathbf{n} > 100$*Thousand Billion/cc*
3, Confinement time: $\tau > 1$*sec* $\tau = W_p / P_{heat}$

FIGURE 2. Three necessary conditions for fusion.

1 . **Research toward the Realization of Thermonuclear Fusion Energy**
- Development of safe and environment-friendly energy source

2 . Nuclear Fusion Research as **Frontier Research**
- Promotion of global research in 100 million °C plasma physics and engineering
- Always leading the frontier of scientific and technological research in the world
- Contribution to the development of scientific research and technology in Japan and world

3 . Infinite **Scientific** Interest
(plasma physics study, non-linear phenomenon, non-local and nonequilibrium physics, plasmas within an atomic nuclei)

4 . Making an Active **International Cooperation**
(Contribution to the globalization)

5 . Contribution to the **Preservation of Global Environment** and **Global Peace**

FIGURE 3. The role of fusion research.

state of matter. In contrast to the Sun which contains hydrogen plasma with its own large gravitational force, the strong magnetic force is necessary to keep the plasma stably with necessary equilibrium in a fusion reactor core on the Earth. The magnetized plasma strongly shows nonlinear and complex features, which leads us to observe a lot of interesting phenomena as a scientific spin-off of the fusion research. Since 98 % of the observable matter in Space is the plasma, understanding of high temperature plasmas helps in our understanding of the general laws of the Universe.

The role of fusion research is summarized in Fig. 3. The ultimate goal of our research is the contribution to the item 5.; i.e. the Preservation of Global Environment and Global Peace.

Figure 4 gives the night view of the Earth, which was taken by NASA. This photo illuminates well how large amount of energy we human beings are continually consuming. It is also well known fact that the average human lifetime is the function

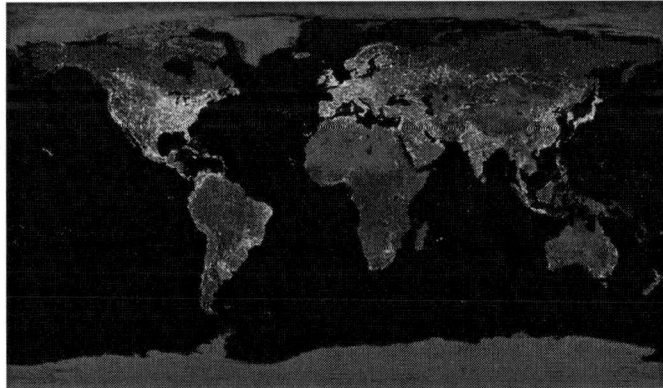

FIGURE 4. Night view of the Earth.

351

World Fusion Activity

International Thermonuclear Experimental Reactor (ITER)

? Site selection : Caderache(France)
? Conducted by 7 parties :
 Japan, EU, USA, Russia, China, Korea, India
? Broader approach hosted by Japan.

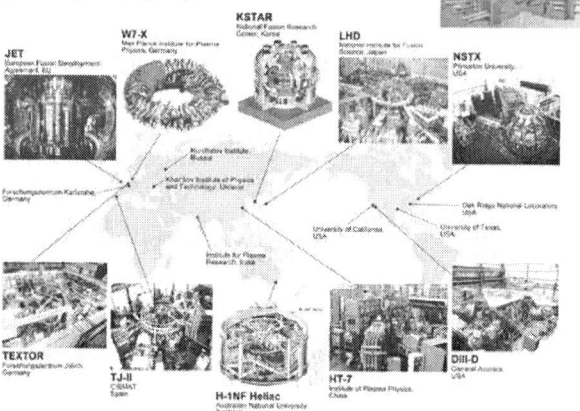

FIGURE 5. World fusion activities.

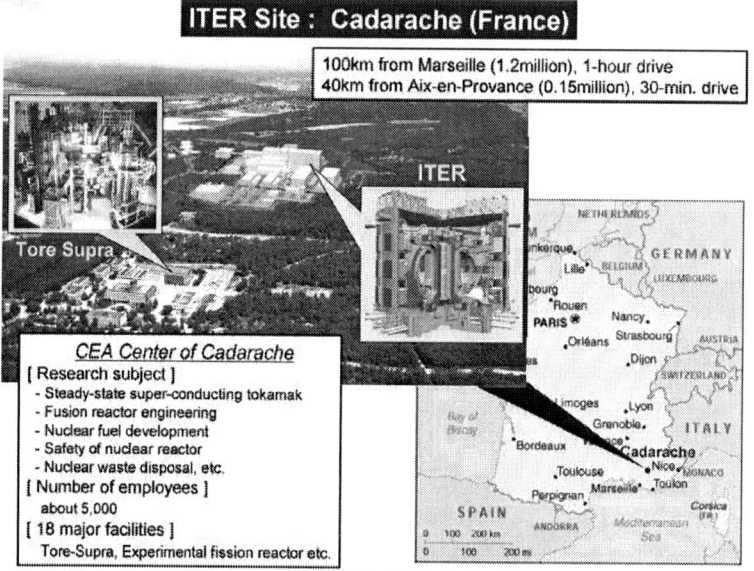

FIGURE 6. THE ITER SITE- CADARACHE

of the energy consumption; while the shortage of energy on a global level is expected after 2050. The realization of the new energy source, i.e. fusion is strongly required.

The World's fusion research activities are highlighted in Fig 5. The experimental reactor project will soon start in Cadarache, France, which is shown in Fig 6. Now, the fusion energy is not anymore the long time dream but has changed to the real target for the fusion researchers. Therefore, we are able to draw up the grand design to reach the final target by defining clearly the critical path.

2. INTRODUCTION TO THE NATIONAL INSTITUTE FOR FUSION SCIENCE (NIFS) IN JAPAN

NIFS is pursuing the integration of science and technology to realize a fusion power plant. The systematization of plasma physics and research and development of reactor relevant engineering are key elements in our strategy. Further, NIFS has been exploring its role as an inter-university research organization while coordinating and executing a variety of excellent collaborative studies together with academic and research institutions from abroad, as well as, in Japan.

The main Mission of NIFS is (i) the experimental study of toroidal plasma confinement using LHD and (ii) theoretical research and computer simulations of the the complex state and nonlinear plasma dynamics such as those seen in high temperature plasmas. These major projects are accompanied by a unique supporting research. Advanced engineering and fusion reactor design studies are strongly promoted. In Fig. 7, objectives of NIFS and its location are shown with the draw chart of the LHD device. The Japanese fusion activity is well highlighted in Fig. 8.

In Japan, the grand design has been newly established at the beginning of 2001 by the Science Council for Science and Technology which is shown in Fig 9. This makes clear the importance of the science and technology to develop fusion energy and has a stratified structure. Eventually, the role of our institute becomes clear in contributing to the science in three basic points.

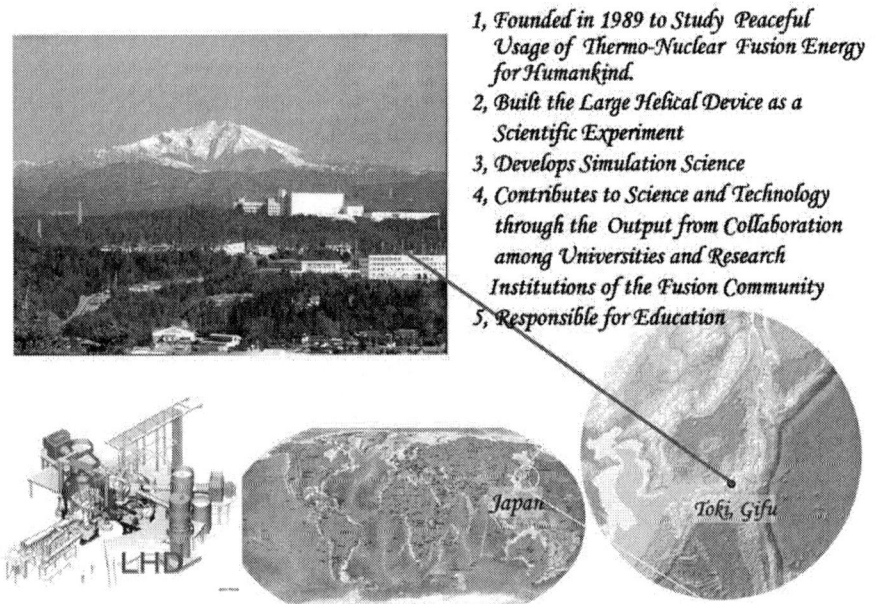

1, Founded in 1989 to Study Peaceful Usage of Thermo-Nuclear Fusion Energy for Humankind.
2, Built the Large Helical Device as a Scientific Experiment
3, Develops Simulation Science
4, Contributes to Science and Technology through the Output from Collaboration among Universities and Research Institutions of the Fusion Community
5, Responsible for Education

FIGURE 7. Objectives of NIFS and its location.

Japanese Fusion Activities

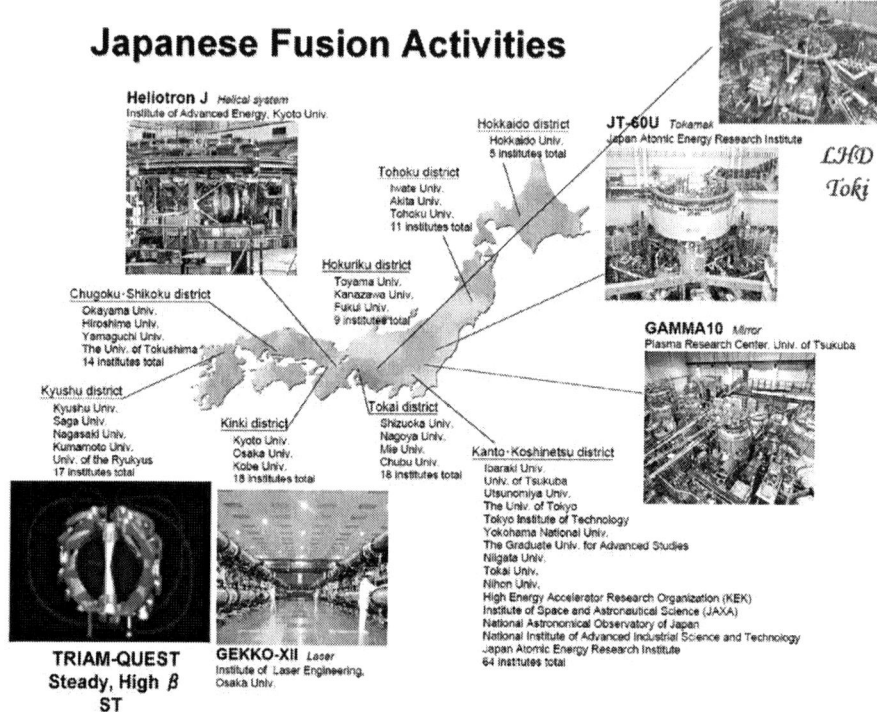

FIGURE 8. Fusion activities in Japan.

Paradigm Shift Required
Stratified Structure of Research towards Fusion Reactor

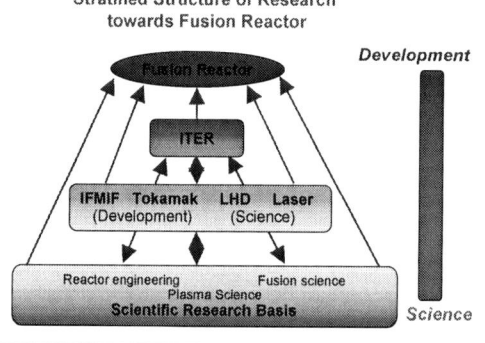

Development

Science

Role of NIFS is clearly suggested by the WG Report
- <u>Keep a close relationship</u> with universities
- <u>Increase collaboration</u> as a center of excellence of fusion research
- <u>Increase educational functions</u> in cooperation with the Graduate University
  for Advanced Studies

FIGURE 9. The grand design of Japanese fusion research.

3. LHD PROJECT

The LHD is a heliotron type fusion device with an intrinsic divertor [4]. Main experimental hall and basic specifications are shown in Fig. 10. It is the largest superconducting fusion device in the world. The major goal of the LHD experiment is to demonstrate the high performance of helical plasmas in a reactor relevant plasma regime. Thorough exploration should lead to the establishment of not only a prospect for a helical fusion reactor but also to a comprehensive understanding of toroidal plasmas. We completed the 9th experimental campaign in FY2005. Diversified studies in LHD have elucidated the broad scope of steady-state high temperature plasmas.

The plasma parameters as well as physical understanding have been progressing steadily since the beginning. They are shortly summarized in Figs 11 and 12. The most highlighted achievement in the last experimental campaign in FY2005 is the discovery of Super-Dense-Core plasma operation accompanied by an *Internal Diffusion Barrier (IDB)*[5], which is shown in Fig. 13. A synergetic effect of highly efficient pumping by means of the Local Island Divertor [6] and core fueling by repetitive pellet injection [7] generates the SDC mode with a central density of $5 \times 10^{20} m^{-3}$. A steep gradient in the density profile is formed by a drastic improvement in particle transport, i.e., IDB. This observation has been done by Profs. N. Ohyabu and A. Komori and their colleagues, by exploring the high density and relatively low temperature reactor core concept operation with more than $5 \times 10^{20} m^{-3}$ and less than 10keV (Fig. 14). This finding has an important meaning equivalent or even more significant than that of H-mode by Prof. F. Wagner in 1982 [8]. It has been said that a helical plasma has an advantage of high density operation and the discovery of the IDB would further emphasize this advantage, much more than was expected.

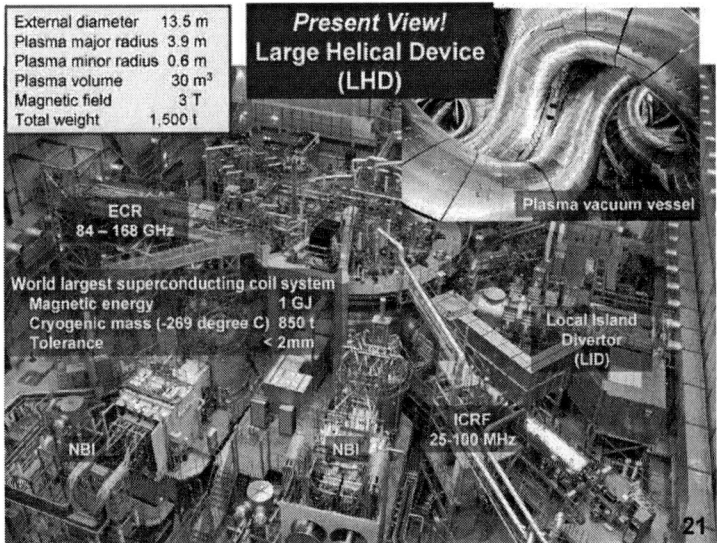

FIGURE 10. Bird-eye view of the LHD experimental hall and basic specifications. Insertion is a view of inside the vacuum vessel in LHD.

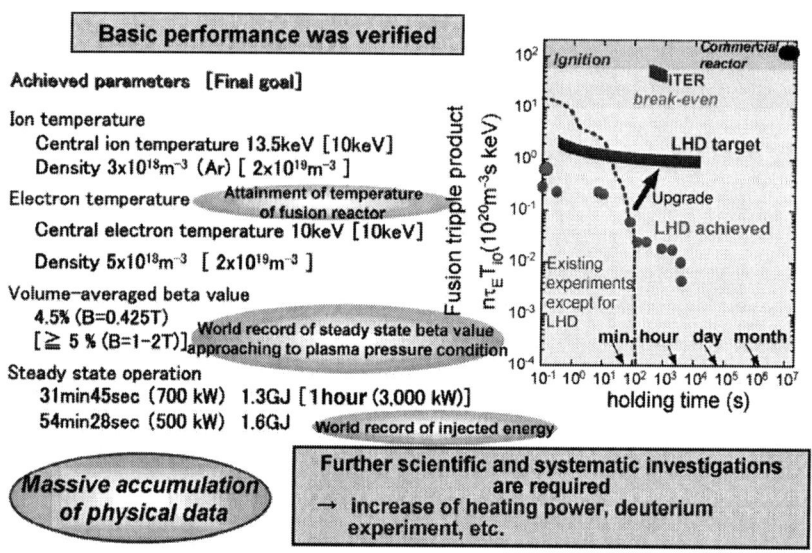

FIGURE 11. Target and achievement of plasma parameters in LHD.

Long pulse operation by means of the ion cyclotron resonant heating (ICRH) has shown progress. The pulse length has been extended to close to one hour (precisely 3268 sec) with a heating power of 500 kW. The total input energy for this discharge reached 1.6 GJ, which is a world record in magnetic confinement experiments [9]. This is summarized in Fig. 15. High beta discharges also have been advanced by exploration of aspect-ratio control and consequently a beta value of 4.5 % has been achieved [10]. The major data obtained recently are shown in Fig. 16.

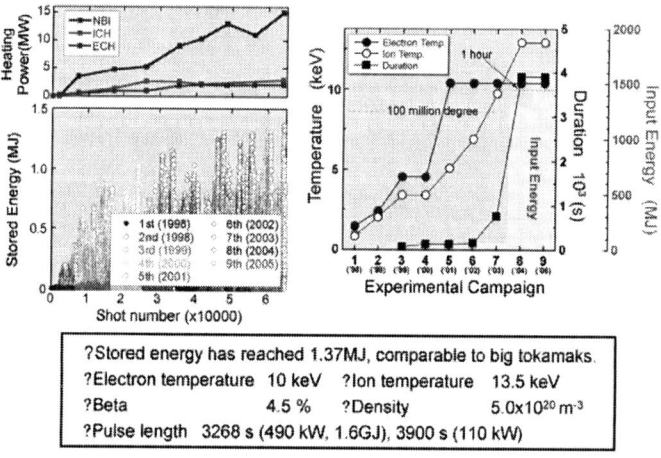

FIGURE 12. Steady progress of plasma parameters in LHD.

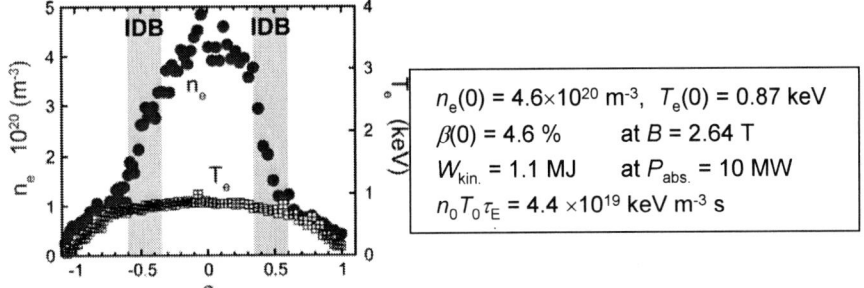

FIGURE 13. Steady progress of plasma parameters in LHD.

LHD produced about 10,000 plasma discharges in FY2005. This high availability of experimental opportunities has indicated a large potential to enable a variety of approaches for scientific research, which is not limited to fusion science in a narrow sense. These activities are made possible by the reliable operation of the superconducting facility. Much effort to improve the cryogenic stability has led to the successful R&D results of the pool-boiling of helium by the

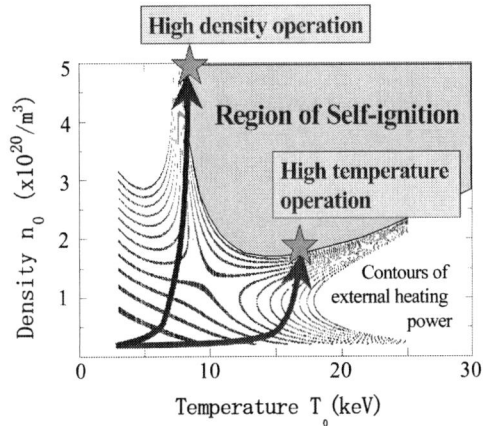

FIGURE 14. Routes to self-ignition.

sub-cooling modification. A consequent upgrade of the helical coil cooling system is scheduled in 2006 as one of the performance improvement plans of LHD [11].

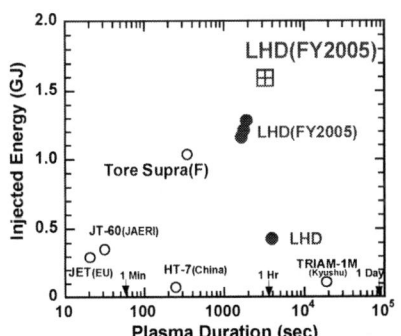

FIGURE 15. LHD has extended the frontier of long pulse plasma experiment.

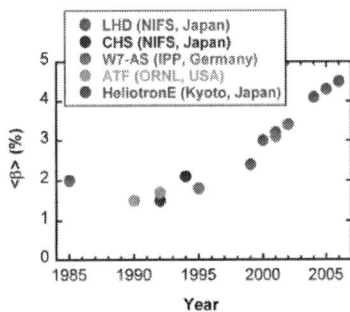

FIGURE 16. Progress of high beta plasmas.

4. SIMULATION SCIENCE AND FUSION ENGINEERING

Computer simulation studies at NIFS are oriented towards the exploration of "Simulation Science" while remaining founded in the large-scale simulation of fusion plasmas such as is the LHD. With the progress of super-computer performances, the paradigm shift of methodology is progressing from the classical scheme of reduction to the system elements to the integration of various interconnecting physics with different time-space scales. This methodology promises to lead to an understanding of the whole structure of natural phenomena. On the way from macro to micro scales, for example, the properties of equilibria with micro-scale effects such as the Hall term, pressure anisotropy, electron inertia, and wave-particle interaction are investigated, in order to understand the basis of extended MHD and consider the closure problems of fluid equations. This is the successful output of the numerical multi-layer re-normalization. Besides sophisticated studies on magnetic fusion plasmas, a variety of theoretical and simulation studies including inertial fusion plasmas, astrophysical plasmas and molecular dynamics have progressed steadily. The activity of the simulation science for the complex fusion plasmas is well highlighted in Fig. 17.

The fusion engineering research center has extended the advanced key-technology for a blanket system, low activation materials and superconducting magnet systems. These activities are closely related to a design study on the Force Free Helical Reactor [12]. The properties of Flibe and vanadium alloys are being intensively investigated.

A variety of coordinated researches is managed by the Coordinated Research Center. Three major activities are: coordinated research with industries and with other institutions in NINS, and an Atomic and Molecular Database. Above activities can be the seeds and rhizomes for interdisciplinary evolution originating from fusion science.

These research activities are integrated into three kinds of collaboration frameworks with their own distinguishing features. The framework of general collaboration research covers a wide spectrum of studies on fusion. The Large Helical Device (LHD) program collaboration research has facilitated participation in the LHD project

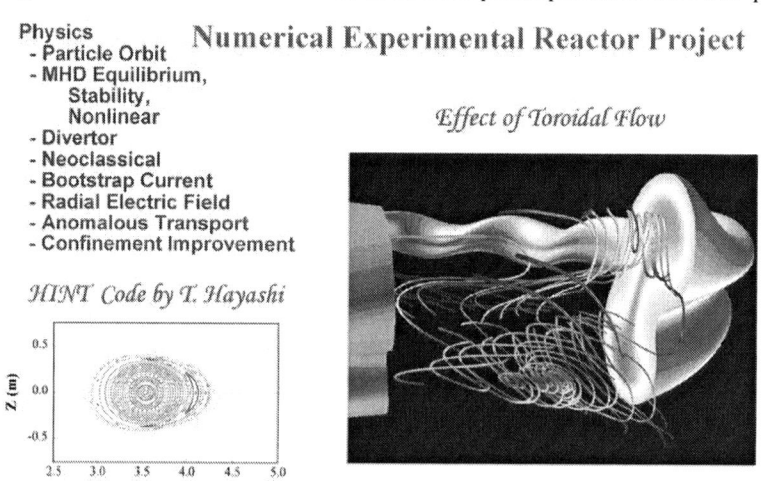

FIGURE 17. Simulation science (complex fusion plasmas).

358

based on the achievements at the universities. The bilateral coordinated collaboration research promotes mutual interaction on an equal footing with affiliated research institutes of the universities. More than 400 collaborating studies have been implemented during the covered period. In conjunction with the logistics to methodically support these three kinds of collaboration frameworks, the environment and infrastructure for efficient collaboration is being improved continuously.

5. DISCUSSIONS

Nowadays, fusion study has become a practical target. This means that this is no more just a dream of human mankind. Once the objective of the fusion research turns to be a real target, we shall be able to proceed to the next step making clear the critical path. ITER project is currently the most important critical path. In Fig. 18, the road map to LHD type DEMO reactor is shown. Integrating the output of LHD and ITER, and in addition to putting together the Numerical Test Reactor project should be sufficient to attempt to realize the DEMO Reactor within the next 30 years.

Nuclear Fusion Research in the World is an effort toward the Realization of Thermonuclear Fusion Energy, developing safe and environment-friendly energy source. It has the aspects of frontier science, since this area is always promoting global research of physics and engineering on 100 million °C plasma, leading the frontier of scientific and technological research and development in the world. Fusion has an infinite scientific interests, which are highlighted as plasma physics study, non-linear

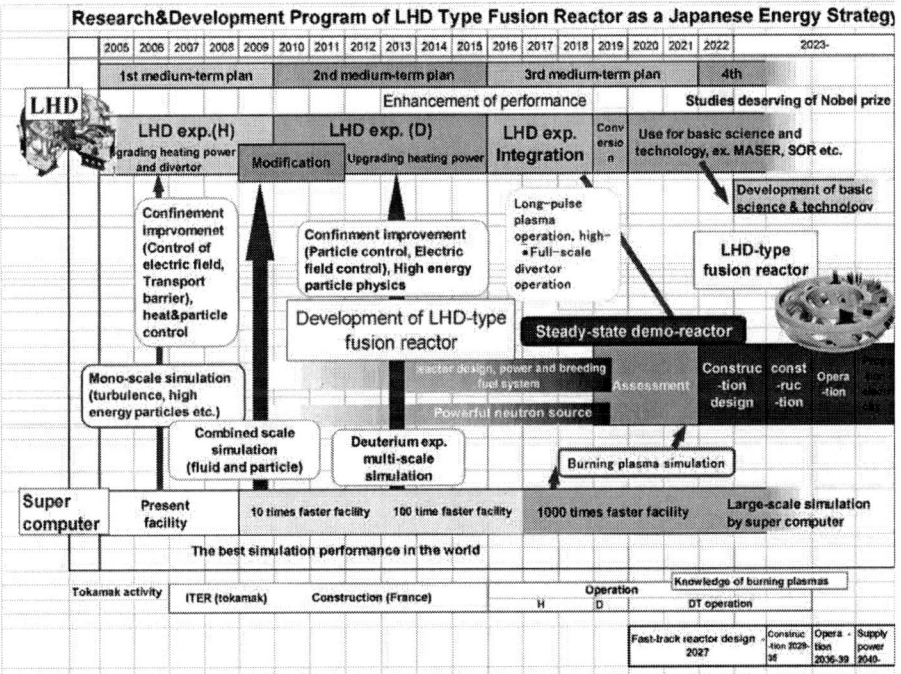

FIGURE 18. The road map to LHD type DEMO reactor.

and complex phenomena, non-local and non-equilibrium physics, plasmas within an atomic nuclei, etc. It is another important point that fusion is making, that is an active international cooperation, which results in a large contribution to the World's globalization. And most importantly, it is the steady Contribution to the Preservation of Global Environment and Global Peace. As it is well known, the dream of astronomers is to find the existence of higher civilization in the Space. The International ALMA Project at the desert in Chile was launched to investigate the life form in the outer Space, which is shown in Fig 19. If the existence of higher civilization could be discovered in tens thousands of light year distance afar, this is the civilization that has existed for tens thousands of years. The fact also proves that the civilization of our planet could continue for the duration as long as theirs. And, they must be actualizing the fusion energy!

FIGURE 19. The International ALMA Project

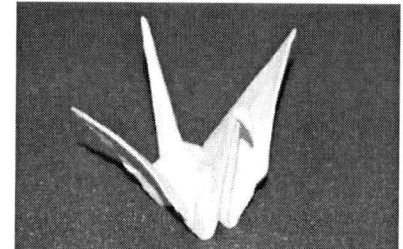

FIGURE 20. The ceramic crane sintered with microwave power.

Finally, I hope to show you one more picture in Fig. 20. This is the ceramic crane sintered with microwave power to heat the electron temperature up to 10 keV. This is one of the very recent results of the spin off of the fusion technology. I am showing it as my concluding slide since the crane is a symbol of Peace in Japan.

ACKNOWLEDGEMENTS

I wish to acknowledge fruitful discussions with Prof. M. M. Skoric and Prof. S. Masuzaki that helped me to complete this paper. I also thank Dr Lj. Hadzievski, Dr. B. Marinkovic, Dr N. Simonovic and Dr. D. Sevic for their efforts to make SPIG2006 a successful event.

REFERENCES

1. H. Bethe, Phys. Rev. **55**, 434-456 (1939).
2. http://www.iter.org/
3. J.D. Lowson, "Some Criteria for a Useful Thermonuclear Reactor", A.E.R.E. report GP/R 1807, December 1955, declassified April 9th 1957
4. O. Motojima et al., Plasma Phys. Control. Fusion **38**, A77-A92 (1996).
5. N. Ohyabu et al., Phys. Rev. Lett. **97**, 055002 (2006).
6. T. Morisaki et al., J. Nucl. Mater. **337-339**, 154-160 (2005).
7. R. Sakamoto et al., Nucl. Fusion **46**, 884-889 (2006).
8. F. Wagner et al., Phys. Rev. Lett. **49**, 1408 (1982).
9. K.Saito et al., presented in International Conference on Plasma Surface Interactions in Controlled Fusion Devices, Hefei, China (2006), submitted to J. Nucl. Mater.
10. O. Motojima et al., Fusion Eng. Des., to be published.
11. S. Imagawa et al., Fusion Eng. Des., to be published.
12. A. Sagara et al., Nucl. Fusion **45**, 258-263 (2005).

Computational Studies and Designs for Fast Ignition

H. Nagatomo[*], T. Johzaki[*], T. Nakamura[*], H. Sakagami[†], and K. Mima[*]

[*]Institute of Laser Engineering, Osaka University, 2-6 Yamada-oka Suita, Osaka 565-0871 JAPAN
[†]National Institute for Fusion Science, Oroshi-cho, Toki, GIFU 509-5292, JAPAN

Abstract. The fast ignition scheme is one of the most fascinating and feasible ignition schemes for the inertial fusion energy. At ILE Osaka University, FIREX (Fast Ignition Realization Experiment) project is in progress. Implosion experiments of the cryogenic target are scheduled in near future. There are two key issues for the fast ignition. One is controlling the implosion dynamics to form high density core plasma in non-spherical implosion, and the other is heating the core plasma efficiently by the short pulse high intense laser. The time and space scale in the fast ignition scheme vary widely from initial laser irradiation to solid target, to relativistic laser plasma interaction and final fusion burning. The numerical simulation plays an important role in demonstrating the performance of the fast ignition, designing the targets, and optimizing laser pulse shapes for the scheme. These all the physics are desired to be self-consistently described. In order to study these physics of FI, we have developed "Fast Ignition Integrated Interconnecting code" (FI³), which consists of collective Particle-in-Cell (PIC) code (FISCOF1D/2D), Relativistic Fokker-Planck with hydro code (FIBMET), and 2-dimensional Arbitrary-Lagrangian-Eulerian (ALE) radiation hydrodynamics code (PINOCO). Those codes are sophisticated in each suitable plasma parameters, and boundaries conditions and initial conditions for them are imported/exported to each other by way of DCCP, a simple and compact communication tool which enable these codes to communicate each others under executing different machines. We show the feature of the FI³ code, and numerical results of whole process of fast ignition. Individual important physics behind the FI are explained with the numerical results also.

Keywords: fast ignition, laser plasma, computational simulation.
PACS: 52.57.Kk, 52.38.-r, 52.65.-y

INTRODUCTION

Fast ignition is a new scheme of laser fusion [1], where a fuel shell is compressed with the use of long pulse (1-20 ns) laser beams to more than 1000 times of the solid density and is heated to create a hot spot at the edge of the compressed fuel by injecting a highly intense short pulse ($\sim$ PW, $\sim$ ps) laser at the moment of maximum compression. In the central ignition scheme, a highly uniform laser irradiation and strict power balance of multi-beam laser system are required to form a hot igniting spot at the center of the compressed fuel. In the fast ignition, such a requirement is relaxed, and all required is to achieve the high-density compression (Figure 1).

CP876, *The Physics of Ionized Gases: 23rd Summer School and International Symposium*, edited by L. Hadžievski, B. P. Marinković, and N. S. Simonović

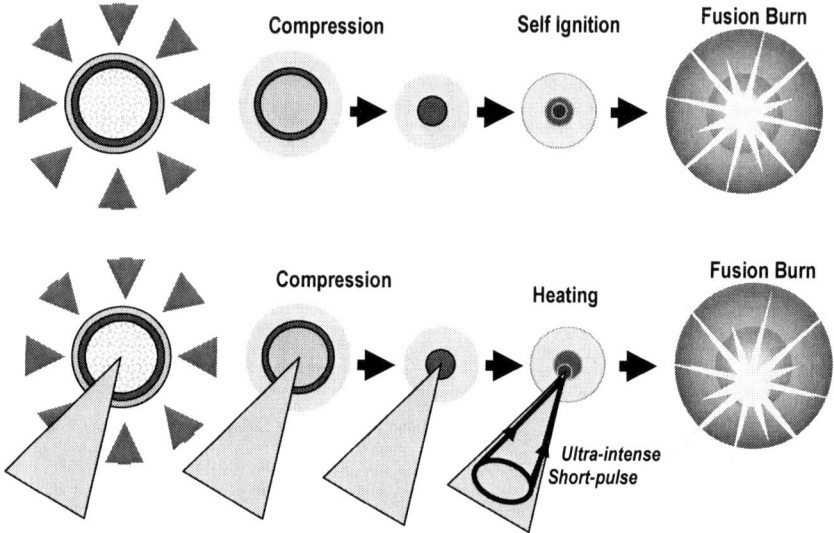

FIGURE 1. Central-hot-spot ignition (above) and fast ignition scheme (below).

The experiment for fast ignition research at Osaka University has been progressing since 2001. At the first stage of the experiments, the neutron yield increased from 10^4 without heating to 10^7, when a 400 J/0.6 ps PW laser was injected into a compressed CD shell. This indicates that the core plasma temperature increases by 500 eV and the energy coupling efficiency between heating laser and core plasma is 20-25% [2]

In the April of 2003, the construction of heating laser of 10 kJ/10 ps/1.06 μm, which is called LFEX (Laser for Fusion Experiment), for FIREX-I (Fast Ignition Realization Experiment) has started [3-4]. Target fabrication and irradiation system of foam cryogenic target are developed as the collaboration program between Osaka University and NIFS (National Institute for Fusion Science) [5]. A foam cryogenic cone target will be imploded and heated as a leading experiment in this year. After the completion of LFEX, we will irradiate a foam cryogenic cone shell target with LFEX in late 2007. The target fabrication technology is further developed to reduce the foam density to less than 20mg/cc by the end of FY2006. If the temperature of heated core plasma reaches higher than 5keV in FIREX-I, we plan to proceed to the FIREX-II as soon as possible.

The numerical simulation plays an important role in demonstrating the performance of the fast ignition, designing the targets, and optimizing laser pulse shapes for the scheme. There are two key issues for the fast ignition. One is controlling the implosion dynamics to form high density core plasma in non-spherical implosion, and the other is heating the core plasma efficiently by the short pulse high intense laser. The time and space scale in the fast ignition scheme vary widely from initial laser irradiation to solid target, to relativistic laser plasma interaction and final fusion burning. These all the physics are desired to be self-consistently described in numerical calculation. However, it is a formidable task to simulate relativistic laser plasma interaction and

362

radiation hydrodynamics in a single computational code, without any numerical dissipation, special assumption or conditional treatment.

Recently, we have developed "Fast Ignition Integrated Interconnecting code" (FI[3]) [6-7] which consists of collective Particle-in-Cell (PIC) code (FISCOF1: Fast Ignition Simulation COde with collective and Flexible particles), Relativistic Fokker-Planck with hydro code (RFP-hydro) code [7], and 2-dimensional Arbitrary-Lagrangian-Eulerian (ALE) radiation hydrodynamics code (PINOCO : Precision Integrated implosion Numerical Observation COde) [8]. Those codes are sophisticated in each suitable plasma parameters, and boundaries and initial conditions for them are imported/exported to each other by way of DCCP [9], a simple and compact communication tool which enable these code to communicate each others in different machines. In this paper, we will present the feature of FI[3] code, and individual codes which consists of FI[3]. In each section, numerical methods and the latest numerical results of these codes which are related to fast ignition are introduced briefly. Finally, fully integrated simulation of fast ignition is presented.

NUMERICAL METHODS

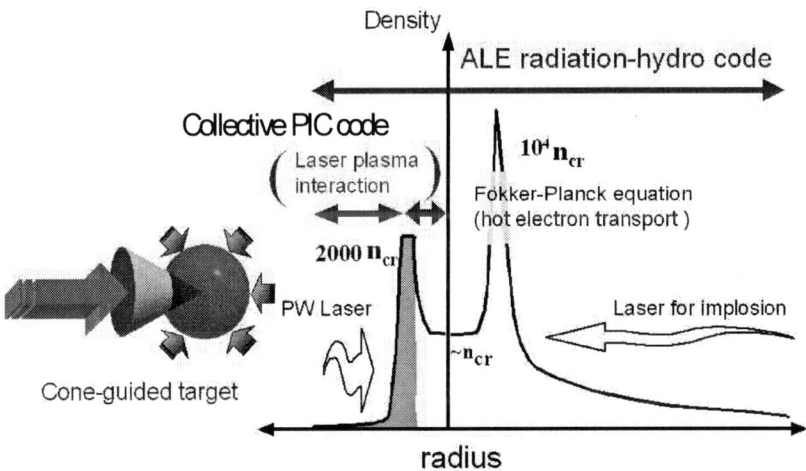

FIGURE 2. Plasma density of fast ignition implosion and computational code which can solve those plasma parameters.

Figure 2 shows the image of plasma density of fast ignition implosion and computational code which can solve those plasma parameters. These codes are best optimized for each plasma parameter regions to avoid undesired numerical dissipations or unwilling huge computing time. At first, cone-guided implosion dynamics is calculated by PINOCO because radiation hydrodynamics is dominant in implosion process. Near the maximum density timing, just before the irradiation of heating laser, the mass density, temperatures, and other profiles calculated by PINOCO are exported to both collective PIC and RFP-hydro code for their initial and boundary conditions.

363

The relativistic laser plasma interaction inside the cone target is simulated by collective PIC code, which exports the time-dependent energy distribution of fast electron to REP-hydro code. The fast electrons calculated by the FISCOF1 or 2 are exported to the RFP-hydro code. Therefore, the core heating process is simulated using both physical profiles of imploded core plasma and fast electron as the boundary conditions. The abstract of the profile data flows are illustrated in Fig.2.

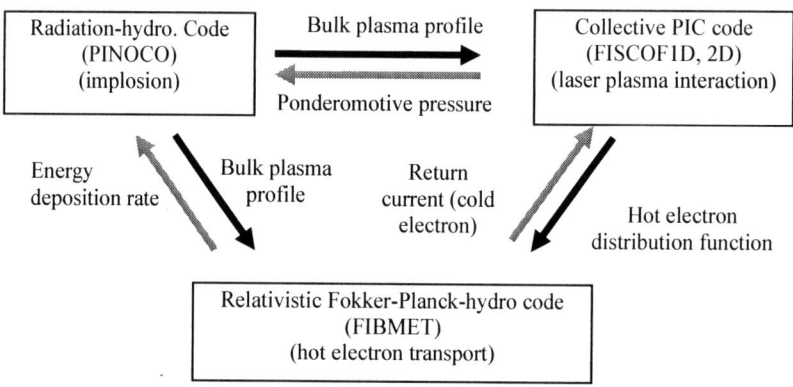

FIGURE 3. Data flow in FI³ system. Black arrows are already executable data flows, and gray arrows are next plan to be considered.

Radiation Hydrodynamics Code (PINOCO)

In PINOCO code, mass, momentum, electron energy, ion energy, equation of states, laser ray-trace, laser absorption, radiation transport, surface tracing and other related equations are solved simultaneously [6, 8, 10]. In the most of the integrated implosion codes except PINOCO, hydrodynamic equations are solved by Lagragian-based ALE method. But they are affected by numerical viscosity easily at rezoning/remapping process. Therefore, we have extended CIP (Cubic Interpreted Polynomial) method [11] into ALE type CIP method, so-call ALE-CIP. This modification has enabled the calculation of large dynamic range of the implosion. Originally, CIP has some characteristics of Lagrangian method, although the fundamental formulas are done for Eulerian coordinates. This CIP method is also employed to track the interface between the different materials clearly also. This tracking system is very useful when multi-material target structures must be considered. The equation of state is based on quotidian equation of state (QEOS) [12] with a fitting formula [13].

In the energy equations, flux-limited Spitzer-Harm type thermal transport model is solved using the implicit 9-point differencing for the diffusion equation with ILUBCG (Incomplete LU Biconjugate Gradient Conjugate) method. For the laser ray-tracing, a simple 1-D ray-tracing method is applied. The radiation transport solver was newly installed. Here, multi-group flux-limited diffusion-type equations are solved with ILUBCG implicit method. In the calculation of opacity and emissivity, LTE (Local Thermal Equilibrium) and CRE (Collisional Radiative Equilibrium) models are prepared for table lookup. Even though we can move the grid points as Lagrangian way in PINOCO simulation, so-called sliding-mesh in which the high resolution region is sliding along the mass center of the target is used in all these simulations. We have to remark that the advantage of the sliding-mesh is not only simple rezoning rule but also better convergence of the iteration method in solving the diffusion equations. Only an implosion code which is based on a high-order scheme such as PINOCO can simulate the problem with the sliding-mesh.

Numerical Example of Implosion Simulation using PINOCO

The formation of high density core plasma is one of the most important issues for FI scheme, as well as heating core plasma problem. For the preliminary study, we have performed the non-spherical implosion with initial perturbation on the target surface to estimate the effect of Rayleigh-Taylor instability [14].

The cone with an opening angle of 30 degree is attached to a spherical shell of polystyrene (ρ=1.06 g/cm^3) which has a uniform thickness of 6 μm. The target is irradiated by uniform laser of which wavelength, energy and pulse width are, λ= 0.53 μm, 6.0 kJ and 1.2 ns (Gaussian, FWHM), respectively. About 70% of the total energy is used until they finish imploding, that is, the effective laser energy is not more than 4.5kJ for each simulation. These target structure and laser pulse shape are not optimize for the cone-guided implosion, because the tailored pulse does not work well in some of them for the existence of the gold cone. If the spherical shell is irradiated by the low intensity laser, the ablated CH plasma hits the surface of the gold cone. That plasma blocks the irradiation of main laser pulse near the cone. For the limitation of computational resources, radiation transport is not included in all these simulations. The configuration of the target is shown in Fig. 4. For the comparison, some spherical implosions without cone target are also carried out. In these simulations, the initial perturbation of mode number ℓ = 0, 6, 12 and 24 are given onto the outer surface of CH ablator, and the amplitudes of the perturbation are 0, 0.1, 0.03, and 0.03 μm respectively. In the table 1, these simulation conditions are summarized. All the cases, the number of computational grid points is fixed to be 300x300 points, which are distributed to the shell target and gold cone region mainly.

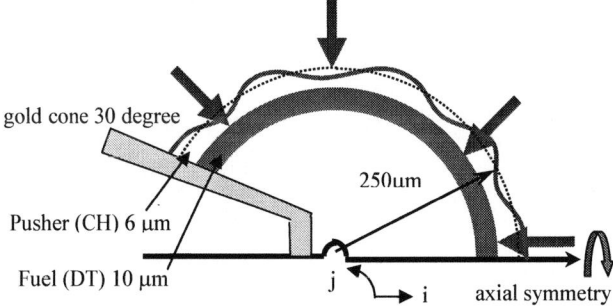

FIGURE 4. Target configuration for FIREX-I experiment, which is spherical CH-DT shell with gold conical target.

Figure 5 shows the contours of the mass density and electron temperature at the timing of maximum mass density in case C-0 (Fig.5 (a)), and in case C-24 (Fig.5 (b)). Although the perturbation at the DT-CH interface is grown strongly due to the RT instability, they keep the almost same implosion velocity and obtained the maximum mass density at almost same timing. In both cases, the hot spot is pushed away from the center of implosion, and that dynamics enable to achieve the high averaged ρR. When the ρR is reached at the maximum, the tip of the gold cone still stay and keeps its shape, which is significant issue for the heating problem of fast ignition. The black lines indicate the contact surface between CH pusher and DT fuel. Around the cone, there are "dead area" where fuel are not accelerated, and not compressed sufficiently because of the existence of the cone. The open angle of the cone, cone material, and thickness of the cone will be optimizing parameters to control the implosion dynamics.

The calculated values of the maximum mass density and areal density ρR of the fuel at the maximum compression time is the essential for the fast ignition scheme. ρR and the maximum mass density are indicated in the table 1. Even though the maximum mass density of DT of case C-0 is lower than that of case S-0, the averaged ρR of case C-0 is higher than that of case S-0. This is caused by the shift of the hot spot and DT fuel drop into the center of the implosion. This fact was reported by our previous work [6] also. If there are initial perturbations, in the spherical implosion cases (S-6, 12 and 24), the averaged ρR are reduced to be 65% of the ideally spherical implosion case of S-0. On the other hand, in case C-6, 12, and 24, the averaged ρR are same level as the S-0 case, or higher. That is, non-spherical implosions are robust under the existence of non-linear hydrodynamic instabilities. In these simulations, the hydrodynamic instabilities of higher harmonic mode are appeared and grow rapidly, and they may affect to the results. For example, those implosions of S-6, 12 and 24 have similar attribute with each other. Even though the analysis of higher mode instability is important, we need more fine computational grids. Also, the radiation transport is not included in this work to save the computing time. The radiation transport may stabilize the RT instability, but the dynamics of the gold cone will be affected by the radiative heating also. Some our previous work suggests that the gold cone inside of the shell was ablated by the irradiation.

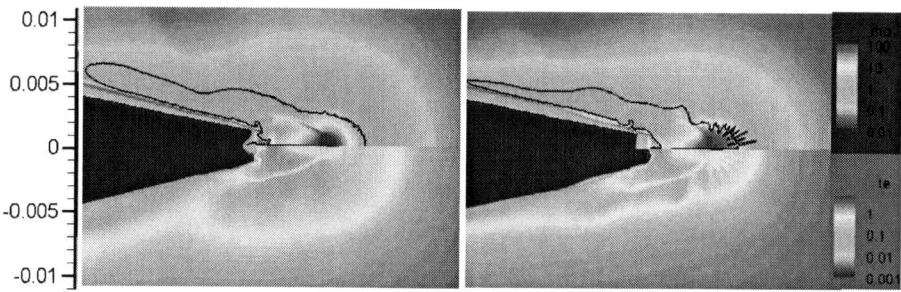

Figure 5. Mass density contours in g/cm^3 (above) and electron temperature contours in keV (below), non-perturbed shell target (a) and a perturbed shell of mode ℓ =24 (b). The black lines are contact surface between CH pusher and DT fuel

TABLE 1. Average ρR and fuel density.

	Spherical target				Cone-guided target			
perturbation mode	0	6	12	24	0	6	12	24
perturbation amplitude (μm)	(0)	0.10	0.03	0.03	(0)	0.10	0.03	0.03
case	S-0	S-6	S-12	S-24	C-0	C-6	C-12	C-24
averaged ρR (g/cm^2)	0.151	N/A	0.0981	0.0935	0.203	0.140	0.151	0.163
the maximum fuel density (g/cm^3)	250	N/A	163	145	133	89.0	101	104
averaged fuel density (g/cm^3)	137	N/A	83.7	77.8	93.1	64.9	69.3	72.2

From these results, we concluded that even if there is hydrodynamic instability, high areal density can be achieved, though high temperature hot spot can not be expected which is required for conventional central ignition scheme. This is the advantage of fast ignition scheme.

Collective PIC code (FISCOF1D and 2D)

For evaluation of fast electron generation due to relativistic laser-plasma interactions (LPI), we use the 1D and 2D collective PIC code (FSCOF1, 2) [11, 12], where collective particles are used to represent many normal particles and then total number of particles and computations are drastically reduced. Even though, the FISCOF1D/2D code enables us to treat a wide range in space and high density region, the exact condition can not be considered for the limitation of computer resources. Therefore, we simulated the small region of relativistic laser plasma interaction and this fast electron bean profile was extended to the realistic scale size in FI^3. This assumption was temporal and will be eliminated after developing accelerated fully parallelized two-dimension Collective PIC code is developed in near future.

Numerical Example of Laser Plasma Interaction using FISCOF2D

Interaction of ultra-intense short pulse laser with cone target is studied using FISCOF2D to understand the characteristics of high energy electrons emitted from the cone target and propagating towards core plasma for fast ignition [16]. Typical cone target geometry is shown in Fig.6. The density of cone target is 100 n_c with pre-plasma whose scale length is 1 micron. The target rear side is surrounded by underdense plasma whose density is 2 n_c which is taken from the result of hydro simulation (PINOCO). The initial electron temperature is set to 10 keV. Ion motion is not taken into account which is considered acceptable for short laser pulse irradiation. The pulse has Gaussian profile in radial direction with focal spot size of 10 micron, and flat profile in time with duration of 150 fs, which irradiates the target from left boundary. The laser peak intensity is 5x10^{19} W/cm^2. Laser pulse is lowest mode of Hermite-Gauss mode, and is focused at the left boundary. Since the Rayleigh length is about 300 μm, most of the laser pulse reaches the target surface without diffraction. For the comparison, planner target with the same laser pulse as the cone target case is performed.

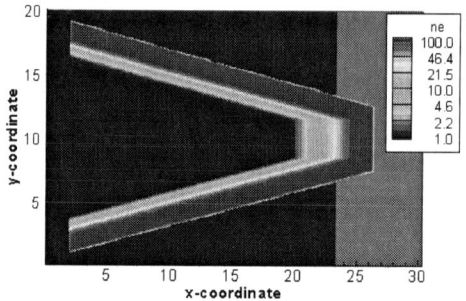

FIGURE 6. The cone target geometry for 2D simulation (initial electron density profile).

The spectrum from cone target is shown in Fig. 7(left). It is found that there are three effective temperatures characterizing the hot electrons. For cone target case, the temperature for high energy component (T_h) is 5.0 MeV, and temperature for electrons of middle range (T_m) is 1.9 MeV. On the other hand, there are two temperatures in plane target. The temperature for high energy component is lower than T_h, and is down to 2.5 MeV. This temperature is explained with ponderomotive energy. The temperatures of low energy component are similar in both cases, but higher energy components are quite different. This is due to two important features of cone geometry, which are laser intensification by cone-guiding and larger interaction area at the cone wing. These two features lead to effective acceleration at cone tip and wing, which will be explained in the following subsections. The absorption rate is about 40 % for cone target, and 20 % for plane target. This higher coupling efficiency from laser to electrons is another advantage

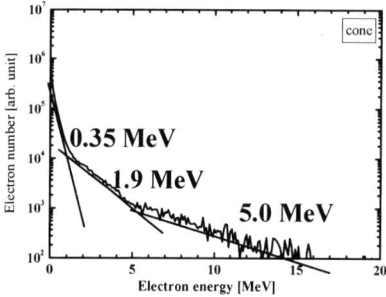 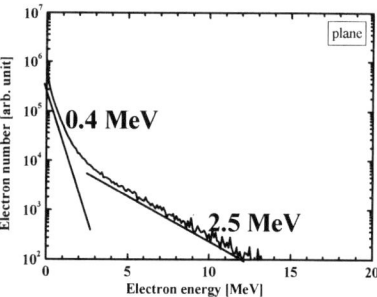

FIGURE 7. Electron spectrum of cone target (left) and electron spectrum of plane target (right) respectively.

Relativistic Fokker-Planck hydrodynamic code (FIBMET)

A 2-D Relativistic Fokker-Planck (RFP) code, FIBMET [8] has been developed for analysis of the fast electron transport and energy deposition processes in dense core plasma, which was coupled with an Eulerian hydrodynamic code to examine core-heating properties. In this code, cold bulk electrons and ions are treated by a 1-fluid and 2-temperature hydro model, and the fast electrons generated by the ignition laser-plasma interactions are treated by the RFP model. In the coupled RFP-hydro code includes magnetic field generated by fast electron current, gradient of plasma resistivity and pressure gradient.

Initial and boundary condition can be imported from PINOCO and FISCOF1D/2D as data flows which are shown in figure 3. Numerical example of FIBMET is written in the latter section "Full integrated Simulation of Fast Ignition".

Distributed Computing Collaboration Protocol (DCCP)

We have three set of codes which have quite different properties in FI^3 project. PINOCO and FIBMET codes are optimized for a vector parallel machine, and FISCOF1D/2D is optimized for a massive scalar parallel machine. Therefore, it is preferable to be combined with each others which use different machines by way of simple and easy communication tool for computational scientists. Since communication in our project is not complex and very straightforward, we design a special lightweight communication protocol, Distributed Computing Collaboration Protocol (DCCP), to transfer data between codes [9].

DCCP is implemented with two kinds of daemon programs and simulation code itself. One of daemon programs is called Communicator that actually transfers data instead of the code, and the other is called Arbitrator that manages communication between Communicators. The code does not send data directly to another code, but only asks the local Communicator, which is running in the background at his site, to transfer his data. The sender code, therefore, does not have to know details of the receiver code, such as IP address of the receiver computer. And then, the sender side Communicator inquires of the Arbitrator where is the remote Communicator that is handling the receiver code, and forwards the requested data to the appropriate Communicator via the Internet. The receiver side Communicator stores the data to storage, and is waiting for a demand. Finally the receiver code requires the data from the receiver side Communicator, and then the Communicator sends the data to the code and communication between two codes has just finished. If the receiver code does not run yet when the sender code sends data to the local Communicator, the Arbitrator orders it to save the data to storage for a future usage. Once the receiver code is invoked, the Arbitrator directs the sender side Communicator to restore and transfer data to the receiver side Communicator, and now the receiver code can get the desired data. Thus the sender code does not also have to take care of a situation whether the receiver code is running or not. Furthermore, if a broadband dedicated line is available between the Communicators, the Arbitrator tells both Communicators to use that dedicated line instead of the Internet and high-speed communication will be performed even both codes do not know about details of network configuration.

TCP/IP based lightweight protocol does not transfer data directly to another code manage information of codes Arbitrator.

FULLY INTEGRATED SIMULATION OF FAST IGNITION

In this section, an example of fully integrated simulation using FI^3 system is presented [17]. This simulation conditions correspond to the GXII experiment at ILE Osaka Univ. [1]. The detail profile of mass density, temperatures, and other conditions of imploded core plasma is imported from PINOCO simulation. The profiles of fast electrons generated in the LPI were evaluated with FISCOF1D that includes both ion and electron motions but not the collision process. The Au cone tip was modelled by the 10µm-thickness plasma with $n_e = 100n_c$. The 60µm-thickness imploded plasma was put behind the cone tip. The simulations were carried out by assuming that the electron density in the rear of cone tip is $n_{e, rear} = 100n_c$, $10n_c$ and $2n_c$. The forward-directed fast electrons were observed behind the cone tip every 20fs. A 750 fs Gaussian pulse ($\lambda_L = 1.06$ µm, $I_L = 10^{20}$ W/cm^2) was assumed as a heating laser. The temporal profiles of intensity $I_{ff}(t)$ and average energy $<E_{ff}>$ of forward-directed fast electrons observed in the three cases are plotted in Fig.3 (a) and (b). The time-integrated energy spectra $f_{ff}(E)$ are plotted in Fig.3 (c). The intensities become maximum at $t \sim 1.25$ps in all cases. The peak intensities are 2.8×10^{19}, 1.7×10^{19} and 0.8×10^{19} W/cm^2 in the cases of $n_{e, rear} = 100n_c$, $10n_c$ and $2n_c$, respectively.

In plasmas, the REB drives the bulk electron return current for keeping the current neutrality. If the bulk electron density is comparable to the REB density, the drift velocity of bulk electron becomes up to the order of light velocity. As the result, strong two-stream instability is induced, which heats up the bulk electrons and inhibits the bulk electron flow. Then, the strong static field is build up, which rapidly slows down the fast electrons and accelerates the bulk return current. Thus, in the low $n_{e, rear}$ case, part of the fast electrons cannot enter the rear side plasma and are confined inside the cone tip, which results in decreasing the peak intensity. After reaching the peak intensity, the REB intensity decreases with the laser pulse in the case of $n_{e, rear} = 100n_c$. Contrary to this, the forward-energy flow is kept constant at the intensity of $\sim 10^{18}$W/cm^2 in the other two cases since the hot electrons confined inside the cone tip (we call them 'sloshing component') are continuously released from there even after the laser irradiation.

Using the imploded-core profile of PINOCO at $t = 1.97$ns, and the time-dependent momentum profiles of REB obtained in the cases of $n_{e, rear} = 2n_c$, $10n_c$ and $100n_c$, we evaluated the core heating properties using FIBMET. The REB source was injected behind the cone tip by assuming the super Gaussian profile with 30µm width. Figure 4 shows the temporal profiles of bulk-electron and ion temperatures averaged over the core region ($\rho > 10$g/cc).

Compared with the results between $100n_c$ and $10n_c$ cases, the temporal evolution of core temperature is different because of the difference in intensity profiles of REBs. However, there are little differences in $\eta_{L \rightarrow e}$ and the spectrum, so that the core temperatures reach the same value (0.45keV) at the end of core heating. The core size ($\int \rho dz = 0.14$ g/cm^2) is smaller than the range of MeV electron (e.g., 0.6g/cm^2 for 1MeV electron), so that the most of fast electrons penetrate the core. The energy coupling from the REB to the core is 22% in both cases and then the coupling from the heating laser to the core is only 4.6%. In the case of $n_{e,rear} = 2n_c$, the beam intensity and the $\eta_{L->e}$ are small compared with the other cases, so that the resultant core temperature is about 0.35 keV.

This result does not exactly agree with the experiment [1] where core plasma was heated to 0.6 keV. We recognize some important physics are still missing, for example, MHD feature in implosion simulation, static field in FIBMET, collisions in FISCOF1D/2D and so on. They will be improved with the further developments of computational codes, and increments of computational resources.

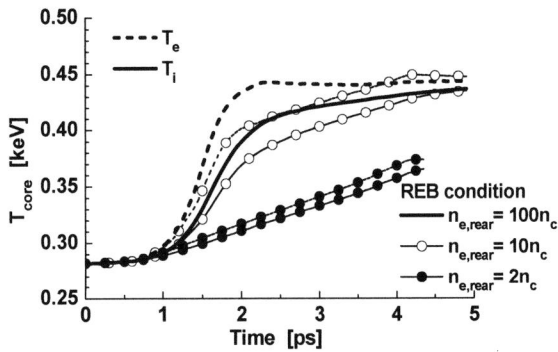

Figure 4. Temporal profiles of bulk-electron and ion temperatures averaged over the dense core region ($\rho > 10g/cc$) obtained for the three different REB conditions ($n_{e,rear} = 2, 10$ and $100\ n_c$).

SUMMARY

Experiments, theoretical and computational researches for the fast ignition scheme have been implemented at ILE Osaka University. In order to simulate the whole process of the complex multi scale plasma physics in the fast ignition, we have developed FI^3 system where three different computational codes are connected to each others. In this paper, we have introduced these latest numerical methods and results for the fast ignition. Though there are some disagreements with the experiments, we study the detail of the physics and improve the numerical codes. This code system can be applicable not only to the IFE but also to the other applications, for example, particle acceleration for medical treatment, and laser nuclei physics.

ACKNOWLEDGMENTS

This work was supported by MEXT, Grant-in Aid for Creative Scientific Research (15GS0214). This simulation was executed on the computers at ILE and Cyber Media Center, Osaka University.

REFERENCES

1. M. Tabak, *et al.*, *Phys. Plasmas* **1**, 1626, (1994).
2. R. Kodama, *et al.*, *Nature*, **418,** 933, (2002).
3. Y. Izawa, et. al., IAEA-CN-116/OV/3-2, (2004).
4. N. Myanaga, *et al*, *J. Phys. IV Frances, EDP Sciences,* **133**, 81-87 (2006).
5. A. Iwamoto, *et al*, *J. Phys. IV Frances*, **133**, 899-901 (2006).
6. H. Nagatomo, *et. al.*, IAEA-CN-116/IFP/07-29, (2004).
7. T. Johzaki, *et al.*, *J. Plasma Fusion Res.*, SERIES, **6,** 341, (2004).
8. H. Nagatomo, *et al.*, *Plasma Physics*, **669**, 253-256, 2003.
9. T. Sakaguchi, *et al.*, *Parallel and Distributed Computing: Applications and Technologies*, LNCS 3320 Heidelberg, Springer-Verlag, 90-93, (2004).
10. H. Nagatomo *et al.*, *Proceedings of IFSA 2001*, Paris: Elsevier, 140-142, (2001).
11. T. Yabe, *et al.*, *J. Comp. Phys.* **169,** 556-593 (2001).
12. R.M. More, *et al.*, *Phys. Fluids*, **31,** (10), 3059-3078 (1988).
13. K. Takami and H. Takabe, *Tech.Rep of Osaka Univ.*. Osaka Univ., **40**, 159 (1990).
14. H. Nagatomo, *et al*, *J. Phys. IV Frances,* **133**, 397-400, (2006).
15. H. Sakagami, *Proceedings of IFSA 2001*, Paris: Elsevier, 434-437 (2003).
16. T. Nakamura, *et al*, *J. Phys. IV Frances*, **133**, 405-408, (2006).
17. T. Johzaki, *et al*, *J. Phys. IV Frances*, **133**, 385-489, (2006).

.

The Peculiar Absorption And Emission Phenomena From Stars To Quasars

E. Danezis[1], L. Č. Popović[2], E. Lyratzi[1], M. S. Dimitrijević[2]

1. University of Athens, Faculty of Physics, Department of Astrophysics, Astronomy and Mechanics,
Panepistimioupoli, Zographou 157 84, Athnes – Greece
2. Astronomical Observatory, Volgina 7, 11160 Belgrade, Serbia

Abstract. The spectra of Hot Emission Stars and AGNs present peculiar profiles that result from dynamical processes such as accretion and/or ejection of matter from these objects. In this paper we analyze DACs and SACs phenomena, which indicate the existence of layers of matter with different physical conditions and we propose that these phenomena can explain the spectral lines peculiarity in Hot Emission Stars and AGNs. We also propose a new model with which we can study the density regions in the plasma surrounding the studied objects, where DACs and SACs of a spectral line are created producing the observed peculiar profiles. Finally, we present some tests to justify the proposed model.

Keywords: Hot Emission Stars, AGNs, DACs, SACs.
PACS: 97.10.Ex, 97.10.Fy, 97.20.Ec, 97.30.Eh, 98.54.Aj

THE SPECTRAL LINES IN ASTROPHYSICAL OBJECTS

It is well known that the absorption spectral lines that we can detect in the spectra of normal stars are an important factor to study physical parameters stellar atmospheres.

In the classical stellar spectra we observe "normal" absorption lines (Fig. 1). However, in the spectra of hot emission stars (Oe and Be stars) we observe spectral lines with complex and peculiar profiles (combination of absorption and emission spectral lines, e.g. P Cygni profiles etc., Fig. 2). The same phenomena also occur in the case of galactic spectra. This means that "normal" galaxies present spectra without emission lines, which are composed spectra of stars from the galaxy. In contrary Active Galactic Nuclei (AGNs) present emission lines (Hα, Hβ) like active Oe and Be stars (Fig. 3). The peculiar lines are always characteristic of the objects with very dynamical processes (accretion, jets, winds etc.). For example, in the case of stars, the peculiar spectral profiles are created in density regions of matter that we can detect quite away from the stellar object (Fig. 4), while in the case of AGNs, accretion, wind (jets, ejection of matter etc.), Broad Line Regions (BLR) and Narrow Line Regions (NLR) are responsible for the construction of the observed peculiar profiles of the spectral lines (Fig. 5).

The spectral lines of hot emission stars and AGNs may be very broad and satellite lines may appear (Fig. 6). An answer for the origin of satellite lines is the matter that exists between the observer and an object (Fig. 7).

CP876, *The Physics of Ionized Gases: 23rd Summer School and International Symposium*,
edited by L. Hadžievski, B. P. Marinković, and N. S. Simonović
© 2006 American Institute of Physics 978-0-7354-0377-2/06/$23.00

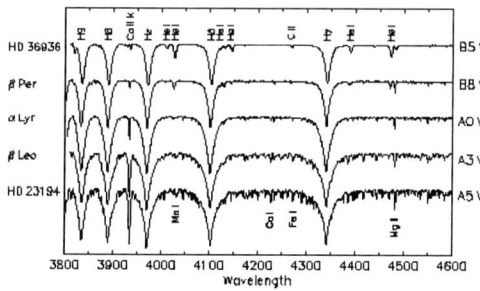

FIGURE 1. Classical Stellar Spectra for different spectral subtype.

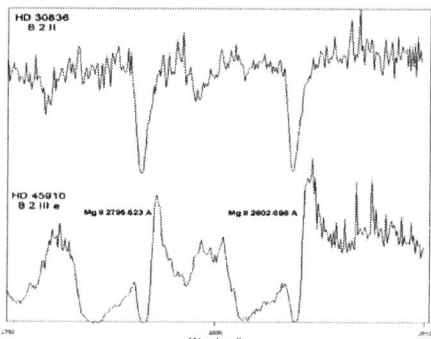

FIGURE 2. Comparison of Mg II resonance lines between the spectrum of a "normal" B star and the spectrum of an active Be star that presents complex and peculiar spectral lines. The combination of an emission and some absorption components construct the P Cygni profile.

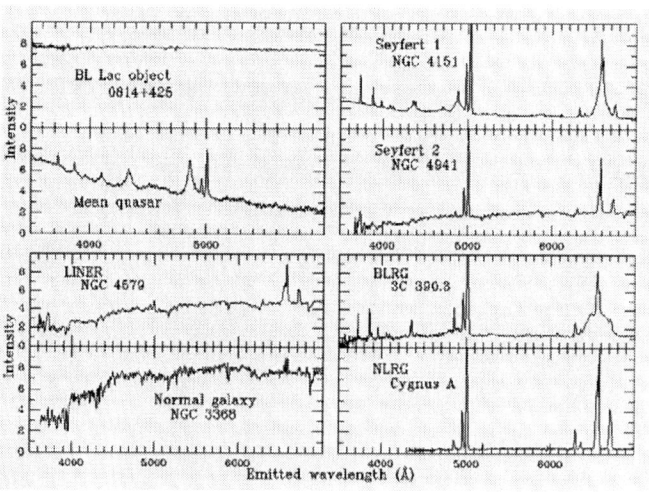

FIGURE 3. Spectra of different types of AGNs.

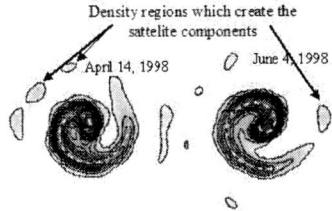

Density regions which create the
sattelite components

April 14, 1998   June 4, 1998

FIGURE 4. Around a Wolf-Rayet star (WR 104) we can detect density regions of matter quite away from the stellar object, able to produce peculiar profiles. (This figure is taken by Tuthill, Monnier & Danchi [1] with Keck Telescope.).

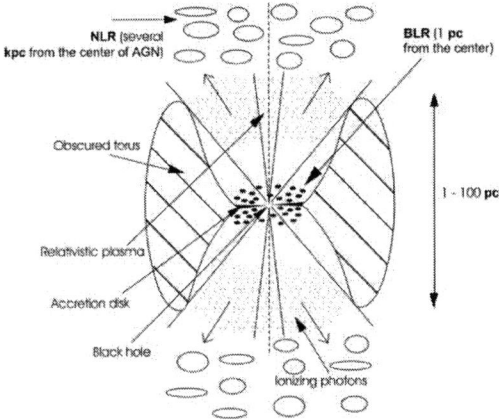

NLR (several
kpc from the center of AGN)

BLR (1 pc
from the center)

Obscured torus

1 - 100 pc

Relativistic plasma

Accretion disk

Black hole

Ionizing photons

FIGURE 5. In the case of AGNs, accretion, wind (jets, ejection of matter etc.), BLR (Broad Line Regions) and NLR (Narrow Line Regions) are the density regions that construct peculiar profiles of the spectral lines.

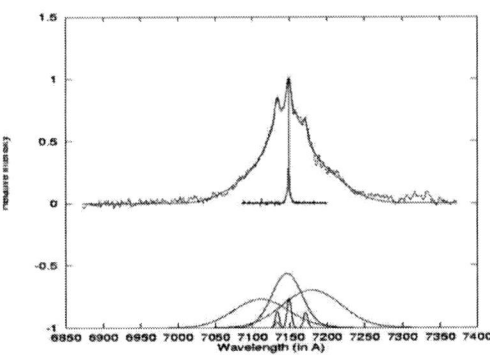

FIGURE 6. Comparison between the observed Hα line of an AGN (III Zw2) with the same line obtained from laboratory plasma. This spectral line is blended with two [NII] satellite lines. The line broadening is a kinematical effect (radial, rotational and random velocities) that arises from the geometry of the emitting region.

375

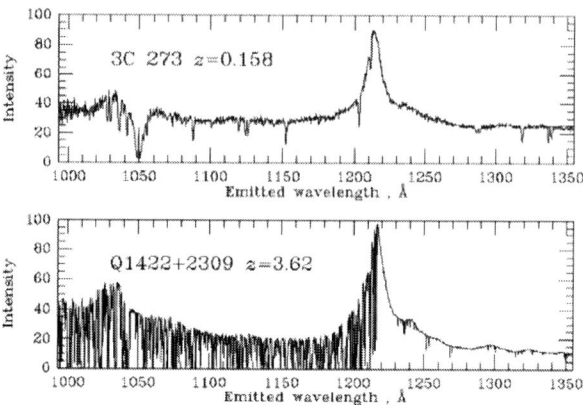

FIGURE 7. Difference between absorption lines in quasar spectra: The absorption lines which are created in the heart of a quasar (up) in comparison with the absorption that can be constructed by matter located between an observer and the quasar.

A MODEL FOR THE REPRODUCTION OF PECULIAR SPECTRAL LINE PROFILES

There are some models to reproduce the observed peculiar profiles in hot emission stars and AGNs. These are several Non-LTE models, which give bad reproduction and several Non-LTE specialized models (e.g. PHOENIX), having very complicated codes that may reproduce the peculiar profiles, but only in some cases.

In order to explain the peculiar profile that we observe in the spectra of hot emission stars and AGNs, Danezis et al. [2, 3] proposed a simple model that is able to explain the structure of the regions that produce these spectral lines. We point out that with this model we can study and reproduce specific spectral lines. This means that we can study specific density regions in the plasma surrounding the studied objects. In order to construct a general model we need to study a number of density regions which produce spectral lines of different ionization potential, meaning different temperature and thus different distance from the studied object.

In order to explain simply our model, we need to explain two similar phenomena, the DACs and SACs phenomena, able to construct peculiar spectral line profiles in some Hot Emission Stars and AGNs.

The DACs Phenomenon

In a stellar atmosphere or disc that we can detect around hot emission stars, an absorption line can be produced in several regions that present the same temperature. From each one of these regions an absorption line arises.

The line profile of each one of these absorption components is a function of a group of physical parameters, as the radial, the rotational, the random velocities and the optical depth of the region that produces the specific components of the spectral line. These spectral lines were named Discrete Absorption Components (DACs) [4].

376

DACs are discrete but not unknown absorption spectral lines. They are spectral lines of the same ion and the same wavelength as a main spectral line, shifted at different $\Delta\lambda$, as they are created in different density regions which rotate and move radially with different velocities [2]. DACs are lines, easily observed, in the spectra of some Be stars, because the regions that give rise to such lines, rotate with low velocities and move radially with high velocities (Fig. 8).

It is very important to point out that we can detect the same phenomenon in the spectra of AGNs, so called Broad Absorption Line Quasars (BAL QSOs). In Fig. 9 one can see the C IV UV doublet of the BAL QSO PG 0946+301.

From the values of radial displacements and the ratio of the line intensities we can detect that the two observed C IV shapes indicate the presence of a DACs phenomenon similar to the DACs phenomenon that we can detect in the spectra of hot emission stars.

The SACs Phenomenon

However, if the regions that give rise to such lines rotate with large velocities and move radially with small velocities, the produced lines have large widths and small shifts. As a result they are blended among themselves as well as with the main spectral line and thus they are not discrete. In such a case the name Discrete Absorption Component is inappropriate and we use only the name Satellite Absorption Components (SACs) [3].

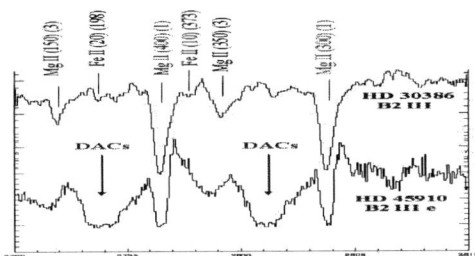

FIGURE 8 Mg II doublet in the UV spectrum of AX Mon, compared with the Mg II lines of the classic B2 star HD30386. In these line profiles we can see the main spectral line and in the left of each one of them a group of DACs.

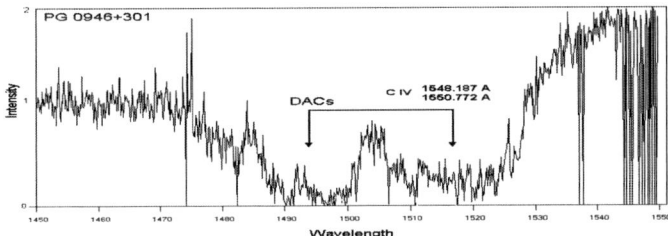

FIGURE 9. C IV UV doublet of the quasar PG 0946+301. We can detect that the two observed C IV shapes indicate the presence of a DACs phenomenon similar to the DACs phenomenon that we can detect in the spectra of hot emission star.

Calculation Of The Peculiar Line Shapes

In the case of SACs phenomenon we need to calculate the line function of the complex line profile.

Recently Danezis et al. [2, 3] proposed a model in order to explain the complex structure of the density regions of hot emission stars and some AGNs, where the spectral lines that present SACs or DACs are created.

The main hypothesis of this model is that the stellar envelope is composed of a number of successive independent absorbing density layers of matter, a number of emission regions and an external general absorption region. By solving the radiation transfer equations through a complex structure, (in more details see [2, 3]) we obtained a function for the line profile, able to give the best fit for the main spectral line and its Satellite Components at the same time:

$$I_\lambda = \left[I_{\lambda 0} \prod_i e^{-\tau_{ai}} + \sum_j S_{\lambda ej} \left(1 - e^{-\tau_{ej}} \right) \right] e^{-\tau_g} \tag{1}$$

where: $I_{\lambda 0}$: is the initial radiation intensity, $S_{\lambda ej}$ is the source function, which, at the moment when the spectrum is taken, is constant and τ is the optical depth in the center of the considered component.

In Eq. (1), the functions $e^{-\tau_{ai}}$, $S_{\lambda ej} \left(1 - e^{-\tau_{ej}} \right)$ and $e^{-\tau_g}$ are the distribution functions of each satellite component and we can replace them with a known distribution function (Gauss, Lorentz, Voigt).

An important fact is that in the calculation of I_λ we can include different geometries (in the calculation of τ) of the absorbing or emitting independent density layers of matter.

The decision on the geometry is essential for the calculation of the distribution function that we use for each component. This means that for different geometries we have different line shapes, presenting the considered SACs.

In the case of rapidly rotating hot emission stars, it is very important to insert in the line function (Eq. 1) the rotational, radial and random velocities of the regions which produce the satellite components. In this case we have to assume the geometry for the corresponding regions.

The Spherical Symmetry Hypothesis

In order to assume the appropriate geometry we took into account the following important facts:

The spectral line profile was reproduced in the best way when one assumes spherical symmetry for the independent density regions. Such symmetry has been proposed by many researchers [5-11].

However, the independent layers of matter, where a spectral line and its SACs are born, could lie either close to the star, as in the case of the photospheric components of the Hα line in Be stars [12, 13], when spherical symmetry is justified, or at a larger distance from the star, where the spherical symmetry can not be justified.

These lead us to conclude that:

1. In the case of independent density layers of matter which lie close to the star we could suppose the existence of a classical spherical symmetry around the star [5-11].

2. In the case of independent density layers of matter which lie at a larger distance from the photosphere, we could suppose the existence of independent density regions such as blobs, which could cover a substantial fraction of the stellar disk and are outwards moving inhomogeneities, spiral streams or CIRs (Corotating Interaction Regions), which may result from non-radial pulsations, magnetic fields or the stellar rotation and are able to make structures that cover a substantial part of the stellar disk [9, 11, 14-23]. These regions, though they do not present spherical symmetry around the star, they form spectral line profiles which are identical with those deriving from a spherically symmetric structure. In such a case, though the density regions are not spherically symmetric, through their effects on the line profiles, they appear as spherically symmetric structures to the observer.

The above mentioned ideas led us to suppose spherical symmetry (or apparent spherical symmetry) around the center of the density regions of matter, where the main spectral line as well as its SACs are born.

So, in the case of spherical symmetry, Eq. 1 takes the following form:

$$I_\lambda = \left[I_{\lambda 0} \prod_i e^{-L_i \xi_i} + \sum_j S_{\lambda ej} \left(1 - e^{-L_{ej} \xi_{ej}} \right) \right] e^{-L_g \xi_g} \tag{2}$$

where: $I_{\lambda 0}$: is the initial radiation intensity, L_i, L_{ej}, L_g: are the distribution functions of the absorption coefficients $k_{\lambda i}$, $k_{\lambda ej}$, $k_{\lambda g}$, ξ is the optical depth in the centre of the spectral line, $S_{\lambda ej}$: is the source function, that is constant during one observation.

Calculation Of The Distribution Functions L

It is known that Be and Oe stars are rapid rotators. This means that we accept that a reason of the line broadening is the rotation of the regions that produce each satellite component. These rapidly rotating density regions may also present radial and random motions. For this reason we search an expression for the distribution function (L) of the spectral line components that has as parameters the rotational, the random and the radial velocities of the spherical region.

The distribution function (L) has the form:

$$L_{final}(\lambda) = \frac{1}{2\lambda_0 z} \int_{-\frac{\pi}{2}}^{\frac{\pi}{2}} \left[erf\left(\frac{\lambda - \lambda_0}{\sqrt{2}\sigma} + \frac{\lambda_0 z}{\sqrt{2}\sigma} \cos\theta \right) - erf\left(\frac{\lambda - \lambda_0}{\sqrt{2}\sigma} - \frac{\lambda_0 z}{\sqrt{2}\sigma} \cos\theta \right) \right] \cos\theta d\theta \tag{3}$$

where λ_0 is the observed wavelength of the center of the spectral line and $\lambda_0 = \lambda_{lab} + \Delta\lambda_{rad}$ where λ_{lab} is the laboratory wavelength and $\Delta\lambda_{rad}$ is the radial displacement, where

$z = \dfrac{V_{rot}}{c}$ and V_{rot} is the rotational velocity of the region which creates the spectral line.

We use this distribution function $L_{final}(\lambda)$ in the line function $e^{-L\xi}$, when the line broadening is an effect of both the rotational velocity of the density region as well as the random velocities of the ions. This means that now we have a new distribution function to fit every satellite component of a complex line profile that present DACs or SACs. We name this function Gauss-Rotation distribution function (GR distribution function).

In Figs. 10-12 we present the fittings of some hot emission stars and AGNs with the proposed model. The thick line presents the observed spectral line profiles and the thin one the best fits. The differences between the observed spectrum and its fit are some times hard to see, as we have accomplished the best fit.

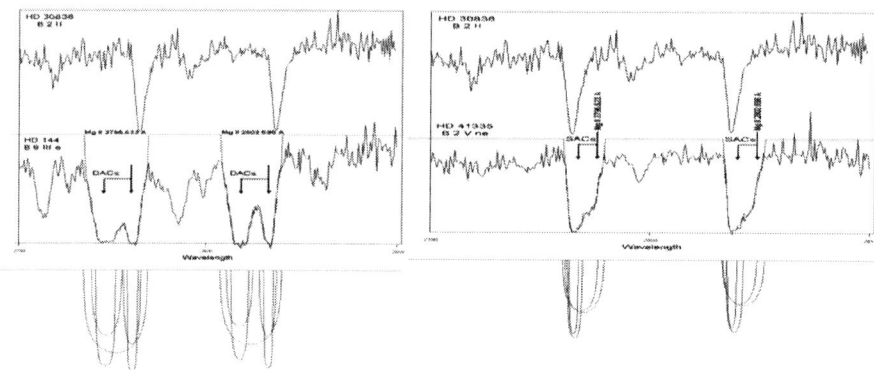

FIGURE 10. Fitting with the proposed model of the Mg II spectral lines of two Be stars that present DACs and SACs, respectively. We point out that we cannot explain and fit these spectral lines with another method. The thick line presents the observed spectral line's profile and the thin one the model fit.

DISCUSSION OF THE PROPOSED MODEL

In order to accept a fit of the complex spectral line as the best, we should apply all the physical criteria and techniques, such as the following:

1. It is necessary to check practically and theoretically the presence of blended lines that can deform the line shape as well as the existence of SACs.
2. The resonance lines as well as all the lines originating in a particular region should have the same number of SACs, depending on the structure of this region, without influence of ionization stage or ionization potential of emitters/absorbers. As a consequence, the respective SACs should have similar or same values of the radial and rotational velocities.
3. The ratio of the optical depths in the centre of two resonance lines has to be the same as the ratio of the respective relative intensities.

4. The proposed line function (Eq. 2) can be used in the case that i=1 and j=1, meaning when we deal with simple, classical spectral lines. This means that we can calculate all the important physical parameters, such as the rotational, the radial and the random velocities, the optical depth and the column density, for all the simple and classic spectral lines in all the spectral ranges.

5. We check the correct number of satellite components that construct the whole line profile. At first we fit using the number of the components that give the best difference graph between the fit and the observed spectral line. Then we fit using one component less than in the previous fit. The F-test between them allows us to take the correct number of satellite components that construct in the best way the whole line profile

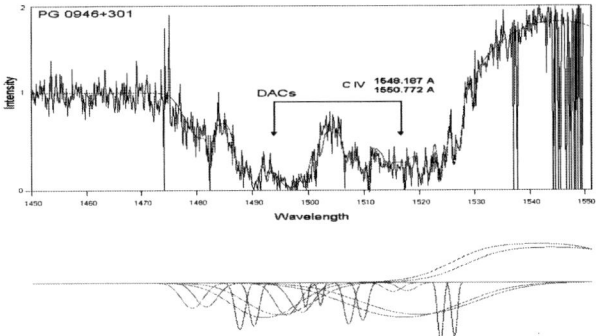

FIGURE 11. Fitting of the C IV UV resonance lines of the BAL QSO PG 0946+301, that present DACs, with the proposed model. We point out that we cannot perfectly fit these spectral lines with another model. The thick line presents the observed spectral line profile and the thin one the model fit.

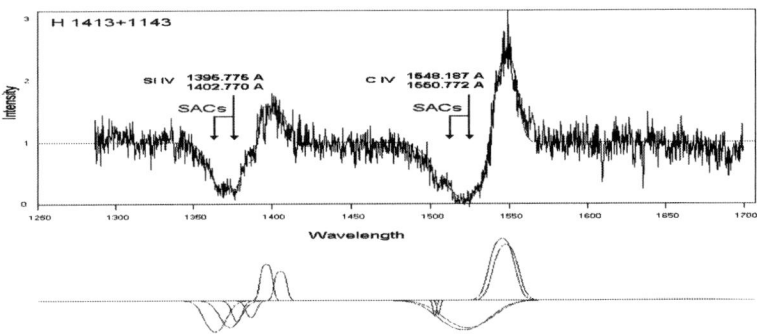

FIGURE 12. Fitting of the Si IV and C IV resonance lines of the BAL QSO H 1413+1143, that present SACs with the proposed model. The SACs phenomenon is able to explain the observed shape. We point out that we cannot perfectly fit these spectral lines with another model. The thick line presents the observed spectral line profile and the thin one the model fit.

TESTING THE MODEL

In order to check the spectral line function (Eq. 2), we calculated the rotational velocity of the He I absorption line at λ 4387.928 Å for five Be stars, using two methods, the classical Fourier analysis and our model. The rotational velocities that we calculate with both methods are almost the same.

We point out that with our model, apart from the rotational velocities, we can also calculate some other parameters as the standard Gaussian deviation (σ), the velocity of random motions of the ions, the radial velocities of the regions producing the studied spectral lines, the full width at half maximum (FWHM), the optical depth, the column density and the absorbed or emitted energy.

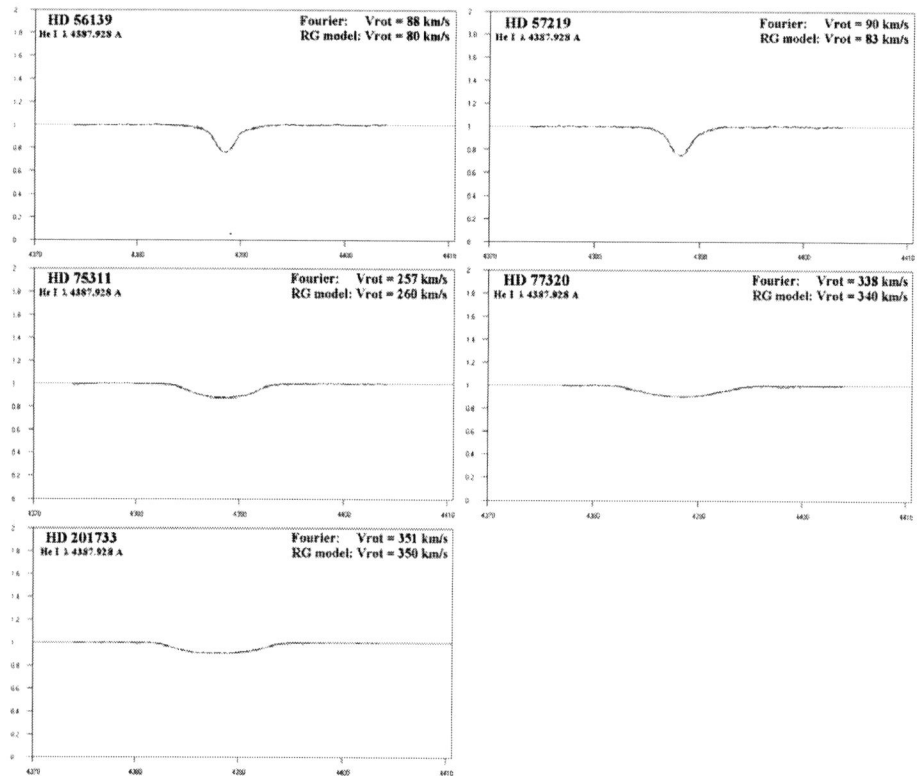

FIGURE 17. The five He I λ 4387.928 Å fittings for the studied Be stars and the measured rotational velocities with both methods. The results are favorable for our model. The thick line presents the observed spectral line profile and the thin one the model fit.

A second test of our model is to calculate the random velocities of the layers that produce the C IV satellite components of 20 Oe stars with different rotational velocities. The values of the random velocities do not depend on the inclination of the rotational axis. As the ionization potential of the regions that create the satellite components for all the studied stars is the same, we expect similar average values of the random velocities for each component for all the studied stars.

We apply the model on the C IV line profiles of 20 Oe stellar spectra taken with the IUE –satellite (IUE Database http://archive.stsci.edu/iue).

We examine the complex structure of the C IV resonance lines ($\lambda\lambda$ 1548.155 Å, 1550.774 Å). Our sample includes the subtypes O4 (one star), O6 (four stars), O7 (five stars), O8 (three stars) and O9 (seven stars). The values of the photospheric rotational velocities are taken from the catalogue of Wilson [24].

After the study of the C IV spectral lines we detect two components in 9 stars, three in 7 stars, four in 3 stars and five in one star. The results that we present in these figures are favourable for our model. The differences between the average values of the random velocities of the satellite components arise from the small variations of the temperature that exist in each one of the regions that produce the satellite components.

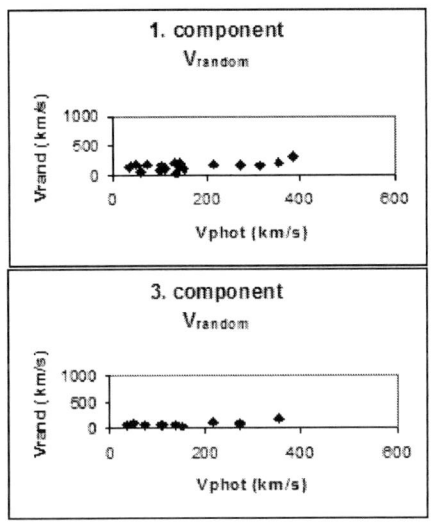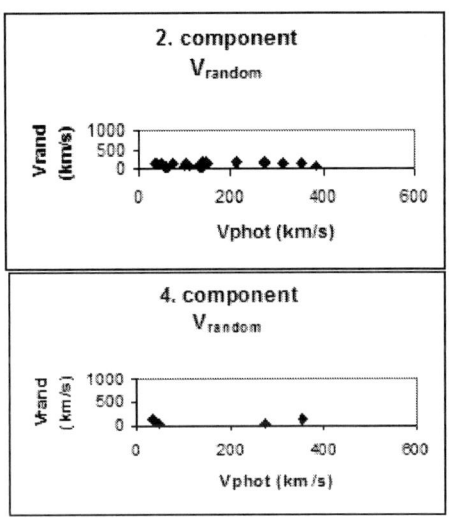

FIGURE 18. Relation between the random velocities and the photospheric rotational velocities of the studied stars.

CONCLUSIONS

We presented an overview of peculiar line profiles from stars to AGNs and a model that can describe peculiar lines.

Here we give some of our conclusions:

1. The peculiar spectral lines in Hot Emission Stars and AGNs are caused mainly by accretion and/or ejection of matter from these objects.
2. Some of the spectral lines peculiarity could be explained by DACs and SACs phenomena, indicating the existence of layers of matter with different physical conditions.
3. The results obtained confirm the assumptions of the proposed model.

ACKNOWLEDGMENTS

This research project is progressing at the University of Athens, Department of Astrophysics, Astronomy and Mechanics, under the financial support of the Special Account for Research Grants, which we thank very much. This work also was supported by Ministry of Science and Environment Protection of Serbia, through the projects "Influence of collisional processes on astrophysical plasma line shapes" and "Astrophysical spectroscopy of extragalactic objects".

REFERENCES

1. P. Tuthill, J. Monnier and W. Danchi, *Nature*, **398**, 487 (1999).
2. E. Danezis, D. Nikolaidis, V. Lyratzi, M. Stathopoulou, E. Theodossiou, A. Kosionidis, C. Drakopoulos, G. Christou and P. Koutsouris, *Ap&SS*, **284**, 1119 (2003).
3. E. Danezis, D. Nikolaidis, E. Lyratzi, L. Č. Popović, M. S. Dimitrijević, E. Theodossiou and A. Antoniou, *Mem. S.A.It.Suppl.*, **7**, 107-113 (2005).
4. B. Bates and D. R. Halliwell, *Mon. Not. R. Astr. Soc.* **223**, 673-681 (1986).
5. H. J. G. L. M. Lamers, R. Gathier and T. P. Snow, *Ap. J.*, **258**, 186 (1982).
6. B. Bates and S. Gilheany, *Mon. Not. R. Astr. Soc.*, **243**, 320 (1990).
7. S. Gilheany, B. Bates, M. G. Catney and P. L. Dufton, *Ap&SS*, **169**, 85 (1990).
8. W. L. Waldron, L. Klein and B. Altner, *nvos.work*, 181(1992).
9. Th. Rivinius, O. Stahl, B. Wolf, A. Kaufer, Th. Gäng, C. A. Gummersbach, I. Jankovics, J. Kovács, H. Mandel, J. Peitz, Th. Szeifert and H. J. G. L. M. Lamers, *A&A*, **318**, 819 (1997).
10. L. S. Cidale, *Ap. J.*, 502, 824 (1998).
11. Markova, N.: 2000, *A&A, Supl. Ser.*, 144, 391.
12. Y. Andrillat and Ch. Fehrenbach, *A&A Supl. Ser.*, **48**, 93-136 (1982).
13. Y. Andrillat, *A&A, Supl. Ser.*, **53**, 319-338 (1983).
14. S. R. Cranmer and S. P. Owocki, *Ap. J.*, **462**, 469 (1996).
15. A. W. Fullerton, D. L. Massa, R. K. Prinja, S. P. Owocki and S. R. Cranmer, *A&A*, **327**, 699 (1997).
16. S. R. Cranmer, M. A. Smith and R. D. Robinson, *Ap. J.*, **537**, 433 (2000).
17. D. J. Mulan, *Ap. J.*, **283**, 303 (1984a).
18. D. J. Mulan, *Ap. J.*, **284**, 769 (1984b).
19. D. J. Mulan, *A&A*, **165**, 157 (1986).
20. R. K. Prinja and I. D. Howarth, *Mon. Not. R. Astr. Soc.*, **233**, 123 (1988).
21. L. Kaper, H. F. Henrichs, J. S. Nichols, L. C. Snoek, H. Volten and G. A. A. Zwarthoed, *A&A Supl. Ser.*, **116**, 257 (1996).
22. L. Kaper, H. G. Henrichs, A. W. Fullerton, H. Ando, K. S. Bjorkman, D. R. Gies, R. Hirata, E. Dambe, D. McDavid and J. S. Nichols, *A&A*, **327**, 281 (1997).
23. L. Kaper, H. F. Henrichs, J. S. Nichols and J. H. Telting, *A. & Ap.*, **344**, 231 (1999).
24. R. E. Wilson, *General Catalogue Of Stellar Radial Velocities,* Washington, Carnegie Institution of Washington Publication 601, (1963).

Plasma diagnostics in the Active Galactic Nuclei environment

S. Ciroi*, G. La Mura*, L.Č. Popović†, D. Ilić** and P. Rafanelli*

*Department of Astronomy, Padova University (Italy)
†Astronomical Observatory Belgrade (Serbia)
**Department of Astronomy, University of Belgrade (Serbia)

Abstract. Active Galactic Nuclei (AGNs) are among the most powerful sources of radiation in the Universe. Their tremendous energy, which can reach values up to 10^{47} erg s^{-1}, makes them as bright as the entire host galaxy or even brighter, and it is produced within a very small spatial region. According to the widely accepted model of AGNs, the engine hidden in the nuclei of active galaxies is a supermassive black-hole (BH) accreting matter and emitting a complicate thermal and non-thermal continuum able to photoionize the gas filling the interstellar medium.

Here we present new results about the physics of plasma in the Broad Line Region (BLR), an ensemble of gaseous clouds believed to orbit the BH at smaller distances than a parsec and to emit bright broad permitted spectral lines. The width of these lines is caused by Doppler broadening because of the clouds motion in the BH gravitational potential. Given its small size, a detailed observation and analysis of the BLR is not possible, and it appears to us as a point-like light source. Therefore, we are forced to use its spectroscopic features to derive its physical parameters. Several indications suggest that BLR clouds are characterized by high values of electron density, higher than 10^9 cm^{-3}, likely as high as 10^{12}-10^{14} cm^{-3}. At such densities it is in principle possible to assume the condition of partial LTE and apply the Boltzmann equation to obtain a rough estimate of the average BLR temperature. We applied this method to the brightest hydrogen emission lines of the Balmer series observed in the optical spectra of 90 type 1 AGNs. The fluxes of Hα, Hβ, Hγ, Hδ and when possible also Hε lines were measured and plotted against the excitation energy featuring the upper level of these transitions to obtain the so-called Boltzmann plot (Popović, 2003). A least-square fit of the points allowed us to successfully determine the temperature of the BLR plasma in most of the targets, obtaining reasonable values typically in the range 10 000-50 000 K.

Moreover, we plotted such temperatures against the FWHM of the ionized gas, which is an indication of the clouds kinematics, confirming that, even with a large spread, there is a clear trend for high velocity clouds to be colder. High velocities are observed in quasars, highly luminous AGNs with very massive BHs, while lower velocities are typical of Seyfert galaxies, less luminous and less massive AGNs. Therefore, in the hypothesis of Keplerian motion around the BH, our results suggest that the relation between the average distance of the BLR and the luminosity of the AGN should be featured by a slope higher than 0.5. This strengthens the recent findings of Kaspi et al. (2000), who, fitting observational data of quasars and Seyfert galaxies, obtained a dependence of the BLR size on the AGN luminosity, expressed as $R_{BLR} \propto L^{0.69}$ in the optical domain.

Keywords: active galaxies, ionization processes, spectroscopy
PACS: 98.54.Cm, 95.75.Fg, 95.85.Kr

INTRODUCTION

Active Galactic Nuclei (AGNs) are extragalactic sources of great interest and great importance in the frame of Universe evolution. Discovered about a century ago, they became known to the scientific community thanks to the work of Carl K. Seyfert, who published in 1943 the optical spectra of six galactic nuclei showing bright and broad

CP876, The Physics of Ionized Gases: 23rd Summer School and International Symposium, edited by L. Hadžievski, B. P. Marinković, and N. S. Simonović emission lines, in place of stellar absorptions, therefore suggesting the presence of some kind of activity inside (Seyfert, 1943). These galaxies were named Seyfert galaxies and several thousands of them were identified and classified during the following decades (see Veron-Cetty and Veron, 2006). Twenty years later, the young astronomer Maarten Schmidt discovered the firts active galaxies at the largest distance known (Schmidt, 1963). These galaxies were named quasars, acronymous for quasi-stellar radio sources, since they were previously identified for their powerful radio emission, but they looked like stars in visible wavelengths. Quasars became quickly fundamental for the cosmological models, because their tremendous amount of emitted energy made them detectable until large distances in the Universe. Several other kinds of active nuclei were also discovered and classified, so that the AGNs family became broader. This forced astronomers to look for a common picture, able to describe the observed different spectroscopic and photometric features of AGNs, the so-called Unified Model (Antonucci and Miller, 1985; Osterbrock, 1993; Urry and Padovani, 1995; Peterson et al., 2000).

According to the model, Seyfert galaxies and quasars are galactic nuclei powered by the same kind of engine, that is a supermassive black-hole (SMBH) accreting matter and emitting a complicate power-law spectrum of energy, extended from gamma-rays to radio wavelengths with the spectral index changing in each spectral window (Figure 1).

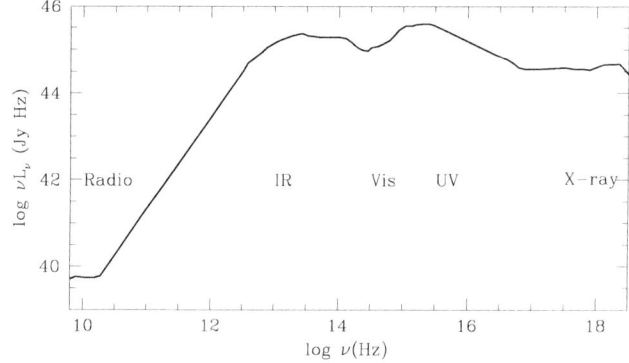

FIGURE 1. The approximate shape of the spectrum for a radio-quiet AGN.

The SMBHs have typical masses ranging from 10^6 up to 10^9 $M_\odot$, and accretion rates between 10^{-2} of the Eddington limit up to Eddington limit and sometimes even more. The total energy budget produced by AGNs goes from 10^{41} up to 10^{47} erg s^{-1}, that is from 10^8 up to 10^{14} $L_\odot$. Taking into account that non-active galaxies show luminosities of 10^{10}-10^{12} $L_\odot$, it is clear that the most powerful AGNs are brighter than their entire host galaxies.

The Unified Model provides that the SMBH and its accretion disc are surrounded by gas strongly ionized by the high energy photons coming from the source. In detail, two regions separated by a thick and obscuring dusty torus, are expected to exist : an ensemble of gaseous clouds located very close to the AGN and responsible for the broad permitted emission lines observed in AGNs spectra, named Broad Line Region (BLR), and a more distant and more extended region of gas from which narrow permitted and forbidden lines are emitted, named Narrow Line Region (NLR). Beacuse of the dusty torus, not always it is possible to observe both BLR and NLR together. When this happens, an AGN is generally classifi ed as Type 1 (Figure 2). On the contrary, when the torus obscures BLR to our view, only NLR is observable and the AGN is classifi ed as Type 2.

Here, we will focus on the physics of the BLR, presenting new recent results about its properties.

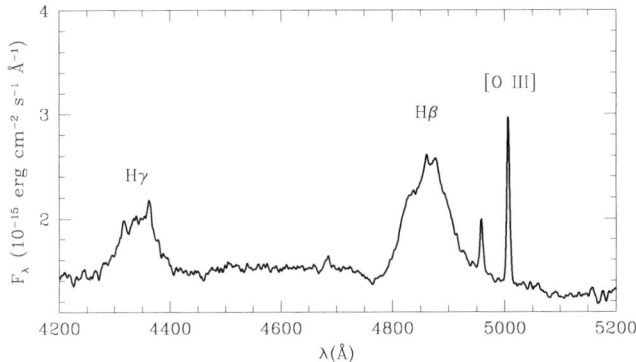

FIGURE 2. Part of the rest-frame optical spectrum of a Seyfert 1 galaxy showing broad Balmer emission lines and narrow forbidden [O III] lines.

MAIN FEATURES OF THE BLR

The Broad Line Region has been study by several authors and from different points of view in the last decades (see e.g. Ferland and Mushotzky, 1982; Osterbrock, 1989; Netzer, 1991; Krolik, 1999; Korista and Goad, 2004). Nevertheless, a clear general understand of its geometry, kinematics and spectroscopic properties is still far to be reached.

Its size is usually confi ned within a parsec, and this prevents any chance for us to obtain a direct view of its spatial structure even with observations from space. Indeed, the parsec scale in an extragalactic source is virtually reachable only through the application of interferometric techniques. Therefore, to date we have only the chance to observe its emission line spectrum in as large as possible samples of Type 1 AGNs, both Seyfert 1 galaxies and quasars, and make some hypotheses and assumptions about its structure and geometry to derive important physical information.

According to a diffuse idea, the BLR is made by high density and high velocity gaseous clouds, photoionized by the central source, randomly moving in the gravitational potential well of the SMBH, and distributed all around the AGN. Nevertheless, several authors are convinced that a flat, disc-like distribution is more reasonable (see e.g. McLure and Dunlop, 2001), since the BLR would be in fact an extension of the accretion disc. This seems to be real in distant quasars, which are usually powerful radio emitters (radio-loud AGNs) whose bright and extended radio jets are expected to confine the BLR on a plane. Anyway, no clear indications have been found of a similar behaviour in Seyfert galaxies (radio-quiet AGNs). Other authors believe that the BLR is made by outflowing clouds in radial motion from the accretion disc.

The high velocity is indicated by the Doppler broadening of the hydrogen and helium emission lines, whose Full Width at Half Maximum (FWHM) span from a few thousands km s^{-1} to values higher than $\sim 10^4$ km s^{-1}. These velocities can be explained only by assuming that the emission is due to a system of clouds, each of one producing the whole complex of observed spectral lines with the Doppler shift associated to its motion. Therefore the emission line profile describes the velocity distribution of the clouds and can be effectively used to derive information about the BLR kinematics. In the hypothesis of Keplerian motion, indeed observed only in one case, NGC 5548 (Peterson and Wandel, 1999), the BLR kinematics is an indicator of the central mass.

The absence of broad forbidden lines, like for example [O III] $\lambda 4363, 4959$ and 5007, which are on the contrary very strong in the Narrow Line Region, suggests that the electron density must be high, at least higher than $\sim 10^{8.5}$ cm^{-3}. It is likely that the density is usually higher than $\sim 10^{10}$ cm^{-3}, since also the ultraviolet intercombination line C III] $\lambda 1909$ is not observed to be broad. What is not known is the upper limit, which is left free to vary up to 10^{14} cm^{-3}.

How much close is on average the BLR to the central engine, is a value which can be obtained even without a spatially resolved image of its structure. In fact, we can take advantage of the variability of the energy source. This variability is observed in each wavelength domain, in radio, as in infrared, visible, ultraviolet and X-rays, but with different timescales which suggest different spatial regions involved. In particular, when periodically observing Type 1 AGNs in visible it is usual to detect a clear luminosity variation of both continuum and broad emission lines, but not in a synchronous way. The emission lines vary later than the continuum, and this time lag is interpreted as the time necessary to light for traveling from the AGN toward the BLR clouds (Korista, 1990). Therefore, the spectroscopic monitoring of AGNs allows to derive light curves of continuum and spectral lines emission, and through the cross-correlation method, to obtain the time lag, and so the average distance of the BLR from the central mass. This technique is known as Reverberation Mapping (RM) and it usually requires long period of observations, like several months up to several years before to estimate the BLR radius (Peterson and Wandel, 2000), which for this reason, could be obtained only for few tens of Seyfert galaxies and quasars until now. Nonetheless, it was demonstrated that Seyfert galaxies have shorter timescales of variability and BLR clouds closer to the central engine than quasars, and much more important that a correlation does exist between the time lag, or the BLR radius, and the continuum luminosity (Figure 3). There was a significant result obtained by Kaspi et al. (2000; 2005), which allowed to much more easily estimate the BLR radius, skipping the extreme time consuming RM technique. Moreover, the RM indicated that ionization degree within the BLR should be stratified in order to justify different time lags for different ionization spectral lines.

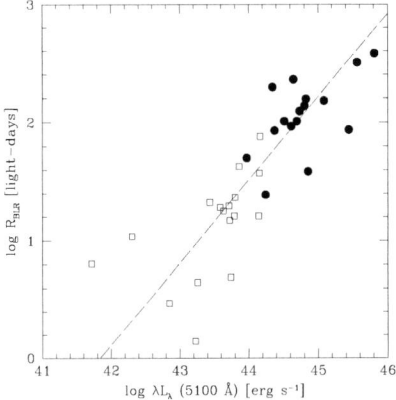

FIGURE 3. The R_{BLR} vs. L_{5100} relation : open squares are Seyfert galaxies, filled circles are quasars.

BLR THROUGH BALMER LINES

Hydrogen Balmer emission lines are doubtless the strongest spectral features of the rest-frame optical spectrum of Type 1 AGNs. Several physical data of the BLR can be extracted by studying in detail the profiles, intensities and ratios of these broad lines. In particular, thanks to fact that Balmer lines belong to the same family of transitions, Popović (2003) had the idea to apply for the first time the Boltzmann Plot (BP) method, well known in laboratory plasma experiments, to a small sample of Seyfert 1 and quasars. He had success in demonstrating that this method is a very promising instrument to gain indications about the existence of Partial Local Thermodynamic Equilibrium (PLTE) in the BLR of most of its AGN sample, and to infer an estimate of the gas average temperature.

Considering a transition from an upper level (u), with energy E_u toward a lower level (l) in the optically thin approximation, the line intensity emitted by a cloud with typical dimension r_* along the line-of-sight can be expressed as follows:

$$I_{ul} = \frac{hc}{\lambda_{ul}} g_u A_{ul} N_u r_* \sim \frac{hc}{\lambda_{ul}} g_u A_{ul} r_* \frac{N_0}{Z} exp\left(-\frac{E_u}{k_B T_e}\right) \tag{1}$$

where N_u, N_0 are the number densities of u-staged and every-excitation emitters, g_u, A_{ul} are the statistical weight and the spontaneous radiative decay probability, and finally Z is the partition function.

In case of PLTE, the upper level population is described by the Boltzmann distribution, that is it depends only on electron temperature T_e. By defining a normalized line intensity:

$$I_n = \frac{\lambda_{ul} I_{ul}}{g_u A_{ul}} \tag{2}$$

and expressing everything by means of log_{10}, we obtain the following linear relation between the intensity of the line and the excitation energy of the upper level of its transition:

$$log_{10}I_n = B - AE_u \qquad (3)$$

where the slope $A = log_{10}e/(k_BT_e)$ depends on the inverse of the average electron temperature (Popović, 2003).

This method has the great advantage to give an estimate of the gas temperature without the need of modelling. But of course it has some limitations, that have been investigated with some tests by Popović (2003) and Ilić (2006).

SAMPLE SELECTION AND DATA ANALYSIS

The first results obtained by Popovč (2003) were recently extended to a larger sample of Type 1 AGNs (La Mura et al., 2006). The public archive of the Sloan Digital Sky Survey Data Release 3 (SDSS DR3) was used to mine all the possible spectra useful for this research. SDSS is definetely one of the most important recent astronomical surveys in optical wavelengths for the huge amount of both photometric and spectroscopic data produced and archived for public use. Spectra cover a range from 3800 Å up to 9000 Å with resolution between 1800 and 2200, and are already reduced and calibrated both in wavelengths and in flux. From this database we have extracted our candidate targets by applying the following criteria : 1) sources must be at redshift $z < 0.3$, in order to include always Hα within the available spectral range; 2) spectra must show the Balmer broad emission lines at least up to $H\delta$, when possible up to $H\varepsilon$; 3) the S/N ratio must be sufficiently high to clearly detect the emission line profiles for each Balmer line considered; 4) spectra must be cleaned from bad pixels, distortions or any other artefact. Such restrictions reduce significantly the number of "good" targets. After the first extraction and subsequest careful check of each single spectrum, we could collect a sample of 90 Type 1 AGNs suitable for the analysis.

The spectra were analyzed by means of the IRAF software packages. Firstly, spectra were corrected for Galactic extinction, using the A(V) values given by the NASA Astrophysical Database (NED) for each target and applying a convenient dust extinction law. In fact the dust absorption causes the reddening of the spectrum changing the Balmer line intensities toward different ratios, and therefore it affects the temperature estimation.

Then, the continuum was subtracted by each spectrum interpolating the wavelength intervals not affected by the numerous emissions with a polynomial function. Each spectrum was de-redshifted to the rest-frame wavelengths and when necessary, the Fe II multiplets contribution was removed by subtracting a template obtained from the spectrum of I Zw 1, a Narrow-Line Seyfert 1 galaxy (see Botte et al., 2004).

Finally, since in type 1 AGNs we observe the contemporary emission from BLR and NLR, this last contribution must be removed. In general, the broad Balmer lines are much brighter than the narrow ones, which however are often visible, especially in Seyfert galaxies (much less in quasars), superimposed to the broad lines and forming composite (broad+narrow) profiles. In order to remove their contributions, we made use of the task NGAUSSFIT, which allows to perform a multi-Gaussian fi tting of complicate profi les (Figure 4). To reduce the number of free parameters, we fi xed the FWHM of narrow Balmer lines to the same value shown by the narrow forbidden lines, like [O III] $\lambda5007$ and [N II] $\lambda6583$, and we considered only one broad profi le. The use of two or more broad components is sometimes reasonable, since the kinematics of the BLR can be more complicate, but it must be done with extreme carefulness since the multi-Gaussian fi t becomes too flexible and arbitrary in spectra with not very high resolution and/or with relatively low S/N ratio. After having obtained a stable solution, for each object the modelled spectrum of its NLR was subtracted, and the residual broad Balmer lines were integrated along their profi les without using a Gaussian model in order to take into account for non-symmetric profi les.

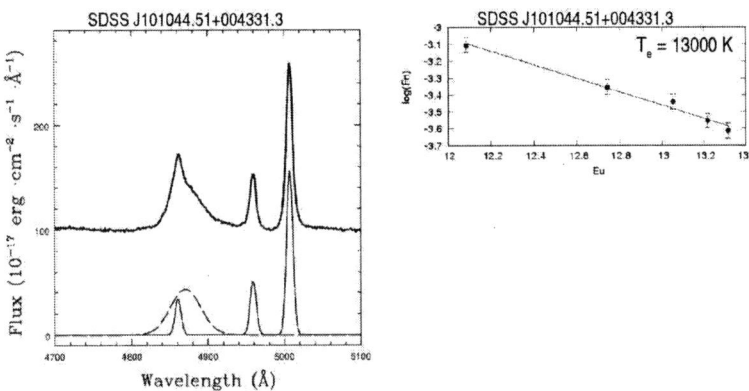

FIGURE 4. An example of multi-Gaussian fit applied to one of the targets (*left*) and the following Boltzmann Plot (*right*).

RESULTS AND CONCLUSIONS

The BP method was successfully applied to 66 out of the 90 spectra. The obtained average electron temperatures range from about 5000 up to 35000 K, with a peak at 20000 K. A few number of AGNs show higher temperatures, between 37000 and 47000 K. The typical indetermination associated to these values is about 30%. In 35 out of 66 AGN we can be confi dent that PLTE exists in their BLR, while in the other 31 AGN the value of the temperature parameter A is too low (T_e higher than 20000 K), and similar BPs can be obtained assuming the Case B recombination. Therefore, for them we cannot be sure about the real existence of PLTE.

The temperatures were plotted against FWHM and FWZI (Full Width Zero Intensity), which are two parameters connected to the kinematics of BLR clouds. It was already suggested by Popović (2003) that there might be a correlation between FWHM (or FWZI) and T_e, but he had a small number of data. Now, the sample is quite larger and we can state that because of the spread a weak correlation exists between these two terms: in fact the correlation parameter is R = -0.44 for *FWHM vs. T_e* and R = -0.47

for *FWZI vs. T_e*. Anyway, a trend is observed showing that lower average electron temperatures of the BLR gas correspond to higher velocity dispersions of the BLR clouds (Figure 5).

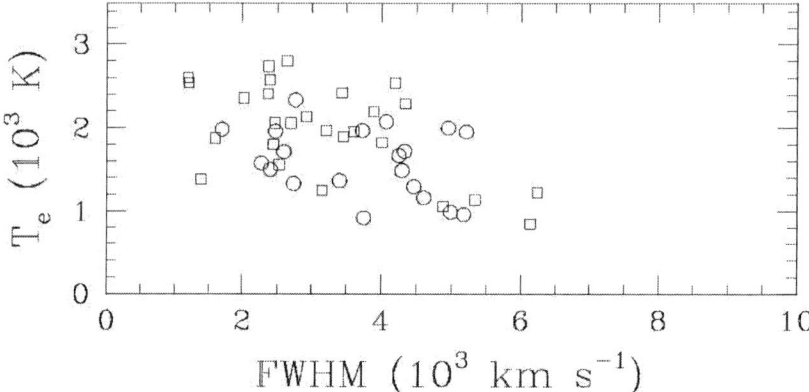

FIGURE 5. The *FWHM vs. T_e* plot for the targets whose BPs were successfully applied: different symbols indicate when PLTE occurs and when we cannot confirm its presence.

We tried to give an explanation to this interesting result on the basis of photoionization calculations and the physical properties of AGN. In particular, taking into account the Saha ionization equation, the ionization degree depends on the electron temperature variation at a fixed value of electron density. Therefore, in principle a lower temperature can be the result of a lower ionization. Moreover, high dipersion velocities of BLR clouds are generally observed in AGN with more massive central BH. But these AGN are also more luminous. Recently, Kaspi et al. (2000; 2005) obtained a fundamental experimental relation between the above mentioned quantities: $R_{BLR} \propto L_{5100}^{0.69}$. This result is generally named *Kaspi's relation* or *BLR size − luminosity relation*, and it was obtained with spectroscopic data of some AGNs (quasars and Seyfert 1 galaxies) by measuring the luminosity of their continuum at rest-frame 5100 Å and estimating the R_{BLR} through the application of the Reverberation Mapping technique. The main consequence of the Kaspi's relation is the great chance to very quickly estimate the BH mass simply measuring the continuum at rest-frame 5100 Å and the FWHM of Hβ in the hypothesis of Keplerian motion in the gravitational potential of the central mass. But, another important product is that the ionizing photons flux scales with luminosity according to $\Phi \propto L^{-0.38}$, and so the ionizing flux decreases with increasing values of luminosity.

To test our idea, we performed a certain number of photoionization models by means of CLOUDY 94 (Ferland, 2000). CLOUDY is a code which requires as input an ionizing source with a spectral energy distribution, the distance between the source and gas, the hydrogen density of gas and eventually the chemical composition of gas, solves the transfer equation, the statistical equilibrium equation and the thermal equilibrium equation, and finally calculate the theoretical emitted spectrum, both continuum and emission lines. For our case (Figure 6), we fixed the values of accretion rate and hydrogen density and we varied the BH mass from $10^6 M_\odot$ up to $10^9 M_\odot$, obtaining as results four models showing decreasing temperature of the gas with increasing values of the FWHM, and a range of temperature compatible with our data. Then, we fixed the BH mass and we tried to explore the effect of changing only the accretion rate, from 0.03 to 0.4 times the Eddington limit: the models show a slightly decreasing temperature with increasing accretion rates. Finally we changed only hydrogen density showing that the temperature can be decreased also increasing the density of gas. In principle, a grid with these models seems to be able to explain the observed $FWHM - T_e$ plot, both the general trend and the observed scatter. We stress that these models do not pretend to give the ultimate interpretation to this plot obtained with the application of the BPs, since additional work must be done for a deeper comprehension of the results. For example, we cannot rule out *a priori* other possible reasons of the scatter, like the varibility of each AGN, which can affect the ionization degree of the BLR clouds and therefore change the temperature estimates.

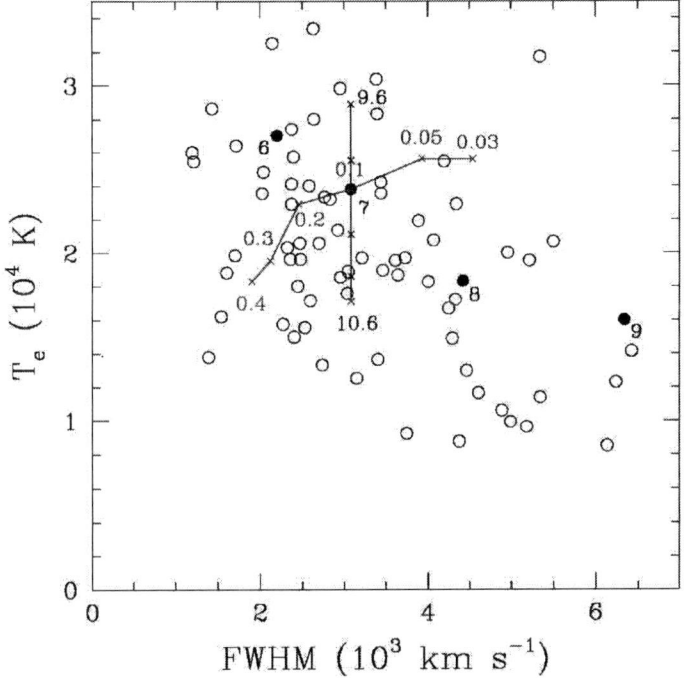

FIGURE 6. Photoionization models aimed to explore the *FWHM* vs. T_e plot: black dots are models with different BH masses; the accretion rate is changed from 3% up to 40% of the Eddington limit and the hydrogen density is varied from $10^{9.6}$ up to $10^{9.6}$ cm^{-3}.

ACKNOWLEDGMENTS

S. Ciroi is very greatful to L.Č. Popović and all the organizers of the 23rd Summer School and International Symposium on the Physics of Ionized Gas for their kind invitation and warm hospitality. L.Č. P. was supported by the Ministry of Science and Environment Protection of Serbia through the project 'Astrophysical Spectroscopy of Extragalactic Objects'.

Data for the present study have been entirely collected at the SDSS database, funding by the Alfred P. Sloan Foundation and the participating Institutions. This research has made use of the NASA/IPAC Extragalactic Database (NED) which is operated by the Jet Propulsion Laboratory, California Institute of Technology, under contract with the National Aeronautics and Space Administration.

REFERENCES

1. R. R. J. Antonucci, and J. S. Miller, *Astrophysical Journal* **297**, 621 (1985)
2. V. Botte, S. Ciroi, P. Rafanelli, and F. Di Mille, *Astronomical Journal* **127**, 3168 (2004)
3. G. J. Ferland, and R. F. Mushotzky, *Astrophysical Journal* **262**, 564 (1982)
4. G. J. Ferland, *Hazy, A Brief Introduction to Cloudy 94*, University of Kentucky Internal Report (2000)
5. S. Kaspi, P. S.Smith, H. Netzer, D. Maoz, B. T. Jannuzi, and U. Giveon, *Astrophysical Journal* **533**, 631 (2000)
6. S. Kaspi, D. Maoz, H. Netzer, B. M. Peterson, M. Vestergaard, B. T. Jannuzi, *Astrophysical Journal* **629**, 61 (2005)
7. K. T. Korista, *Publications of the Astronomical Society of the Pacific* **102**, 1351 (1990)
8. K. T. Korista, and M. R. Goad, *Astrophysical Journal* **606**, 749 (2004)
9. J. H. Krolik, *Active galactic nuclei : from the central black hole to the galactic environment*, Princeton University Press (1999)
10. D. Ilić, L.Č. Popović, E. Bon, E.G. Mediavilla, and V.H. Chavushyan, *astro-ph/0607253*, accepted for publication in MNRAS (2006)
11. G. La Mura, L.Č. Popović, S. Ciroi, P. Rafanelli, and D. Ilić, submitted to ApJ (2006)
12. R. J. McLure, and J. S. Dunlop, *Monthly Notices of the Royal Astronomical Society* **327**, 199 (2001)
13. H. Netzer, *Broadline Region Models and the Luminosity/magnitude Relation for Active Galactic Nuclei* in Variability of Active Galaxies **377**, 107 (1991)
14. D. E. Osterbrock, *Astrophysics of gaseous nebulae and active galactic nuclei*, University Science Books (1989)
15. D. E. Osterbrock, *Astrophysical Journal* **404**, 551 (1993)
16. B. M. Peterson, and A. Wandel, *Astrophysical Journal* **521**, L95 (1999)
17. B. M. Peterson, and A. Wandel, *Astrophysical Journal* **540**, L13 (2000)
18. B. Peterson, B. Wilkes, P. Murdin, *Active Galaxies: Unified Model* in Encyclopedia of Astronomy and Astrophysics (2000)
19. L. Č Popović, *Astrophysical Journal* **599**, 140 (2003)
20. M. Schmidt, *Nature* **197**, 1040 (1963).
21. C. K. Seyfert, *Astrophysical Journal* **97**, 28 (1943)
22. C. M. Urry, and P. Padovani, *Publications of the Astronomical Society of the Pacific* **107**, 803 (1995)
23. M.-P. Veron-Cetty, and P. Veron, *Quasars and Active Galactic Nuclei (12th Ed.)* (2006)

Effect of Magnetic Topology on Edge Plasma Behavior in LHD Heliotron

S. Masuzaki, T. Morisaki, M. Kobayashi, T. Watanabe, N. Ohyabu, A. Komori, O. Motojima and the LHD Experimental Group

National Institute for Fusion Science, Oroshi 322-6, Toki 509-5292, JAPAN

Abstract. The Large Helical Device (LHD) is the largest heliotron-type super-conducting fusion experimental device. One of the features of the heliotron configuration is its unique edge magnetic field topology. There exist an intrinsic stochastic layer just outside of the last closed flux surface (LCFS), residual islands embedded in the stochastic layer, whisker structures, laminar layers and intrinsic divertor structure (helical divertor). That contrasts to 'onion-skin' like magnetic field structure in poloidal divertor tokamaks scrape-off layer (SOL). The edge field line structure can be characterized by Kolmogorov length and field line connection length from wall to the other wall. In the stochastic layer, Kolmogorov length is much longer than connection length. Typical connection length of field lines in the stochastic layer and laminar layers are several kilometers and below several tens of meters, respectively, and these structures co-exist. Therefore, the radial profile of field lines connection length becomes complex contrasting to that in poloidal divertor tokamaks SOL. Plasma transport in the LHD edge region has been studied experimentally by using Langmuir probes and Thomson scattering method, and numerically by using three dimensional plasma and neutral transport codes. In this paper, the physical basis of the heliotron configuration, especially the characteristics of the edge magnetic field topology is presented. Understandings of plasma transport in such a unique magnetic field structure are discussed.

Keywords: LHD, heliotron, field line structure, edge plasma, edge modeling.
PACS: 52.25.-b, 52.25.Fi, 52.25.Xz, 52.55.Hc, 52.65.-y

1. INTRODUCTION

In toroidal plasma confinement systems using a magnetic field, both toroidal and poloidal magnetic fields are necessary to have an equilibrium in which the plasma pressure is balanced by the magnetic forces. There are two types of toroidal plasma confinement systems. One uses toroidal coils and toroidal plasma current to generate toroidal and poloidal magnetic field, respectively, such as tokamak. The other type does not use plasma current, but only use external coil systems such as stellarator and heliotron. In the latter type, magnetic structure is non-axisymmeric, three dimensional in nature. The shape of the magnetic surface rotates helically with a constant pitch. Therefore, this type is called helical devices in this paper.

Helical devices are free from current drive system and instabilities induced by plasma current such as disruption. That means helical devices have an advantage over tokamak in steady state operation in principle. The heliotron configuration is one of

CP876, The Physics of Ionized Gases: 23rd Summer School and International Symposium,
edited by L. Hadžievski, B. P. Marinković, and N. S. Simonović

the types of helical devices, and has been intensively studied at Kyoto University and National Institute for Fusion Science in Japan. The Large Helical Device (LHD) is a heliotron-type device, and is the world largest helical device with super-conducting coils [1]. One of the major experimental goals in LHD is to achieve improvement of plasma confinement and sustainment of steady state high performance plasma with edge plasma control. As described later, heliotron configuration has a unique magnetic topology outside the closed flux surfaces region, and its effect on plasma behavior has been investigated in LHD to achieve the experimental goal.

In the next section, physical basis of heliotron configuration including the unique edge magnetic topology is briefly described. The plasma behavior in the LHD edge region is described in section 3.

2. PHYSICAL BASIS OF THE HELIOTRON CONFIGURATION

2.1. Characteristics of the Heliotron Configuration and Design of LHD

Heliotron-type devices have pole number l=2 helical coils and poloidal field coils (see Figure 1(a)). The toroidal mode number of the helical coils is 5 and thus the toroidal mode number of LHD (m) is 10. The directions of the helical coils currents are same, and the coils produce both the poloidal and the toroidal components simultaneously. This is a major difference between the heliotron and standard stellarator configurations. In the case of LHD, three pairs of poloidal field coils are installed. These Poloidal field coils

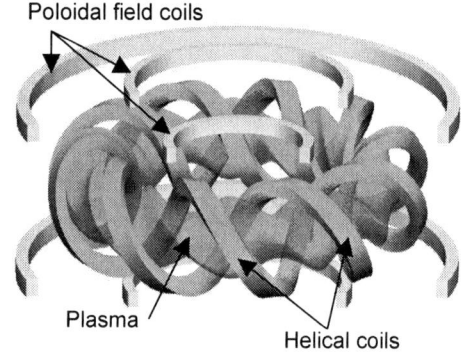

FIGURE 1. Coil system and plasma in LHD heliotron.

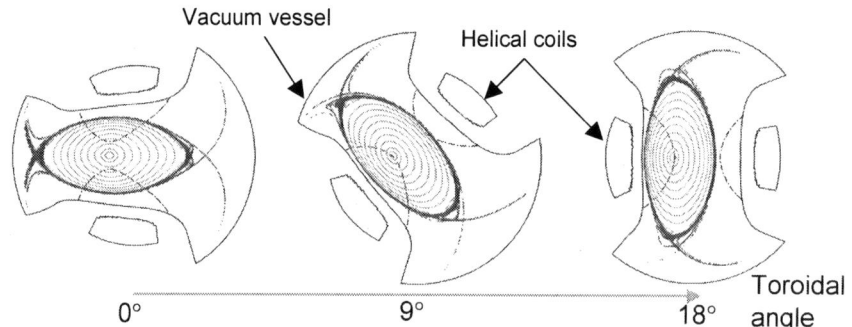

FIGURE 2. Poincaré plots of field lines in typical poloidal cross sections in LHD.

396

generate the vertical, quadrapole, and hexapole fields, and control the radial position and shape of the magnetic surface. Poincaré plots of field lines that have more than 1.2-toroidal-turn connection lengths in typical poloidal cross-sections in LHD are shown in Figure 2. The magnetic surfaces have roughly elliptical shape. There are four divertor legs like in the double-null poloidal divertor configuration in tokamaks. This divertor structure (helical divertor) is naturally equipped without divertor coils in the heliotron configuration, and is a feature of the configuration.

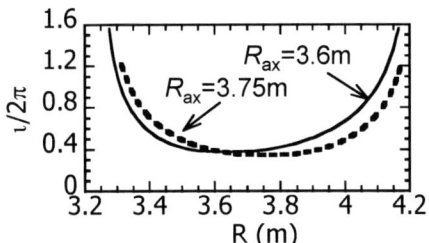

FIGURE 3. Rotational transform of LHD heliotron.

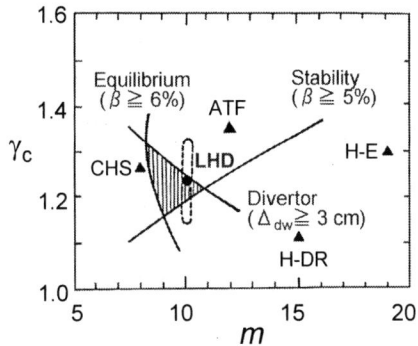

FIGURE 4. Schematic γ_c-m plot for LHD design optimization [2]. Other heliotron-type devices are plotted in the same figure. Relevant physics requirements are fulfilled in the shaded area. Δ_{div} is diveror-wall clearance.

Figure 3 shows the rotational transform of LHD for two typical operational magnetic configurations. R_{ax} is the major radius of magnetic axis. In contrast to the standard tokamak configuration, the rotational transform increases with increasing minor radius, and it is larger than 1 at the last closed flux surface (LCFS). This is also one of the main characteristics of the heliotron configuration.

Physical studies carried out to optimize LHD design from the viewpoints of high β, good high-energy particle confinement and sufficient divertor-wall clearance [2]. Figure 4 depicts pitch parameter γ_c ($\equiv (m/l)(a_c/R_c)$, where a_c and R_c are minor and major radii of the helical coils) and m space in heliotron-type devices. Relevent physics requirements, indicated in the figure, are fulfilled in the shaded area. In the LHD case, m number is determined to be 10. The pitch parameter γ_c is adjustable by the multi-layer operation of the helical coils in LHD. As the result of the studies, the parameters for the design of LHD were determined as in table.1.

TABLE 1. Major parameters of the helical coils in LHD

Parameter	
l: pole number	2
m: toroidal mode number	10
R_c: major radius of the helical coils	3.9m
a_c: minor radius of the helical coils	0.975m
γ_c: pitch parameter, $(m/l)(a_c/R_c)$	1.25
α^*: pitch modulation, $\theta = (m/l)\phi + \alpha^*\sin[(m/l)\phi]$	0.1
B_0: field strength	~ 3T

397

2.2. Edge Magnetic Topology in LHD

In heliotron-type devices, the edge magnetic structure is more complicated than the scrape-off layer (SOL) in axisymmetric poloidal divertor tokamaks, and the helical divertor exists without additional coils as shown in Figure 2. The magnetic structure in LHD under vacuum condition has been investigated using the calculations of magnetic field line tracing, and detailed descriptions of them are in the references [3,4].

2.2.1. Edge Magnetic Structure

In straight helical systems, a clearly defined separatrix exists due to their helical symmetry as in the axisymmetric divertor tokamaks. It separates the closed surfaces region from the open field line region. However, in the toroidal helical systems, there is no clear separatrix, and two types of open field line layers exist between the LCFS and the residual X-points due to the toroidal effect. That complicates the edge magnetic structure in this system. The edge magnetic structure in LHD is illustrated in Figure 5(a). With an increase of minor radius, the poloidal mode numbers of the natural island layers which exist in the peripheral region decrease and the widths of the islands increase. Eventually, the island layers overlap, and the stochastic field structure appears (stochastic layer in Figure 5(a)). When the field lines in this region approach the residual X-point, they are folded and stretched by radial movement of the X-point and high local rotational transform ($\iota/2\pi = 5$ at the residual X-point) and shear on the torus outboard or inboard side. This generates a structure of multiple layers ('whisker' in Figure 5(a)). The field lines in the region surrounded by these whiskers ('laminar' in Figure 5(a)) are connected to the wall with relatively short connection lengths. In Figure 5(b), the profile of calculated connection lengths of magnetic field lines (L_c) is shown as a function of the starting position (4.5 m < R < 5 m) on the mid-plane ($Z = 0$) at the horizontally elongated cross-section for the case of the magnetic axis position (R_{ax}) of 3.75 m. In this calculation, the LCFS is defined as the flux surface which is kept after 40 toroidal circulations ($\sim$ 1 km) of field line tracing. Connection lengths of magnetic field lines, L_c, just outside of the LCFS are over a few km ($\sim$ 100 toroidal circulations), and this region is the stochastic layer.

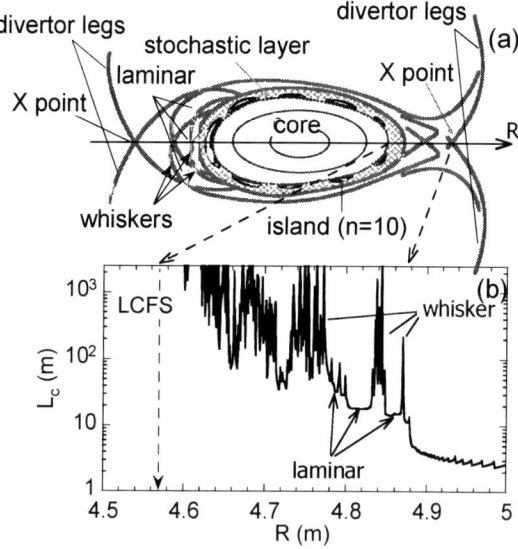

FIGURE 5. (a) Schematic view of the edge magnetic topology in LHD heliotron. (b) Radial profile of the connection length of field lines in the LHD edge region.

398

At larger R, field lines with long L_c, over several hundred meters, discretely appear, that is, whiskers. In this region, there are regions of relatively short field lines with a flat L_c profile (laminar in Figure 5(b)). The field lines in the laminar layer connect to the divertor plates, but do not connect to the stochastic layer. The field lines with long L_c in the edge surface layers finally reach the divertor plates through the residual X-point. Outside of the residual X-point, the toroidal component of the magnetic field is decaying rapidly and becomes smaller than the poloidal component. Therefore, the connection lengths of field lines from the residual X-point to the divertor plates are very short compared to L_c, only several meters. This is in marked contrast to an axisymmetric divertor in tokamaks.

The features of the edge magnetic structure in LHD heliotron can be summarized as following: (1) Existence of 2 types of open field line layers between LCFS and the residual X-point; the stochastic layer and the whiskers. (2) Much longer connection lengths of magnetic field lines in the stochastic layer than that in SOL in an axisymmetric divertor in tokamaks. (3) In the divertor region, the poloidal component of the magnetic field is larger than the toroidal component and thus the field lines length from the residual X-point to the divertor plates are very short compared to L_c.

2.2.2 Intrinsic Helical Divertor

The positions of the divertor traces at the divertor plates were calculated using magnetic field line tracing. A random walk process was included in the calculation in order to investigate the effects of cross-field transport [3,4]. At every step of distance λ, the position of the field line deviates by δ in the plane perpendicular to the field lines with random azimuthal directions. Two thousand start points of simulated particles were distributed just inside LCFS. Figure 6 shows the results of that calculation for the cases of $R_{ax} = 3.6$ m and 3.75 m. The contour plots of the number of field lines connected to the divertor region, that is divertor traces, are indicated on the plane of the toroidal and the poloidal angles. The difference in the profile caused by

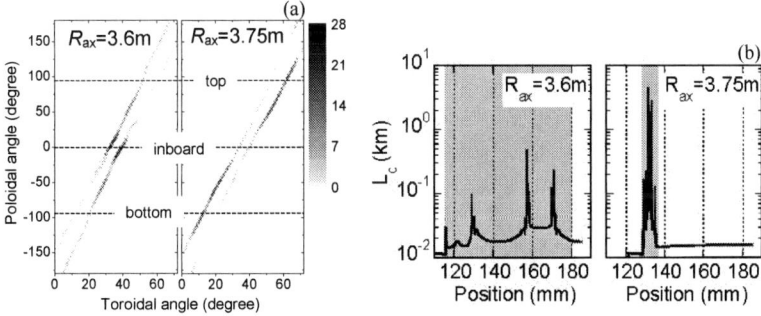

FIGURE 6. (a) Calculated divertor traces for $R_{ax} = 3.6$ m and $R_{ax}=3.75$ m. They are contour plots of the number of test particles starting in the vicinity of the LCFS, which are deposited on the divertor in the plane of the toroidal and poloidal angles. In these cases, the field line tracing simulates the particle diffusion by using a random walk process ($\lambda = 0.2$m, $\delta = 0.6$mm, $D \sim 0.3$ m^2/s)[5]. (b) L_c profiles of the field lines connected to a divertor plate for $R_{ax}=3.6$m and 3.75m. Hatched regions are divertor leg.

the difference in the position of the magnetic axis is evident. The three dimensional structure of the helical divertor in LHD is clearly shown in the contour plots. Strong non-uniformity appears in both profiles. This means that the particle deposition is also non-uniform even in the helical direction. This particle deposition profile depends on the magnetic configuration, such as R_{ax}. In the case of R_{ax} = 3.6 m, the particle deposition is concentrated in the torus inboard side. On the other hand, the deposition is concentrated in the top and bottom region for R_{ax} = 3.75m. Figure 6(b) shows L_c profiles on a torus-inboard side divertor plate. The profile has some peaks correspond to whiskers in the edge region, and the structure is drastically changed with R_{ax}

3. THE PLASMA BEHAVIOR IN THE LHD EDGE REAGION

3.1. Edge Plasma Profile

The degree of stochasticity of the edge magnetic field structure in various vacuum magnetic configurations in LHD was quantitatively estimated, using the Kolmogorov length (L_k) as a measure. In the stochastic region, magnetic field lines present chaotic trajectories. A flux tube there deforms its shape and the circumference d of the tube increases exponentially [6], which is described as $d(s) = d_0 \exp(s/L_K)$ where d_0 and s are initial value of the circumference and length of the flux tube, respectively. Since L_K is the e-folding length of the exponential increase of the circumference, it can be a good measure of the degree of stochasticity [7]. In order to obtain L_K, 100 field lines were traced for 50 toroidal turns, i.e. ~1200 m, from the circularly distributed starting points whose diameter was 1 mm on the poloidal plane. Then the circumference d of the flux tube was measured every toroidal turn, which resulted in the averaging effect

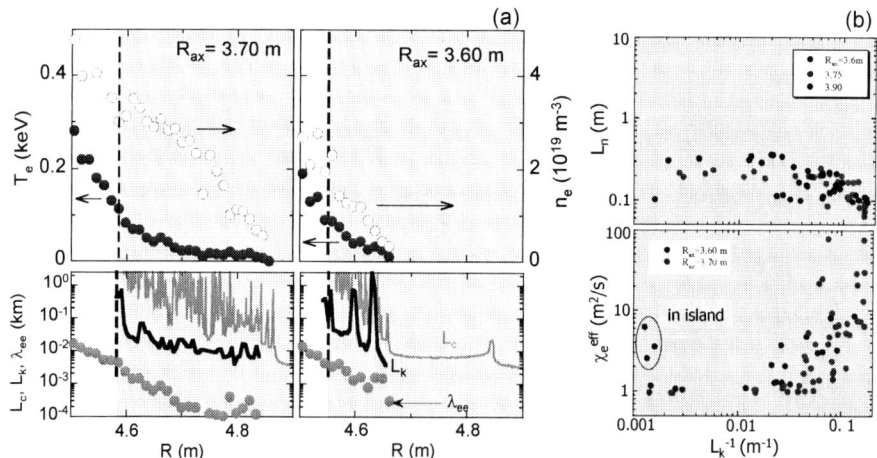

FIGURE 7. (a) Electron density (n_e), temperature (T_e), connection length (L_c), Kolmogorov length (L_k) and collision mean free path (λ_{ee}) profiles for Rax=3.6m and 3.7m configurations. (b) Radial electron heat conductivity χ_e^{eff} and e-folding lengths of electron density profiles as a function of the inverse Kolmogorov length L_K [6].

over one toroidal turn in measuring d. If the field line hits the wall during the trace, the calculation was stopped at that point. Small bundles of starting points were distributed every 5 mm from just inside LCFS to the X-point on the mid-plane at the poloidal cross-section where plasmas are horizontally elongated. These bundles were exactly on the line of sight of the Thomson scattering measurement.

It is found that the edge electron temperature profile changes its gradient at the position where the degree of stochasticity begins to increase (L_k begins to decrease) several cm outside the LCFS (see Figure 7 (a)). In the profiles no distinguished change of plasma parameters was seen at the LCFS, which suggests the existence of a region just outside the LCFS where the confinement performance is relatively as good as that in the closed region. The radial electron heat conductivity was estimated using a simple energy balance equation and it was confirmed that the heat conductivity is strongly affected by the degree of stochasticity (see Figure 7 (b)).

Numerical study of plasma transport properties in the edge region using the 3 dimensional edge transport code EMC3-EIRENE [8,9] has been started. Figure 8(a) and (b) show L_c profile and the electron temperature profile on a horizontally elongated poloidal cross-section obtained by the 3D modeling, respectively, for the case of $R_{ax} = 3.75$ m, $n_u = 2 \times 10^{19}$ m^{-3}, $\chi_\perp = 0.6$ m^2/s, $D_\perp = 0.3$ m^2/s. The power flow to the SOL, P_{SOL}, is 4MW in this case. The pattern of long L_c region is clearly reflected on the T_e profile, appearing as high temperature. This means an energy flow channel through long flux tubes, which approach to the LCFS. The radial T_e profile calculated by the 3D code is good agreement with the profiles measured by the Thomson scattering system, as shown in Figure 8 (c) [10].

3.2. Relationship between Edge and Divertor Plasmas

The relation between edge and helical divertor plasmas has been investigated using Langmuir probes embedded in a divertor plate, Thomson scattering and interferometer, in particular electron density n_e and temperature T_e behaviors, by line-averaged density $\overline{n}_e$ scan ($\overline{n}_e \sim 1\text{-}8\times10^{19}$ m^{-3}) [11]. Figure 9 shows typical results of this investigation. Experimental data show that both n_e and T_e in the divertor plasma

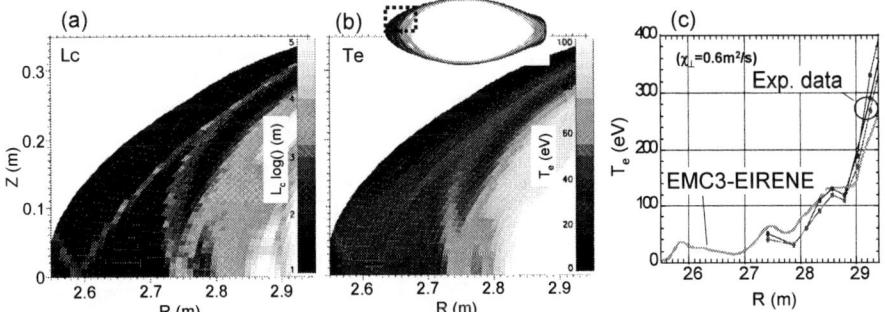

FIGURE 8. (a) L_c and (b) T_e profiles in a horizontally elongated poloidal plane (see insertion) obtained by EMC3-EIRENE for the case of $R_{ax} = 3:75$ m; $n_u = 2\times10^{19}$ m^{-3}, $\chi_\perp = 0.6$ m^2/s, $D_\perp = 0.3$ m^2/s and $P_{SOL} = 4$MW [10]. (c) Radial profiles at the inner mid-plane of T_e obtained by the 3D modelling and the experiments for $n_e \gg 2\times10^{19}$ m^{-3} and input power~ 4MW.

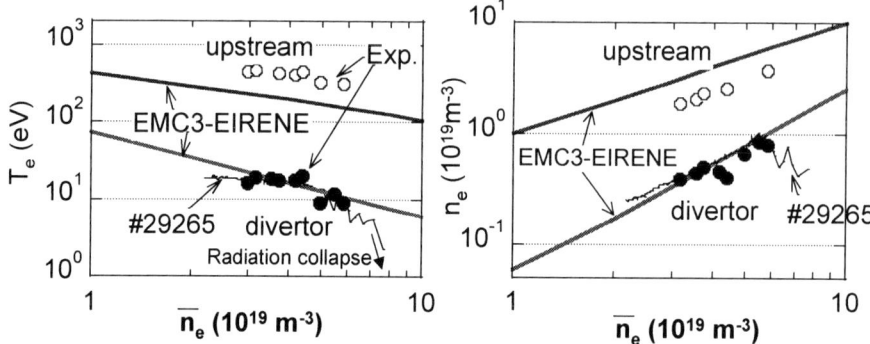

FIGURE 9. T_e and n_e at near the LCFS (upstream, open circles) and the divertor (closed circles) plasmas in $R_{ax}=3.75$m configuration as functions of operational density (line averaged density from the center chord). The NBI power range is 4 – 5 MW. Open and closed circles are data from different discharges at the timing of the maximum stored energy. Thin lines show the time evolution of a discharge (#29265). Bold lines are numerical results [10, 11].

largely drop comparing to them at upstream region (near the LCFS) and thus plasma pressure drops by almost 2 orders of magnitude in the divertor plasma. With increasing $\overline{n}_e$, T_e decreases and n_e increases both at upstream and divertor. Nevertheless, high-recycling regime where n_e in the divertor increases rapidly with n_u increase has not been clearly observed. Divertor-detachment like behaviors of n_e and T_e are observed at high-density region. However that is not stable, and plasma radiatively collapses at last. 3D modeling data that was calculated under the same condition as previous section show good agreement with experimental data [10]. In this calculation, impurities are not taken into the account and thus the large drops of T_e and n_e are due to transport. The 3D modeling shows the existence of counter-flows in the edge region (see Figure 10). A possible cause of the drastic drop of plasma pressure is perpendicular momentum loss due to friction between counter-flows. The very long L_c in the stochastic layer is considered a possible reason of the one order drop of T_e.

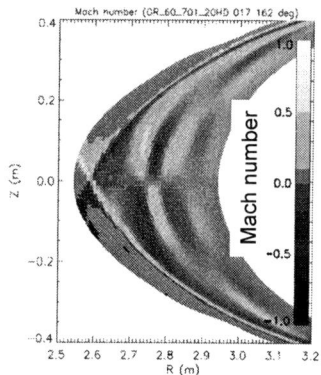

FIGURE. 10 Mach number profile in the edge region [10].

3.3. Relationship Between Divertor Heat and Particle Deposition Profiles and Magnetic Structure

The relationship between the divertor particle flux profile and magnetic field lines topology on divertor plates has been studied using Langmuir probe arrays and thermocouples embedded in the divertor plates [5]. Figure 11 shows that the ion saturation current (I_{sat}) profiles (corresponding to the ion flux profile) have peaks at the position where long field lines connect. Profiles of electron density and

402

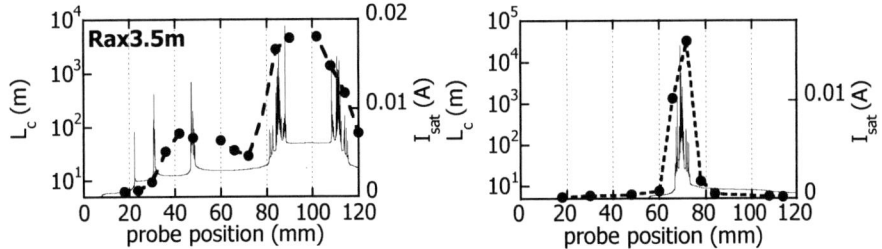

FIGURE 11. Profiles of field line connection lengths ~L_c, solid lines and ion saturation current ~I_{sat}, closed circles and thin lines, on a torus inboard side divertor plate for R_{ax}=3.5m and 3.9m configurations.

temperature on the divertor plate are similar to the ion saturation current profile and thus the heat flux profile also similar to the I_{sat} profile. These results suggest that the long field lines that approach the LCFS are the main heat and particle transport channels from the stochastic layer to the divertor, and this consideration is consistent with that in section 3.1.

Measurements of the heat and the particle depositions using Langmuir probe arrays and thermocouples embedded in the divertor plates at some typical positions show that the deposition profiles qualitatively agree with the predicted profiles by field lines tracing calculation (see Fig. 6(a)). This result indicates that the prediction of the heat and the particle deposition profiles is possible in the complex edge magnetic topology in the heliotron configuration.

With increasing plasma pressure (β), the magnetic surfaces are modified. R_{ax} shifts outward due to the Shafranov shift, and the magnetic surfaces in the periphery are destroyed. Thus, the shape of the LCFS varies with increasing the β and thus the edge magnetic topology also changes. In high-β experiments ($\beta > 1\%$), the heat and the particle deposition profiles on a divertor plate and the poloidal-toroidal distribution largely modified. Field lines tracing calculations under finite β conditions using the HINT code [12] is under preparation, and detailed investigations of the heat and the particle deposition on the divertor in high-β plasmas will be conducted. That is essential for the investigation of the helical divertor in a heliotron-type reactor.

4. SUMMARY AND PROSPECT

Physical basis of the heliotron configuration is described especially pertaining to the edge magnetic topology. Experimental and numerical investigations of the edge plasma behavior in LHD heliotron were explained, and the effect of magnetic topology on edge plasma behavior in LHD heliotron was discussed. The main channel of the heat and the particle transport from the LCFS to the divertor is long field lines ($L_c >$ several hundred meters) connecting these regions. It is observed experimentally that both electron density and temperature in the divertor plasma considerably drop compared to them at the LCSF, and high recycling regime has not been observed clearly. The result of numerical investigation using 3D transport code EMC3-EIRENE agrees well with the experimental observation. A possible reason of the drop of the electron temperature is the long L_c. The existence of counter-flows in the stochastic layer is expected from the result of numerical study, and the friction between counter-flows is considered a main cause of the large momentum loss. Plasma flow measurement in the stochastic layer is necessary to confirm the existence of the counter-flows, and a "Mach probe" measurement is in preparation. In high β plasmas, magnetic topology is modified. To investigate the effect of magnetic topology on edge plasma behavior in high β plasmas, a new field lines tracing code taking into account the equilibrium has been developed.

In some tokamaks such as Tore Supra and TEXTOR, the ergodic divertor experiments have been conducted. The magnetic topology in the ergodic divertor is similar to that in the heliotron edge region, and some common plasma behaviors have been observed. In DIII-D tokamak, the edge ergodize experiment to control ELM has been conducted. Common understandings in tokamaks and LHD heliotron are expected.

In LHD, the Internal Diffusion Barrier (IDB) has been discovered under the Local Island Divertor (LID) configuration [13]. Very high electron density ($\sim 5 \times 10^{20}$ m^{-3}) and high plasma pressure (central $\beta > 4\%$) have been obtained inside the IDB. A necessary condition is high neutral particle pumping efficiency, and the LID configuration provides it [14]. A problem of the configuration is very small wet area, and it is difficult to use this configuration in steady state high input power operation. Now we prepare to closure the helical divertor to provide high pumping efficiency, and sustain the IDB plasma in steady state [15].

REFERENCES

1. A. Iiyoshi et al., *Nucl. Fusion* **39**, 1245-1256 (1999).
2. for example, K. Yamazaki et al., "Physics studies on helical confinement configurations with *l*=2 continuous coil systems", in Plasma Physics and Controlled Nuclear Fusion Research (Proc. 13th Int. Conf. Washington DC, 1990), IAEA-Cn-53/C-4-11.
3. N. Ohyabu et al., *Nucl. Fusion* **34**, 387-400 (1994).
4. T. Morisaki et al., *Contr. Plasma Phys.* **40**, 266-270 (2000).
5. S. Masuzaki et al., *Nucl. Fusion* **42**, 750-758 (2002).
6. T. Morisaki et al., *J. Nucl. Mater.* **313-316**, 548-552 (2003).
7. A.B. Rechester and M.N. Rosembluth, *Phys. Rev. Lett.* **40**, 38-41 (1978).
8. Y. Feng et al., *Contr. Plasma Phys.* **44**, 57-69 (2004).
9. D. Reiter et al., *Fusion Science and Technology* **47**, 172-186 (2005).
10. M. Kobayashi et al., presented in 17th International conference on Plasma Surface Interactions in Controlled Fusion Devices (2006), submitted to *J. Nucl. Mater.*
11. S. Masuzaki et al., *J. Nucl. Mater.* **313-316**, 852-856 (2003).
12. T. Hayashi et al., *Phys. Fluids B* **2**, 329-337 (1990).
13. N. Ohyabu et al., *Phys. Rev. Lett.* **97**, 055002 (2006).
14. S. Masuzaki et al, presented in 17th International conference on Plasma Surface Interactions in Controlled Fusion Devices (2006), submitted to *J. Nucl. Mater.*
15. S. Masuzaki et al., to be published in *Fusion Science and Technology.*

Relativistic Laser Acceleration Of Electrons Along Solid Surfaces

Z. M. Sheng, Y. T. Li, M. Chen, Y. Y. Ma, X. H. Yuan, M. H. Xu, Z. Y. Zheng, W. X. Liang, Q. Z. Yu, Y. Zhang, F. Liu, Z. H. Wang, Z. Y. Wei, Z. Jin, and J. Zhang

Institute of Physics, Chinese Academy of Sciences, Beijing 100080, China

T. Nakamura and K. Mima

Institute of Laser Engineering, Osaka University, Osaka 565-0871, Japan

Abstract. Recent experimental and theoretical studies on surface electron emission will be presented. A collimated fast electron beam was observed along the target surface irradiated by intense laser pulses up to 20TW when the laser is incident with large angles such as over 45 degree. Numerical simulations suggest that such an electron beam is formed due to the confinement of the surface quasistatic electric and magnetic fields. Meanwhile, an acceleration process similar to the inverse-free-electron-laser is found to occur and is responsible for the generation of the most energetic electrons. A general formula for electron angular distributions accounting for the quasistatic electric and magnetic fields is given. In certain conditions, quasi-monoenergetic electron beams are also produced. These results are of interest for potential applications of laser-produced electron beams and helpful to the undersanding of the cone-target physics in the fast ignition related experiments.

Keywords: Fast electron beam, surface emission, fast ignition, cone target.
PACS: 52.25.Tx, 52.38.Kd, 52.57.Kk

INTRODUCTION

Generation of various energetic particle beams, such as electron beams, positron beams, ion beams, and neutron beams from relativistic laser plasma interactions on table-top facilities has attracted significant attention in the past years [1,2]. This is mainly owing to the development of the chirped-pulse-amplification (CPA) technology, which enables one to obtain ultrashort laser pulses at the power up to the petawatt level currently and with focused intensities over 10^{20}W/cm^2. These energetic particles can find applications in the fast ignition of fusion targets, advanced particle accelerators, medical applications, ultrashort fast diffraction, advance radiation sources, nuclear waste separation and transmutation, radiography, and so on. Among the different particle beams, the generation and transport of fast electrons are the most fundamental processes in the intense laser interactions with plasmas [3-6]. They

CP876, The Physics of Ionized Gases: 23rd Summer School and International Symposium,
edited by L. Hadžievski, B. P. Marinković, and N. S. Simonović
© 2006 American Institute of Physics 978-0-7354-0377-2/06/$23.00

determine other secondary processes such as high energy ion, hard x-ray, and neutron emissions.

A variety of mechanisms for fast electron generation effective in different conditions have been proposed, such as resonance absorption [7], vacuum heating [8], acceleration by laser ponderomotive force and J×B force [9], stochastic heating and acceleration [10], and direct laser acceleration of electrons in a laser self-focusing channel [11]. At certain conditions, the energy conversion efficiency as high as 40% is found from the laser interaction with solid targets. The emission direction of fast electrons, however, depends not only upon the acceleration mechansim, but also upon the transporting process when overdense plasma is involved. A few experiments and simulations suggest that the produced electron beams often have a larger angular spread, unfavorable for applications. For example, the fast ignition concept requires energetic electrons to transport as effectively as possible through the long-scale-length plasma surrounding the ignition fuel pellet and deposit their energies as much as possible into its center [12]. Efforts have been made by use of proper target design, for example, the hollow cone target [13], which is found to be helpful to increase the heating efficiency to the target core. In some other experiments, it is found that there is no significant cone target effect [14]. It is thus necessary to study this problem further more.

In this article, we present experimental and theoretical studies on the laser interaction with solid targets with large incident angles. This is comparable to the laser interaction with cone-shell in the cone target geometry. It shows that electrons can be accelerated collimatedly along the solid surface provided the preplasma is small enough. We demonstrate that an acceleration mechanism similar to the inverse-free-electron-laser appears in such an interaction geometry, which contributes to these most energetic electrons. Those with relatively low energies follow the emission angle determined by the laser incident angle, the experienced electrostatic and magnetic fields. Finally, in the laser interaction with solid wire and slice targets, we demonstrate that quasi-monoenergetic electron bunches can be produced directly by laser ponderomotive-force acceleration.

EXPERIMENTAL OBSERVATION OF THE SURFACE EMISSION OF FAST ELECTRONS

The experiments were carried out using the Xtreme Light II (XL-II) laser system at the Institute of Physics, Chinese Academy of Sciences. The laser system can deliver laser pulses with an energy up to 0.6 J in 30 fs at 800 nm. The amplified spontaneous emission was measured to be $\sim 10^{-5}$. The laser pulse was focused by an f/3.5 off-axis parabolic mirror onto a 30 μm thick aluminum foil. The focal spot size was monitored by an x-ray pinhole or a knife-edge camera. The diameter of the focus was ~ 10 μm at the full width at half maximum (FWHM). The experimental setup is illustrated schematically in Fig. 1 [15]. The angular distribution of fast electrons was measured by a stack array of imaging plate (IP) at a distance of 5.5 cm radially from the focus. The size of a piece of IP is 30x50mm. Most of the 2π space in the laser incident plane was covered by the stack array except about 25° radial angle left for laser incidence.

The electron energy range was chosen by different thick aluminum filters in front of the first IP layer and aluminum filters between the adjacent layers. Imaging plate, a photo-stimulated luminescence detector with a large dynamic range, excellent linearity, and high sensitivity, is very suitable for x-ray and charged particle measurements in laser-plasma experiments [16]. IP is sensitive to ion, x-ray, and electron. However, thick aluminum filters over 400 μm set in front of the first IP layer can block the ions generated in the experiments. Comparison of the signal intensity on IP with and without a 1500 G magnetic field shows that the contribution of x rays is < 5%. Thus, the signal recorded by the IP is mainly from fast electrons.

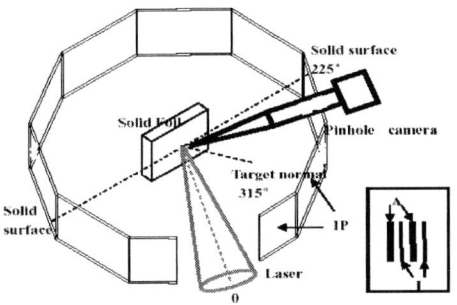

FIGURE 1. Schematic of experimental setup. The angular distribution of fast electrons was measured by a stack array of imaging plates.

Figures 2(a)-2(c) show the angular distributions of the fast electrons in the polarization plane for three laser incidence angles of 22.5°, 45°, 70°, respectively. 0° corresponds to the laser incident direction. Each data point in the polar diagrams was obtained by integrating the signal intensities on IP. One can see that in the polarization plane some fast electrons are distributed between the target normal and the laser specular directions in front of the target. This is similar with previous theoretical and experimental observations form different groups [17,18]. The most striking aspect of our measurements is the presence of part of fast electrons ejected along the front target surface, marked as "surface electrons" in the figures [In Fig. 2(a) an inset with enlarged scale is used to show them clearly]. In particular, the component of the surface fast electrons increases with the laser incidence angle. It even dominates the fast electron emission, the peak reaching ~5 times greater than the component of the hot electrons close to the target normal for the case of 70° incidence angle. Moreover, in this case the surface electron beam is also well-collimated with a emission cone angle less than 15° (the full width at half maximum).

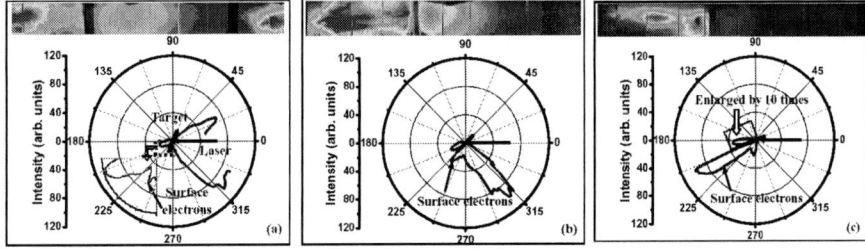

FIGURE 2. Angular distributions of the fast electrons with energies greater than 300 keV in the laser incident plane for the p-polarized laser pulse at an intensity of 1-2×10^{18} W/cm^2 for three different incidence angles of 22.5° (a), 45° (b) and 70° (c).

The distributions of the transmitted fast electrons are also shown in Fig. 2. The intensities of the hot electrons behind the targets are much less that in front of the targets. To show the signals clearly, the intensities of the transmitted electrons have been artificially multiplied by a factor of 10 in Fig. 2(c). One can see that the peak of the transmitted electron beam is deflected slightly to the rear target surface for the 22.5° incidence angle. This deflection, however, become much obvious for the 45° and 70° incidence angles, nearly paralleling with the rear surface. This agrees with the recent theoretical proposal that the surface magnetic field tends to confine fast electrons at the front surface [19].

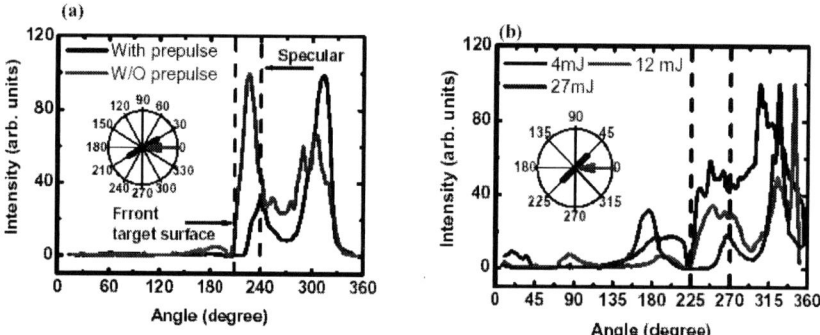

FIGURE 3. Angular distributions of fast electrons with and without a prepulse at the incident angle 60° (a), and the distributions for different prepulse energies at the incident angle 45°(b).

To find out the generation conditions for the surface electron emission, s-polarized and circularly polarized laser pulses, as well as different target materials (CH and Al), were also tried at different laser incident angles in the experiments, respectively. The results were similar to those shown in Fig. 2. To check the role of the electron density scale length, a prepulse with a duration of 200 ps, split from the main beam before the compressor, was used to create a preplasma. The separation between the prepulse and the main pulse was ~0.5 ns. The size of the prepulse focal spot was ~20 μm in diameter. The main laser intensity was set to be $1-2 \times 10^{18}$ W/cm^2, similar to that in Fig. 2. Figure 3(a) shows the angular distributions of fast electrons with and without a 36 mJ prepulse at an incident angle 60°. The emission peak shifts to the laser specular direction and the electron number close to the front target surface decrease almost to noise level after introducing the prepulse. Figure 3(b) shows the results for three different prepulse energies at an incident angle 45°. The surface fast electrons are affected very little when using the smallest prepulse energy 4 mJ. However, after increasing the prepulse energy, the surface emission switches to laser specular direction. For the case of 27 mJ, the peak has fully shifted to the specular direction. The electron number close to the surface also decreases to zero accordingly. In our experiments the energy of the intended prepulse must be lower than ~10 mJ to form the surface beam. In the experiments, the surface electron beam is easy to occur at large incident angles. This is probably because the reflecting surface $n_{cr}\cos\theta$ appears at small densiteis for large incident angles and thus the effect of the preplasma is reduced.

SIMULATIONS OF SURFACE EMISSION OF FAST ELECTRONS

To understand the formation of the surface fast electron jets, numerical simulations have been conducted with our two-dimensional (2D) fully relativistic particle-in-cell (PIC) code. In the PIC simulations a p-polarized laser pulse with an irradiance of 2-5x10^{18}W/cm^2 is incident at 70° onto an 8 n_c, 4λ_0 thick plasma slab with a sharp boundary, where n_c and λ_0 are the critical density and the laser wavelength respectively. The diameter of the laser focus is 10λ_0 with a Gaussian profile. The laser electric field is in the Y-direction. Figure 4(a) shows typical trajectories of fast electrons at the front target surface. One can see that some fast electrons in the focus move along the target surface in an oscillating form, instead of ejecting into the target region. After running away from the focus they have deviated from the initial directions, resulting in a surface fast electron jet.

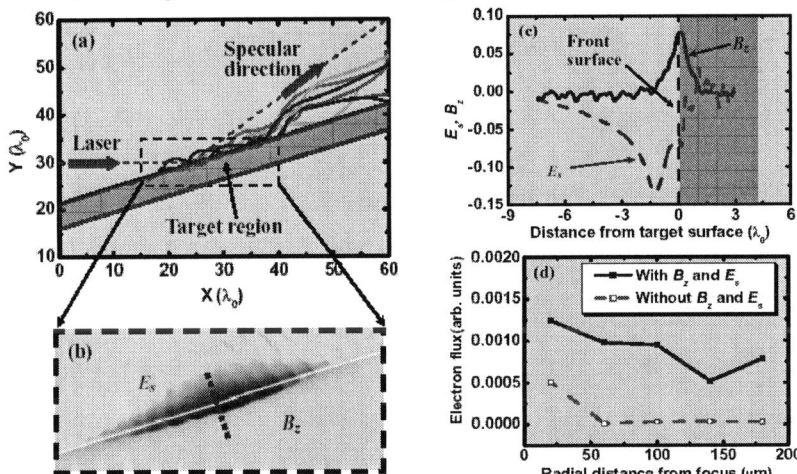

FIGURE 4. (a) Typical fast electron trajectories at the front target surface; (b) The distributions of the static electric field (E_s) in front of the surface and the magnetic field (B_z) inside the target; (c) The fields along the dashed line in (b) at 250 fs. (a)-(c) are obtained from the 2D PIC simulations. (d) Electron flux in the front target surface vs the radial distance from the focus for fast electrons with energies >300 keV, obtained from MC simulations.

This novel phenomenon can be explained as follows. When the fast electrons generated by $J \times B$ heating or vacuum heating in a plasma with a steep density gradient are accelerated into the target bulk at the early stage, a magnetic field will be induced by the fast electron current itself around the front surface. Figure 4(b) shows the spatial distribution of the magnetic field B_z inside the target, and the electrostatic field E_s [where $E_s = (E_x^2 + E_y^2)^{1/2}$] due to charge separation in front of the target, in the dashed box region in Fig. 4(a). The fields are normalized by the incident laser amplitude. Both the magnetic and electric fields are located in the skin layers near the surface. Part of the fast electrons generated in the interaction region will be reflected back to the vacuum by B_z. However, the negative sheath field E_s, whose peak position

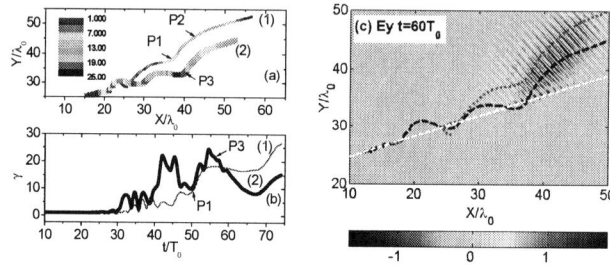

FIGURE 5. Selected two electron trajectories [labeled with (1) and (2)] and their energy changes along the trajectories (a) and with time (b). The color bar in (a) shows the relativistic factor of the electrons. Frame (c) shows snapshots of the laser field together with the two electron trajectories.

slightly shifts to the vacuum relative to that of the B_z [see Fig. 4(c)], will push them back to target again. This push-pull process will lead to the enhancement of the surface electron current and the magnetic field. Therefore a flow of fast electrons along the target surface in an oscillating form will be produced self-consistently due to the confinement of the surface magnetic fields and electric fields. Thus a fast electron jet along the target surface will be generated.

The oscillations of the fast electrons along the surface in the focus are very similar with the electron motion in a laser self-focusing channel, where electrons make betatron oscillations forced by the self-generated quasistatic electric and magnetic fields [11,20]. The oscillating electrons can gain a significant amount of energy from the laser fields when the electron oscillation frequency is resonant with the laser frequency, similar to the inverse-free-electron-laser acceleration. In our simulations we also identify a fraction of surface electrons are accelerated additionally during their oscillations [20]. Figure 5 shows two typical electron trajectories and their energy change with time. Typically they oscillate for a few periods along the surface before leaving the interaction region.

In other simulations with a preplasma added in front of the sharp-boundary plasma slab, no surface fast electrons are observed. This agrees with the experimental measurement for the case with large preplasma. A large scale preplasma will spoil the field structure and lead to disappearance of the surface electron beams. Thus both the experimental and simulated results show that surface quasistatic magnetic and electric fields are essential to the formation of surface electron beams.

To further illustrate the effect of the surface magnetic field and the sheath electric field on fast electron emission, a 3D MC code ITS3.0 has been used to simulate the electron propagation [21]. In the simulations a fast electron beam with a cone angle of $30°$ and with an exponential distribution of $\sim \exp(E/kT)$ is incident at $70°$ onto a $30\mu m$ thick Al target, where E and kT are the energy and effect temperature of fast electrons, respectively. Here we take kT=300keV, which is comparable to our experimental results. The diameter of the electron source is 5 μm. The values of the magnetic fields and the electric fields obtained from the PIC simulation were used as the field input. Figure 4(d) shows the dependence of the electron flux of the fast electrons with energies >300 keV on the radial distance from the focus in the plane of the front target surface. The flux has been normalized to one source particle. It is obvious that the electron flux without magnetic field B_z and electric field E_s is local and rapidly decays to zero when fast electrons transport away from the source. However, the fast electrons can flow along the front surface to far distance from the source with much higher flux when B_z and E_s are considered.

EFFECT OF QUASI-STATIC ELECTRIC AND MAGNETIC FIELDS ON ANGULAR DISTRIBUTIONS OF FAST ELECTRONS

For the electrons emitted in a different direction from the front surface, a formula has been derived before to describe the angular distributions [18]. To derive this formula, one assumes specular reflection of the incident laser and the solid target is modeled with some quasistatic electric fields around the front surface. The quasistatic manetic fields have been neglected. However, from simulations [19,20], it is obvious that the quasistatic magnetic fields play a significant role in determing the fast electron emission. Therefore, it is necessary to reconsider this problem.

We consider a geometry that a planar laser pulse is incident at angle α onto a solid target, where quasistatic electric (Ex) and magnetic (Bz) fields are induced near the target surface region during the interaction. To get the fast electrons emission directions, one can start with the Lagrange function for the electron motion as before [18,22]. Alternatively, by use of momentum and energy conservations between the fast electrons and absorbed photons, one can get the relation in a simpler way. As it is well known the component of the total canonical momentum along the target surface is always conserved. Thus one has:

$$N_e p_\parallel = N_p \hbar k_\parallel - N_e (q \delta A_{0y} / c) \tag{1}$$

where N_e and N_p are electron density and photon density, and δA_{0y} is the vector potential change associated with quasistatic magnetic fields. The number of the interacting photons per electron, N_p / N_e, can be found from the energy conservation equation:

$$N_p \hbar \omega = N_e (\gamma - 1) mc^2 + N_e q \delta \varphi. \tag{2}$$

Here the contribution of the quasistatic electric field to the electrons is included. We define the emission angle θ as: $\sin \theta = p_y / p$, $\tan \theta = p_y / p_x$ and assume $p_z = 0$ and $\gamma = \sqrt{1 + (p_x^2 + p_y^2) / m^2 c^2}$. From Eqs. (1) and (2), making use of $\hbar k_\parallel = (\hbar \omega / c) \sin \alpha$, we can obtain the emission direction of electrons [22]:

$$\sin \theta = \frac{\gamma - 1 + \hat{q} \delta \Phi}{(\gamma^2 - 1)^{1/2}} \sin \alpha, \tag{3}$$

where $\delta \Phi = |e| [\delta \varphi - \delta A_{0y} / \sin \alpha] / mc^2$ is the normalized potential variation and $\hat{q} = q / |e|$ is the normalized charge. Compared with the relation given in [18], one may find that it is formally the same as the relation we have gotten before if one take $\hat{q} = -1$ for electrons, except for the definition of $\delta \Phi$. Here the effect of the quasistatic magnetic field has been taken into account, which plays in an opposite way with the quasistatic electric field.

According to PIC simulations, when an ultrashort intense laser pulse obliquely incident onto a solid target, negative static electric fields and positive static magnetic fields are generated in front of the target surface. Therefore one has $E_x = -\partial \varphi / \partial x < 0$ and $B_z = \partial A_{0y} / \partial x > 0$. One notes that $\hat{q}\delta\Phi$ in Eq. (3) is a path integral for individual particles. The formula (3) suggests that more emission regions of fast electrons than those predicted in Ref. 18. It can describe electrons emitted in different directions, corresponding to different $\hat{q}\delta\Phi$.

Usually it is difficult to get an explicit expression for $\delta\Phi$. Some attempt has been done in Refs. 18 and 19. Because of the dependence on the electron trajectories, in numerical simulation we can only give the approximate value of $\delta\Phi$. Numerical simulations show that individual electrons can experience various $\delta\Phi$ values either positive or negative. Figure 6 shows a simulation result by use of a 1D3V PIC code accommodated in the Lorenz-boosted frame for oblique incidence of laser pulses [18,23]. The target is composed of a high density region at $5n_c$ with width $d = (3 \sim 8)\lambda$ and preformed plasma which decreases exponentially with scale length $L = 0 \sim 3\lambda$ from the high density platform. Following the field distributions at $t = 65\tau$ given in Fig. 6(a), one estimates that the value for the electrons moving to the left is $\delta\Phi \approx -0.2$. In another way, one can get the minimum $\delta\Phi$ by fitting the numerical simulations as shown in Fig. 6(b) with Eq. (3), which gives $\delta\Phi_{min} = -1.0$.

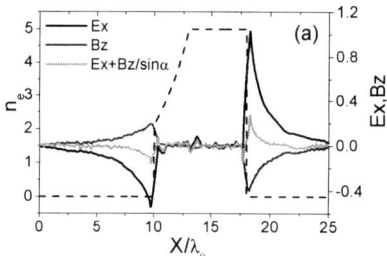 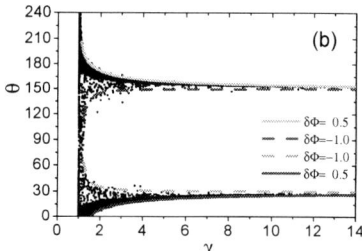

FIGURE 6. (a) Initial density profile for the simulation (given by the dashed line), induced quasistatic magnetic field (Bz) and electric field (Ex) (normalized by $m\omega c/e$ and smoothed with neighbor points) after the laser has reflected away from the target found in 1D PIC simulation ($t = 65\tau$). The combined value Ex+Bz/sinα is also shown; (b) Angular distributions of electrons outside of the target after the interaction of a laser pulse (with p-polarized pulse) with a solid target. The solid lines give the fit curves with Eq. (3) for different values of $\delta\Phi$. The incident laser pulse takes a sine-square profile with peak amplitude $a_0 = 2.0$, pulse duration $t_0 = 15\tau$, and incident angle $\alpha = 30^0$.

GENERATION OF DENSE QUASI-MONOENERGETIC ELECTRON BEAMS

In a similar interaction geometry as shown above, even quasi-monoenergetic electron bunches can be generated in the case with large incident angles of the laser pulse. They are produced by the ponderomotive force of a relativistic intense laser pulse directly during its interaction. The mechanism is studied both analytically and by multi-dimensional PIC simulations [24,25]. In the following, we show typical results found when the laser pulse interacts with wire or slice targets by using 2D PIC simulations. The wire target is a thin uniform plasma strip with width D=0.1λ_0 (in the y-direction) and length L=5 λ_0 (in the x-direction). The incident laser has a Gaussian profile transversely with a beam waist of 2µm and it propagates along the wire or slice target as ploted in Fig. 7(a).

Figure 7(a) snapshot of electrons emitted from the slice target. Electrons near the peak of transverse laser field are pulled out from the target to vacuum and are accelerated by **V×B** mechanism. The positive field peaks pull electrons toward the negative y-direction and negative peaks pull electrons toward the positive y-direction. Since both positive peaks and negative peaks are in one-wavelength spacing, the emitted electrons appear as separated bunches both upside and downside. The trajectories of selected electrons are given in Fig. 7(b). They show that electrons are expelled from the target surface by the laser beam. Some of the trajectories appear similar to the ponderomotive force scattering in vacuum [26]. Note that these energetic electrons usually emit with a small angles against the x-direction.

It shows that the obtained energy spectrum has a quasi-monoenergetic peak around 11MeV as shown in Fig. 7(c). For comparison, we have also studied the interaction of the same laser pulse with a plasma slice with a transverse width D=20λ_0. The resulting electron energy spectrum shows an exponential distribution as usually seen. Even though the quasi-monoenergetic electron beams are found with wire or slice target with the laser pulse propagating along the wire axis, such electron beams can be found in the laser interaction with plane targets provided the incident angle is large enough.

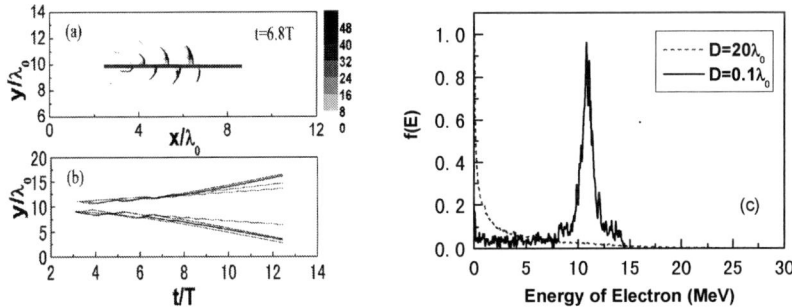

FIGURE 7. (a) Energy density distribution $\mathcal{M}_e$ in space for the initial density of the wire target at $n_0=30n_c$; The solid line marks the initial wire target; (b) The trajectories of some selected electrons; (c) Electron energy spectra obtained for different target widths D=0.1λ_0 and 20λ_0. The incident laser pulse is with the amplitude $a_0=8.54$, transverse width 2λ_0 and pulse duration of 5 laser cycles.

SUMMARY

We have studied experimentally and theoretically the electron emission along the solid surface. Both experiments and simulations suggest that such emission can be observed provided the preplasma is controlled to be small enough. In this case, large quasistatic electric and magnetic fields can be induced around the solid surface, which tends confine large number of fast electrons at the surface. At the beginning, fast electrons are generated mainly due to the vacuum heating and JxB acceleration. At later stage, when the quasistatic electric and magnetic fields are established at the front surface, additional acceleration mechanism similar to the inverse-free-electron laser acceleration may occur for some electrons.

An analytical formula for the electron emission direction accounting for the induced quasistatic fields is found, which can describe partially the electron emission directions found in experiments.

When the laser field is relativistic-intense, ponderomotive force acceleration may become very important. In this case, quasi-monoenergetic electron bunches can be produced provided that the laser pulse is grazingly incident to the solid targets.

ACKNOWLEDGMENTS

This work was supported in part by the National High-Tech ICF Committee of China, the National Science Foundation of China, the National Key Laboratory of High Temperature and High Density Plasma, and the JSPS-CAS Core-University Program on Plasma and Nuclear Fusion.

REFERENCES

1. G. Mourou, T. Tajima, and S. Bulanov, Rev. Mod. Phys. 78, 309 (2006).
2. J. Zhang, Y. T. Li, Z. M. Sheng, Z. Y. Wei, Q. L. Dong, X. Lu, Appl. Phys. B 80, 957 (2005).
3. M. H. Key et al., Phys. Plasmas 5, 1966 (1998).
4. J. R. Davis et al., Phys. Rev. E 56, 7193 (1997); A. R. Bell et al., Phys. Rev. Lett. 91, 035003 (2003).
5. M. Borghesi et al., Phys. Rev. Lett. 83, 4309 (1999); L. Gremillet et al., Phys. Rev. Lett. 83, 5015 (1999).
6. M. Honda, J. Meyer-ter-Vehn, and A. Pukhov, Phys. Plasmas 7, 1302 (2000).
7. D. W. Forslund, J. M. Kindel, Kenneth Lee, E. L. Lindman, and R. L. Morse, Phys. Rev. A 11, 679-683 (1975).
8 F. Brunel, Phys. Rev. Lett. 59, 52-55 (1987).
9 W. L. Kruer and K. Estabrook, Phys. Fluids 28, 430-432 (1985)
10 Z. M. Sheng et al., Phys. Rev. Lett. 88, 055004 (2002)
11 A. Pukhov, Z. M. Sheng, and J. Meyer-ter-Vehn, Phys. Plasmas 6, 2847-2854 (1999)
12. M. Tabak et al., Phys. Plasmas 1, 1626 (1994).
13. R. Kodama et al., Nature (London) 412, 798 (2001) ; R. Kodama et al., ibid. 418, 933 (2002).
14. P. A. Norreys et al., Phys. Plasmas 11, 2746 (2004) ; S. D. Baton et al., Plasma Phys. Controlled Fusion 47, B777 (2005).
15. Y. T. Li et al., Phys. Rev. Lett. 96, 165003 (2006).
16. K. A. Tanaka et al., Rev. Sci. Instrum., 76, 013507 (2004).
17. Y. Sentoku et al., Phys. Plasmas 6, 2855 (1999); Y.T. Li et al., Phys. Rev. E 64, 046407 (2001); L. M. Chen et al., Phys. Rev. Lett., 87, 225001 (2001).
18. Z.-M. Sheng et al., Phys. Rev. Lett. 85, 5340 (2000).
19. Y. Sentoku et al. Phys. Plasmas, 11, 3083 (2004) ; T. Nakamura et al., Phys. Rev. Lett., 93, 265002 (2004).
20. M. Chen et al., Opt. Express 14, 3093 (2006).
21. J. A. Halbleib et al., "ITS 3.0: Integrated TIGER Series of Coupled Electron/Photon Monte Carlo Transport Codes", SAND91-1634, (March 1992).
22. M. Chen, Z. M. Sheng, J. Zhang, Phys. Plasmas 13, 014504 (2006).
23. A. Bourdier, Phys. Fluids 26, 1804 (1983); R. Lichters, J. Meyer-ter-Vehn, and A. Pukhov, Phys. Plasmas 3, 3425 (1996).
24. Y. Y. Ma, W.W. Chang, Y. Yin et al., Chin. Phys. Lett 18,1628 (2001).
25. Y. Y. Ma et al., submitted for publication (2006).
26. F.V. Hartemann et al., Phys. Rev. E 58, 5001 (1998); ibid. 51, 4833 (1995).

Some Routes in Forming $C_3H_n^+$ Ions and Deuterated Variants under Interstellar Conditions

Igor Savić[*] and Dieter Gerlich[¶]

[*] *Department of Physics, Faculty of Science and Mathematics, University of Novi Sad, Trg Dositeja Obradovića 4, 21000 Novi Sad, Serbia*
[¶] *Department of Physics, Technische Universität Chemnitz, 09126 Chemnitz, Germany*

Abstract. Laboratory experiments on hydrogenation and deuteration of C_3^+, C_3H^+ and $C_3H_2^+$ in collisions with H_2 and HD have been performed from room temperature down to 15 K using a 22-pole ion-trap. At room temperature C_3^+ reacts slowly with H_2 but the reactivity increases with decreasing temperature. It has been shown that the association reaction $C_3^+ + H_2 \rightarrow C_3H_2^+ + h\nu$ can compete with the exothermic reaction $C_3^+ + H_2 \rightarrow C_3H^+ + H$. In collisions of C_3^+ with HD, formation of C_3D^+ is slightly favored over C_3H^+ formation. A pronounced competition between various channels has been detected for deuterated variants of the $C_3H^+ + H_2$ system. Most surprising is that formation of C_3HD^+ is over one hundred times faster then formation of $C_3H_2^+$ in collisions of C_3H^+ and HD. An tentative explanation is that the H-HD exchange takes place via an open-chain H_2CCCH^+ intermediate. Reactions of $C_3H_2^+$ and $C_3H_3^+$ with H_2 are very slow. The formation of $C_3H_2D^+$ or $C_3HD_2^+$ and finally C_3HD via dissociative recombination has been discussed. The reaction $C_3H_3^+ + HD \rightarrow C_3H_2D^+ + H_2$ can be ignored in astrochemical models since the reaction rate at 15 K is very small; however, quite efficient routes have been found starting from C_3^+ and proceeding via deuterated C_3H^+ to $C_3H_2^+$ and $C_3H_3^+$. The new reaction rate coefficients are recommended to be included in astrophysical databases. Nonetheless it is still unclear how to explain the large abundance of C_3H_2 and larger hydrocarbons and their deuterated variants observed in cold interstellar clouds.

Keywords: laboratory astrochemistry, low temperature collisions, hydrogenation, deuteration, ISM: molecules, ions, C_3H_2, C_3HD, $C_3H_2D^+$

PACS: 52.20.Hv, 82.20.Tr, 82.30.Fi, 95.30.Ft, 98.38.-j

INTRODUCTION

Within the last 30 years, different chemical models have been formulated for predicting the formation, processing and destruction of interstellar molecules. In all of them, gas phase reactions and surface processes are of key importance.

In general, gas phase ion-molecule reactions have no barriers [1] and due to the long range ion-induced dipole attraction their reaction rate coefficients are usually larger then those for neutral-neutral reactions. This is the reason why already early simulations (e.g. [2], [3]) of dense or diffuse clouds and accretion disks, using mainly ion-molecule reactions were quite successful in explaining the observed molecular abundances. The final step in the synthesis of neutrals coming from ion-molecule

CP876, *The Physics of Ionized Gases: 23rd Summer School and International Symposium*,
edited by L. Hadžievski, B. P. Marinković, and N. S. Simonović

reactions is their recombination with electrons. Today's chemical reaction networks are very sophisticated, complex and specialized. In addition to ion-molecule reaction, they include a variety of other types of reactions. In order to numerically solve the problem one has to select the most important processes for predicting the evolution of interstellar clouds. For this precise reaction rate coefficients are needed and, therefore, dedicated measurements performed under conditions of astrophysical relevance are necessary.

For some ion-molecule reactions the rate coefficients are temperature independent but there are also many exceptions. Changes especially occur at low temperatures. Especially important for astrochemistry are radiative association reactions [4], near thermoneutral reactions and reactions slowed down by small barriers along the reaction path.

Almost 80 % of the detected interstellar molecules contain one or more carbon atoms. One problem in astrochemistry is to find chemical pathways leading from small molecules like C, C_2, C_3, C_2H or C_3H to those complex molecular structures which are suspected to be synthesized in inter- or circumstellar regions. It has been pointed out recently [5] that even for small molecules such as C_4H and C_3H_2 our present understanding of the chemical route is not sufficient to explain observations and that new formation routes must be found to explain abundances and correlations of molecules in PDRs.

It has been observed that in cold ISM environments the DX/HX abundance ratio can be a factor of 10^4 larger than the D/H elemental ratio. Understanding isotope fractionation is of central importance for modeling deuterium enrichment in interstellar molecules. It is generally accepted that most important ways in producing deuterated molecules are collisions between ions and HD, D or D_2. The difference in zero-point vibrational energy of the molecules determines the endo- or exothermicity of an H-D exchange reaction. Usually reactions proceed much faster in the exothermic direction than in endothermic one; however, an activation barrier can complicate the situation. New low-temperature measurements [6], [7], [8] have shown that there can be significant deviations from simple models, due to small barriers or bottlenecks of the potential energy surfaces. In addition, symmetry selection rules play a central role in replacing one ore more atoms in a group of identical ones by isotopes and vice versa [9]. This together with conservation of the total nuclear spin can have a huge influence [10].

A typical example for an unsolved problem in today's reaction models is the very large deuterium fractionation of c-C_3H_2 in TMC-1. For the C_3HD/C_3H_2 ratio, a value of 0.048 has been reported in [11]; other observations lead to ratios between 0.08 and 0.16, depending on the position [12]. Theoretical predictions stay below 0.01 [13], [14], [15]. In order to reproduce observed C_3HD/C_3H_2 ratio, it has been proposed in [16] that the reaction

$$C_3H_3^+ + HD \rightarrow C_3H_2D^+ + H_2 \qquad (1)$$

may occur rapidly at low temperatures and that C_3HD is formed via dissociative recombination of $C_3H_2D^+$. This has been excluded by quantum chemical calculations [17] since high transition barrier have been found.

The detailed understanding of reaction dynamics at very low energies is very important from a fundamental point of view. It is interesting what happens if the total energy varies by a small amount near zero. One consequence is that the number of partial waves contributing to the collision complex formation becomes smaller and smaller and finally only s-waves need to be considered. In reactions where H atoms are involved this effect play a role already at collision energies in the meV range. In this limit, quantum-mechanical calculations (but also statistical models) become simplified. Dynamical effects such as resonances may play a role. At low energy collision energies, other effects can occur because excitation of rotational or fine structure states can be the dominant contribution to the total energy. In general, a low total energy leads to long collision complex lifetimes enabling very unlikely processes to occur like radiative association, rearrangement via internal tunneling, or non-adiabatic mixing.

EXPERIMENTAL AND TYPICAL RESULTS

The results presented in this paper have been obtained using a variable temperature rf 22-pole ion trapping apparatus. The machine is fully described in [18] and here only a brief description is given. Primary ions are generated externally in a storage ion source [19]. Then a pulsed bunch of ions is mass selected in a first quadrupole mass filter (operated in this work in the mass selective mode) and injected into a 22-pole ion trap with low kinetic energy. In the 22-pole trap, ions are confined in radial direction by an effective potential. This is created by applying two opposite phases of an rf generator to the two sets of 11 electrodes. In the axial direction the trap can be closed by small potential barriers created by voltages applied to the gate electrodes at both ends of the trap. In order to achieve temperatures as low as 5 K, the trap is mounted onto a closed cycle refrigerator and surrounded by a second thermal shield held at ~50 K. The temperature is usually measured using a carbon resistor, a calibrated diode or a hydrogen gas thermometer. Buffer and reactant gases introduced by cooled tubes are in thermal equilibrium with the surrounding walls. The stored ions are efficiently coupled to the temperature of the buffer gas via collisions. In the present case a short and intense pulse of He is introduced into the trap. The ions can be stored for times varying from milliseconds to seconds. During this time, they react with the neutral target gas. Usually only few thousands primary ions are used per pulse in order to avoid space charge effects and saturation of the Daly detector. After a given reaction time, the remaining primary and the newly formed product ions are extracted from the trap by a pulsed voltage applied to the exit electrode. They are mass analyzed in the second mass filter and finally counted using a Daly type detector. For improving the statistics and eliminating fluctuations, the procedure of ion formation, trapping, reaction and analysis is repeated very often for each mass of interest and for typically ten different storage times.

All primary ions were produced in the storage ion source by electron bombardment (collision energy ≤ 40 eV) of allene ($CH_2=C=CH_2$, Aldrich, 97 %). Until recently it was unclear how C_3^+ really looks like. Recent calculations [20] indicate a cyclic structure with the lowest states being perturbed by the Jan-Teller effect. Since C_3H^+, $C_3H_2^+$ and $C_3H_3^+$ have cyclic and noncyclic isomers, several tests, based on reactions in the trap, have been performed in order to determine the possible presence of excited isomers. The reactivity of the linear and cyclic ions is usually different. In most studies

417

performed with ions produced in the storage ions source, only mono-exponential decay have been observed, one exception being $C_3H_2^+$. This indicates that the storage ion source produces predominantly the lowest energy isomer.

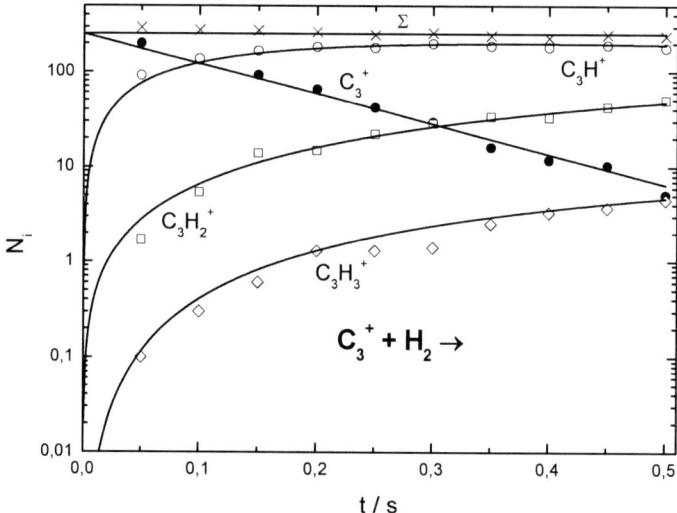

FIGURE 1. Typical experimental results. Shown is the averaged number of primary and product ions trapped per filling, N_i, as function of the storage time t. Within less than 100 ms the C_3^+ ions are cooled to the ambient temperature. They react with H_2 (number density 4.1×10^9 cm^{-3}) via a cascade of reactions forming C_3H^+, $C_3H_2^+$ and $C_3H_3^+$. The solid lines are solutions of an adequate reaction rate system.

In the present experiments, both H_2 and HD has been used as neutral reactant. The H_2 had a purity of 99,99990 % (Messer-Griesheim) while HD (Cambridge Isotope Laboratories Inc.) had a purity of 98 % with 2 % of H_2 and D_2 [7]. The number density of target gas is determined with an ion gauge calibrated with respect to a spinning rotor gauge. In the present work the error of the absolute rate coefficients is estimated to be 20%, mainly caused by fluctuations of the density.

Fig. 1 shows a typical experimental result measured at 44 K. Plotted is the number of trapped ions, N_i, as a function of the storage time t which is identical with the reaction time. The analogue reactions with HD are more complex. Typical experimental results for these reactions are presented in [21] and [22].

RESULTS AND DISCUSSION

Ion-molecule reactions between hydrocarbons and H_2 have been studied often. The reaction

$$C_3^+ + H_2 \rightarrow C_3H^+ + H \qquad (2)$$

is exothermic by 197 kJ mol^{-1} [23]. Therefore, the reaction rate coefficients measured at room temperature, 1.8×10^{-10} [24], 2.5×10^{-10} [23], 3×10^{-10} [25] and 4.6×10^{-10} cm^{3}s^{-1} [21] are surprisingly low. In addition, a value of 2×10^{-9} cm^{3}s^{-1} has been measured at 10 K [26]. Our recent temperature dependent measurements [21] have approved that the reaction becomes faster with decreasing temperature. Below

44 K a constant value of 1.7×10^{-9} cm^{3}s^{-1} is reached. Within our experimental error of 20 %, this is in accordance with the Langevin capture rate coefficient. The temperature dependence of the rate coefficient can be approximated with the function $k = \alpha (T/300K)^{\beta}$ with the parameters $\alpha = 4.7 \times 10^{-10}$ cm^{3}s^{-1} and $\beta = -0.69$ for temperatures between T = 44 and 300 K.

It is surprising that the association reaction

$$C_3^+ + H_2 \rightarrow C_3H_2^+ + h\nu \tag{3}$$

competes with the exothermic reaction (2). At 20 K almost 5 % of the collision complexes are stabilized by emission of a photon. For the temperature range T = 20 - 44 K the coefficients are $\alpha = 5.7 \times 10^{-12}$ cm^{3}s^{-1} and $\beta = -1.02$ while reaction rate coefficient $k = 8.5 \times 10^{-11}$ cm^{3}s^{-1} have been measured at temperature of 15 K [21].

The reaction of C_3^+ with HD has one more product channel since formation of C_3H^+, C_3D^+ and C_3HD^+ is possible:

$$C_3^+ + HD \rightarrow C_3H^+ + D \tag{4}$$
$$C_3^+ + HD \rightarrow C_3D^+ + H \tag{5}$$
$$C_3^+ + HD \rightarrow C_3HD^+ + h\nu \tag{6}$$

The reaction rate coefficient for reaction (4) is 9.3×10^{-10} cm^{3}s^{-1} at 15 K. It is slightly higher than the value for reaction (5) of 7.6×10^{-10} cm^{3}s^{-1}. For the reaction (6) 5.9×10^{-11} cm^{3}s^{-1} has been measured [21]. As can be seen, reaction of C_3^+ with HD is slightly slower then reaction with H_2 and even slower with D_2 (1.3×10^{-9} cm^{3}s^{-1} [27]). This is in accordance with the mass dependence of the capture cross section.

Within a simple classical picture, reaction between an ion and HD favors H abstraction because the separation of the center of charge and the center of mass orients HD such that the H atom preferentially points toward the ion (especially in the rotational ground state). The rate coefficients measured for reactions (4) and (5) are in contradiction with this simple model. Therefore, it can be concluded that hydrogen abstraction does not occur in a direct way. This conclusion can be corroborated by the temperature dependence of reaction (2). Namely, the observed hydrogen abstraction temperature dependence can be explained by formation of a long-lived collision complex since several stable strongly bound $C_3H_2^+$ ion isomers exists. The lifetime of such complex must be long enough for rearrangement. The possible structures of $(C_3H_2^*)^+$ [28] indicate that the hydrogen molecule is first weakly connected to the C_3^+ ion, which may have to go through a special geometry (may be linear) in order to brake the hydrogen bond and to form the linear C_3H^+ product. The fact that the reaction probability is much smaller at room temperature than at low temperatures can be taken as a hint that the initially formed intermediate state has only a small binding energy. Indication that some of the collision complexes are really long lived is that the radiative association also contributes with more than 3 %.

The first results using flow tube technique on reactions between C_3H^+ and H_2 have been discussed in detail in [25], [29], [30], [31]. The high pressure flow tube conditions lead to saturated three-body association and significant suppression of hydrogen abstraction. This strong competition in $C_3H_3^+$ and $C_3H_2^+$ formation has been verified by pressure dependent measurements over a wide range of pressures in an 80 K ion trap [4]. Since the lifetime of intermediate complex is temperature dependent, decrease of $C_3H_2^+$ products with decreasing the temperature has lead to the wrong conclusion that the reaction

$$C_3H^+ + H_2 \rightarrow C_3H_2^+ + H \tag{7}$$

is endothermic by 4 kJ mol^{-1} [30]. Later ab initio calculations got the same value [28] or an even higher endothermicity of 7 kJ mol^{-1} [32]. It has been proposed in [32] that more studies between 10 and 300 K are needed to give some guidance to ab initio calculations and that has been done in the 22-pole trapping machine [27], [21]. The results on reaction (7) and reaction

$$C_3H^+ + H_2 \rightarrow C_3H_3^+ + h\nu \tag{8}$$

at temperatures between 10 and 50 K and with n-H$_2$ and p-H$_2$ have been reported in [27] and within the range 15 – 300 K for n-H$_2$ in [21]. There is a good overall agreement between these experiments. At 300 K rate coefficients for reactions (7) and (8) sum up to 2 % of collision rate but they reach almost 20 % at 20 K. For the hydrogen abstraction reaction (7), the parameters α and β are 1.5×10^{-10} cm^{3}s^{-1} and + 0.09 respectively in the the temperature range 15 – 44 K while for 44 – 300 K they are 1.4×10^{-11} cm^{3}s^{-1} and – 1.05 [21]. The rate coefficient for radiative association (8) increases from 1.5×10^{-12} cm^{3}s^{-1} at T = 72 K to 6×10^{-11} cm^{3}s^{-1} at T = 15 K. The use of p-H$_2$ instead of n-H$_2$ speeds up radiative association but it slows down hydrogen abstraction due to competition between these two channels. This is mainly due to the colder conditions prepared with the non-rotating H$_2$. Therefore, all results [27], [21] clearly show that hydrogen abstraction in C$_3$H$^+$ + H$_2$ collisions is not endothermic.

Using HD instead of H$_2$ in collisions with C$_3$H$^+$ has revealed a lot of unexpected results. Measured reaction rate coefficients at temperature of T = 15 K for reactions

$$C_3H^+ + HD \rightarrow C_3D^+ + H_2 \tag{9}$$
$$C_3H^+ + HD \rightarrow C_3HD^+ + H \tag{10}$$
$$C_3H^+ + HD \rightarrow C_3H_2^+ + D \tag{11}$$
$$C_3H^+ + HD \rightarrow C_3H_2D^+ + h\nu \tag{12}$$

are 5.6×10^{-11} cm^{3}s^{-1}, 4.6×10^{-10} cm^{3}s^{-1}, 3×10^{-12} cm^{3}s^{-1}, and 3.2×10^{-11} cm^{3}s^{-1}, respectively [21]. In comparison with reactions with H$_2$ [27], [21], and D$_2$ [26], the overall fastest reaction is C$_3$H$^+$ + HD, reaching almost 50 % of the collision rate at 15 K. That can be explained by the fact that HD is a more efficient cooler than the homonuclear hydrogen molecules where nuclear spin restrictions restrict rotational transitions. In addition it has to be emphasized that C$_3$H$^+$ + HD collisions lead almost exclusively to deuterated hydrocarbons, a very important results for interstellar chemistry. Note however that the major product, C$_3$HD$^+$, cannot be easily converted into C$_3$H$_2$D$^+$.

For reaction (10) and (11) an incredible large isotope effect has been discovered. Formation of C$_3$HD$^+$ is 150 times more probable than formation of C$_3$H$_2$$^+$! That fact can not be explained based on differences in zero point energies. It is probably caused by an H-HD switching reaction. In this case the addition of the HD on one side of the linear intermediate complex leads to the loss of the H atom bound on the other side. In order to explain qualitatively the competition between hydrogen abstraction and radiative association, the isotope effect and total reactivity a simple model potential has been proposed in [27] and [21].

Careful measurements on reaction (1) have been performed recently [22]. It has been found that the rate coefficient for deuteration of C$_3$H$_3$$^+$ in collisions with HD is smaller then 1×10^{-16} cm^{3}s^{-1}. The reaction rate coefficient of 5×10^{-13} cm^{3}s^{-1} has been measured for the reactions C$_3$H$_2$$^+$ + H$_2$ → C$_3$H$_3$$^+$ + H and C$_3$H$_2$$^+$ + D$_2$ → C$_3$HD$^+$ + HD at 10 K, while value of 1.7×10^{-14} cm^{3}s^{-1} have been obtained for the $C_3H_2^+ + D_2 \rightarrow C_3D_2^+ + H_2$ reaction [27], [26]. Collisions of $C_3H_3^+$ and $C_3H_2^+$ with HD or D_2 are not of importance for astrochemical models since these reactions are very slow. On the other hand they are interesting from fundamental point of view since exothermic reactions are possible. Small but finite rate coefficients measured at low temperature indicate that there must be some ways to circumvent the barrier or to tunnel through it.

CONCLUSIONS

In the focus of this contribution are low-temperature ion trap measurements of reactions involving $C_3H_n^+$ (n = 0 - 3) ions and hydrogen and deuterated variants. The relevance for the carbon chemistry of interstellar clouds has been discussed. We conclude that chemical reaction networks have to be revised since important reactions are missing or are included with wrong rate coefficients especially concerning their temperature dependence. Since branching ratio in distributing the hydrogen or deuterium atom from and HD molecule is almost 1:1 for $C_3^+ + HD$ collisions and completely different ratios are obtained for the competing products in $C_3H^+ + HD$ collisions, branching ratios have to be checked individually for each reaction. Including the new rate coefficients into models, especially those for reactions with C_3^+ at low temperature will increase the number density of hydrocarbons in dense interstellar clouds. The data presented in [21], [22] and here are not yet complete for correctly describing the correlations between different hydrocarbons and their deuterated variants. Therefore more dedicated experiments on this subject have to be done. It is sure that the most interesting reactions are $C_3HD^+ + H_2 \rightarrow C_3H_2D^+ + H$ or $C_3HD^+ + H \rightarrow C_3H_2D^+ + h\nu$ since on this way C_3HD^+ can be again pushed into the game. In the addition to the low temperature reactions with H_2, HD and D_2, experiments on collisions of $C_mH_n^+$ with H and D atoms have to be performed. With this aim, a special trapping apparatus has been constructed in our laboratory [33].

From the failure of the related theory we conclude that a much higher accuracy must be postulated from the quantum chemical calculations, especially in the vicinity of asymptotic regions or at transition states. Of special interest for the deuteration are zero-point energies in all critical regions. Assuming that all the details of the potential energy surface are accurately known, detailed dynamical calculations can be performed.

ACKNOWLEDGMENTS

Financial support by the Deutsche Forschungsgemeinschaft (DFG) is gratefully acknowledged, especially via the Forschergruppe FOR 388 "Laboratory astrophysics". I.S. thanks the Department of Physics, Faculty of Science and Mathematics, University of Novi Sad for financial support of the participation of conference.

REFERENCES

1. Gerlich, D., *J. Chem. Soc. Faraday Trans.* **89**, 2199-2208 (1993).
2. Herbst, E., and Klemperer, W., *ApJ* **185**, 505-534 (1973).
3. Black, J. H., and Dalgarno, A., *ApJ* **184**, L101-L104 (1973).
4. Gerlich, D., and Horning, S., *Chem. Rev.* **92**, 1509-1539 (1992).
5. Teyssier, D., Fosse, D., Gerin, M., Pety, J., Abergel, A., and Roueff, E., *A&A* **417**, 135-149 (2004).
6. Gerlich, D., and Schlemmer, S., *Planetary and Space Science* **50**, 1287-1297 (2002).
7. Asvany, O., Savić, I., Schlemmer, S., and Gerlich, D., *Chem. Phys.* **298**, 97-105 (2004).
8. Asvany, O., Schlemmer, S., and Gerlich, D., *ApJ* **617**, 685-692 (2004).
9. Gerlich, D., *J. Chem. Phys.* **92**, 2377-2388 (1990).
10. Gerlich, D., Windisch, F., Hlavenka, P., Plašil, R., Glosik, J., *Phil. Trans. R. Soc. Lond. A*, accepted (2006).
11. Turner, B. E., *ApJS* **136**, 579-629 (2001).
12. Bell, M. B., Avery, L. W., Matthews, H. E., Feldman. P.A., Watson, J. K. G. Madden, S. C. and Irvine, W. M., *ApJ* **326**, 924-930 (1988).
13. Millar, T. J., Bennett, A., Herbst, E. *ApJ* **340**, 906-920 (1989).
14. Roberts, H., Millar, T. J., *A&A* **361**, 388-398 (2000).
15. Roberts, H., Millar, T. J., *A&A* **364**, 780-784 (2000).
16. Howe, D. A., Millar, T. J., *Mon. Not. R. Astron. Soc.* **262**, 868-880 (1993).
17. Talbi, D., Herbst, E., *A&A* **376**, 663-666 (2001).
18. Gerlich, D., *Physica Scripta* **T59**, 256-263 (1995).
19. Gerlich, D., "Inhomogeneous rf Fields: a Versatile Tool for the Study of Processes with Slow Ions", in *State-Selected and State-to-State Ion-Molecule Reaction Dynamics, Part 1: Experiment*, edited by Ng, C. Y. and Baer, M., Adv. in Chem. Phys. LXXXII, 1992, p.p. 1-176.
20. Rosmus, P. (private communication)
21. Savić, I., Gerlich, D., *Phys. Chem. Chem. Phys.* **7**, 1026-1035 (2005).
22. Savić, I., Schlemmer, S., Gerlich, D., *ApJ* **621**, 1163-1170 (2005).
23. Hansel, A., Richter, R., Lindinger, W., *Int. J. Mass Sprectrom. Ion Proc.* **94**, 251-260 (1989).
24. Bohme, D. K., Wlodek, S., *Int. J. Mass Spectrom. Ion Proc.* **102**, 133-149 (1990).
25. Herbst, E., Adams, N. G., Smith, D., *ApJ* **269**, 329-333 (1983).
26. Sorgenfrei, A., „Ion-Molekül-Reaktionen kleiner Kohlenwasserstoffe in einem gekühlten Ionen-Speicher", Ph.D. Thesis, Albert-Ludvigs-Univesität Freiburg 1994.
27. Sorgenfrei, A. and Gerlich, D. "Ion-Trap Experiments on $C_3H^+ + H_2$: Radiative Association vs. Hydrogen Abstraction" in *Molecules and Grains in Space*, edited by I. Nenner, AIP Conference Proceedings 429, New York: American Institute of Physics, 1994, pp. 505-513.
28. Wong, M. W., Radom, L., *J. Am. Chem. Soc.* **115**, 1507-1514 (1993).
29. Herbst, E., Adams, N. G., Smith, D., *ApJ* **285**, 618-621 (1984).
30. Smith, D., Adams, N. G., Ferguson, E. E., *Int. J. Mass Spectrom. Ion Proc.* **61**, 15-19 (1984).
31. Smith, D. and Adams, N. G., *Int. J. Mass Spectrom. Ion Proc.* **76**, 307-317 (1987).
32. Maluendes, S. A., McLean, A. D., Yamashita, K., Herbst, E., *J. Chem. Phys.* **99**, 2812-2820 (1993).
33. Luca, A., Borodi, G., and Gerlich, D., "Interactions of ions with Hydrogen atoms", Progress report in XXIV ICPEAC 2005, Rosario, Argentina, July 20-26, 2005, edited by Colavecchia, F.D., Fainstein, P.D., Fiol, J., Lima, M.A.P., Miraglia, J.E., Montenegro, E.C., and R.D.Rivarola

Microlensing signatures in spectra of quasars: X-ray radiation

Predrag Jovanović and Luka Č. Popović

Astronomical Observatory Belgrade, Volgina 7, 11160 Belgrade, Serbia

Abstract. Gravitational microlensing is a very useful tool for investigating the innermost part of quasars, especially for studying a relativistic accretion disk around a massive black hole (BH) supposed to exist in the center of every quasar. Here we present a short overview of our recent investigations of the gravitational microlensing influence on detected X-ray radiation from accretion disks of quasars. We set our focus to the analysis of the Fe $K\alpha$ spectral line and the X-ray continuum variations due to gravitational microlensing.

The disk emission was analyzed by numerical simulations, based on a ray-tracing method in a Kerr metric, taking into account only photon trajectories reaching the observer's sky plane. The influence of microlensing on a standard accretion disk was studied using three types of a microlensing model: point-like microlens, straight-fold caustic and quadruple microlens (microlensing pattern).

Our results show that gravitational microlensing can produce significant variations and amplifications of the Fe $K\alpha$ line and X-ray continuum flux and that even very small mass objects could produce such changes. These deformations of the X-ray radiation depend on both the disk and microlens parameters and they are significantly larger than the corresponding effects on the optical and UV emission lines and continua, due to the smaller dimensions of the X-ray emitting region. Although gravitational microlensing is an achromatic effect, it can induce wavelength dependent variations of the X-ray continuum.

Keywords: gravitational lenses, active galaxies, quasars, black holes, accretion disks, X-ray radiation, Fe $K\alpha$ line
PACS: 95.30.Sf, 98.62.Sb, 98.54.-h, 04.70.-s, 98.62.Mw, 78.70.En

INTRODUCTION

Quasars are the most powerful sources of X-rays. Except in X-rays, they are intensively emitting in a wide spectral band, from γ-rays to radio waves. Some of them are so bright that they can be seen at a distance of more than 12 billion light years, being the most distant observed objects. They were discovered in the 1960s as radio sources seemed to be associated with "stars", and were called quasi-stellar radio sources or quasars. This theory of quasar origin was soon rejected because they had spectra similar to the nuclei of Seyfert galaxies. Nowadays, it is clear that quasars are a type of Active Galactic Nuclei (AGN). Active galaxies have a small, often highly variable and very bright core. Variations in the X-ray radiation of AGN can be caused, among other things, by gravitational microlensing, especially in the case of the gravitationally macrolensed quasars [1, 2, 3].

The X-rays of AGN are generated in the innermost region of an accretion disk around a central super-massive black hole (BH). An emission line, Fe $K\alpha$, has been observed at 6-7 keV in majority of AGN (see e.g. [4, 5]). This line is probably produced in a very compact region near the black hole [5, 6, 7] and can bring essential information

CP876, *The Physics of Ionized Gases: 23rd Summer School and International Symposium*,
edited by L. Hadžievski, B. P. Marinković, and N. S. Simonović

about the plasma conditions and the space-time geometry around the black hole. Thus, it seems clear that the Fe Kα line can be strongly affected by microlensing and recent observations of several lens systems support this idea [1, 2, 3].

In this paper we present a short overview of our recent investigations of the gravitational microlensing influence on detected X-ray radiation from relativistic accretion disk around a massive black hole supposed to exist in a quasar's center. For this purpose, we set our focus to the analysis of the Fe Kα spectral line and the X-ray continuum variations due to gravitational microlensing. The initial assumption was the existence of a super massive black hole ($10^7 - 10^9$ $M_\odot$), surrounded by an accretion disk that radiates in X-rays, in the center of all types of AGN. Accretion disks could have different forms, dimensions and emission, depending on the type of central BH, being rotating (Kerr metric) or non-rotating (Schwarzschild metric). The disk emission was analyzed by numerical simulations, based on a ray-tracing method in a Kerr metric, taking into account only photon trajectories reaching the observer's sky plane (see [8] and references therein). The influence of microlensing on a standard accretion disk was studied using three types of a microlensing model: point-like microlens, straight-fold caustic and quadruple microlens (microlensing pattern).

VARIATIONS OF THE Fe Kα LINE AND THE X-RAY CONTINUUM DUE TO MICROLENSING

As mentioned before, to study the effects of microlensing on a compact accretion disk we used ray tracing method considering only those photon trajectories which reach the observer's sky plane at infinity. The amplified brightness with amplification $A(X,Y)$ for the continuum is then given by [9]:

$$I_C(X,Y;E_{obs}) = I_P(E_{obs}, T(X,Y)) \cdot A(X,Y), \qquad (1)$$

and for the Fe Kα line by:

$$I_L(X,Y;E_{obs}) = I_P(E_0 \cdot g(X,Y), T(X,Y)) \cdot \delta(E_{obs} - E_0 \cdot g(X,Y)) \cdot A(X,Y), \qquad (2)$$

where I_P is an emissivity law of the disk, $T(X,Y)$ is the temperature, X and Y are the impact parameters which describe the apparent position of each point of the accretion disk image on the celestial sphere as seen by an observer at infinity; E_0 is the line transition energy ($E_0^{Fe\ K\alpha} = 6.4$ keV) and $g(X,Y) = E_{obs}/E_{em}$ is the energy shift due to relativistic effects (E_{obs} is the observed energy and E_{em} is the emitted energy from the disk). The total observed flux for the continuum and the Fe Kα line is then given by:

$$F(E) = \int_{image} [I_C(X,Y;E) + I_L(X,Y;E)]d\Omega, \qquad (3)$$

where $d\Omega$ is the solid angle subtended by the disk in the observer's sky and the integral extends over the whole emitting region.

The influence of gravitational microlensing on the Fe Kα line was analyzed by adopting averaged values of disk parameters from the study of 18 Seyfert I galaxies [4]:

inclination $i = 30°$ and emissivity index $q = 2.5$. For the inner radius we took $R_{in} = R_{ms}$, where R_{ms} is the radius of the marginally stable orbit, that corresponds to $R_{ms} = 6\,R_g$ (R_g is gravitational radius: $R_g = GM/c^2$) in the Schwarzschild metric and to $R_{ms} = 1.23\,R_g$ in the Kerr metric with angular momentum $a = 0.998$. Taking into account that for the adopted emissivity index ($q = 2.5$), emission is concentrated in the innermost part of the disk, we assumed that the outer radius is $R_{out} = 20\,R_g$.

The amplified line profiles were computed for different locations of a point-like microlens with projected Einstein ring radius $ERR = 10\,R_g$, in respect to the center of the disk. Figure 1 shows the results for Schwarzschild (left) and Kerr (right) metrics in the case of a point-like microlens, but similar results were obtained also for the case of a straight-fold caustic [8, 10]. Several outstanding changes of the line shape with the location of the microlens, and consequently with the transit of a microlens across the disk, are shown in Figure 1. These changes are reflected in the number of peaks, their relative separation and asymmetrical enhancement of the line profile. For both metrics the amplification has a maximum for the approaching part of the rotating disk and consequently, the amplification affects mainly the blue part of the line. This asymmetrical enhancement induced by microlensing is stronger in Schwarzschild than in Kerr metric.

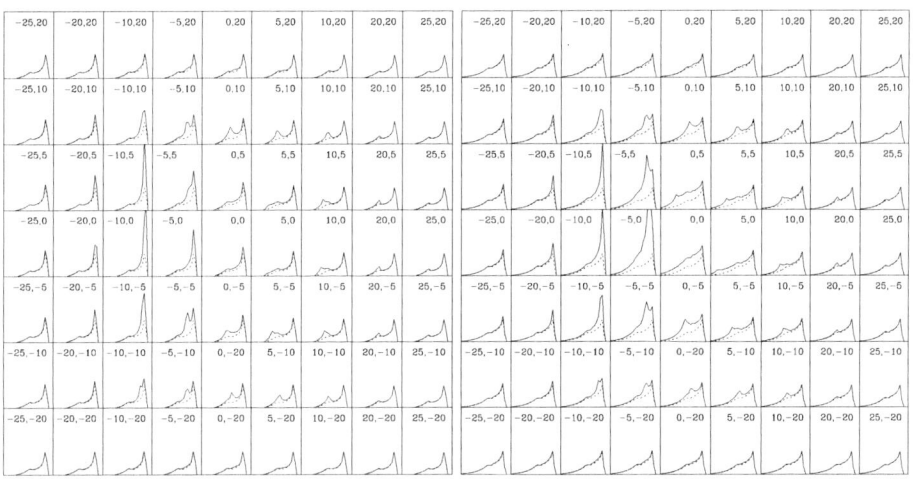

FIGURE 1. The line profiles for different positions of a point-like microlens [8]. The calculations were performed for Schwarzschild metric (left) and for Kerr metric with $a = 0.998$ (right). The adopted disk parameters are: $R_{in} = 6\,R_g$, $R_{out} = 20\,R_g$, $i = 35°$ and q=2.5. The ERR of the microlensing object is 10 R_g. The relative intensity (y-axis) is in the range from 0 to 3 and the energy (x-axis) from 0.1 to 10 keV. The numbers at the top of the figures are coordinates (impact parameters) of the microlens center expressed in R_g.

In Figure 2 we present the variations of the total X-ray emission spectra (continuum + Fe $K\alpha$ line) during a straight fold caustic crossing (A_0=1, β=1 and ERR=50 R_g) for Schwarzschild and Kerr metrics, respectively. The radial dependence of the emissivity is related to the black body radiation law (see [9]). The sizes of the continuum and line emission regions are the same, $R_{inn} = R_{ms}$ and $R_{out} = 20\,R_g$.

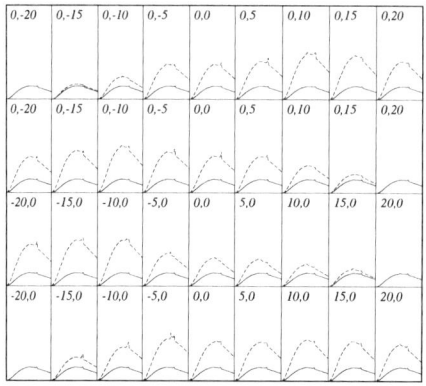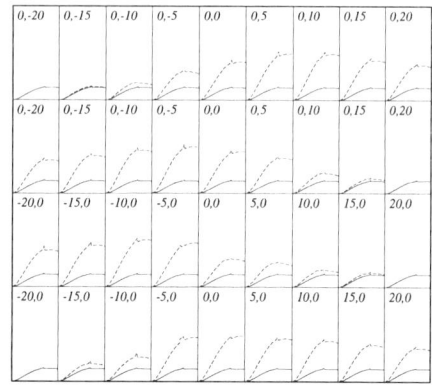

FIGURE 2. Simulations of the behavior of the X-ray continuum and Fe Kα line variations due to microlensing by a caustic in the case of Schwarzschild (left) and Kerr (right) metrics. The parameters of the caustic are: A_0=1, β=1 and ERR=50 R$_g$. In the first and second rows we present the caustic crossing perpendicular to the rotating axis in both directions and in the third and fourth rows we show the caustic crossing along the rotation axis in both directions, also. The radii of the continuum and the Fe Kα line emission accretion disks are the same: $R_{in} = R_{ms}$ and $R_{out} = 20$ R$_g$. The unperturbed and normalized emission correspond to solid and dashed lines, respectively. The relative intensity ranges from 0 to 7 (y-axis) and the energy interval from 0.1 to 10 keV (x-axis).

The obtained results demonstrate that even microlenses with small masses, corresponding to $ERR \sim 10\ R_g$, can induce significant changes in the iron line and X-ray continuum fluxes, due to the small dimensions of the X-ray emitting region [8, 11]. These changes depend on both the disk and microlens parameters [10, 12].

WAVELENGTH DEPENDENT VARIATIONS OF THE X-RAY CONTINUUM AMPLIFICATION

When discussing the influence of gravitational microlensing on the spectra of lensed quasars, the color index and the amplified flux features are usually used as indicators of the microlensing [9]. Taking into account that the emitters at different radii in the accretion disk have different temperatures and make different contributions to the observed continuum flux at a given wavelength, the exact shape of the amplification as a function of wavelength (or energy) for a partly microlensed disk can be calculated. During a caustic crossing, microlensing effects would depend on the location of the emitters and, consequently, would induce a wavelength dependence of the amplification. This dependence is easily noticeable in the spectra of Figures 1 and 2. In Figures 3 and 4 we present the amplification as a function of the observed energies for an accretion disk with inner radius $R_{in} = R_{ms}$ and outer radius $R_{out} = 30\ R_g$, assuming caustic crossing along the disk's rotation axis. The black body (Figure 3) and the modified black body (Figure 4) emissivity laws, for both Schwarzschild and Kerr metrics, are considered.

As it can be seen in Figures 3 and 4 the amplification is different for different

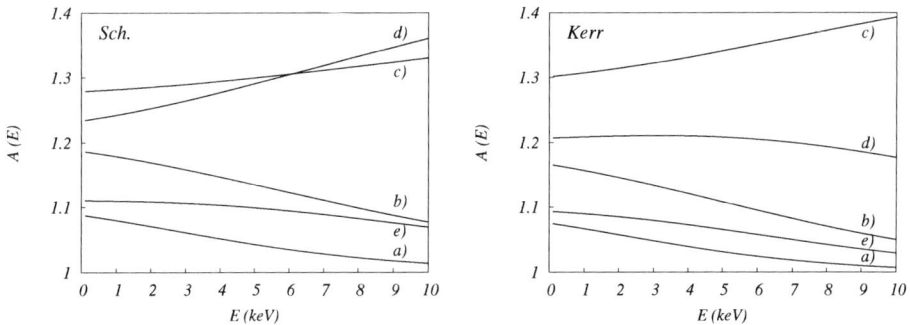

FIGURE 3. Microlensing amplification as a function of emitted energies (the chromatic effect of microlensing). The calculations were performed for caustic with $ERR = 50\ R_g$ crossing along the rotation axis (Y=0) for the following positions on X-axis: a) X=20 R_g; b) X=10 R_g; c) X=0 R_g; d) X=-10 R_g and e) X=-20 R_g. The radii of the disk are $R_{inn} = R_{ms}$, $R_{out} = 30\ R_g$. The black body emissivity law in both Schwarzschild (left) and Kerr metrics (right) is assumed.

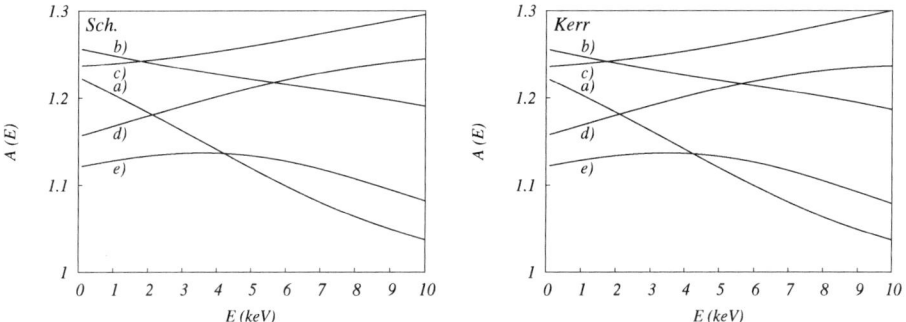

FIGURE 4. The same as in Figure 3, but for the "modified" black body emissivity law.

observed energies. It is higher for the hard X-ray continuum when the caustic crosses the central part of the disk [9]. Depending mainly on the caustic location and on the chosen emissivity law, the difference of the amplification in the energy range 0.1 - 10 keV can be significant (e.g. $\sim 20\%$ for very small mass microlenses, $ERR = 50\ R_g$, see Figures 3 and 4). This effect could induce a noticeable wavelength dependent variability of the X-ray continuum spectrum during a microlensing event (of even a 30%), providing a tool to study the innermost regions of accretion disks [9, 13].

MICROLENSING OF GRAVITATIONALLY LENSED QUASARS

In the aim to discuss the lack of correlation between line and continuum variability in the recently observed gravitationally lensed quasars (MG J0414+0534 [1], QSO 2237+0305 [2], and H1413+117 [3]) in context of the microlensing hypothesis, we modeled the behavior of the X-ray continuum and the Fe Kα line during a microlensing event for different sizes of the continuum and the Fe Kα line emission regions. We assumed that the Fe Kα line was formed in the innermost part of the disk ($R_{in} = R_{ms}$; $R_{out} = 20\ R_g$) and that the continuum (emitted in the energy range between 0.1 keV and 10 keV) was mainly originated from a larger region ($R_{in} = 20\ R_g$; $R_{out} = 100\ R_g$). The simulations were made in the case of image A of quasar QSO 2237+0305, assuming microlensing by a magnification pattern generated by a population of low mass deflectors (with mean mass $< m >= 0.35\ M_\odot$), distributed randomly in a rectangular region in the lens plane, significantly larger than the considered region in the source plane. The size of modeled magnification pattern was 1ERR×2ERR, that corresponds to 334.63 R_g × 669.26 R_g in the source plane for a black hole of mass $M_{BH} = 10^9 M_\odot$ (Figure 5 left). For numerical reasons, the size of the microlens magnification map was given in pixels: 1000×2000 (1pix=0.33463 R_g in source plane). As one can see from Figure 5 (left), the microlensing pattern structures are comparable with a compact X-ray accretion disk.

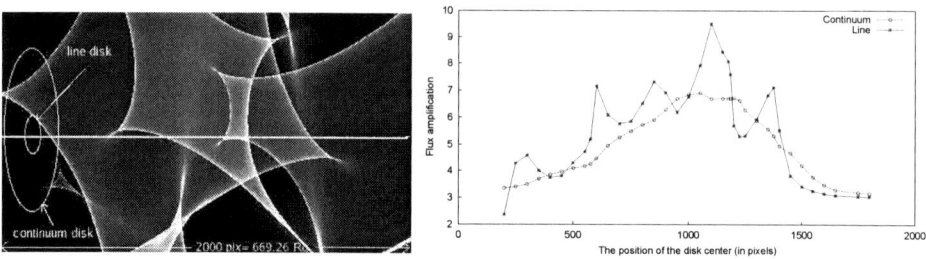

FIGURE 5. *Left:* microlensing map of QSO 2237+0305A image with 1ERR×2ERR (1000 pix ×2000 pix=334.63 R_g × 669.26 R_g) on a side and scheme of the projected disk with outer radius $R_{out} = 20\ R_g$ and 100 R_g for the Fe Kα line and the X-ray continuum, respectively. The straight line presents the path of the center of the disk (left side of the pattern corresponds to 0 pix). *Right:* the amplification of the Fe Kα line and the X-ray continuum total flux for different positions of the disk center on the straight line in the left figure [9].

To explore the line and the continuum variations we moved the disk center along the microlensing map shown in Figure 5 (left) from left to the right, i.e. from 0 to 2000 pixels. In Figure 5 (right) we presented the corresponding total line (marked by asterisks) and X-ray continuum flux (marked by open circles) variations. As one can see in Figure 5 (right), there is a global correlation between the total line and continuum flux during the complete path. However, the total continuum flux variation is smooth and has a monotonic change, while the total line flux varies very strongly and randomly. In fact, during some parts of the microlensing event we found that the total Fe Kα line flux changed, while the continuum flux remained nearly constant (e.g. the position of the disk center between 1000 and 1200 pixels). This and the shapes of the line and continuum total flux amplification indicated that the observed microlensing amplification of the Fe Kα in three lensed quasars could be explained if the line was originated in the innermost part of the disk and the X-ray continuum in some larger region [14, 15, 16].

WORK IN PROGRESS: PRELIMINARY RESULTS

For the purpose of analyzing the microlensing time scales and time dependent response of amplification due to microlensing in quasar's spectra emitted in the X-ray, UV and optical bands, we numerically simulated a straight-fold caustic crossing along an accretion disk that was stratified into three parts:

1. An innermost part that emits X-ray continuum (1.24 Å- 12.4 Å or 1-10 keV). The inner radius is taken to be $R_{in} = R_{ms}$ and outer radius is $R_{out} = 80\ R_g$.
2. An UV emitting part of the disk (that contributes to the emission from 1000 Å– 3500 Å), with $R_{in} = 100\ R_g$ and $R_{out} = 1000\ R_g$
3. An outer optical part of the disk with $R_{in} = 100\ R_g$ and $R_{out} = 2000\ R_g$ that emits in the wavelength band from 3500 Å until 7000 Å.

The corresponding variations of the total flux for QSO 2237+0305 are given in Figure 6, and one can see that the microlensing time scales are different for different spectral bands. The phase of the maximal amplification in the X-ray band is on the order of several months, while the variations of the UV/optical emission regions are on the order of several years. Also, a higher amplification in the X-ray band than in the UV/optical bands is present, so one can expect that the Fe Kα line and X-ray continuum amplification due to microlensing can be significantly larger than the corresponding effects in the optical and UV emission lines and continua [13].

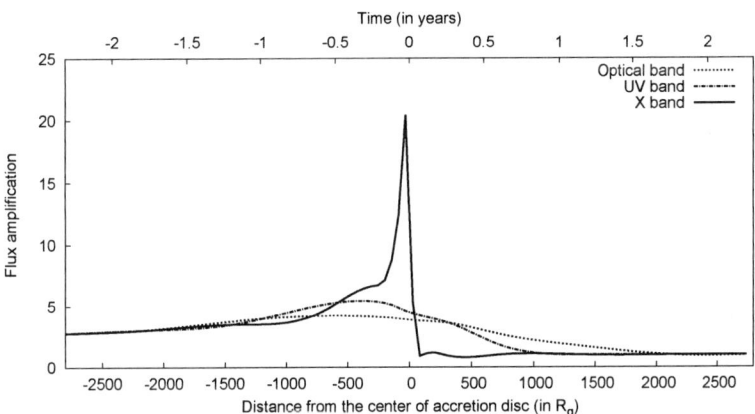

FIGURE 6. The variations of normalized total continuum flux in optical (3500-7000 Å), UV (1000-3500 Å) and X (1.24-12.4 Å i.e. 1-10 KeV) band due to microlensing by a caustic crossing over accretion disk in Schwarzschild metric. Time scale corresponds to the QSO 2237+0305 gravitational lens, where redshifts of microlens and source are $z_l = 0.04$ and $z_d = 1.69$, respectively [13].

CONCLUSIONS

Here we summarize the several interesting results of our investigations of gravitational microlensing influence on X-ray radiation from accretion disks of quasars.

1. Even microlenses with very small masses can produce significant variations and amplifications of the Fe Kα line and X-ray continuum fluxes. These deformations of the X-ray radiation depend on both the disk and microlens parameters.
2. Although microlensing is an achromatic effect, it can induce wavelength dependent variations of the X-ray continuum of $\sim 30\%$ due to the radial distribution of temperature in an accretion disk.
3. Microlensing can satisfactorily explain the lack of a correlation between the X-ray continuum and the Fe Kα line, observed in some gravitational lens systems.
4. Microlensing time scales analysis showed that the variations of the X-ray continuum are the fastest ($\sim$ several months), and those of the optical and UV continua are weaker and much slower ($\sim$ several years). Also, the Fe Kα line and X-ray continuum amplifications due to microlensing are expected to be significantly larger than the corresponding effects on optical and UV emission lines and continua.

According to the presented results, one can conclude that monitoring of gravitational lenses in the X-ray spectral band may help us to understand the physics and to reveal the structure of the innermost parts of quasars, particularly their relativistic accretion disks.

ACKNOWLEDGMENTS

This work is a part of the project 146002: "Astrophysical Spectroscopy of Extragalactic Objects" supported by the Ministry of Science and Environment of Serbia.

REFERENCES

1. G. Chartas, E. Agol, M. Eracleous, G. Garmire, M. W. Bautz, N. D. Morgan, *ApJ*, **568**, 509 (2002)
2. X. Dai, G. Chartas, E. Agol, M. W. Bautz, G. P. Garmire, *ApJ*, **589**, 100 (2003)
3. G. Chartas, M. Eracleous, E. Agol, S. C. Gallagher, *ApJ*, **606**, 78 (2004)
4. K. Nandra, I. M. George, R. F. Mushotzky, T. J. Turner & T. Yaqoob, *ApJ*, **477**, 602 (1997)
5. A. C. Fabian, K. Iwasawa, C. S. Reynolds, A. J. Young, *PASP*, **112**, 1145 (2000)
6. K. Iwashawa, A. C. Fabian, A. J. Young, H. Inoue, C. Matsumoto, *MNRAS*, **306**, L19 (1999)
7. K. Nandra, I. M. George, R. F. Mushotzky, T. J. Turner, T. Yaqoob, *ApJ*, **523**, 17 (1999)
8. L. Č. Popović, E. G. Mediavilla, P. Jovanović, J. A. Muñoz, *A&A*, **398**, 975 (2003)
9. L. Č. Popović, P. Jovanović, E. G. Mediavilla, A. F. Zakharov, C. Abajas, J. A. Muñoz, G. Chartas, *ApJ*, **637**, 620 (2006)
10. L. Č. Popović, P. Jovanović, E. G. Mediavilla, J. A. Muñoz, *A&AT*, **22**, 719 (2003)
11. L. Č. Popović, P. Jovanović, *POBeo*, **73**, 215 (2002)
12. L. Č. Popović, E. G. Mediavilla, J. A. Muñoz, M. S. Dimitrijević, P. Jovanović, SerAJ, **164**, 53 (2001)
13. P. Jovanović, *PASP*, **118**, 656 (2006)
14. P. Jovanović, *Mem. S. A. It.*, **7**, 56 (2005)
15. P. Jovanović, L. Č. Popović, M. S. Dimitrijević, *POBeo*, **76**, 205 (2003)
16. P. Jovanović, L. Č. Popović, *PASRB*, **5**, 195 (2005)

AUTHOR INDEX

A

Adamov, M. G., 333
Alzetta, G., 62
Aubrecht, V., 338
Aucouturier, M., 191
Azria, R., 112

B

Bartlova, M., 338
Baudon, J., 28
Bennion, I., 169, 216
Bertin, M., 112
Bibić, N., 209
Bocvarski, V., 28
Bolorizadeh, M. A., 41
Brezinsek, S., 235
Brunger, M. J., 41
Burgdörfer, J., 143

C

Cáceres, D., 112
Campbell, L., 41
Cartaleva, S., 62
Cartwright, D. C., 41
Chen, M., 405
Ciroi, S., 385
Coelho, P. J., 224
Colao, F., 309
Colombeau, B., 181
Cowern, N. E. B., 181
Cvejanović, D., 72
Czarnetzki, U., 260

D

Danezis, E., 373
Deiss, C., 143
Dimitrijević, M. S., 373
Domaracka, A., 112
Dreher, J., 169
Dubov, M., 169, 216

Ducloy, M., 28
Dujko, S., 51
Dyatko, N. A., 15

F

Fantoni, R., 309
Folkard, M., 3

G

Gans, T., 260
Gerlich, D., 415
Gigosos, M. A., 294
González, M. A., 294
Gozzini, S., 62
Graham, W. G., 250
Grauer, R., 169
Grucker, J., 28
Gwilliam, R. M., 181

H

Hey, J. D., 235
Hoffmann, S. V., 3
Horváth, G., 284

I

Ilić, D., 385
Ingolfsson, O., 117
Ionikh, Y. Z., 15
Ivković, M., 301

J

Jin, Z., 405
Johzaki, T., 361
Jovanović, P., 423
Jovicevic, S., 309

431